Next-Generation Plant-based Foods

David Julian McClements • Lutz Grossmann

Next-Generation Plant-based Foods

Design, Production, and Properties

David Julian McClements
Department of Food Science
University of Massachusetts
Amherst, MA, USA

Lutz Grossmann
Department of Food Science
University of Massachusetts Amherst
Amherst, MA, USA

ISBN 978-3-030-96763-5   ISBN 978-3-030-96764-2   (eBook)
https://doi.org/10.1007/978-3-030-96764-2

© The Editor(s) (if applicable) and The Author(s), under exclusive license to Springer Nature Switzerland AG 2022

This work is subject to copyright. All rights are solely and exclusively licensed by the Publisher, whether the whole or part of the material is concerned, specifically the rights of translation, reprinting, reuse of illustrations, recitation, broadcasting, reproduction on microfilms or in any other physical way, and transmission or information storage and retrieval, electronic adaptation, computer software, or by similar or dissimilar methodology now known or hereafter developed.

The use of general descriptive names, registered names, trademarks, service marks, etc. in this publication does not imply, even in the absence of a specific statement, that such names are exempt from the relevant protective laws and regulations and therefore free for general use.

The publisher, the authors and the editors are safe to assume that the advice and information in this book are believed to be true and accurate at the date of publication. Neither the publisher nor the authors or the editors give a warranty, expressed or implied, with respect to the material contained herein or for any errors or omissions that may have been made. The publisher remains neutral with regard to jurisdictional claims in published maps and institutional affiliations.

This Springer imprint is published by the registered company Springer Nature Switzerland AG
The registered company address is: Gewerbestrasse 11, 6330 Cham, Switzerland

David Julian McClements dedicates this book to his daughter Isobelle McClements and wife Jayne Farrell
Lutz Grossmann dedicates this book to his family and all the genius minds who created the knowledge for this work. Let's make the world a better place through positive science.

Acknowledgments

The authors thank numerous people for their expert advice and support during the writing of this book. Several people kindly read first drafts of the chapters and provided constructive feedback. We give special thanks to Erin Rees Clayton and Priera Panescu of the Good Food Institute who provided insightful comments and suggestions on several chapters. We thank Celia Homyak of Ripple Foods, Ester Ferrusola Pastrana of New Roots Foods, Alexandra Dallago of Eat Just, and Imtiaz Billah of Beyond Meat who provided feedback on the plant-based milk, dairy, egg, and meat chapters, respectively. We thank Hualu Zhou from the University of Massachusetts who provided comments on several chapters. We also thank Nicole Negowetti of the Plant Based Foods Association who provided insightful comments on the concluding chapter. This feedback certainly helped us to improve our book. Any errors remaining are our own.

Several people also provided expert advice on specific topics. We thank Zack Henderson of Sensient Technologies for providing insights into plant-based food colors. We also thank all the individuals and companies who kindly allowed us to use photographs of their analytical equipment or food products. Marc Johnson of Texture Technologies Corporation provided photographs of the texture analysis equipment, Philip Rolfe of Netzsch Instruments provided photographs of the Dynamic Shear Rheology equipment, Pascal Bru of Formulaction provided images of the Turbiscan, and Ken Wendt of HunterLab provided photographs of color theory and colorimeters. We also thank the sales department of Malvern Panalytical for providing photographs of particle characterization equipment, including static and dynamic light scattering. We thank Impossible Foods, Beyond Meat, Just Eat, and Revo Foods for providing images of their plant-based food products and Perfect Day for providing images of their cellular agriculture dairy products. In addition, we thank Coperion, C. Gerhardt Analytical Systems, Albert Handtmann Holding, Bühler Group, and ProXES for providing us images of food processing equipment. All brand names appearing in this book are for product identification purposes only and all information presented is available to the general public. No endorsement or criticism was intended, nor was endorsement or criticism implied of products not mentioned. For the purpose of clarity, product category names like 'plant-based milk' and 'plant-based meat' are used throughout the book but we want to highlight that some countries limit the use of such terms that are derived from animal-based foods for labelling of plant-based foods.

The authors thank Daniel Falatko from Springer for accepting our proposal for this book and for all his help in bringing the book to fruition.

The authors thank all of their students, visiting scholars, and postdocs who have carried out research and development on plant-based foods as well as involving the authors in thoughtful discussions on these products. Especially, we thank Hualu (Lu) Zhou, Zhiyun (Kevin) Zhang, Yunbing (Shelly) Tan, Kanon Kobata, Giang Vu, Hung (Harry) Pham, Carlos Woern, Xiaoyan Hu, Dawn Wannasin, Cheryl Chung, Eric Tsin, Minqi Wang, and Kanokporn (Ye) Leethanapanich. We also thank our colleagues at the University of Massachusetts for their valuable insights and collaboration on plant-based foods, including Eric Decker, Lili He, Charmaine Koo, Jiakai Liu, Amanda Kinchla, Allissa Nolden, David Sela, Hang Xiao, Yeonhwa Park, and Jean Alamed.

Finally, the authors thank their families and friends for their constant support and encouragement throughout the writing of this book. Special thanks go to David Sela and Matthew Moore for all the excellent advice they gave on plant-based foods over beers at Progression microbrewery in Northampton, Massachusetts. Special thanks also goes to Jayne Farrell and Isobelle McClements who encouraged one of us (DJM) to start writing the book.

Contents

1	**The Rise of Plant-Based Foods**		1
	1.1	Introduction	1
	1.2	Environmental and Sustainability Reasons for Consuming Plant-Based Foods	2
		1.2.1 The Inefficiency of Animals as Foods	4
		1.2.2 Impact of Animal Foods on the Environment	5
		1.2.3 Improving the Efficiency of Food Production	6
		1.2.4 Establishment of Planetary Boundaries	8
		1.2.5 Ensuring Biodiversity	10
	1.3	Ethical Reasons for Consuming Plant-Based Foods	11
	1.4	Health Reasons for Consuming Plant-Based Foods	12
	1.5	The Importance of Taste	13
	1.6	Opportunities for the Food Industry	15
	1.7	Other Sources of Alternative Proteins	18
	1.8	Conclusions	19
	References		19
2	**Properties and Functionality of Plant-Based Ingredients**		23
	2.1	Introduction	23
	2.2	Proteins	24
		2.2.1 Protein Structure	24
		2.2.2 Protein Extraction and Refinement	34
		2.2.3 Protein Ingredients	37
		2.2.4 Protein Characterization	41
	2.3	Carbohydrates	43
		2.3.1 Carbohydrate Structure	44
		2.3.2 Carbohydrate Isolation	45
		2.3.3 Carbohydrate Ingredient Properties	45
		2.3.4 Carbohydrate Characterization	48

	2.4	Lipids	49
		2.4.1 Lipid Structure	50
		2.4.2 Lipid Isolation	52
		2.4.3 Lipid Ingredients	53
		2.4.4 Lipid Characterization	57
	2.5	Other Additives	58
		2.5.1 Salts	58
		2.5.2 Colors and Flavors	59
		2.5.3 pH Controllers	61
		2.5.4 Crosslinking Agents	61
		2.5.5 Preservatives	62
		2.5.6 Micronutrients and Nutraceuticals	62
	2.6	Ingredient Functionality	63
		2.6.1 Solubility	63
		2.6.2 Molecular Binding Interactions	64
		2.6.3 Fluid-Holding Capacity	65
		2.6.4 Thickening	66
		2.6.5 Gelling	70
		2.6.6 Binders and Extenders	73
		2.6.7 Emulsification	74
		2.6.8 Foaming	75
		2.6.9 Melting and Crystallization	75
		2.6.10 Nutrition	76
		2.6.11 Gastrointestinal Fate	76
		2.6.12 Miscellaneous Functions	78
	2.7	Ingredient Utilization	78
	2.8	Minimally-Processed Ingredients	79
	2.9	Conclusions and Future Directions	80
	References		81
3	**Processes and Equipment to Create Plant-Based Foods**		**89**
	3.1	Introduction	89
	3.2	Molecular Approaches to Structuring Plant-Based Ingredients	90
		3.2.1 Biopolymer Phase Separation	90
		3.2.2 Gelation	95
		3.2.3 Phase Transitions	97
	3.3	Advanced Particle Technologies	97
		3.3.1 Types of Advanced Particles	97
		3.3.2 Applications of Advanced Particle Technologies	101
	3.4	Mechanical Processing Methods	104
		3.4.1 Size Reduction	104
		3.4.2 Separation and Fractionation Methods	110
	3.5	Structure Formation Methods	116
		3.5.1 Extrusion	118
		3.5.2 Shear Cell	124
		3.5.3 Additive Manufacturing	127

	3.6	Thermal Processing Methods	134
		3.6.1 Blanching	134
		3.6.2 Inactivation of Antinutrients	135
		3.6.3 Pasteurization and Ultra-High Temperature (UHT) Treatments	137
	3.7	Fermentation Methods	141
		3.7.1 Enzymatic Fermentation	141
		3.7.2 Microbial Fermentation	141
	3.8	Process Design Examples: Soy, Oat, and Nut Milk	143
	3.9	Conclusions and Future Directions	145
		References	146
4	**Physicochemical and Sensory Properties of Plant-Based Foods**		**155**
	4.1	Introduction	155
	4.2	Appearance	156
		4.2.1 Factors Affecting Appearance	156
		4.2.2 Modeling and Prediction of Appearance	162
		4.2.3 Major Factors Impacting the Appearance of Plant-Based Foods	168
		4.2.4 Measurement of the Appearance of Plant-Based Foods	174
		4.2.5 Color Attributes of Plant-Based Foods	176
	4.3	Texture	177
		4.3.1 Fluids	177
		4.3.2 Solids	183
		4.3.3 Practical Considerations	197
	4.4	Stability	198
		4.4.1 Gravitational Separation	199
		4.4.2 Particle Aggregation	202
		4.4.3 Phase Separation	206
		4.4.4 Chemical Degradation	207
		4.4.5 Microbial Contamination	209
		4.4.6 Quantification of Stability	209
	4.5	Fluid Holding and Cookability Properties	211
	4.6	Partitioning, Retention and Release Properties	213
		4.6.1 Partitioning Phenomena	213
		4.6.2 Retention and Release Processes	216
	4.7	Oral Processing and Sensory Attributes	218
		4.7.1 Oral Processing	218
		4.7.2 Sensory Evaluation	219
	4.8	Conclusions	222
		References	222
5	**Nutritional and Health Aspects**		**227**
	5.1	Introduction	227
	5.2	Macronutrients	228
		5.2.1 Proteins	229

		5.2.2	Lipids	237
		5.2.3	Carbohydrates	244
	5.3	Micronutrients		249
		5.3.1	Vitamins	251
		5.3.2	Minerals	253
	5.4	Nutraceuticals		256
	5.5	Gastrointestinal Fate: Digestibility, Bioavailability, and Fermentability		258
	5.6	Impact of Diet on the Gut Microbiome		261
	5.7	Nutritional Studies Comparing Plant- and Animal-Based Diets		263
	5.8	Evolution, Genetics and Meat Consumption		264
	5.9	The Agricultural Revolution and Meat Consumption		267
	5.10	Improving Healthiness of Plant-Based Foods		267
		5.10.1	Fortification	268
		5.10.2	Reformulation: Reduced Fat, Salt, Sugar, and Digestibility	269
		5.10.3	Agricultural and Processing Approaches	274
	5.11	Microbiological and Chemical Toxins		274
	5.12	Conclusions		275
	References			275
6	**Meat and Fish Alternatives**			285
	6.1	Introduction		285
	6.2	Properties of Meat and Fish		289
		6.2.1	Muscle Structure and Composition	289
		6.2.2	Appearance	292
		6.2.3	Textural Attributes	294
		6.2.4	Cooking Loss and Heat-induced Changes	294
		6.2.5	Flavor Profile and Oral Processing	297
	6.3	Ingredients for Formulating Plant-based Meat Analogs		298
		6.3.1	Plant Proteins	299
		6.3.2	Lipids	302
		6.3.3	Binders	306
		6.3.4	Coloring Agents	310
		6.3.5	Flavoring Agents	312
	6.4	Processing Methods		315
		6.4.1	Protein Texturization	315
		6.4.2	Plant-based Meat Preparations	317
		6.4.3	Plant-based Meat Products	318
	6.5	Key Properties		320
		6.5.1	Color	321
		6.5.2	Texture	322
		6.5.3	Fluid Holding	323
		6.5.4	Flavor	325
		6.5.5	Nutritional Value	326
		6.5.6	Environmental Sustainability	329

	6.6	Future Directions	330
	References		331
7	**Eggs and Egg Products**		**341**
	7.1	Introduction	341
	7.2	Properties of Hen's Eggs	342
		7.2.1 Composition and Structure	342
		7.2.2 Processing	345
		7.2.3 Physicochemical Properties	345
		7.2.4 Functional Properties	348
		7.2.5 Flavor	353
	7.3	Plant-Based Egg Analogs	354
		7.3.1 Composition and Structure	354
		7.3.2 Processing	356
		7.3.3 Physicochemical Properties	358
		7.3.4 Functional Properties	364
		7.3.5 Characterization of Egg Analogs	367
		7.3.6 Commercial Egg Analogs	368
	7.4	Comparative Nutrition, Sustainability, and Ethics of Eggs and Egg Analogs	370
	7.5	Egg Products	372
		7.5.1 Emulsified Products: Mayonnaise and Salad Dressings	372
		7.5.2 Thickened and Gelled Products: Custards, Flans, and Quiches	379
		7.5.3 Foamed Products: Meringues, Mousses, and Soufflé	380
		7.5.4 Baked Products: Cakes, Cookies, and Pastries	381
		7.5.5 Advantages of Egg Analogs to Food Manufacturer	382
	7.6	Conclusions and Future Directions	383
	References		384
8	**Plant-Based Milk and Cream Analogs**		**389**
	8.1	Introduction	389
	8.2	Attributes of Cow's Milk	390
		8.2.1 Composition and Microstructure	390
		8.2.2 Processing	392
		8.2.3 Physicochemical and Sensory Properties	392
		8.2.4 Functional Versatility	394
		8.2.5 Nutritional Profile	395
	8.3	Production of Plant-Based Milk Analogs	395
		8.3.1 Plant Tissue Disruption Approaches	396
		8.3.2 Emulsification Approaches	398
	8.4	Physicochemical Attributes	406
		8.4.1 Appearance	407
		8.4.2 Texture	410
		8.4.3 Stability	416

8.5	Sensory Attributes	425	
8.6	Nutritional Attributes	426	
8.7	Nutritional Fortification	431	
8.8	Environmental Impact: Life Cycle Analysis	433	
8.9	Conclusions and Future Work	435	
References		435	

9 Dairy Alternatives – Cheese, Yogurt, Butter, and Ice Cream 443

- 9.1 Introduction ... 443
- 9.2 History of Plant-Based Cheeses ... 444
- 9.3 Animal-Based Cheeses ... 444
  - 9.3.1 Raw Materials ... 452
  - 9.3.2 Cheese Production ... 453
  - 9.3.3 Production of Cheese Varieties ... 455
  - 9.3.4 Key Physicochemical Properties ... 457
- 9.4 Plant-Based Cheese Ingredients ... 465
  - 9.4.1 Polysaccharides ... 465
  - 9.4.2 Proteins ... 467
  - 9.4.3 Fats ... 468
- 9.5 Production of Plant-Based Cheese ... 470
  - 9.5.1 Overview of Production Methods ... 471
  - 9.5.2 Fractionation Route ... 475
  - 9.5.3 Tissue Disruption Route ... 480
- 9.6 Sustainability and Health Considerations ... 487
  - 9.6.1 Greenhouse Gas Emissions ... 487
  - 9.6.2 Health Aspects ... 489
- 9.7 Other Dairy Alternatives ... 491
  - 9.7.1 Yogurt ... 491
  - 9.7.2 Ice Cream ... 500
  - 9.7.3 Whipping Cream ... 503
  - 9.7.4 Butter ... 505
- 9.8 Future Considerations ... 507
- References ... 508

10 Facilitating the Transition to a Plant-Based Diet ... 523

- 10.1 Introduction ... 523
- 10.2 Research ... 523
  - 10.2.1 Ingredient Innovation ... 523
  - 10.2.2 Food Quality Design ... 525
  - 10.2.3 Nutritional Implications ... 526
  - 10.2.4 Environmental Impacts ... 526
  - 10.2.5 Socioeconomic Impacts ... 528
- 10.3 Education ... 528
- 10.4 Consumer Awareness ... 529
- 10.5 Government Support ... 529

10.6	Food Systems Approaches	531
10.7	Final Thoughts	531
	References	533

Appendix: Analysis of Plant-Based Ingredients and Foods 535

Index 559

About the Authors

David Julian McClements was born in the north of England but has lived in California, Ireland, France, and Massachusetts since then. He is currently a distinguished professor in the Department of Food Science at the University of Massachusetts where he specializes in the areas of food design and nanotechnology. He has written numerous books, published over 1200 scientific articles, been granted several patents, and presented his work at invited talks around the world. He is currently the most highly cited author in the food and agricultural sciences. He has received awards from numerous scientific organizations in recognition of his achievements and is a fellow of the *Royal Society of Chemistry, American Chemical Society, and Institute of Food Technologists.* His research has been funded by the United States Department of Agriculture, National Science Foundation, NASA, the Good Food Institute, and the food industry.

Lutz Grossmann is currently an assistant professor in the Food Science Department at the University of Massachusetts Amherst. He grew up in Baden-Württemberg (Germany) and graduated in Food Science and Engineering from the University of Hohenheim in Stuttgart. His research focuses on facilitating a sustainable food system transition by designing holistic approaches to increase the consumption of plant- and microbial protein-rich foods. He is especially interested in combining up- and downstream processing technology with physicochemical, technofunctionality, and engineering concepts to create food textures that are nutritious, sustainable, and tasty. His research has been funded by the United States Department of Agriculture and The Good Food Institute.

Chapter 1
The Rise of Plant-Based Foods

1.1 Introduction

There has been a surge of interest by many consumers in moving towards a more plant-based diet. An increasing number of consumers are adopting vegan (no animal products), vegetarian (no meat, but still some dairy and eggs), or pescatarian (no meat, but some fish) diets. More commonly, however, consumers are adopting a flexitarian diet where they still eat meat but are trying to reduce the total amount they consume. Indeed, a survey reported by the Plant-based Foods Association in the United States indicated the number of people identifying as an omnivore, a flexitarian, a vegetarian, or a vegan were 65%, 29%, 4% and 2%, respectively (www.plantbasedfoods.org). People give a variety of reasons for excluding or decreasing animal-based foods in their diet, with the main ones being health, sustainability, and animal welfare (Fox & Ward, 2008; Fresan et al., 2020; Stoll-Kleemann & Schmidt, 2017). As flexitarians currently make up the largest percentage of consumers interested in adopting plant-based foods, and the largest potential untapped market is still omnivores, there has been a focus on creating plant-based foods that accurately simulate the look, feel, and taste of animal-based ones. These products can then easily be incorporated into a person's diet without making any large lifestyle changes. Animal-based foods include a diverse range of products, such as meat (beef, lamb, pork, chicken, burgers, sausages, nuggets), seafood (fish, shrimps, scallops), dairy (milk, cream, cheese, yogurt), and egg (scrambled egg, mayonnaise, flans) products. Many of these animal-based foods are inherently complex materials whose physicochemical properties, sensory attributes, and nutritional profiles depend on the type and structural organization of the different components they contain. For instance, the proteins in muscle foods are organized into complex nested fibrous structures, which have a pronounced impact on the desirable look, feel, and mouthfeel of these meat and fish products. The food ingredients found in plants (such as proteins, carbohydrates, and lipids) usually have very different molecular features and functional attributes than those found in animals. Consequently, creating plant-based foods that accurately simulate the desirable attributes of animal-based ones is often extremely challenging. The purpose of this book is to highlight the fundamental science and technology behind the formation of next-generation high-quality plant-based foods. By "next-generation", we mean foods that are specifically designed to mimic the properties of existing animal-based foods, such as meat, seafood, egg, and dairy products. We do not consider natural plant-based foods, such as fruits, vegetables, nuts, cereals, and legumes, or traditional plant-based foods prepared from these ingredients, such as tempeh, tofu, and seitan. We acknowledge that these are an important component of any healthy plant-based diet, but they are beyond the scope of the current work.

In this chapter, we provide an overview of the primary consumer motivations for adopting a plant-based diet. We also highlight the opportunities for the food industry created by the increasing demand for next-generation plant-based foods.

1.2 Environmental and Sustainability Reasons for Consuming Plant-Based Foods

One of the most urgent and compelling reasons for reducing the amount of animal products in the human diet is the negative impact of raising animals for food on the environment. The rearing of cows, pigs, sheep, chickens, and fish for foods is

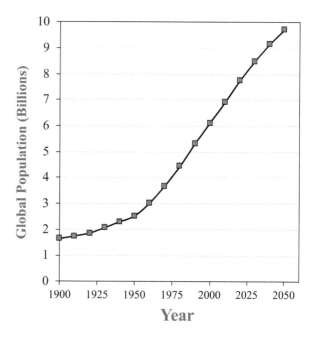

Fig. 1.1 The global population is predicted to increase to around 10 billion by 2050, meaning that more food will need to be produced to feed these extra people. (Data from *Our World in Data*. ourworldindata.org)

depleting the planet's resources, polluting the land, water, and air, and decreasing the biodiversity of native species (Poore & Nemecek, 2018). Left unchecked, this problem is only going to worsen. The global population is continuing to rise and is expected to reach nearly 10 billion by 2050, meaning that enough food will need to be produced by this date to feed another 3 billion people (Fig. 1.1). Moreover, people in many parts of the world are becoming wealthier and living longer, which means they want to consume a higher-quality diet. The consumption of animal products, particularly meat, is often associated with a more affluent lifestyle. As an example, there has been an 80% increase in the number of animals eaten globally over the past two decades. It is also predicted that there will be another 50% increase in meat consumed globally by 2050 if current trends continue.

The website *Our World in Data* (ourworldindata.org) run by Drs. Hannah Ritchie and Max Roser from the University of Oxford in the United Kingdom contains data on global meat production over the past few decades (Fig. 1.2). This data shows that there has been a rapid increase in the amount of meat produced to fulfill the world's rising demand. The world consumes almost four times as much meat now as it did 50 years ago. Indeed, around 340 million metric tons of meat were produced in 2018, comprising around 69 billion chickens, 1.5 billion pigs, 574 million sheep, and 302 million cattle. Thus, for every person on the planet, almost 10 animals died per year to feed them meat. Of course, this depends on the economic resources and cultural habits of the people living in different parts of the world. On average, Americans eat much more meat than people in most other regions of the world. In 2017, the average American ate over 124 kg (273 lb.) of meat a year, which is equivalent to about a quarter of a cow, one pig, or 37 chickens. If we want to reduce

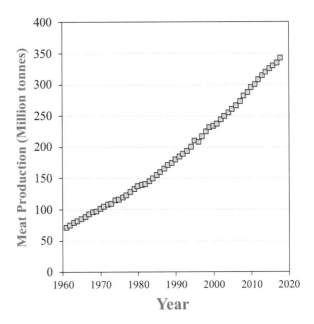

Fig. 1.2 The amount of meat produced each year has increased steeply over the past few decades. (Data from *Our World in Data*. ourworldindata.org)

the negative impacts of eating animal-based foods on our environment, it will be important that the food industry has viable alternatives to these products that consumers actually want to eat.

1.2.1 The Inefficiency of Animals as Foods

One of the major problems with eating animals is that they are not a particularly efficient means of converting the planet's scarce resources into food. The amount of food protein that can be produced on the same area of land for different kinds of livestock animals is compared to that of plants in Fig. 1.3. This data shows that it would be much better for us to eat plant-based foods directly, rather than using them to feed animals that we then eat. This is because of the low feed conversion efficiency of animals, *i.e.*, they need to use some of the energy and resources they consume to grow and maintain organs that are not typically used as meat (such as hooves, cartilage, and bones). Moreover, some of the energy from the food they consume is used to power the biological processes needed for the animals to stay alive, such as breathing, thinking, moving, and maintaining their body temperature.

If one just looks at the feed conversion efficiencies, then one would think that it is never a good idea to consume meat. But one has to be careful since in some cases, animals are converting something that we cannot eat (grass) into something that is highly nutritious (meat and milk). For food security reasons, there are, therefore, some arguments for raising livestock on land that humans cannot easily cultivate for plant-based foods, such as land where it is impractical to grow arable crops. Even

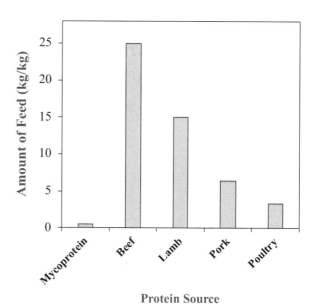

Fig. 1.3 Comparison of protein yield per unit area for animal- and plant-based protein-rich foods. Much more food protein can be produced by plants than by livestock. (Data extracted from Alexander et al. 2017)

so, the fact that animals are not particularly efficient at turning feed into food is a good argument for converting much of the land currently used to produce livestock into land for producing agricultural crops. However, we acknowledge that this is not always possible, and there are good economic and cultural arguments that can be made for still rearing some livestock in many parts of the world (Houzer & Scoones, 2021).

1.2.2 Impact of Animal Foods on the Environment

In those regions where it is feasible to replace animals with agricultural crops, there are very strong arguments for doing so. The fact that livestock, particularly cattle, are not particularly efficient at converting feed into food means that much more land, water fertilizers, and pesticides are needed to produce the same amount of food as could be produced from eating plant-based foods directly. Moreover, there is much more pollution and greenhouse gas emissions associated with livestock than agricultural crops. The magnitude of this effect can be seen in studies on the environmental impacts of the modern food supply. A study by researchers at the University of Oxford in the United Kingdom compared the environmental impacts of around 38 thousand farms that produced 40 different agricultural products from around the world (Poore & Nemecek, 2018). This study compared the environmental impacts of producing a wide range of animal- and plant-based foods. The impact of different protein-rich foods, including a plant-based one (tofu) and several animal-based ones (beef, lamb, pork, and poultry) on markers of environmental effects, such as land use, water use, greenhouse gas emissions, and acidification, are compared in Fig. 1.4. This data clearly shows that the production of the plant-based protein is much less damaging to the environment than the animal-based ones, requiring less land and water to produce, and causing much less pollution and greenhouse gas emissions. A number of other studies have also come to similar conclusions about the overall environmental benefits of switching from an animal-based to a plant-based diet (Xu et al., 2021). Most of these studies have considered switching to a broad plant-based diet, such as one containing more fruit, vegetables, cereals, and nuts, rather than introducing more next-generation plant-based foods into the diet. However, there is also some evidence from life cycle analyses that swapping specific animal-based foods (such as meatballs or burgers) with plant-based alternatives can also have benefits on the environment (Saerens et al., 2021; Saget, Costa, Santos, Vasconcelos, Gibbons, et al., 2021a; Saget, Costa, Santos, Vasconcelos, Styles, & Williams, 2021b).

It was recently reported by the Good Food Institute that the global food system causes around 34% of total global greenhouse gas emissions, with half of this being attributable to protein production (GFI, 2021a). This amount exceeds the total emissions from all sectors in the United States combined, which highlights the magnitude of the problem. These greenhouse emissions are linked to a variety of sources,

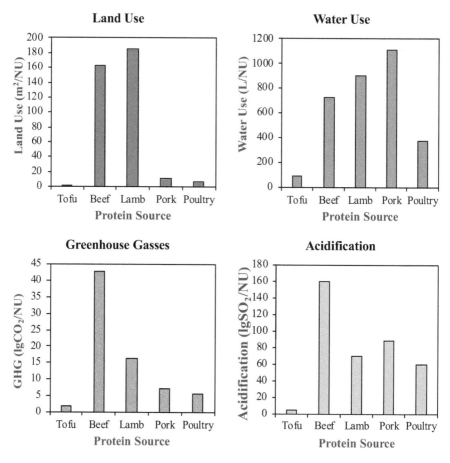

Fig. 1.4 Comparison of the environmental impacts of different protein sources. Raising livestock for food is much worse for the environment than growing agricultural crops, such as soybeans for tofu. (Data from Poore and Nemecek 2018)

including deforestation for livestock and feed production, emissions from feed crops, and direct emissions, including methane and nitrous oxide.

1.2.3 Improving the Efficiency of Food Production

The environmental impact of raising animals for food depends on the manner in which livestock are raised and killed. There are huge variations depending on the animals and farming practices employed. The domesticated animals used for livestock production have been bred over centuries to be highly efficient at producing as much meat as possible in as short a time as possible. Moreover, modern industrial farms often involve concentrated animal feeding operations (CAFOs), where large numbers of animals are confined into a relatively small space, and their feed is carefully designed to ensure maximum growth rates. Somewhat surprisingly, the animals raised for meat on large industrial farms often have less environmental impact (on some important measures) than those free to roam across green pasture lands. This was highlighted in a paper published by a group of researchers based in California and Oregon (Swain et al., 2018), who concluded "modern, intensive livestock systems, especially for beef, offer substantially lower land requirements and greenhouse gas emissions per kilogram of meat than traditional, extensive ones." For instance, the carbon footprint of factory-farmed beef production ranged from 9 to 42 units (kg CO_{2-eq} kg^{-1}), whereas that of pastoral farmed beef (where the cattle were free to roam around) was 12–129 units. Similarly, the land use for the factory-farmed beef (15–29 m^2 kg^{-1}) was considerably lower than that for the pastoral version (286–420 m^2 kg^{-1}). These authors argued that modern factory farming practices could help protect the environment, especially in places like the Brazilian Amazon, since less land and resources would be needed to produce the same amount of meat. The reason that less water and feed are needed is because the animals grow much faster and reach their full size in a shorter time. As a result, there would be less pressure to cut down the rainforests, which are essential for maintaining a healthy global climate.

Having said this, there are serious ethical and environmental problems associated with CAFOs. In particular, there are concerns about animal welfare and the generation of millions of tons of manure, which pollutes the surrounding land, air, and water. Indeed, the sewage lakes associated with CAFOs greatly reduce the quality of life and health of the people living near them. If people consumed less meat as part of their diet, there would be fewer concerns with producing livestock using traditional methods. As a result, a greater fraction of livestock animals could be raised under conditions that improved their welfare without causing excessive pollution. In addition, newer regenerative agricultural practices may be able to reduce greenhouse gas emissions and even sequester some of them from the air, thereby reducing global warming (Kleppel, 2020). Moreover, these kind of farming practices can lead to improvements in soil quality and less pollution, which are essential for ensuring sustainable and resilient agricultural systems (Kwon et al., n.d.).

Researchers are trying to develop innovative methods to reduce the negative impacts of livestock on our environment. A major contribution of cows to global warming is their tendency to produce methane in their stomachs when they digest grasses and other feed, which is then released into the atmosphere through belches or flatulence. Scientists are currently working to reduce the methane emissions from cows, thereby decreasing their negative impact on global warming (Schlossberg, 2020). For instance, it was found that adding relatively small quantities of a crimson seaweed (*Asparagopsis*) grown off the coast of Australia to a cow's diet could reduce methane production by around 98%. The seaweed does this because it contains a natural compound (bromoform) that blocks the conversion of hydrogen and carbon to methane inside the cow's stomach. Moreover, growing the seaweed extracts greenhouse gasses from the air, thereby having a positive effect on the climate. For these reasons, seaweed farms and processing facilities are being set up to extract this methane-reducing compound from a sustainable resource, so that they can provide ingredients that farmers can incorporate into their animal feed. If these ingredients can be produced economically in sufficient quantities, they could have an important impact on global warming. However, it still has to be established that these ingredients do not damage the cow's health or alter meat quality and yield. Again, switching to a more plant-based diet would more effectively alleviate many of the current environmental problems associated with livestock production.

Fish is another important source of high-quality protein in the human diet, as well as a good source of other important nutrients, such as omega-3 fatty acids, vitamins, and minerals. However, overfishing of wild fish populations is depleting the oceans of this valuable resource (FAO, 2020; GFI, 2019). Moreover, climate change is altering fish migration patterns, with profound effects on the fishing industry and coastal communities (FAO, 2018; Lavelle, 2015). Wild fish may also contain appreciable levels of toxins, such as mercury or persistent organic pollutants. The rapidly growing aquaculture industry alleviates some of these issues but has its own challenges, including the need for protein-rich resources to feed the fish and environmental pollution, such as eutrophication (DeWeerdt, 2020; White, 2017). There are also substantial losses in aquaculture due to fish diseases, such as sea lice in salmon, which contribute to food waste and economic losses (DeWeerdt, 2020). Finally, there are concerns that the antibiotics and pesticides used to tackle these diseases may contaminate fish and the environment.

1.2.4 Establishment of Planetary Boundaries

A comprehensive study of the impact of different foods on the health of people and the planet was carried out by the EAT-Lancet commission led by Professor Walter Willet from the School of Public Health at Harvard University (Willett et al., 2019). The researchers recommended that one of the most effective means of creating a healthier and more sustainable food production system was to increase the amount of plant-based foods in the human diet, while reducing the level of animal foods consumed, especially red meat and processed meat (Willett et al., 2019). The report states:

> Transformation to healthy diets by 2050 will require substantial dietary shifts. Global consumption of fruits, vegetables, nuts and legumes will have to double, and consumption of foods such as red meat and sugar will have to be reduced by more than 50%. A diet rich in plant-based foods and with fewer animal source foods confers both improved health and environmental benefits.

The authors of the EAT-Lancet commission also stated that "Global food production threatens climate stability and ecosystem resilience and constitutes the single largest driver of environmental degradation and transgression of planetary boundaries". It is clear that a radical transformation of the global food system is required to help combat problems associated with climate change and food

1.2 Environmental and Sustainability Reasons for Consuming Plant-Based Foods

sustainability. The commission made a number of concrete suggestions about how to achieve this goal. In particular, they set several "universal scientific targets for the food system that apply to all people and the planet". To this end, the commission established a set of guidelines to define how a healthy food supply can be produced without transgressing planetary boundaries and damaging the planet. Some of the boundaries proposed are summarized in Table 1.1. Individuals, governments, and industries can apply this framework to work towards a more environmentally sustainable food supply, so that we can feed future generations without damaging the planet. Unfortunately, we are already approaching or exceeding the proposed boundaries, and this is likely to worsen unless individuals, companies, and governments act soon (Fig. 1.5).

Table 1.1 Targets for key food and agriculture processes that need to be reached to help produce a healthy diet for all without damaging our planet – set by the EAT-Lancet Commission

Earth system process	Control variable	Target boundary
Climate change	GHG emissions	5 gigatonnes $CO_{2\text{-eq}}$/year
Land-system change	Cropland use	13 million km^2
Freshwater use	Water use	2500 km^3/year
Nitrogen cycling	Nitrogen application	90 teratonnes/year
Phosphorous cycling	Phosphorous application	8 teratonnes/year
Biodiversity rate	Extinction rate	10 species per year

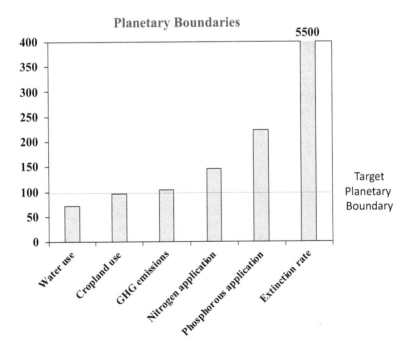

Fig. 1.5 Comparison of the current situation with the target planetary boundaries (expressed as a percentage of target). The data shows that a number of planetary boundaries are already being exceeded and need to be reversed. (Data from Willett et al. 2019)

The EAT-Lancet commission came up with a specific set of actions that should be adopted to meet the proposed targets, which involved some radical changes in the way that foods are produced and consumed (Willett et al., 2019):

- *Dietary changes*: People should eat less meat and more healthful plant-based foods like fruits, vegetables, whole grains, nuts, and seeds.
- *Increase yields and efficiency*: The agricultural, transportation, and food production systems should be improved to sustainably produce healthy foods using fewer resources (land, water, fertilizers, pesticides, and fossil fuels).
- *Reduce waste*: The amount of food currently wasted during production, storage, and consumption should be reduced.

Substantial changes in the agricultural, transportation, and food manufacturing sectors are therefore required to achieve these goals, as well as in the way foods are marketed and sold. This will require policy changes at both the national and international levels so as to promote affordable, accessible, and sustainable healthy foods, such as taxes, subsidies, regulations, grants, and education programs. These policy changes should be designed to encourage the agriculture and fishing industries to produce a diverse range of healthy and sustainable foods, rather than just increasing quantities and lowering costs. There should also be greater emphasis in the agricultural and food sectors on developing and adopting technological and management innovations. Improved farming practices like regenerative agriculture will be critical, as well as advanced technologies such as gene editing, cellular agriculture, alternative proteins, biotechnology, nanotechnology, sensors, robotics, automation, and big data (McClements, 2019; WEF, 2019). The food and agricultural industries, therefore, need to be radically transformed. Indeed, the authors of a recent study in *Science* magazine concluded that a rapid and radical transformation of the global food supply was critical to meeting the Paris Agreement climate goals of limiting global warming to 2 °C or less (Clark Michael et al., 2020).

In this book, we focus on the development of plant-based alternatives to traditional animal-based products, such as meat, fish, eggs, and dairy, which will make an important contribution to ensuring this transformation is successful.

1.2.5 Ensuring Biodiversity

Finally, the land used to raise the agricultural crops needed to feed cattle is displacing other animals, leading to a dramatic decrease in global biodiversity. To produce more food for humans, rainforests, savannas, and grasslands are being converted to crop and pasture lands, which is threatening many plant and animal species with extinction. The World Wildlife Fund estimates that over 60% of the mammals, birds, fish, and reptiles on the planet have been lost due to human activities since 1970, with food production being a major contributor to this problem. These findings were confirmed by an extensive 2021 report on the impact of the modern food system on biodiversity loss by Chatham House, also known as the Royal Institute of International Affairs, which is based in London (Benton et al., 2021). The authors found that the average rate of species extinction is orders of magnitude higher now than at any time during the past 10 million years. This timeframe is 50 times longer than modern humans have been on the planet. The conversion of natural ecosystems into farmland for pasture or crop production was identified as the main driver for this biodiversity loss, which itself was driven by the demand for ever cheaper and more abundant foods, especially animal-based ones. The authors of this report highlighted the urgent need for societal, policy, and economic changes to create a more sustainable food supply, with special emphasis again on switching from animal- to plant-based foods.

In short, numerous important studies have concluded that we should eat less meat if we are serious about addressing global warming and other environmental issues.

1.3 Ethical Reasons for Consuming Plant-Based Foods

A major reason that some people do not consume animals is because of ethical reasons – they are concerned about animal welfare and other undesirable consequences of raising animals for food (Alvaro, 2017a; Ursin, 2016). Indeed, a recent study reported that animal welfare was the top concern of Americans aged 18–30 (Feldmann et al., 2021). These people believe that animals have certain rights, so it is wrong to confine and kill them to provide food for humans. Many consumers exhibit a form of cognitive dissonance - they dissociate living animals from the foods on their plates. Few people in most developed countries would be willing to slaughter a live animal for food. However, many people are happy to buy a hamburger, chicken nuggets, or a beef steak. As mentioned earlier, nearly 70 billion animals are killed each year for food, with most of them being chickens. The way these animals are bred, raised, and slaughtered varies greatly depending on the nature of the farm. Some animals live on farms or in pasture lands where they are free to roam, whereas others are confined to tiny spaces where they may not even be able to turn around. Increasingly, more and more animals are being raised on large industrial-scale farms so that vast quantities of animal-based foods, such as meat, eggs, and milk, can be produced cheaply (Rossi & Garner, 2014). However, the practices used to raise and slaughter animals on these farms have adverse effects on animal welfare, the environment, zoonotic disease transmission, and rural communities.

As well as being highly confined, many livestock animals are kept under unnatural lighting conditions and made to stand on hard or uneven surfaces that make them uncomfortable or distressed (Rossi & Garner, 2014). They may also be packed so closely to other animals that it promotes mental distress, fighting, and injury. To avoid these problems, farmers may cut off the beaks of chickens, horns of cattle, or tails of pigs. Many livestock animals are bred to rapidly produce large yields rather than for their comfort or wellbeing. Moreover, they are fed foods that are not normal to them. Animals living under these highly unnatural conditions are often more susceptible to disease and deformity, further reducing their quality of life. Philosophers have used virtue ethics to argue that it is unethical to consume meat when similar or better plant-based alternatives are available that do not involve harming animals (Alvaro, 2017a, 2017b). They argue that a temperate, just, fair, and compassionate person would not want to cause distress to another living creature that can feel pain and distress. However, other philosophers have used virtue ethics to argue that, in some cases, a virtuous person would be justified in including meat in their diet (Bobier, 2021). This argument is partially based on the fact that even growing plants for foods causes some harm to animals, *e.g.,* pesticides, the clearing of land, and the mechanical harvesting of crops can cause injury or death to animals, birds, and fish. (It should be noted that this could be avoided by altering agricultural practices to prevent these effects). Followers of this philosophical position, which are known as virtuous new omnivores, do not support the consumption of animal foods produced in "factory farms". Instead, they believe that it is ethical to consume animals that do not feel pain (such as insects) or animals that have died naturally or from other causes (such as roadkill), as this would mean that there would be less demand for livestock animals and therefore less suffering. However, it is unlikely that roadkill would be a sufficient or acceptable source of meat for most people. Insects, on the other hand, may be a viable source of protein in people's diet, provided they can overcome their feelings of disgust and neophobia (La Barbera et al., 2018; Tan et al., 2015). It should be noted that lobsters were once considered to be a food for the poor in Europe and North America in the seventeenth and eighteenth centuries, but then became a fashionable and desirable food item for many people (Spanier et al., 2015), which highlights the possibility of changes in food acceptance.

Overall, it seems that if people believe that animals have some rights, they should either stop eating meat and other animal products, or, at least, they should only buy them if they had been treated more ethically.

1.4 Health Reasons for Consuming Plant-Based Foods

Another major reason many people adopt a more plant-based diet is that they believe it is healthier for them (Corrin & Papadopoulos, 2017; Fox & Ward, 2008). Plant- and animal-based diets differ in the types and amounts of macronutrients (proteins, lipids, and carbohydrates), micronutrients (vitamins and minerals), and nutraceuticals (such as carotenoids, prebiotics, dietary fibers, bioactive peptides, and omega-3 fatty acids) they contain. They also differ in their rate and extent of digestion within the gastrointestinal tract depending on their composition, structure, and degree of processing. Moreover, they differ in the levels of potentially negative nutrients and other constituents they contain, such as salts, sugars, saturated fats, and toxins. Finally, plant- and animal-based foods have different impacts on the gut microbiome of humans, which may influence their health (Toribio-Mateas et al., 2021).

Therefore, there may be appreciable differences in the nutritional and health implications of consuming omnivore, vegetarian, or vegan diets. However, the relationship between diet and health is extremely complex, and nutritional and medical scientists are still trying to understand the relative advantages and disadvantages of different kinds of diet. Even so, studies suggest that healthful plant-based diets, particularly those rich in fruits, vegetables, legumes, nuts, whole grains, tea, and coffee appear to have various health benefits over diets containing substantial amounts of animal foods, especially red and processed meats (Hu et al., 2019). In contrast, unhealthful plant-based diets, which contain large amounts of refined grains, potatoes, sweets, desserts, snacks, fruit juices, and sweetened beverages may have detrimental effects on health. The potential health impacts of switching from an omnivore to a plant-based diet are discussed in more detail in a later chapter (Chap. 5).

As well as promoting human health, switching to a more plant-based diet may also reduce the risk of human disease. A large proportion of the antibiotics administered in many countries are to ensure the health of livestock animals rather than for the treatment of human diseases, which is leading to an increase in antibiotic resistance. As a result, some of the antibiotics that are currently effective at treating human diseases may not work in the future, which could have serious health implications for the general population (Ma et al., 2019; Mthembu et al., 2021). Another potential health problem is associated with the rearing of animals on large industrial-scale farms. Having livestock animals living in close proximity to each other increases the risks of zoonotic diseases, *i.e.*, the chance for a disease arising in livestock and then being transmitted to humans. The devasting global consequences of viral infections arising from zoonotic transmissions, such as swine flu, avian flu, and COVID-19, highlight the potential severity of these diseases on human health and the global economy. Consequently, there is a strong argument not to eat animals or their products if they are raised on factory farms because this puts us all at an increased risk of illness or death from zoonotic diseases (Jones, 2021).

1.5 The Importance of Taste

Although consumers may be driven to change their dietary habits due to ethical, environmental, or health reasons, it is important that plant-based foods are also desirable, affordable, convenient, and readily available, otherwise a sufficiently large proportion of the general population will not adopt them. Indeed, the International Food Information Council (IFIC) carries out an annual consumer survey to identify the main factors that drive consumer preferences for foods and beverages. In 2021, the major drivers (with % importance) were taste (82%), cost (66%), health (58%), convenience (52%), and environmental sustainability (31%) (Fig. 1.6). The relative importance of these drivers has remained about the same for the past decade or so. These surveys highlight the fact that if a food is not tasty and affordable, it is unlikely to be adopted by consumers no matter how healthy or

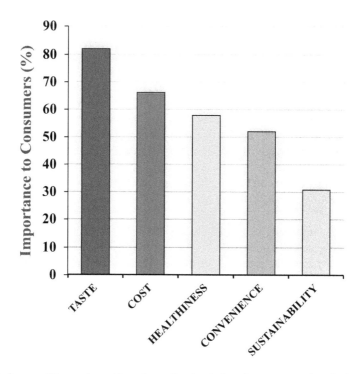

Fig. 1.6 Survey of the purchase drivers impacting the choice of consumers to buy foods or beverages (1014 American consumers, March 2021). (Data from the 2021 FOOD & HEALTH SURVEY by the International Food Information Council)

sustainable it is. Indeed, a survey carried out by The Good Food Institute reported that the most common reasons that people give for not eating more plant-based meat products are taste and cost (GFI, 2021b). Other reasons that consumers reported are that they are too processed, contain too many ingredients, or do not seem as nutritious as animal-based foods. The plant-based food industry has therefore devoted considerable resources to creating high quality and inexpensive products that consumers actually want to incorporate into their diet. Indeed, this has been one of the major achievements of the food industry over the past decade or so – there are now a wide range of delicious, affordable, and convenient plant-based foods available to consumers. Many of these foods are designed to accurately mimic the physicochemical and sensory attributes of conventional animal-based products, such as meat, fish, eggs, or dairy products. This is because many consumers are already familiar with these products and may have an emotional attachment to them. Moreover, as mentioned earlier, the biggest potential market for plant-based foods is the growing flexitarian one. The creation of plant-based foods requires a comprehensive understanding of numerous disciplines, including agriculture, food chemistry, chemical engineering, materials science, analytical chemistry, engineering, nutrition, microbiology, human physiology, sensory science, psychology, and marketing. Considerable progress has already been made for some types of processed

1.6 Opportunities for the Food Industry

Fig. 1.7 Plant-based burgers sizzling on the grill. (Image kindly supplied by Impossible Foods (Redwood City, CA), with permission)

meat products, such as plant-based burgers (Fig. 1.7). Further advances in this area will therefore depend on improving our understanding of each of these sub-disciplines, as well as successfully integrating them together. Many of the current generation of plant-based foods are highly processed products that contain a large number of different ingredients. Consequently, there is a need to carry out research and development that can simplify the ingredients and processes required to produce these foods, while also maintaining food quality. In addition, many of the current generation of products have not been specifically designed to have better or comparable nutritional profiles as animal-based ones, and so further research is also required in this area.

1.6 Opportunities for the Food Industry

The development of plant-based foods, as well as those assembled from other sources of alternative proteins, provide a number of economic opportunities for investors and for the food industry. The increasing focus of many governments on achieving net-zero emissions means that a growing number of consumers are looking for more climate-friendly and sustainable sources of proteins. In principle, these products can be manufactured at lower costs than traditional animal-based ones, leading to increased sales and profits. The World Economic Forum has highlighted the enormous opportunities for innovation within the food industry, including in the alternative protein space (WEF, 2019). Indeed, there has been a rapid growth in the number of established food companies introducing plant-based product lines, as well as the formation of numerous start-up companies creating new kinds of plant-based products (GFI, 2021b).

The Good Food Institute, a non-profit organization based in Washington D.C. whose mission is to promote the transition to a more sustainable and healthy food supply by stimulating research in alternative proteins, reviewed the growth of plant-based foods in its State of the Industry Report: Plant-Based Meat, Eggs, and Dairy for 2020 (GFI, 2021b). The annual investment in plant-based foods has increased substantially over the past decade as more consumers incorporate plant-based foods into their diet (Fig. 1.8). Indeed, there was a 27% growth in the plant-based food market in the United States in 2020, with the total value reaching around $7 billion. The growth of this segment of the market was reported to be twice that of the overall retail food market. One of the fastest growing segments in this market was the plant-based meat segment, which saw sales increased by around 45%.

The plant-based food market is comprised of a broad range of products that are usually designed to mimic the properties of animal-based products, such as meat, fish, eggs, and dairy (Fig. 1.9). Currently, plant-based milks make up the biggest percentage of this segment of the food industry, with plant-based meats being second. For many plant-based products, there is still considerable room to grow, which is partly because of the difficulty in accurately simulating the physicochemical and sensory attributes of the original animal-based products. As the science develops, the quality of these products will increase, and their cost will decrease, which is likely to spur further growth in these sectors.

Despite their substantial growth over the past few years, plant-based foods still only make up a relatively small fraction of the total food market (Fig. 1.10).

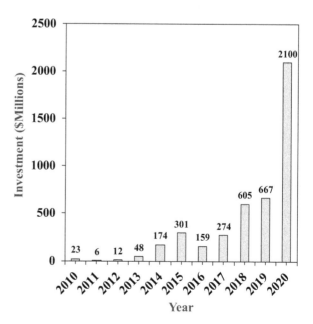

Fig. 1.8 Global annual investment in plant-based companies from 2010 to 2020. (Data from the Good Food Institute 2021b)

1.6 Opportunities for the Food Industry

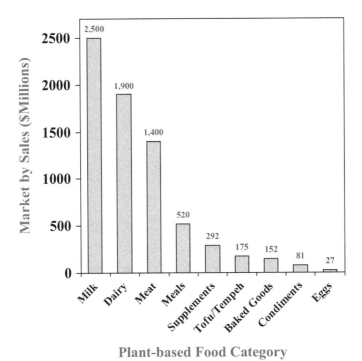

Fig. 1.9 Retail sales of plant-based foods by category. (Data from the Good Food Institute 2021b)

Plant-based milks have been the most successful product in this respect, currently making up around 15% of the total market for milk and milk-like products (GFI, 2021b). In contrast, plant-based meat, cheese, and egg only make up 1.4, 1.1, and 0.4% of the total market, which means there is considerable room for growth. One of the main challenges is to make high-quality and affordable versions of conventional meat, cheese, and egg products. Over half the households in the U.S. have already purchased some form of plant-based foods, so if these kinds of products can be produced, they will likely be successful. Moreover, the people who tend to buy plant-based products tend to be from higher income brackets and are younger than average consumers (GFI, 2021b), which suggests they have sufficient resources to buy these products and that the category will continue to be important in the future. Indeed, The Good Food Institute has reported that the global plant-based meat market alone has been projected to be worth around $12 to $380 billion by the 2030s.

Another potential advantage to food manufacturers in replacing animal-based foods with plant-based ones is to reduce fluctuations in the supply and price of key food ingredients (Grizio & Specht, 2021). Many animal products are susceptible to volatility in their price and availability due to disruptions in the supply chain. For instance, Avian flu, Swine flu, and Covid-19 led to significant disruptions in the supply of chicken, pork, and beef. Consequently, if food manufacturers replaced animal-derived ingredients with plant-based alternatives, then they may be able to

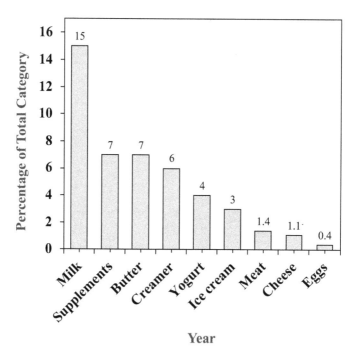

Fig. 1.10 Percentage of the total category in the U.S. retail market for different plant-based foods. (Data from the Good Food Institute 2021b)

reduce these price and supply fluctuations. Moreover, plant-based ingredients are often cheaper than animal-based ones, which could lead to economic advantages. Finally, it is often easier to handle and store plant-based ingredients than animal-based ones since the latter are highly perishable.

1.7 Other Sources of Alternative Proteins

In this book, we focus on the use of plants as an alternative source of dietary proteins to animals. However, there are also various other sources of alternative proteins that are being explored for their potential to replace animal proteins in foods (McClements, 2019). Insects are being utilized as a protein-rich food source. Insects can be caught in the wild or raised in large insect farms. There are a large number of edible insects around the world that can be consumed by humans. Indeed, over 2 billion people already regularly eat insects as part of their diet. Many types of insects are rich sources of proteins, lipids, dietary fibers, vitamins, and minerals and can therefore be part of a healthy diet. However, many consumers in developed countries do not currently find insects to be a desirable food source. Other ethical and sustainable sources of alternative proteins that can be used to formulate plant-based meat, fish, egg, and dairy analogs are fermentation and cell cultivated products. In the case of fermentation, proteins and other ingredients are typically produced by yeast, fungi, or bacteria in fermentation tanks. DNA sequences for specific animal proteins can be inserted into these microbes so that they can be programmed to produce these proteins, which can then be isolated, purified, and used as food ingredients. Alternatively, whole microbial cells can be cultivated and then consumed as a protein-rich source. Meat and fish tissues can also be grown in fermentation tanks by incubating living cells (taken from an animal that does not need to be killed) in a suitable nutrient and growth media under appropriate environmental conditions. These tissues can then be used to make meat- or fish-like products. The application of insects, fermentation, and cell cultivated products is beyond the scope of this book but is likely to be an important alternative source of proteins in the future.

1.8 Conclusions

There is growing evidence of the negative impacts of diets containing high levels of animal-based foods (particularly those from cows) on the environment and human health. Moreover, there are serious ethical concerns linked to animal welfare when we raise and slaughter animals for foods. For these reasons, many consumers are switching from an omnivore to a vegan, vegetarian, pescatarian, or flexitarian diet, which involves consuming less or no meat. Producing plant-based foods requires much less land and water than raising livestock, as well as causing considerably less pollution and biodiversity loss. If we are serious about feeding the billions of new people that will be born over the next 30 years or so, without seriously damaging our planet, we have to make dramatic changes in the way we produce and distribute our foods.

The authors of the EAT-Lancet commission state that "a diet rich in plant-based foods and with fewer animal source foods confers both improved health and environmental benefits". There is therefore compelling evidence for eating less meat. However, people will not incorporate plant-based foods into their diet if they are not delicious, affordable, convenient, healthy, and safe. In this book, we describe the ingredients and processes that can be used to create next-generation plant-based foods that meet these criteria. In particular, we focus on the production of plant-based analogs of meat, seafood, egg, and dairy products because these are currently the most appealing to most consumers. We hope the knowledge presented in this book will stimulate more research and development in this area and facilitate the transition to a more sustainable, healthier, and ethical food supply.

References

Alexander, P., Brown, C., Arneth, A., Dias, C., Finnigan, J., Moran, D., & Rounsevell, M. D. A. (2017). Could consumption of insects, cultured meat or imitation meat reduce global agricultural land use? *Global Food Security, 15*, 22–32.

Alvaro, C. (2017a). Ethical veganism, virtue, and greatness of the soul. *Journal of Agricultural & Environmental Ethics, 30*(6), 765–781.

Alvaro, C. (2017b). Veganism as a virtue: How compassion and fairness show us what is virtuous about veganism. *Future of Food-Journal on Food Agriculture and Society, 5*(2), 16–26.

Benton, T., Bieg, C., Harwatt, H., Pudasaini, R., & Wellesley, L. (2021). *Food system impacts on biodiversity loss: Three levers for food system transformation in support of nature* (pp. 1–75). Chatham House.

Bobier, C. A. (2021). What would the virtuous person eat? The case for virtuous omnivorism. *Journal of Agricultural & Environmental Ethics, 34*(3).

Clark Michael, A., Domingo Nina, G. G., Colgan, K., Thakrar Sumil, K., Tilman, D., Lynch, J., Azevedo Inês, L., & Hill Jason, D. (2020). Global food system emissions could preclude achieving the 1.5° and 2°C climate change targets. *Science, 370*(6517), 705–708.

Corrin, T., & Papadopoulos, A. (2017). Understanding the attitudes and perceptions of vegetarian and plant-based diets to shape future health promotion programs. *Appetite, 109*, 40–47.

DeWeerdt, S. (2020). Cultivating a sea change: Can aquaculture overcome its sustainability challenges to feed a growing global population? *Nature, 588*, S60–S62.

FAO. (2018). *Impacts of climate change on fisheries and aquaculture: Synthesis of current knowledge, adaptation and mitigation options. In.* Food and Agriculture Organization of the United Nations.

FAO. (2020). *The state of world fisheries and aquaculture 2020. Sustainability in action.* Food and Agriculture Organization of the United Nations.

Feldmann, D., Thayer, A., Hanna, M., Dashnaw, C., & Hansen, T. (2021). Influencing young America to act. In *Cause and social influence* (pp. 1–5).

Fox, N., & Ward, K. (2008). Health, ethics and environment: A qualitative study of vegetarian motivations. *Appetite, 50*(2–3), 422–429.

Fresan, U., Errendal, S., & Craig, W. J. (2020). Influence of the socio-cultural environment and external factors in following plant-based diets. *Sustainability, 12*(21).

GFI. (2019). *An ocean of opportunity: Plant-based and cultivated seafood for sustainable oceans without sacrifice* (pp. 1–39). The Good Food Institute.

GFI. (2021a). *Global food system transition is necessary to keep warming below 1.5°C* (pp. 1–12). The Good Food Institute.

GFI. (2021b). *State of the industry report: Plant-based meat, eggs, and dairy* (pp. 1–85). Good Food Institute.

Grizio, M., & Specht, L. (2021). *Plant-based egg alternatives: Optimizing for functional properties and applications* (pp. 1–22). The Good Food Institute.

Houzer, E., & Scoones, I. (2021). *Are livestock always bad for the planet? Rethinking the protein transition and climate change debate* (pp. 1–74). PASTRES.

Hu, F. B., Otis, B. O., & McCarthy, G. (2019). Can plant-based meat alternatives be part of a healthy and sustainable diet? *JAMA – Journal of the American Medical Association, 322*(16), 1547–1548.

Jones, B. (2021). Eating meat and not vaccinating: In defense of the analogy. *Bioethics, 35*(2), 135–142.

Kleppel, G. S. (2020). Do differences in livestock management practices influence environmental impacts? *Frontiers in Sustainable Food Systems, 4*.

Kwon, H., Liu, X. Y., Xu, H., & Wang, M. C. (n.d.). Greenhouse gas mitigation strategies and opportunities for agriculture. *Agronomy Journal*.

La Barbera, F., Verneau, F., Amato, M., & Grunert, K. (2018). Understanding Westerners' disgust for the eating of insects: The role of food neophobia and implicit associations. *Food Quality and Preference, 64*, 120–125.

Lavelle, M. (2015). Collapse of New England's iconic cod tied to climate change. *Science, 1*.

Ma, Z. X., Lee, S. Y., & Jeong, K. C. (2019). Mitigating antibiotic resistance at the livestock-environment interface: A review. *Journal of Microbiology and Biotechnology, 29*(11), 1683–1692.

McClements, D. J. (2019). *Future foods: How modern science is transforming the way we eat.* Springer.

References

Mthembu, T. P., Zishiri, O. T., & El Zowalaty, M. E. (2021). Genomic characterization of antimicrobial resistance in food chain and livestock-associated salmonella species. *Animals, 11*(3).

Poore, J., & Nemecek, T. (2018). Reducing food's environmental impacts through producers and consumers. *Science, 360*(6392), 987.

Rossi, J., & Garner, S. A. (2014). Industrial farm animal production: A comprehensive moral critique. *Journal of Agricultural & Environmental Ethics, 27*(3), 479–522.

Saerens, W., Smetana, S., Van Campenhout, L., Lammers, V., & Heinz, V. (2021). Life cycle assessment of burger patties produced with extruded meat substitutes. *Journal of Cleaner Production, 306*, 127177.

Saget, S., Costa, M., Santos, C. S., Vasconcelos, M. W., Gibbons, J., Styles, D., & Williams, M. (2021a). Substitution of beef with pea protein reduces the environmental footprint of meat balls whilst supporting health and climate stabilisation goals. *Journal of Cleaner Production, 297*, 126447.

Saget, S., Costa, M., Santos, C. S., Vasconcelos, M., Styles, D., & Williams, M. (2021b). Comparative life cycle assessment of plant and beef-based patties, including carbon opportunity costs. *Sustainable Production and Consumption, 28*, 936–952.

Schlossberg, T. (2020). An unusual snack for cows, a powerful fix for climate. In *Washington Post* (Vol. November 27, 2020). Washington Post.

Spanier, E., Lavalli, K. L., Goldstein, J. S., Groeneveld, J. C., Jordaan, G. L., Jones, C., Phillips, B. F., Bianchini, M. L., Kibler, R. D., Diaz, D., Mallol, S., Goni, R., van Der Meeren, G. I., Agnalt, A. L., Behringer, D. C., Keegan, W. F., & Jeffs, A. (2015). A concise review of lobster utilization by worldwide human populations from prehistory to the modern era. *ICES Journal of Marine Science, 72*, 7–21.

Stoll-Kleemann, S., & Schmidt, U. J. (2017). Reducing meat consumption in developed and transition countries to counter climate change and biodiversity loss: A review of influence factors. *Regional Environmental Change, 17*(5), 1261–1277.

Swain, M., Blomqvist, L., McNamara, J., & Ripple, W. J. (2018). Reducing the environmental impact of global diets. *Science of the Total Environment, 610*, 1207–1209.

Tan, H. S. G., Fischer, A. R. H., Tinchan, P., Stieger, M., Steenbekkers, L. P. A., & van Trijp, H. C. M. (2015). Insects as food: Exploring cultural exposure and individual experience as determinants of acceptance. *Food Quality and Preference, 42*, 78–89.

Toribio-Mateas, M. A., Bester, A., & Klimenko, N. (2021). Impact of plant-based meat alternatives on the gut microbiota of consumers: A real-world study. *Food, 10*(9).

Ursin, L. (2016). The ethics of the meat paradox. *Environmental Ethics, 38*(2), 131–144.

WEF. (2019). *Innovation with a purpose: The role of technology innovation in accelerating food systems transformation* (pp. 1–42). World Economic Forum.

White, P. (2017). *Aquaculture pollution: An overview of issues with a focus on China, Vietnam, and the Philippines*. World Bank Group.

Willett, W., Rockstrom, J., Loken, B., Springmann, M., Lang, T., Vermeulen, S., Garnett, T., Tilman, D., DeClerck, F., Wood, A., Jonell, M., Clark, M., Gordon, L. J., Fanzo, J., Hawkes, C., Zurayk, R., Rivera, J. A., De Vries, W., Sibanda, L. M., … Murray, C. J. L. (2019). Food in the Anthropocene: The EAT-Lancet Commission on healthy diets from sustainable food systems. *Lancet, 393*(10170), 447–492.

Xu, X., Sharma, P., Shu, S., Lin, T.-S., Ciais, P., Tubiello, F. N., Smith, P., Campbell, N., & Jain, A. K. (2021). Global greenhouse gas emissions from animal-based foods are twice those of plant-based foods. *Nature Food, 2*(9), 724–732.

Chapter 2
Properties and Functionality of Plant-Based Ingredients

2.1 Introduction

In this chapter, we focus on the ingredients that can be used to formulate plant-based foods. The creation of these products requires careful selection and utilization of an appropriate combination of functional plant-derived ingredients. These ingredients should lead to a final product with physicochemical, functional, and sensory attributes closely simulating those of the animal-based food it is designed to replace, such as a meat, fish, egg, or dairy product. Moreover, these ingredients should have several other attributes if they are going to be used to formulate successful commercial products, including regulatory approval, low cost, reliability of supply, consistency of performance, ease of use, abundance, and label friendliness. Ideally, a plant-based product should be formulated using the least number of ingredients possible, and all the ingredients should be label-friendly, as many consumers state that they are reluctant to adopt the current generation of plant-based foods because they either contain too many ingredients or ingredients that are synthetic or undesirable (GFI, 2021). Different kinds of functional ingredients are required to formulate plant-based foods with the desired look, feel, smell, taste, sound, and mouthfeel, including gelling agents, binders, thickeners, emulsifiers, foaming agents, colorants, flavors, vitamins, minerals, nutraceuticals, buffering agents, sizzling agents, and preservatives. Each category of plant-based food (*e.g.*, meat, fish, egg, or dairy analogs) requires a different set of ingredients to obtain the expected quality attributes of that specific product. The formulation of a high-quality plant-based product requires knowledge of the nature and interactions of the different kinds of functional ingredients, as well as how their behavior changes during food processing, storage, and preparation. In this chapter, we therefore provide an overview of the properties and functionality of the most common functional ingredients used to formulate plant-based foods. These ingredients include proteins, carbohydrates, lipids, and various additives.

2.2 Proteins

Plant proteins are widely used as functional ingredients in plant-based foods because of their functional versatility. In this section, we use the term "plants" to cover species from terrestrial sources (such as soybeans, peas, wheat, and corn), as well as from marine and microbial sources (such as algae and microalgae). These latter sources are technically not plants, but they are often used as functional ingredients in plant-based foods. Plant proteins exhibit a broad range of functional attributes that makes them suitable for formulating plant-based foods, they act as binders, structure formers, thickeners, gelling agents, emulsifiers, foaming agents, fluid holders, and nutrients (Loveday, 2019, 2020). For instance, plant proteins may be used as emulsifiers to stabilize the fat droplets in plant-based meat, fish, egg, and milk, as gelling agents in plant-based meat, fish, egg, and cheese, or as foaming agents in plant-based whipped cream or ice cream. Choosing a suitable plant protein or a combination of plant proteins for a particular application is therefore critical (McClements & Grossmann, 2021a). Numerous kinds of plant proteins are available that have different molecular, physicochemical, functional, and nutritional attributes (Tables 2.1a and 2.1b). It should be noted that the functionality of a protein depends on its biological origin, the way it was isolated, its thermal history, and the environmental conditions it is used under.

2.2.1 Protein Structure

In general, proteins consist of linear chains of amino acids linked together by peptide bonds (Brady, 2013). There are 20 standard proteinogenic amino acids commonly found in nature that are used to assemble these polypeptide chains. In general, the structure of proteins is defined at different hierarchical levels (Fig. 2.1):

- *Primary structure*: This refers to the number, type, and sequence of amino acids in the polypeptide chain.
- *Secondary structure*: This refers to the presence of local regions within the polypeptide chain that have some structural organization, such as α-helices, β-sheets, or β-turns. The other regions of the polypeptide chain are assumed to have a disorganized structure.
- *Tertiary structure:* This refers to the overall 3D configuration a single polypeptide chain adopts in a particular environment.
- *Quaternary structure:* This refers to the supramolecular structure that some proteins are found in. These structures consist of one or more types of proteins (or other molecules) linked together by physical and/or chemical bonds.

Ultimately, the primary structure of a protein is governed by the DNA of the organism that produced it. Proteins differ in their primary structures because they are designed through evolutionary pressures to serve different functions in nature, such

2.2 Proteins

Table 2.1a Molecular and physicochemical attributes of major proteins in meat, eggs, and milk

	Protein % in Fraction	M_W (kDa)	pI	T_m (°C)	Conformation
Meat proteins					
Connective tissue					
Collagen	50-90	300	5–8	60–70	Fibrous
Muscle					
Myosin	29	480	~ 5.3	40–60	Fibrous-globular
Sarcoplasmic	29	20–100	Varies	50–70	Globular
Hemoglobin	3	67	6.8	67	Globular
Myoglobin	1	17–18	6.8–7.2	79	Globular
Actin	13	43	~ 5.2	70–80	Globular
Egg proteins					
Egg yolk			6.0		
Livetins (α, β, γ)	12	33–203	4.3–7.6 (5.3)	83.3	Globular
Phosvitin	7	35	4.0	80.0	Globular
HDL	12	400	4.0	72–76	Colloidal assemblies with lipids & phospholipids (4-20 nm)
LDL	68		3.5	72–76	Colloidal assemblies with lipids & phospholipids ($\approx$ 30 nm)
Egg white			4.5		
Ovalbumin	58	45	4.6	85	Globular
Conalbumin	13	80	6.6	63	Globular
Ovomucoid	11	28	3.9	70	Globular
Ovoglobulins	8	30–45	5.5–5.8	93	Globular
Lysozyme	3.5	14.6	10.7	78	Globular
Ovomucin	1.5	210	4.5-5.0	–	Globular
Milk proteins					
Caseins			4.6	N/A	Flexible
α_{S1}–casein	39	23.6		N/A	Flexible
α_{S2}–casein	10	25.2		N/A	Flexible
β–casein	36	24.0		N/A	Flexible
κ–casein	13	19.0		N/A	Flexible
Whey			5.2		
β–lactoglobulin	51	18.4	5.4	72	Globular
α–lactalbumin	19	14.2	4.4	35 & 64[a]	Globular
BSA	6	66.3	4.9	64	Globular
Immunoglobulins	12	Range	Range	Range	Globular
Lactoferrin	1–2	78	8–9	70 & 90[a]	Globular

[a]The lower and higher thermal denaturation temperatures of the α-lactalbumin and lacterferrin represeant the apo- (calcium or iron free) and holo- (calcium or iron bound) forms, respectively
Key: M_W molecular weight, pI isoelectric point, T_m thermal denaturation temperature. Adapted from McClements & Grossmann, 2021b

Table 2.1b Molecular and physicochemical properties of the major proteins in selected plant sources

	Protein in fraction (%)	M_W (kDa)	pI	T_m (°C)	Conformation
Soy					**Hydrophilic Multimers**
Globulins					
β-conglycinin (7S)	17–24	150–200	5	80	Globular
Glycinin (11S)	36–51	300–380	4.5	93	Globular
Pea					**Hydrophilic Multimers**
Globulins	55–80		4.5	75–79	
- Legumin (11S)		360			Globular
- Vicilin (7S)		150			Globular
- Convicilin		280			Globular
Albumins	18–25				Globular
- Albumin (2S)		50	6.0	110	Globular
Lentil			4.5	120	**Hydrophilic Multimers**
Globulins	51	15–92			Globular
- Legumins	45	14–92			Globular
- Vicilins	4	20–82			Globular
Albumins	17	20–82			Globular
Glutelins	11	17–46			Globular
Prolamins	4	17–64			Globular
Chickpea			4.5	90	**Hydrophilic Multimers**
Globulins	74	15–92			Globular
Albumins	16	20–82			Globular
Prolamins	0.5	17–64			Globular
Lupin			4.5	79–101	**Hydrophilic Multimers**
Globulins	75	150–216	5.6–6.2	103	Globular
Albumins	25		4.3–4.6		Globular
Canola		14–59	4.5	84–102	**Hydrophilic Multimers**
Globulins	60				Globular
Albumins	20				Globular
Glutelins	15–20				Globular
Prolamins	2–5				Globular
Corn Zein			6.4	89	**Hydrophobic Multimers**
α-zein	75–85	19–24			Packed helices
β-zein	10–15	14–15			Packed helices
γ-zein	5–10	16–27			Packed helices

Key: M_W molecular weight, *pI* isoelectric point, T_m thermal denaturation temperature. Adapted from McClements & Grossmann, 2021b

2.2 Proteins

as enzymes, transporters, signaling molecules, and structure formers. The number, type, and sequence of the amino acids in the polypeptide chain determine their configuration in the present natural environment. There is a tendency for protein molecules to adopt a configuration that minimizes their free energy in a particular environment, which involves adopting a configuration that maximizes the number of favorable molecular interactions and minimizes the number of unfavorable ones. These molecular interactions include van der Waals, steric, electrostatic, hydrogen bonding, and hydrophobic interactions, as well as configurational entropy effects. The relative importance of these interactions depends on the primary structure of the protein, as well as on the environmental conditions, such as pH, ionic strength, and temperature.

In general, proteins may adopt a variety of tertiary structures, which are typically classified as globular, flexible, or fibrous depending on the overall configuration of the polypeptide chain (Fig. 2.2). Globular proteins have tight compact structures, which are roughly spheroid. Flexible proteins have relatively disordered structures with a high degree of conformational mobility. Fibrous proteins, which are typically formed from polypeptide chains that adopt helices, are relatively stiff and extended. The importance of these different structural motifs is discussed later in this section, as they impact the ability of plant proteins to simulate the functionality of some animal proteins.

The conformation that a protein tends to adopt in its natural environment is known as the *native state*, which is usually the configuration of the polypeptide chain that gives the lowest free energy (Fig. 2.3). The native structure of a protein governs its biological functions, such as enzyme activity, signaling, transporting, motility, mechanical properties, or structure formation. After proteins have been isolated from their natural environment, the tertiary and quaternary structure may change because of an alteration in the balance of molecular interactions. Moreover, polypeptide chains may be prevented from reaching their lowest free energy state in a particular environment due to the presence of kinetic energy barriers. In this case, a protein may be trapped in one or more *denatured states* because it cannot jump over a kinetic energy barrier and reach the native state. This phenomenon is important for the functionality of proteins in foods. For instance, the globular proteins in raw eggs are in their native state before cooking but they unfold and aggregate after cooking, thereby remaining in a denatured state that leads to the desirable semi-solid structure of cooked eggs.

Many plant proteins are relatively large globular proteins that are found in nature in the form of supramolecular structures consisting of numerous proteins that may be similar or different and are linked together by physical or chemical interactions (Fig. 2.4). The type and number of proteins in these supramolecular structures depends on their botanical origin and how they were isolated, as some extraction procedures may disrupt the linkages between the individual proteins. The native state of the individual globular proteins may also be disrupted during the extraction process. As a result, the functional properties of proteins depend on their native and aggregation states, and so a plant protein from the same source may have different functional attributes depending on how it was isolated.

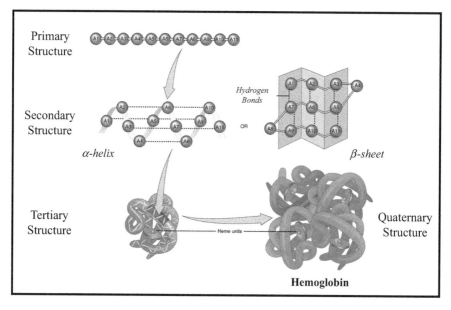

Fig. 2.1 Primary, secondary, tertiary and quaternary structures of a globular protein (hemoglobulin). (Adapted from Wikimedia Commons OpenStax College under CC BY 3.0 (https://creativecommons.org/licenses/by/3.0))

The tertiary structures of the proteins found in animals (Fig. 2.2) are often very different from those found in plants, which tend to be mainly large globular proteins (Fig. 2.4). For instance, flexible proteins like the caseins found in milk or fibrous proteins like collagen found in meat are rarely found in plants. As a result, it is more difficult to simulate their functional attributes using plant proteins. Having said this, some important functional proteins found in animals do have globular structures (such as β-lactoglobulin in milk or ovalbumin in egg), which are therefore more like the globular proteins found in plants (such as 11S glycinin from soybeans). Consequently, it is often easier to simulate their functional attributes. However, the molecular weights, surface chemistries, and thermal stabilities of globular plant proteins are different to those of globular animal proteins, which can still be challenging if these proteins are replaced with plant proteins.

The quaternary structures of the proteins found in animals are also often very different from those found in plants. For instance, the muscle tissues in meat and fish consist of bundles of fibrous proteins encased in a sheath of connective tissue comprised of collagen triple helices (Fig. 2.6) (Tornberg, 2013; Tornberg et al., 2000). The unique molecular architecture of the proteins in muscles and connective tissue plays a critical role in determining their functional attributes, such as the tenderness and juiciness of meat products. It is extremely challenging to mimic the complex molecular architecture of the proteins in meat and fish using the globular proteins derived from plants, which makes it challenging to simulate their desirable physicochemical and sensory attributes. Similarly, the production and

2.2 Proteins

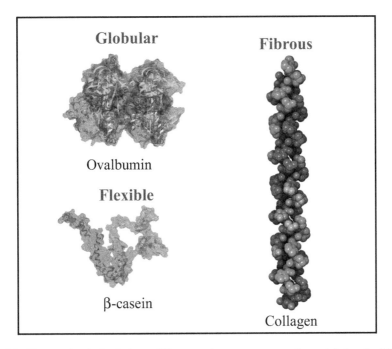

Fig. 2.2 The proteins in foods have different tertiary structures, such as globular, flexible, or fibrous depending on the conformations of their polypeptide chains. Here, we show the structure of three animal proteins that fall into these categories. The original structures the collagen triple helix (3BOS) and ovalbumin tetramer (1OVA) were taken from the NIH Protein Database. The original structure of the β-casein (CSN2) was taken from the AlphaFold Protein Structure Database. Optimized protein structures were kindly prepared by Jeff Sanders using Schrödinger Release 2021–4: Maestro, Schrödinger, LLC, New York, NY, 2021

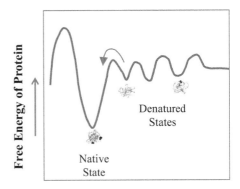

Fig. 2.3 Proteins can adopt different configurations depending on their free energy and the presence of any energy barriers. If the energy barriers are too high, they may not be able to move to the lowest free energy state (native state)

characteristics of dairy products like milk, yogurt, and cheese are governed by the nature of the casein molecules within them (Chandan & Kilara, 2013; Fox et al., 2016). Caseins are fairly disorganized and flexible molecules that contain both

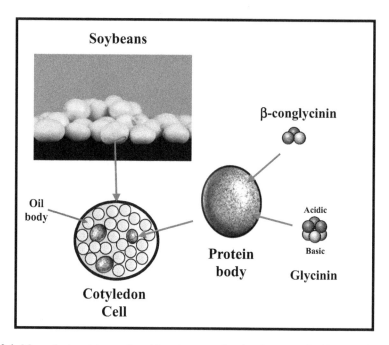

Fig. 2.4 Many plant proteins are found in nature as molecular clusters packed into protein bodies inside plant cells. This schematic diagram shows the proposed organization of the proteins in soybeans. (Image of soybeans taken from CSIRO under CC BY 3.0 (https://creativecommons.org/licenses/by/3.0/))

non-polar and polar regions, with a number of phosphate groups in the polar regions (Farrell et al., 2002). As a result, they are amphiphilic charged molecules that can interact with their neighbors through a combination of hydrophobic, calcium-mediated, and electrostatic interactions to create the unique colloidal particles (casein micelles) found in milk. Moreover, these casein micelles undergo partial disassembly and reassembly during the formation of yogurt and cheese, which leads to the generation of a 3D network that contributes to the desirable textural attributes of these products. Again, it is hard to mimic the complex behavior of these flexible disorganized casein molecules using rigid globular plant proteins, which makes it challenging to create accurate plant-based dairy products. Potential strategies for simulating the functional properties of animal proteins using plant proteins are discussed later in this book, in the chapters on different plant-based foods.

Nonetheless, knowledge of the molecular characteristics of animal proteins is useful when designing plant-based analogs using plant proteins. For this reason, the general features of the different types of functional attributes that are related to the protein structure found in animals and plants is given here (Brady, 2013; Damodaran, 2021).

Globular proteins: The polypeptide chains in globular proteins fold into compact structures that have limited flexibility (Fig. 2.2). The hydrophobic effect is usually the dominant force that favors the adoption of this type of structure, *i.e.,* the

2.2 Proteins

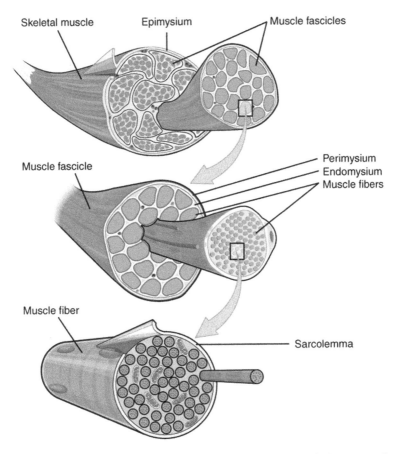

Fig. 2.5 The muscle tissues in meat and seafood have complex hierarchical structures that are often difficult to simulate using globular plant proteins. Muscle tissues contain bundles of protein-rich fibers packed into larger fibrous structures. (*Attribution*: Wikimedia OpenStax, CC BY 4.0 (https://creativecommons.org/licenses/by/4.0))

tendency for the polypeptide chains to reduce the contact area between non-polar amino acid side groups and water. As a result, the majority of non-polar amino acids tend to form a hydrophobic interior, while the majority of polar amino acids tend to form a hydrophilic exterior. However, there may still be some non-polar groups located at the exterior of the proteins, which contributes to their surface hydrophobicity. Various other kinds of molecular interactions may also play a role in determining the structural organization of globular proteins, such as hydrogen bonds (α-helix, β-sheet, and β-turn), electrostatic forces (attractive and repulsive), van der Waals attraction, and disulfide bonds. The presence of these different molecular interactions in a particular protein depends on the type and sequence of amino acids it contains.

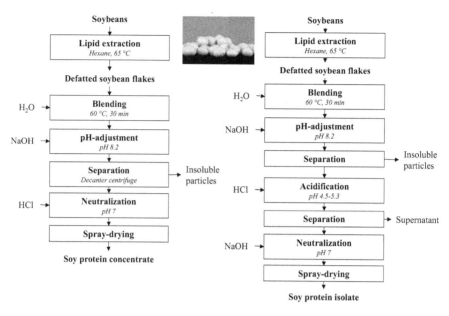

Fig. 2.6a Soy protein powders can be produced to yield different protein contents. Concentrates typically have a protein content of around 60 to 80%, whereas isolates have a protein content >88 %. The process may include additional washing steps. (See text for details). (Image of soybeans taken from CSIRO under CC BY 3.0 (https://creativecommons.org/licenses/by/3.0/)). Adapted from (Puski et al., 1987) and (Johnson, 1999)

The molecular properties of globular proteins influence their functional attributes, such as their solubility, emulsifying, foaming, and gelling properties. For instance, the solubility of globular proteins in different solutions depends on their molar mass, conformation, surface hydrophobicity, and electrical properties. The water-solubility normally increases as their surface hydrophobicity decreases and their electrical charge increases.

Most globular proteins exhibit surface activity since they have both polar and non-polar regions on their surfaces. As a result, they have a propensity to adsorb to oil-water or air-water interfaces so they can be utilized as emulsifiers or foaming agents. Many globular proteins may also be utilized as heat-set gelling agents because they unfold when heated above their thermal denaturation temperature. Consequently, some of the non-polar and sulfhydryl groups that were originally located in the interior of the proteins are exposed to the aqueous phase, which promotes protein aggregation through hydrophobic attraction and disulfide bonding. The globular proteins found in milk (such as β-lactoglobulin) and egg (such as ovalbumin) exhibit this type of gelling behavior, which plays an important role in determining their functional attributes. Many of the globular proteins found in plants (such as soy or pea protein) also can unfold and aggregate when they are heated, which means that they can also be used as heat-set gelling agents (Tables 2.1a and 2.1b).

2.2 Proteins

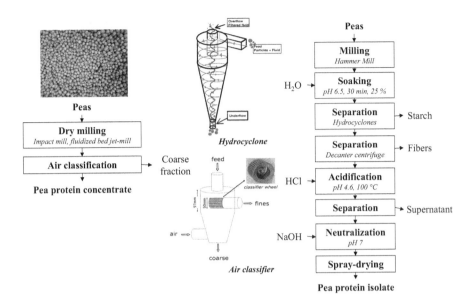

Fig. 2.6b Pea protein concentrates can be obtained using a dry milling and air classification approach to separate the starch granules (around 20 μm) from the protein bodies (around 1–3 μm). Pea protein isolates (86 % protein in dry matter) are produced using wet-extraction and isoelectric point precipitation. (See text for details). (Image of pea taken from: David Adam Kess, under CC BY-SA 4.0 (https://creativecommons.org/licenses/by/4.0/)). Adapted and reprinted with permission from Elsevier and CC BY 4.0 (http://creativecommons.org/licenses/by/4.0/) from Durango-Cogollo et al. (2020), Pelgrom et al. (2013), Schutyser et al. (2015) and Salome et al. (2007)

As mentioned earlier, many plant proteins exist in the form of supramolecular assemblies in their natural environment that are held together by physical or covalent bonds (Fig. 2.4). The functional performance of these proteins is influenced by the extent of dissociation of these supramolecular assemblies during the extraction and purification processes used to create plant-based ingredients. Moreover, the temperatures at which plant proteins undergo thermal denaturation are often different from those of animal proteins (higher or lower), which is important when trying to create plant-based foods that behave the same as animal-based ones during cooking (Tables 2.1a and 2.1b). For instance, it would be desirable to use a globular plant protein that had a similar thermal denaturation temperature as egg proteins when creating egg analogs. Plant proteins also have different isoelectric points than animal proteins, which can influence their ability to function in certain kinds of food products. As an example, it may be important to select a plant protein with a similar isoelectric point as casein when creating yogurt or cheese analogs that rely on protein aggregation when the pH is reduced. For these reasons, it is usually necessary to choose plant proteins that simulate the properties of the animal proteins in the animal-based food products they are designed to replace. On a final note, most of the plant proteins utilized to create plant-based foods are globular proteins, including soy, pea, potato, mung bean, and rice proteins (Sha & Xiong, 2020).

Flexible proteins: Some food proteins have disordered flexible conformations, with the polypeptide chain rapidly altering its structure in solution, such as the caseins from milk and the gelatin from meat and fish (at high temperatures) (Fig. 2.2). For example, gelatin is obtained by acid or alkaline hydrolysis of the collagen proteins obtained from the waste products generated by the meat and fish industry, such as skin, tendons, ligaments, and bones. It is a highly flexible and disordered molecule when it is heated above its helix-coil transition temperature. However, it forms helical structures when cooled below this temperature, which can then form crosslinks with neighboring gelatin molecules through hydrogen bonding. An important functional characteristic of gelatin in many food products is its ability to form transparent reversible cold-set gels. There are only a few plant proteins that have similar flexible structures like casein or gelatin, which means it is often challenging to simulate the desirable physicochemical and sensory properties of foods where these molecules play an important role, such as yogurt, cheese, meat, and desserts. However, polysaccharides derived from plant sources, either on their own or in combination with plant proteins, may be able to mimic some of the desirable features of flexible animal proteins. For instance, some polysaccharides can form transparent gels when cooled below a certain temperature, such as agar, carrageenan, gellan gum or agarose, although the gel-sol transition temperatures are often too high (Amici et al., 2001; Rhein-Knudsen et al., 2017).

Fibrous proteins: The desirable physicochemical and sensory properties of several animal-based foods is due to the presence of fibrous proteins. In particular, the collagen present in meat and fish products has a rigid rod-like structure due to the triple helix formed by three intertwined polypeptide chains (Fig. 2.2). Collagen is present in the skin, bones, hooves, and/or connective tissue of animals, such as pigs, cows, and fish. The triple helix structure is disrupted when it is heated above the helix-coil transition temperature due to the weakening of the hydrogen bonds and the increase in conformational entropy effects, which causes a softening in the textural attributes of meat and fish. The muscle tissues in meat and fish contain actin and myosin proteins, which are assembled into complex fibrous structures that influence the texture of these foods. It is difficult to find plant proteins that naturally have fibrous structures and assemble into such 3-D structures, which makes it challenging to mimic the properties of meat and fish with plant-based ingredients. However, some plant proteins (such as glutenin from wheat or soy) can be induced to form fibrous structures through processing methods (Grabowska et al., 2016; Mattice & Marangoni, 2020b) and some mycoproteins (mycelia) naturally have filamentous structures that mimic the proteins in meat (Delcour et al., 2012). Several processing and physicochemical methods that can be used to create fibrous meat-like structures from plant proteins are discussed in a later chapter, such as extrusion, shear cell, and controlled phase separation methods (Chap. 3).

2.2.2 Protein Extraction and Refinement

Protein-rich ingredients can be isolated from various kinds of plant materials using a combination of different processing methods. Ingredients containing different protein concentrations, fractions, and purities can be obtained depending on the sophistication of the isolation methods used, including flours (<60% protein), concentrates (60–90% protein), and isolates (<90% protein) (Loveday, 2019, 2020; Sha & Xiong, 2020). Ideally, the isolation method should produce proteins that retain their functional properties, which often requires that they do not become denatured or aggregated during the process. Moreover, it is often necessary to remove or deactivate components that interfere with the functional performance of specific proteins (such as carbohydrates, lipids, minerals, enzymes, and antinutritional factors). One of the challenges with the current generation of plant protein ingredients is that they are often a sidestream of other processes, such as the extraction of oil or starch from plant materials. Consequently, the isolation process has not been optimized to retain the desirable functional properties of the proteins. Instead, it has been designed to increase the yield of oil or starch. In the future, it will be important to optimize the extraction of all the different functional ingredients from plants to improve their functionality and sustainability.

In general, the process used to isolate proteins from plant materials has to be tailored to their specific characteristics. However, there are some operation units that are commonly used to isolate plant proteins. Typically, the plant material will be cleaned and then any outer coating removed (such as husks, shells, or skins). In some cases, a defatting stage may be used to remove most of the lipids from oil-rich plant materials that have been milled before (Fig. 2.6a). Traditionally, solvent extraction was used for this purpose, but this may lead to loss of protein functionality. Consequently, alternative methods are being developed as discussed in Sect. 2.4.2. Moreover and if required, the plant material may be softened by heating in water. The plant material may also be heated to deactivate any endogenous enzymes that could cause undesirable changes in ingredient quality, *e.g.,* enzymes that promote oxidation or hydrolysis. The softened material may then be mechanically disrupted to reduce the particle size and increase its surface area, *e.g.,* using a high-shear blender. Exogenous hydrolytic enzymes can be added to break down cell wall materials, thereby facilitating the release of the proteins (and oil bodies). Polysaccharides may also be removed by alkaline extraction and selective precipitation using concentrated alcohol solutions. These procedures often result in a protein-rich slurry containing a mixture of soluble and non-soluble proteins. The non-soluble ones can be separated by centrifugation, gravitational settling, or filtration methods. The soluble proteins can then be collected by adjusting the pH close to their isoelectric point or by adding salt to cause them to precipitate from solution. The precipitated protein fraction can then be collected, washed, dried, and ground into a powder that can be used as a food ingredient. In some cases, more sophisticated processes may be used to obtain particular fractions of proteins that may have specific functionalities, such as by ultrafiltration (Ratnaningsih et al., 2021).

In the remainder of this section, we describe several protein extraction and isolation techniques used for common plant proteins: soy, pea, gluten, and potato protein.

Many plant-based foods are formulated using soy concentrates or isolates (Fig. 2.6a). Soy protein concentrates (50–80% protein) are produced using various approaches. The simplest approach is to use de-fatted soybean flakes/flours as a starting material, which are normally obtained from soybeans by de-hulling, flaking/milling, extracting with hot hexane to remove the lipids, and then drying. The dried flakes are then re-dispersed in an alkaline solution to solubilize the proteins. The insoluble fraction is separated using a decanter and then the soluble fraction is spray-dried to yield a soy protein concentrate (Johnson, 1999). Other processes use ethanol extraction, calcium precipitation, air classification, filtration steps, and enzyme treatments to yield soy protein concentrates that are low in antinutritional compounds (Konwinski, 1992; Nardelli, 1994; Singh, 2006, 2007; Thomas, 2001). Soy protein isolates (> 90% protein) are also produced from defatted soybean flakes/flours that have been solubilized under alkaline conditions. However, instead of taking the soluble fraction and subjecting it to a drying step, the pH of the supernatant is adjusted to the isoelectric point of the soy proteins (~ pH 4 to 5), which results in their precipitation (Puski, 1987). Protein aggregation and yield might be enhanced by heating the protein solution. The agglomerated proteins are subsequently separated from the dispersion, neutralized, and spray-dried to yield a soy protein isolate (Segal & Green, 2017).

Pea protein concentrates and isolates follow the same principles but are produced using slight variations (Fig. 2.6b). Peas are naturally low in fat, *e.g.*, dried split peas contain less than 4% fat (USDA, 2021), but they are rich in starch, which can be separated by centrifugal forces. The starch is present as granules with diameters ranging from about 2–40 µm, which is considerably smaller than the protein bodies that have diameters of around 1–3 µm (Ratnayake et al., 2002; Schutyser et al., 2015). The size and density differences of these two types of natural particles are used to separate the starch granules from the protein bodies by dry or wet fractionation. Dry fractionation utilizes air classification to separate the particles in pea flours using a classifier wheel (Pelgrom et al., 2013). The fine fraction that is rich in protein bodies passes through the wheel, whereas the coarse fraction that is rich in starch granules separates to the bottom. The protein content in the fine fraction is around 40–60% (Schutyser et al., 2015). In contrast, pea protein isolates are usually produced using hydrocylones to separate and classify the starch granules from the protein bodies. The hydrocyclones typically produce a protein-rich fraction with a protein content of 40–50%. This fraction is then separated from the dietary fibers using a decanter, precipitated at pH 4.6 at elevated temperatures, and then dried to yield protein contents of around 86% or higher (Salome, 2007).

Wheat proteins are obtained by utilizing the self-aggregating tendency of gluten molecules in water (Pojic et al., 2018). Wheat grains are milled into flours, mixed with water, and then incubated to induce gluten agglomeration. The gluten particles are separated from the starch slurry using decanters, centrifuges and/or screens, then dried to produce a powder that can be used in foods.

Several methods are available for extracting proteins from potatoes (Zhang et al., 2017). Potato proteins can be obtained from the potato juice generated by grinding potatoes in a sodium sulfite solution to prevent oxidation. The potato juice has around 3% of protein and the protein can be recovered by different processes. The dispersion can be clarified using decanter centrifuges, and additional protein can be recovered from further separation processes such as continuous conical centrifugal sieves and hydrocyclones. For example, a common method is to concentrate the potato juice by filtration and subsequently adjust the pH to between 4.8 to 5.6. The solution is then jet-cooked at 102–115 °C and the coagulated protein is separated using decanter centrifuges and dried to yield a powder with around 90% protein (Grommers & van der Krogt, 2009). Other methods can also be used to obtain specific potato protein fractions, such as various filtration or chromatography methods (Giuseppin, 2008; Waglay et al., 2019).

2.2.3 Protein Ingredients

As mentioned earlier, plant protein ingredients used in the food industry come in various forms including flours, concentrates, or isolates, which differ in the total amount of protein they contain (Loveday, 2019, 2020; Sha & Xiong, 2020). Flours are often formed by grinding and sieving certain dried fractions of a plant (such as soy, pea, or wheat). As a result, they usually contain relatively high levels of starches, dietary fibers, and minerals, as well as proteins, which impacts their functional performance (Sharan et al., 2021). The amount of protein in an isolate (> 90%) is higher than that in a concentrate (60–90%), which is higher than that in a flour (< 60%). For academic studies, isolates are often used because of their high purity, but, in commercial food applications, they are often too expensive. The type and concentration of non-protein fractions in isolates or concentrates may have a major impact on their functional performance. For this reason, it is often important to be able to purchase protein ingredients with well-defined compositions and/or to measure the composition of any ingredients prior to use. It should be noted that sometimes less purified ingredients do exhibit good functionality, which can improve their sustainability because less energy and other resources to extract and purify them is required (Kornet et al., 2021).

In some applications, plant protein ingredients are utilized in the form of texturized vegetable proteins (TVPs), which are preformed protein-rich particulates designed to mimic the structural and textural attributes of products such as ground meat, sausages, and burgers. These products are usually prepared by extruding flours or concentrates obtained from soybeans or other plant sources (such as wheat, chickpeas or oats) into specific sizes and shapes (like chunks, nuggets, flakes, or fibers) (Zhang et al., 2019). Typically, TVPs only contain around 50–70% protein, with the remainder being mainly starch, dietary fibers, and lipids.

Currently, a major challenge when formulating plant-based foods is the lack of protein ingredients with reliable functional performances, which is the result of several factors. First, there are variabilities in the type and amounts of proteins present in the crops they are extracted from, which also depends on the plant breed used and the growing conditions. Hence, there is a need to better understand and control crop breeding and growing. Second, there are many kinds of plant proteins within a particular plant, each with different molecular and functional attributes. Consequently, a protein ingredient may behave differently depending on which fractions it contains. Moreover, the presence of starches, dietary fibers, lipids, and minerals in a plant protein ingredient may alter its functional performance. Third, the native structure of plant proteins can be disrupted by the processes used to create powdered ingredients. Changes in the protein structure can have a profound effect on their stability and functionality, such as solubility, emulsifying, foaming, binding, thickening, and gelling properties. Consequently, it is important to use a protein ingredient where the level of denaturation of the proteins is known and controlled. Fourth, the aggregation state of plant proteins may be altered during the purification processes, which again can have a pronounced impact on their functionality.

The molecular and physicochemical attributes of several common animal and plant proteins are summarized in Tables 2.1a and 2.1b. In general, these attributes depend on the biological function of the proteins in their natural environment, as well as the isolation and other processing operations used to convert them into food-grade protein ingredients. Some of the common challenges encountered when selecting good quality plant proteins for application in plant-based foods are highlighted here:

- *Off flavors* – A number of plant proteins have unpleasant flavor profiles because they contain undesirable aromas, which may be naturally present or formed by chemical reactions after harvesting (Bangratz & Beller, 2020; Rackis et al., 1979). Moreover, some plant proteins have undesirable tastes, being perceived as bitter or astringent, which is often due to the presence of phytochemicals that are associated with them, such as saponins, phenolic acids, and flavanols (Bangratz & Beller, 2020; Sharan et al., 2021). Soy and lupin proteins are often considered to have unpleasant profiles, which limit their application in some foods (Tables 2.2a and 2.2b). Researchers are therefore trying to identify the different kinds of undesirable aromas and tastes in plant proteins and develop effective strategies to reduce them (Sharan et al., 2021). For instance, plant breeding methods can be used to change the composition of plant proteins and the phytochemicals associated with them, thereby reducing undesirable flavors. Alternatively, processing operations can be used to remove or deactivate these off-flavors, such as fractionation, membrane processing, or chromatography. Finally, it may be possible to add other ingredients into these products that can mask the undesirable flavors, either by binding to receptors on the tongue (flavor blocking) or to the undesirable proteins or phytochemicals (flavor maskers).
- *Poor solubility* – The poor water-solubility of many plant proteins limits their application. The solubility of proteins may be due to their inherent molecular characteristics or due to changes in these characteristics during their isolation, purification, and processing. The solubility of proteins is usually classified by the Osborne scheme, which was first proposed almost a century ago (Osborne, 1924): albumins are soluble in water; globulins are soluble in salt solutions; prolamins are soluble in alcohol solutions; and glutelins are soluble in weakly alkaline or acidic solutions. The relative amounts of these different protein classes in a particular plant protein ingredient depend on its origin (Tables 2.1a and 2.1b). However, plant proteins can be fractionated, which can enrich specific protein classes and improve their functional attributes or create ingredients with specific

2.2 Proteins

Table 2.2a Examples of different types of plant proteins available to formulate plant-based foods, along with information about their abundance, cost, physicochemical, functional, and nutritional attributes. It should be noted that these values often depend on how the proteins are isolated and processed

	Protein content	PDCAAS	Allergen risk	Commercial stage	Flavor	Functionality	Cost	GCV
Soy	>30%	>0.8	Serious	Commodity	Undesirable	Excellent	Low	Abundant
Pea	20–30%	0.6–0.79	Medium	Large	Acceptable	Good	Low	Medium
Wheat	10–20%	0.40–0.59	Serious	Large	Acceptable	Excellent	Very low	Abundant
Canola	20–30%	>0.8	Serious	Small	Acceptable	Good	–	High
Chickpea	20–30%	0.40–0.59	Medium	Large	Acceptable	Good	High	High
Fava bean	20–30%	0.40–0.59	Mild	Small	Acceptable	OK	High	Medium
Lentil	20–30%	0.40–0.59	Medium	Small	Acceptable	–	High	High
Lupin	>30%	0.40–0.59	Serious	Small	Undesirable	OK	–	Medium
Mung bean	20–30%	0.40–0.59	Medium	Large	Fairly neutral	Good	High	Medium
Navy bean	20–30%	0.6–0.79	Medium	R&D	Fairly neutral	OK	–	Low
Peanut	20–30%	0.40–0.59	Serious	Small	Fairly neutral	Good	–	High
Sunflower	20–30%	0.6–0.79	Very mild	Small	Acceptable	Good	High	High
Almond	20–30%	0.20–0.39	High	Small	Fairly neutral	OK	High	Medium
Corn	10–20%	0.20–0.39	Very mild	Small	Fairly neutral	Low	High	Abundant
Oat	10–20%	0.6–0.79	Mild	Small	Fairly neutral	Low	–	Medium
Potato	5–10%	>0.8	Very mild	Large	Acceptable	Good	High	Abundant
Quinoa	10–20%	0.6–0.79	Mild	R&D	Acceptable	Good	High	Low
Rice	5–10%	0.40–0.59	Very mild	Large	Fairly neutral	OK	Medium	Abundant
Sorghum	5–10%	0.20–0.39	Very mild	R&D	Fairly neutral	Poor	–	High

Key: *GCV* global crop volume, *PDCAAS* protein digestibility-corrected amino acid score. This data was adopted from the Good Food Institute report "Plant Protein Primer" (www.gfi.org)

Table 2.2b Key attributes of plant proteins for application in plant-based foods

Ranking	Protein Conc.	PDCAAS	Allergen risk	Commercial stage	Flavor	Functionality	Cost ($/kg protein)	GCV (MMT)
Excellent	>30%	>0.8	Mild and low % of people at risk	Commodity	Flavorless	High functionality	<$2	>100
Good	20–30%	0.6–0.79	↕	Large	↕	↕	$2–4	10–99
OK	10–20%	0.40–0.59	↕	Small	Acceptable	↕	$5–9	1–9
Low	5–10%	0.20–0.39	↕	Start-up	↕	↕	$10–19	0.1–0.9
Poor	<5%	<0.20	Severe and high % of people at risk	R&D	Objectionable	Low functionality	>$20	<0.1

Key: *GCV* global crop volume, *PDCAAS* protein digestibility-corrected amino acid score. The scheme for comparing the different protein attributes was adapted from the Good Food Institute report "Plant Protein Primer" (www.gfi.org)

functional characteristics. As mentioned earlier, the solubility of a protein may also be influenced by its denaturation and aggregation state, which can be altered by the processing and storage conditions. For these reasons, it is often useful to optimize the extraction and purification processes used to control the denaturation and aggregation state of the proteins. For instance, the solution and environmental conditions employed, such as pH, salt type, solvent, and temperature, can be selected based on the specific attributes of the protein, such as isoelectric point, molecular weight, surface hydrophobicity, and thermal denaturation temperature.

- *Inconsistency* – Another major hurdle to the utilization of plant protein ingredients in commercial products is their batch-to-batch variability. This variability may arise due to differences in the source materials, extraction methods, processing operations, or storage conditions used, as these factors influence the composition, denaturation state, and aggregation state of the proteins. For these reasons, many companies are trying to optimize their extraction and processing operations to obtain higher quality and more consistent plant protein ingredients.
- *Purity* – As well as proteins, plant protein ingredients may also contain various other substances, such as starches, dietary fibers, sugars, lipids, minerals, and phytochemicals, which impact their functional performance. The type and concentration of these other substances are often not clearly specified by ingredient suppliers and so it is difficult to determine their potential effects on ingredient performance in different applications. Consequently, there is a need to create plant protein ingredients that have more clearly controlled and specified compositions. In some cases, it may be desirable to include combinations of different

components in plant protein ingredients to improve their performance. For instance, an ingredient containing a dietary fiber-protein mixture may perform better than one that only contains proteins in some applications. Nevertheless, this would have to be established on a case-by-case basis.

There are a number of reasons why many existing plant protein ingredients do not exhibit the desired functional characteristics. For instance, soybeans are usually cultivated and processed to have a high yield and an efficient extraction of soybean oil. As a result, they often do not have an optimum protein content and these proteins are often damaged during oil isolation, thereby losing their functionality. With the growing importance of plant-based foods, a number of scientists are developing varieties of soybeans and other crops that contain much higher protein concentrations than normal, thereby increasing their potential as a source of protein ingredients (GFI, 2021). Moreover, companies are developing more gentle methods of processing agricultural commodities so as to prevent the denaturation and aggregation of the functional proteins during isolation. For instance, enzymes are being used to selectively break down the cellular structure of plant materials so as to release the proteins (Bychkov et al., 2019; Sari et al., 2015). It is therefore likely that a new generation of higher quality protein ingredients will become available in the future due to advances in crop breeding and isolation procedures.

Rather than being extracted from plants, plant protein ingredients can also be produced using microbial fermentation approaches (Celik & Calik, 2012; Rasala & Mayfield, 2015). In this kind of cellular agriculture, the DNA sequence that encodes for a particular protein is inserted, for example, into a microbial plasmid, which is then introduced into a microbial species that is able to undergo fermentation under commercial conditions, such as a suitable bacteria, yeast, or fungi. The microorganisms are incubated in an aqueous solution containing a nutrient blend that encourages their growth and reproduction (such as sugars, vitamins, and minerals). They are then kept under environmental conditions that promote their growth (such as controlled oxygen, temperature and light levels). As they grow and divide, they generate the plant proteins encoded within the plasmid DNA. The plant proteins produced are then separated from the microbial cells using a suitable isolation method, purified, and then used as functional ingredients. One of the current drawbacks of this approach is its relatively low yields and high costs. More research is therefore needed to optimize this process so as to economically produce the amounts of protein needed for applications in plant-based foods.

2.2.4 Protein Characterization

The molecular characteristics of proteins are typically characterized using a range of different analytical techniques (Kessel & Ben-Tal, 2018). In this section, a brief overview of some of the most widely used methods is given:

Primary structure: The number, type, and sequence of the amino acids in a protein can be determined using various approaches. The type of amino acids can be determined by hydrolyzing the protein using a strong acid or base to break all the peptide bonds. The relative amounts of the different amino acids present can then be determined using chromatography. Some amino acids may chemically degrade in the presence of strong acids or bases, and therefore different methods are needed to quantify them. The amino acid sequence can be assessed by selective fragmentation of the polypeptide chain using enzymes to produce a series of relatively short peptides (10–20 amino acids). A chemical (the Edman reagent) that binds only to the amino end of the peptides is then added. This terminal amino acid can then be cleaved from the rest of the peptide chain and identified. This procedure is repeated multiple times to determine the full amino acid sequence of each peptide. By cleaving the original protein at different positions on the polypeptide chain using different enzymes it is possible to work out the overall amino acid sequence of the entire protein from that of the peptides. More recently, advanced mass spectrometry methods have been developed to provide information about the amino acid sequence of proteins (see Appendix). In this case, the proteins are partially hydrolyzed to form peptides either before or during mass spectrometry, depending on the approach used. Insights into the number of amino acids in a protein can be obtained by measuring its molecular weight, which is typically carried out using electrophoresis (*e.g.*, SDS-PAGE), chromatography (*e.g.*, size exclusion), or light scattering methods (*e.g.*, laser diffraction).

Secondary structure: Information about the secondary structure of proteins, such as the fractions of α-helix, β-sheet, β-turn and disordered regions, can be obtained using spectroscopic methods that are sensitive to the local ordering of the polypeptide chains. The most commonly used methods are circular dichroism (CD) and Fourier Transform infrared (FTIR) spectroscopy.

Tertiary and quaternary structure: The tertiary and quaternary structure of proteins is typically obtained using X-ray diffraction analysis of crystallized proteins or NMR analysis of proteins in solution. Insights into protein structure can also be obtained using advanced electron microscopy methods. Finally, computer modeling methods are becoming increasingly accurate at predicting the tertiary structures of proteins from their primary structures. Many of the methods used to determine the tertiary structure of proteins require sophisticated and expensive equipment that is run by highly trained operators. As a result, it is not widely utilized within food companies. Moreover, some insights into the conformation of proteins (*e.g.*, native *versus* denatured state) can be obtained by using more widely available and affordable equipment. For instance, the conformational changes of globular or fibrous proteins can be determined by measuring changes in the heat flow with temperature using different scanning calorimetry (DSC) or by measuring changes in fluorescence emission spectra with temperature using fluorescence spectroscopy instruments. These devices are often useful for establishing the native state of globular plant proteins prior to utilization. As an example, the heat flow *vs.* temperature curves measured for a native and denatured plant protein are shown in Fig. 2.7. An endothermic peak is clearly observed for the native protein, which is due to the

2.3 Carbohydrates

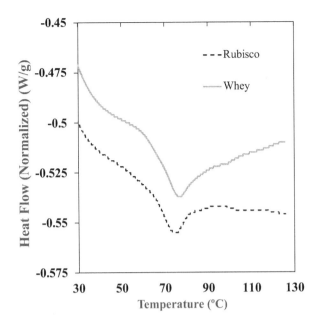

Fig. 2.7 DSC profiles of plant (Rubisco) and animal (whey protein) globular proteins during heating (pH 7). When they were cooled, no peaks were observed, indicating that they were irreversibly denatured. (Data kindly provided by Yunbing Tan)

unfolding of the protein molecules. In contrast, no thermal transitions are observed for the denatured proteins. As the denaturation state of proteins can play an important role in determining their functionality in plant-based foods, it is usually important to measure this using an appropriate analytical tool.

2.3 Carbohydrates

Carbohydrates are another important functional ingredient used to formulate plant-based foods. They play an important role in controlling the appearance, texture, flavor, and cookability of these products. Carbohydrates are organic molecules consisting of a single monosaccharide or numerous monosaccharides linked together by glycosidic bonds (Brady, 2013; Huber & BeMiller, 2021). Animal-derived foods typically contain relatively low levels of carbohydrates. For instance, the USDA FoodData Central on-line database reports that most meat, fish, egg, and milk have carbohydrate contents of 0, 0, 2.4, and 4.8%, respectively (fdc.nal.usda.gov). In contrast, many plants contain relatively high levels of carbohydrates. For example, the same database reports that wheat, corn, soybeans, and peas have carbohydrate levels of around 76, 74, 11 and 15%, respectively (on a wet basis). These carbohydrates are synthesized in plants from sunlight and carbon dioxide by *photosynthesis*.

Biologically, they play numerous roles in plants, such as energy sources, protectors, structure builders, and signalers. In plant-based foods, carbohydrate ingredients can be used for their flavoring, coloring, thickening, gelling, structuring, emulsifying, and fluid holding properties. In this section, we provide a brief overview of the molecular structure and properties of carbohydrates, as well as their use as functional ingredients.

2.3.1 Carbohydrate Structure

Carbohydrates differ from each other based on the type, number, sequence, and bonding of the monosaccharides present (Williams & Phillips, 2021), which is shown schematically in Fig. 2.8. Based on the number of monosaccharides (n) they contain, they are categorized as monosaccharides ($n = 1$), disaccharides ($n = 2$), oligosaccharides ($n = 3$–10), or polysaccharides ($n > 20$). Monosaccharides and disaccharides ("simple sugars") tend to be white crystalline substances that are perceived as sweet by human beings. Many oligosaccharides have prebiotic properties, which means they can stimulate the growth of bacteria in the colon. Polysaccharides and carbohydrates derived from plants in general differ in their structural, physicochemical, and physiological attributes, which influences their performance as functional ingredients in plant-based foods. Molecularly, carbohydrates have different molar masses (low to high), structures (linear or branched), electrical characteristics

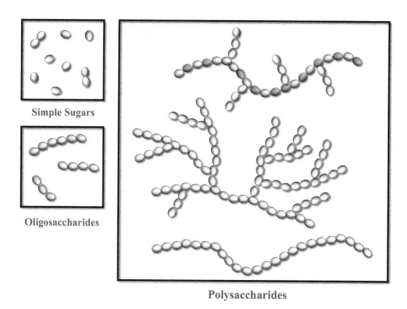

Fig. 2.8 Carbohydrates differ from each other depending on the type, number, and sequence of monosaccharides they contain

(anionic, neutral, and/or cationic), and polarities (polar and/or non-polar). These molecular differences lead to differences in their physicochemical and functional attributes, such as their solubility, thickening, gelling, binding, fluid-holding, and emulsifying characteristics (Williams & Phillips, 2021). Differences in the molecular properties of carbohydrates also lead to differences in their behavior within the human gut, such as their digestibility and fermentability, which leads to differences on their impacts on human nutrition and health (see Chap. 5).

2.3.2 Carbohydrate Isolation

In this section, we focus on the isolation of polysaccharides from plant materials because they are usually the most important carbohydrates used as functional ingredients in plant-based foods. Commercially, starches are normally obtained from maize, wheat, tapioca, or potato using a combination of milling, separation, and drying methods (Mitchell & Hill, 2021). The precise method used depends on the botanical origin of the plant material used. The extraction procedure is usually designed to ensure that the starch granules remain in a native state, with their morphology and partly crystalline structure intact. Hydrocolloids, such as agar, alginate, carrageenan, cellulose, guar gum, pectin, and xanthan gum are derived from a variety of different biological sources, including microbial fermentation, seaweed, apple and citrus pomace, cotton, and seeds (Phillips & Williams, 2021) (Table 2.3). Consequently, the extraction and other purification procedures have to be tailored to the nature of the source material. Nevertheless, there are some procedures that are commonly used in the isolation of different kinds of hydrocolloids, such as hydrolysis, separation, concentration, recovery, washing, drying, and milling. In particular, acid, alkali or enzymatic hydrolysis is often used to break down non-polysaccharide components in plant materials, while incubation in concentrated alcohol solutions (*e.g.,* ethanol) is often used to recover the polysaccharides by selectively precipitating them from solution. The polysaccharides can then be washed, dried, and milled to create a powder that can be used as a food ingredient. In some cases, the polysaccharides may be chemically or enzymatically derivatized during the production process to modify their functional attributes, *e.g.,* demethylation of pectin, methylation of cellulose, or phosphorylation of starch.

2.3.3 Carbohydrate Ingredient Properties

When formulating plant-based foods, it is important to select the most appropriate combination of carbohydrate-based ingredients to obtain the physicochemical, sensory, and nutritional attributes required in the final product. Sugars, such as glucose, fructose, or sucrose, are commonly utilized as ingredients for their sweetness (Huber & BeMiller, 2021). Sugars may also play other roles in foods, such as

Table 2.3 Molecular characteristics of some plant-based polysaccharides that can be used as functional ingredients in foods

Name	Source	Main structure type	Major monomer	Function
Agar	Algal	Linear	β-D-Galactopyranose	Structuring, thickening, gelling (Ca^{2+}), stabilizing
Alginate	Algal	Linear	β-D-Mannuronic acid	Structuring, thickening, gelling (Ca^{2+}), stabilizing
Carrageenan	Algal	Linear/helical	Sulfated Galactan	Structuring, thickening, gelling (K^+ or Ca^{2+}), stabilizing
Chitosan	Mushroom	Linear	2-amino-2-deoxy-β-D-glucose	Structuring, gelling (polyphosphate)
Guar gum	Seeds	Linear with side chains	D-mannose and D-galactose	Thickening, stabilizing
Gum Arabic	Acacia sap	Branched coil domains on protein scaffold	Galactose	Emulsification, stabilizing
Inulin	Plants or bacteria	Linear with occasional branches	β-D-fructose	Prebiotic, thickening
Locust bean gum	Seeds	Linear with side chains	D-mannose and D-galactose	Structuring, thickening, stabilizing
Methyl cellulose	Wood pulp	Linear	Methylated glucose	Thickening, stabilizing, gelling (heat-set)
Pectin (citrus)	Plant cell walls	Highly branched coil	Glucuronate (backbone)	HM: Structuring, gelling (sugar + heat), stabilizing LM: Structuring, gelling (Ca^{2+}), stabilizing
Pectin (beet)	Sugar beet pulp	Branched coil with protein	Glucuronate (backbone)	Structuring, emulsifying, gelling (sugar + heat; Ca^{2+} or laccase), stabilizing
Starch granules	Corn, potato, rice, wheat	Branched and linear	α-D-glucose	Thickening, gelling, binding, stabilizing
Tara gum	Plant seeds	Linear with side chains	D-mannose and D-galactose	Thickening, stabilizing
Xanthan gum	*Xanthomonas campestris* exudate	Linear/helical (high MW)	β-D-glucose (backbone)	Structuring, thickening, stabilizing

Note: In practice, within each category, commercial polysaccharide ingredients vary in their molecular and physicochemical properties, which influences their functional attributes
Adapted from McClements and Grossmann (2021b) with permission.

controlling water activity or acting as bulking agents. Certain kinds of oligosaccharides, such as fructo- and galacto-oligosaccharides, are utilized as prebiotic ingredients in foods to stimulate the growth of beneficial bacteria in the human colon, thereby providing health benefits (Holscher, 2017). Polysaccharides, such as starch, modified starches, modified celluloses, guar gum, locust bean gum, pectin, xanthan gum, and gum arabic are commonly used for their thickening, gelling, emulsifying, binding, and fluid-holding properties (Williams & Phillips, 2021). However, they may also have important health effects depending on their digestibility and fermentability. For instance, rapidly digested starches may lead to an undesirable spike in blood glucose levels, whereas dietary fibers may have health benefits like reducing constipation, cholesterol levels, and colon cancer (see Chap. 5). The selection of an appropriate polysaccharide or combination of polysaccharides to create the desired physicochemical, functional, and nutritional attributes in a plant-based food is therefore critical.

A summary of the different kinds of carbohydrates that can be used as functional ingredients in plant-based foods is given in Table 2.3. A more detailed description of the functional attributes of polysaccharides is provided in Sect. 2.6.

There are a number of factors to consider when purchasing carbohydrate-based ingredients for utilization in plant-based foods, especially polysaccharides. Commercial sugar ingredients are typically available with well-defined and consistent compositions and functionalities. In contrast, the functionality of polysaccharide ingredients often depends on their botanical origin, the isolation, purification, and processing operations used to convert them into ingredients, and the storage conditions they experience during their lifetime (such as time, temperature, light, and oxygen levels). Consequently, the functional performance of a polysaccharide ingredient with a specific name (*e.g.*, "pectin") may vary considerably depending on these factors. In this case, it is important to select an ingredient with the most appropriate functional attributes for the intended purpose, as well as to ensure that its properties are reliable from batch-to-batch. Many of the reasons for these batch-to-batch variations are the same as those discussed for proteins. For instance, there may be differences in the type of polysaccharides present in the starting materials used to fabricate the ingredients, such as their molecular weights, compositions, and structures. Moreover, the isolation and purification procedures may have selectively enriched certain polysaccharide fractions with specific molecular characteristics, or they may have altered some of the original molecular characteristics. In addition, there may be different types and amounts of impurities remaining in the final ingredients, such as other carbohydrates, proteins, lipids, and minerals. To ensure the reliable functional performance of carbohydrate-based ingredients, it is often useful to establish standardized characterization methods, which will be discussed in the following section.

In addition to the technofunctional properties of the polysaccharides, various other factors also need to be considered, such as their regulatory status, cost, ease of use, reliability, and label friendliness. For instance, some consumers are avoiding products containing carrageenan because they believe this seaweed-derived ingredient has negative health effects. This is causing some food manufacturers to avoid using carrageenan as a functional ingredient in their products. Moreover, some consumers may avoid products that contain chemically modified polysaccharides, such as methylcellulose, because they only want to consume products created from "all-natural" ingredients. In this case, manufacturers are trying to identify alternative ingredients that provide the same functionality. Ideally, these ingredients should also be sustainable and healthy.

2.3.4 Carbohydrate Characterization

Knowledge of the molecular characteristics of the carbohydrates used to formulate plant-based foods is important for ensuring they provide the desired functional attributes in the final product. As mentioned earlier, the composition and properties of sugars is usually well-known and consistent because they are highly purified ingredients. In contrast, the composition and performance of polysaccharide-based ingredients often vary considerably from batch-to-batch or between different suppliers because these attributes depend on their natural origin (such as species and growing conditions), as well as the extraction, purification, and other processing operations used to convert them into food ingredients. For this reason, it is usually important to have appropriate information about the molecular characteristics of the polysaccharides in such food-grade ingredients. A variety of analytical instruments are available to provide information about the composition and structure of polysaccharides, which have been reviewed in detail elsewhere (Alba & Kontogiorgos, 2021; Nielsen, 2017; Ren et al., 2019). In this section, we provide a brief overview of these methods.

The overall composition of an ingredient (such as the carbohydrate, protein, fat, ash, and moisture content) can be determined by standardized proximate analysis methods. The total carbohydrate content within an ingredient can be determined using chemical methods, such as the phenol-sulfuric acid method. Commercial polysaccharide-based ingredients often contain different polysaccharide fractions with different molecular and functional characteristics. Information about the number of fractions present can be obtained by measuring their molecular weight profiles using chromatography methods (such as gel permeation chromatography or high-performance liquid chromatography) or electrophoresis methods (such as polyacrylamide gel electrophoresis, carbohydrate gel electrophoresis (PACE)). If a single peak or band is obtained, then one can assume the ingredient mainly contains one type of polysaccharide. Conversely, if multiple peaks are observed, then it can be assumed that there are different fractions present.

The molecular weight of a polysaccharide has a major impact on many of its functional attributes, including solubility, thickening, gelling, and structure formation. The average molecular weight or the full molecular weight distribution of polysaccharides can be determined using various methods, including high performance liquid chromatography, gel permeation chromatography, mass spectrometry, light scattering, and rheological measurements. The monosaccharide composition of polysaccharides is usually determined by acid hydrolysis to break all of the glycosidic bonds followed by chromatography or electrophoresis analysis. It is more difficult to establish the nature of the glycosidic bonds, the degree and position of branching, the monosaccharide sequence, and the presence of any side groups (*e.g.,* such as methyl groups, carboxyl, amino, or sulfate groups). Typically, a combination of different analytical methods is required to obtain this information, such as chemical derivatization, enzymatic hydrolysis, infrared spectroscopy, Raman spectroscopy, mass spectrometry, and/or nuclear magnetic resonance methods.

Information about the conformation of polysaccharide molecules in solution can be obtained using analytical techniques such as electron microscopy, atomic force microscopy, circular dichroism, light scattering, and viscosity measurements.

When using polysaccharide-derived ingredients to formulate plant-based foods it is often important to use one or more of these techniques to characterize their properties so as to ensure they have the desired functional attributes.

2.4 Lipids

Lipids are another important class of functional ingredient used to formulate plant-based food products (McClements & Grossmann, 2021b). These ingredients are selected to mimic the desirable physicochemical, functional, and sensory attributes that animal lipids normally provide to meat, seafood, egg, and dairy products. It is therefore important to understand the nature and behavior of the lipids in animal foods when designing plant-based foods that accurately mimic their properties. Animal lipids contribute to the desirable appearance, mouthfeel, and taste of animal-based products, as well as to their nutritional profiles. The contribution of these lipids to the overall quality attributes of these products depends on their type, concentration, and structural organization. The lipids in animals are comprised of different kinds of molecules that are insoluble in water but soluble in organic solvents, such as triacylglycerols, diacylglycerols, monoacylglycerols, free fatty acids, phospholipids, sterols, and oil-soluble vitamins (Gunstone, 1996). The lipids in animal-based foods may be present in a variety of forms, such as bulk lipids (as in lard), biological cells (such as the adipose tissue in meat and fish), and colloidal particles (such as the fat globules in milk or lipoproteins in egg). It is typically important to mimic the composition and structural organization of these animal-based lipids, as well as their physicochemical and functional attributes, when creating plant-based alternatives.

The lipids used as ingredients in plant-based foods come from a variety of sources, including avocado, canola, cocoa, coconut, corn, safflower, sesame, soybean, and sunflower (Sha & Xiong, 2020). Many of these lipids contain high levels of unsaturated fatty acids and are therefore fluid at room temperature (such as avocado, canola, corn, safflower, sesame, soybean, and sunflower oil), whereas some contain appreciable levels of saturated fatty acids and are therefore semi-solid at room temperature (cocoa butter and coconut oil). It is therefore important to select a lipid that has the required crystallization/melting properties for the desired application. For instance, in plant-based milks and eggs it may be better to utilize a lipid phase that is fluid at room temperature, whereas for a plant-based cheese formulation it is better to use one that is partially crystalline to provide the required mechanical properties. Plant-based lipids usually come in a bulk form and so it may be necessary to convert them into emulsions or other forms for the formulation of many plant-based foods.

2.4.1 Lipid Structure

Triacylglycerols, which are also referred to as triglycerides, are the dominant form of lipids found in animal and plant products. They are comprised of three fatty acid chains bound to three hydroxyl groups on a glycerol molecule *via* ester bonds (Brady, 2013). The fatty acids may have different numbers of carbon atoms, different numbers and positions of double bonds, different isomeric forms (cis or trans), and different attachment points on the glycerol molecule (Fig. 2.9). The nature of the fatty acids used to assemble a triacylglycerol molecule influences its physicochemical, functional, and nutritional attributes in foods. For this reason, the food industry often controls the fatty acid profiles of natural lipids using different approaches. For instance, genetic engineering and crop breeding methods can be used to create plants with different lipid profiles. In addition, the lipids can be separated into fractions with different fatty acid profiles based on their melting points using fractionation methods. This typically involves holding the lipids at a particular temperature to crystallize a specific fraction of the triacylglycerols and then removing the crystals by filtration, sedimentation, or centrifugation. It also has to be mentioned that the position of the fatty acids on the glycerol molecules in natural lipids is usually not random. A process called inter-esterification, which can be achieved using chemical or enzymatic methods, can be used to randomize the position of the fatty acids, which leads to changes in the physicochemical and nutritional properties of the lipids. In particular, the solid fat content *versus* temperature profile is altered by this process, which is important in some applications. Hydrogenization is another process, which is usually carried out by heating the lipids in the presence of hydrogen gas and a catalyst, can be used to reduce the degree of unsaturation of fatty acids, thereby making them more solid-like and less prone to oxidation. However, there have been health concerns with using this process to modify the physicochemical properties of lipids because it increases the percentage of saturated and trans-fatty acids present at the same time (if not carefully controlled), which has been linked to an increased risk of cardiovascular diseases (Chap. 5).

Monoacylglycerols and diacylglycerols have a similar structure to triacylglycerols but they only have one or two fatty acids attached to the glycerol molecule, respectively. Like in triacylglycerols, the nature and position of the fatty acid molecules varies depending on the origin and processing of the lipids. Monoacylglycerols and diacylglycerols may be used as functional ingredients in plant-based foods, *e.g.*, to create solid-like textures or as emulsifiers. In addition, they are naturally formed within the human gut due to the action of lipases on ingested triacylglycerols. The monoacylglycerols and free fatty acids formed during lipid digestion combine with bile salts and phospholipids in the small intestine to create mixed micelles that transport these digested products to the epithelium cell walls, where they can be absorbed. The mixed micelles also play an important role in promoting the absorption of other hydrophobic molecules, such as oil-soluble vitamins and nutraceuticals.

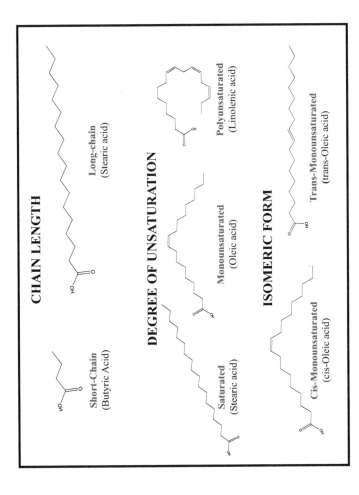

Fig. 2.9 The triacylglycerol molecules in animal or plant lipids consist of three fatty acids attached to a glycerol backbone. The chain length, degree of unsaturation, and isomeric form of the fatty acids influence their functionality and health effects. (Chemical structures were kindly drawn by Yuting Wang)

Phospholipids used in foods often consist of two fatty acids and a phosphate group attached to a glycerol molecule. The nature of the fatty acids, such as their chain length and unsaturation, vary according to their biological origin. In addition, different kinds of functional groups may be attached to the phosphate group, such as choline (PC), ethanolamine (PE), serine (PS), and inositol (PI). Phospholipids can be used as emulsifiers to stabilize oil-in-water emulsions, or they can be used to assemble liposomes that encapsulate and deliver bioactive agents. The functional performance of phospholipids depends on the nature of fatty acids and functional groups they contain.

Animal products typically contain cholesterol, which is a sterol molecule that is essential for human health, but is also linked to diseases under certain circumstances, such as heart disease. In contrast, plants contain phytosterols and phytostanols that have a similar structure to cholesterol but have been shown to have health benefits, such as reducing cholesterol levels. Oil-soluble vitamins are also lipids that are found in foods and that are critical to human health, which include vitamins A, D, E, and K (Chap. 5). Some of these vitamins can be obtained directly from certain plant-based foods (such as vitamins A (as precursor), E and K) but some of them are usually obtained primarily from animal foods (such as vitamin D). As a result, it may be necessary to fortify plant-based foods with vitamin D or to take vitamin D supplements to avoid suffering from a vitamin deficiency.

2.4.2 Lipid Isolation

Edible lipids can be isolated from a variety of oil-rich plant and algal sources, including algae, canola, coconut, corn, cottonseed, flaxseed, olives, palm, rapeseed, safflower, and sunflower. Typically, these materials are seeds or other oil-rich parts of the plants (like mesocarp or germs) (Rani et al., 2021). Traditionally, the industrial extraction of edible oils from oilseeds and other oil-rich plant materials is carried out using solvent extraction methods, which is sometimes preceded by a mechanical pressing step (Rosenthal et al., 1996). The mechanical pressing step is useful for plant-materials that contain large quantities of oil that can easily be expressed by applying an external pressure, such as olives. If an additional solvent extraction step is applied, hexane is typically used as an organic solvent in these processes because of its high extraction efficiency and good recovery properties (Rani et al., 2021). Solvent extraction methods usually involve a number of steps, including cleaning to remove extraneous matter; removing hulls, shells or other outer coatings; grinding to decrease the particle size and increase the surface area; cooking to soften the plant material and enhance solvent penetration; soaking in organic solvent to extract the oil; separation of organic solvent from aqueous slurry; evaporation of organic solvent from the oil phase. Once the oil has been isolated it may be further processed to remove any undesirable impurities, such as free fatty acids, phospholipids, pigments, and pro-oxidants. Moreover, it may also be processed to improve its functional properties using methods such as winterization, fractionation, interesterification, or hydrogenation. Solvent extraction processes have been reported to be highly efficient, leading to oil yields and solvent recoveries of more than 95% (Rosenthal et al., 1996). For this reason, they are widely used around the world in the commercial production of edible oils (Rani et al., 2021). However, there are concerns with the safety and negative environmental effects of using organic solvents for this purpose, as well as the costs required to recover them and their adverse effects on oil quality. For these reasons, there has been interest in developing alternative extraction methods, such as enzyme-assisted, microwave-assisted, ultrasound-assisted, and supercritical fluid extraction methods, which each have their own advantages and disadvantages (Rani et al., 2021).

One of the most promising alternative approaches for oil extraction are enzyme-assisted aqueous-based methods because they do not require an organic solvent. These methods typically involve a series of steps, such as cleaning, crushing, heating, soaking, enzyme hydrolysis, separation, and fractionation (Rani et al., 2021). An important step in these processes is the utilization of specific enzymes that selectively hydrolyze the polysaccharides and proteins around the oil bodies where the oil is normally stored in the plant, since this facilitates the release of the oil. The main advantages of aqueous extraction methods are that there are no harsh chemicals involved, they do not require solvent recovery processes, they are safer, and it is possible to isolate both the oil and protein fractions in a functional form.

2.4.3 Lipid Ingredients

The ability of lipids to crystallize and form semi-solid 3D networks plays an important role in determining the textural attributes of many animal-based products, such as meat, cheese, and ice cream. The solid fat content (SFC) *versus* temperature profile of lipids is important in these kinds of products (Fig. 2.10), which is strongly influenced by the types of fatty acids in the triacylglycerol molecules. The melting point of triacylglycerols increases with increasing number of carbon atoms and decreasing number of double bonds. The animal fats obtained from meat and milk often contain high amounts of long-chain saturated fatty acids, which means they tend to have relatively high melting points, so they are partially solidified at room temperature (Fig. 2.11). Fat melting and crystallization play an important role in determining the quality attributes of a number of common animal-based foods, including the melting of cheese, the spreading of butter, the texture of ice cream, and the foaming of whipped cream. The formulation of plant-based alternatives to these products therefore requires the utilization of plant-derived lipids that can provide similar melting/crystallization attributes.

A major difficulty in mimicking the desirable crystallization/melting behavior of some animal fats is that most plant fats have relatively low melting points because they consist of high levels of unsaturated fatty acids (Table 2.4). As a result, these fats tend to be liquid-like at room temperature and so cannot reproduce the desirable functional attributes exhibited by animal fats that contain high levels of saturated

Fig. 2.10 The solid fat content-temperature profile of edible fats depends on the fatty acid composition, which depends on their biological origin. Plant-based fats are designed to mimic the profiles of animal-based ones

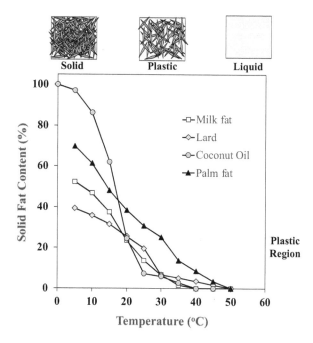

fatty acids. It is possible to increase the melting point of unsaturated fats using hydrogenation, which involves attaching hydrogen atoms to the double bonds as explained before. However, the utilization of hydrogenated plant oils (especially partially hydrogenated ones) in foods is usually undesirable due to the fact that they can contain appreciable levels of saturated and trans-fatty acids at the same time (if the process is not carefully controlled), which have been associated with an elevated risk of cardiovascular diseases (Hu et al., 2001). Consequently, the manufacturers of plant-based foods tend to utilize plant oils that naturally contain relatively high levels of saturated fatty acids (but little or no trans-fatty acids), like cocoa butter, coconut oil or palm fat (Fig. 2.10). It should be noted that the concentration of saturated fatty acids in coconut oil is relatively high (Table 2.4), which has led to some health concerns from nutrition scientists (Ludwig et al., 2018). However, the saturated fatty acids in coconut oil are mainly medium chain ones (8–14 carbon atoms), which may behave differently from a nutritional perspective than the long chain ones (16–18 carbon atoms) found in most animal fats.

More generally, the fatty acid profiles of triacylglycerols influence their nutritional effects (Chap. 5). Observational studies and clinical trials suggest that unsaturated fatty acids (particularly omega-3 polyunsaturated ones) are beneficial to human health, whereas saturated ones are detrimental (Saini & Keum, 2018; Shahidi & Ambigaipalan, 2018). These effects depend on the precise type and amounts of fatty acids consumed, as well as what they replace in the diet. Livestock animals (such as cows, sheep and pigs) usually contain relatively low amounts of polyunsaturated fatty acids, whereas marine animals (such as fatty fish) contain relatively

2.4 Lipids

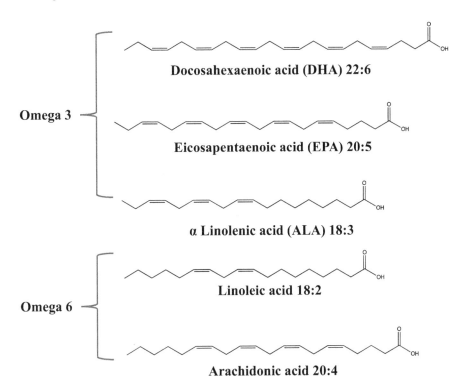

Fig. 2.11 Examples of different kinds of unsaturated fatty acids found in animal- and/or plant-based lipids. (Images kindly provided by Izzy McClements (drawn by ChemDraw))

high amounts. Several common unsaturated fatty acids found in animal- and/or plant-based foods are highlighted in Fig. 2.11. Fish oils typically contain high levels of docosahexaenoic acid (DHA) and eicosapentaenoic acid (EPA), which are long-chain omega-3 fatty acids that are claimed to have health benefits, like reducing heart and neurodegenerative diseases. The triacylglycerols isolated from plant sources usually contain relatively high amounts of omega-6 (like palmitoleic acid) or omega-9 (like oleic acid) monounsaturated acids or omega-6 (like linoleic acid) polyunsaturated fatty acids, rather than omega-3 ones (Fig. 2.11). Nevertheless, certain kinds of plant-based oils do contain relatively high amounts of an omega-3 fatty acid known as alpha-linolenic acid (ALA), such as canola, flaxseed, soybean, and walnut oils (Rajaram, 2014). These oils can therefore be utilized as a source of omega-3 fatty acids when formulating plant-based foods. Having said this, the health benefits of ALA are typically not as potent as those of EPA and DHA. For this reason, algae oil, which can be rich in DHA, can be used as an alternative source of omega-3 fatty acids in these products. In addition, modern gene editing methods, such as CRISPR, are being utilized to create agricultural commodities (such as soybeans) that contain high levels of healthy omega-3 fatty acids (GFI, 2021).

Table 2.4 Fatty acid compositions of common plant and animal fats

Fatty Acid	Beef	Pork	Poultry	Salmon	Milk	Egg	Canola	Coconut	Corn	Olive	Palm	Peanut	Sunflower	Soybean
C4:0	–	–	–	–	3	–	–	–	–	–	–	–	–	–
C6:0	–	–	–	–	2	–	–	–	–	–	–	–	–	–
C8:0	–	–	–	–	–	–	–	8	–	–	–	–	–	–
C10:0	–	–	–	–	3	–	–	7	–	–	–	–	–	–
C12:0	–	–	–	–	4	–	–	48	–	–	–	–	–	–
C14:0	3	2	1	3	12	–	–	16	–	–	–	–	–	–
C16:0	27	27	22	11	26	23	4	9	13	10	44	13	7	10
C16:1	11	4	6	5	3	3	–	–	–	–	–	–	–	–
C18:0	7	11	6	4	11	6	2	2	3	2	4	3	3	4
C18:1	48	44	37	25	28	41	56	7	31	78	40	38	14	23
C18:2	2	11	20	5	2	21	26	2	52	7	10	41	75	51
C18:3	–	–	1	5	–	–	10	–	1	1	–	–	–	7
C18:4	–	–	–	2	–	–	–	–	–	–	–	–	–	–
C20:1	–	–	1	–	–	1	–	–	–	–	–	–	–	–
C20:4	–	–	–	7	–	–	–	–	–	–	–	–	–	–
C20:5	–	–	–	5	–	–	–	–	–	–	–	–	–	–
C22:4	–	–	–	2	–	–	–	–	–	–	–	–	–	–
C22:5	–	–	–	7	–	–	–	–	–	–	–	–	–	–
C22:6	–	–	–	17	–	–	–	–	–	–	–	–	–	–
Other	2	1	6	2	6	2	2	1	0	2	2	1	4	5
SFA	37	40	29	18	61	29	6	90	16	12	48	16	10	14
MUFA	59	48	44	30	31	43	56	7	31	78	40	38	14	23
PUFA	2	11	21	50	2	23	36	2	53	8	10	41	75	58

Key: SFA, MUFA, and PUFA refer to saturated, monounsaturated, and polyunsaturated fatty acids, respectively. Adapted from McClements and Grossmann (b, McClements & Grossmann, 2021b) with permission.

2.4 Lipids

A major hurdle to incorporating unsaturated fatty acids (especially polyunsaturated ones) into foods is their tendency to degrade due to oxidative reactions when the product is processed, stored, or utilized (McClements et al., 2021). These lipid oxidation reactions result in the generation of a series of volatile products that lead to an aroma profile that consumers find unpleasant ("rancid") and that may also exhibit toxicity (Arab-Tehrany et al., 2012; McClements et al., 2021; Nogueira et al., 2019). Oxidation may also lead to color fading for some unsaturated pigments, like carotenoids (Qian et al., 2012). To overcome these problems, scientists have developed a variety of strategies to inhibit lipid oxidation in foods. These strategies include reducing the exposure of the lipids to conditions that promote oxidation reactions such as heat, oxygen, light, and pro-oxidants, adding preservatives that inhibit oxidative reactions such as chelating agents and antioxidants, and controlling the structural organization of the lipids within the foods (Jacobsen, 2015; Jacobsen et al., 2013; McClements & Decker, 2018). In the case of plant-based foods, it is important to use preservatives derived from plants, preferably all-natural ones such as botanical extracts, because this is more label-friendly ('clean labeling'). The creation of healthier plant-based foods in the future may therefore rely on the utilization of these strategies to protect polyunsaturated plant oils from oxidation.

In summary, it is important that the lipid ingredients used to formulate plant-based foods exhibit the required physicochemical, functional, and nutritional attributes. For instance, they should have an appropriate flavor profile, solid fat content – temperature dependence, rheology, chemical stability, and nutritional profile. These parameters depend on the type and amounts of lipids present within them. In addition, it is important to consider other factors, such as their regulatory status, cost, reliability, and sustainability. Environmentally conscious consumers may avoid products containing palm oil because the methods used to grow and cultivate palm trees may cause environmental damage and a loss of biodiversity. Health conscious consumers may avoid foods containing coconut oil because of the high levels of saturated fats. Some consumers may not consume foods formulated from oils that have been derived from genetically modified crops, such as genetically-engineered soybeans. Consequently, manufacturers must take these issues into account when selecting an appropriate lipid-based ingredient to use in plant-based foods. Moreover, it is important to have good analytical tools to determine the quality and properties of the lipid ingredients used to formulate plant-based foods.

2.4.4 *Lipid Characterization*

A number of characteristics of lipid ingredients may be determined to assess their appropriateness for a particular application, including the type of lipid classes present, their fatty acid profiles, their quality, their solid fat content, and their susceptibility to oxidation. The different kinds of analytical methods available to characterize these properties have been reviewed in detail elsewhere (Christie & Han, 2010; Nielsen, 2017). For this reason, only a brief overview is given in this section. The type of lipid classes within a lipid ingredient, such as the relative amounts of triacylglycerols, diacylglycerols, monoacylglycerols, free fatty acids, phospholipids, and cholesterol, can be determined using chromatography methods, such as thin layer chromatography (TLC) or high performance liquid chromatography (HPLC). The fatty acid profile of lipids can be determined by saponification, methylation, and then analysis by gas chromatography (GC). The position of the fatty acids on the glycerol molecule can be determined by various chemical, enzymatic, and spectroscopic (NMR) methods. The quality of lipids can be determined by measuring their acid value (which provides a measure of the amount of free fatty acids present) and their saponification number (which provides a measure of the average molecular weight of the fatty acids). The solid fat content *versus* temperature profile of lipids can be determined using dilatometry, differential scanning calorimetry, or NMR methods. The susceptibility of lipid ingredients to oxidation can be determined by measuring primary (peroxide values or conjugated dienes) and secondary (thiobarbituric acid reactive substances and aldehydes) reaction products, usually as a function of time. The lipid oxidation reaction can be accelerated by heating and/or adding oxygen or other prooxidants to more rapidly predict the potential oxidative stability of a lipid during long-term storage. Valuable information about the chemical structure of lipids can also be obtained using NMR, infrared, and mass spectrometry methods.

2.5 Other Additives

Plant-based foods also contain a variety of other additives to improve their quality attributes, extend their shelf lives, or enhance their nutritional profiles (Sha & Xiong, 2020). A number of the most important ones are briefly discussed in this section.

2.5.1 Salts

Salts are added to plant-based foods to enhance their flavor profiles, as well as to control their physicochemical attributes. After ingestion, the mineral ions in foods may mix with or dissolve within the saliva in the mouth. They may then diffuse to the salt receptors located in the taste buds on the tongue, which leads to a signal that is sent through the nerve cells to the brain, where it is perceived as saltiness. The quality of the saltiness perception depends on the type and concentration of the mineral ions present. Consequently, it is important to carefully select an appropriate mineral-rich ingredient to provide an appropriately perceived saltiness in plant-based food products. The minerals may come from crystalline salt (such as table or sea salt) or may be an integral part of another ingredient (such as soy sauce). The mineral ions in salts also play several other important roles in plant-based foods.

2.5 Other Additives

They screen the electrostatic interactions between charged substances, which will decrease the electrostatic repulsion between similarly charged substances or increase the attraction between oppositely charge substances. The addition of salts can therefore alter the interactions between proteins, polysaccharides, and phospholipids in plant-based foods, which can influence their solubility, thickening, gelling, and emulsifying properties. High levels of salts also influence the water activity of foods, which impacts their chemical stability and susceptibility to microbial growth. It should be noted that salts may also be present in the protein ingredients used to formulate plant-based foods because of the methods used to isolate them (such as salting out or pH adjustments).

2.5.2 Colors and Flavors

Colorings and flavorings are often incorporated into plant-based foods to enhance their aesthetic appeal. These additives are often needed to make plant-based products look and taste like the animal foods they are designed to simulate. They also serve as important indicators for proper use and handling of foods by consumers. For instance, meat (like beef) is expected to turn from a reddish color in its raw state to a brownish color after it is cooked. Different manufacturers have taken different approaches to simulate the color of meat products. The company 'Impossible Foods' (Redwood City, CA) uses leghemoglobin, an iron-binding protein originally identified in the roots of soybean plants, to simulate the color of meat (see Chap. 6). This heme protein undergoes fairly similar color changes as the myoglobin found in real meat. Other companies have used natural colors, such as beet extract, to provide a reddish color to plant-based meat analogs. In the case of plant-based egg analogs, the yellowish color of real eggs can be simulated using natural plant pigments, such as curcumin (from turmeric) or carotenoids (from carrots or other sources). Often these additives are strongly hydrophobic molecules and so specially designed delivery systems are required to incorporate them into plant-based foods, such as emulsions or nanoemulsions (Zhang et al., 2020). Typically, the manufacturer will first characterize the optical properties of the animal-based product whose appearance is to be simulated, such as the tristimulus color coordinates ($L^*a^*b^*$). Then, they will select one or more plant-based pigments to match the desired lightness and color. Several of the most common pigments used to formulate plant-based foods are discussed in more detail in Chap. 4.

A number of different flavoring agents may also be utilized to simulate the flavor profile of specific animal-based products, including salts, sugars, spices, herbs, and flavor extracts (Sha & Xiong, 2020). In this case, the food manufacturer will try to identify the key volatile and non-volatile substances in an animal-based food that are responsible for its unique flavor profile. The overall flavor profiles of these foods may be the result of multitudes of different substances that were naturally present in the original food or that were produced during food processing, storage, and cooking, including salts, sugars, organic acids, amino acids, peptides, lipids, and their degradation products and derivatives. Then, the manufacturer will try to identify plant-based sources of these substances that can be isolated and converted into food ingredients, or they will try to chemically synthesize them. In addition, controlled chemical reactions, such as the Maillard, caramelization, or enzymatic reactions, can be used to produce particular flavor profiles. For example, yeast extracts are often used to provide "meaty" flavors (umami) in plant-based foods (Sha & Xiong, 2020; Watson, 2019). In addition, advanced fermentation technologies are being used to create a range of flavors that can be used in these products. Different flavor profiles can be produced by altering the microbes (such as bacteria, molds, and yeast), substrates (such as soy, wheat, or peas), and fermentation conditions (such as time, temperature, and humidity) used. Many plant-based ingredients naturally contain off-flavors, such as earthy, chalky, greeny, astringent, or bitter notes. In this case, it is often necessary to include flavor blocking agents that interact with the receptors on the tongue or flavor masking agents that interact with the off-flavors themselves, thereby inhibiting the ability of the off-flavors and receptors to come together (Fig. 2.12).

More comprehensive discussions of the physical and chemical basis of the color and flavor of plant-based foods are given in Chap. 4. Additional details about the colors and flavors used in specific plant-based foods are given in the chapters on meat, seafood, egg, and dairy analogs.

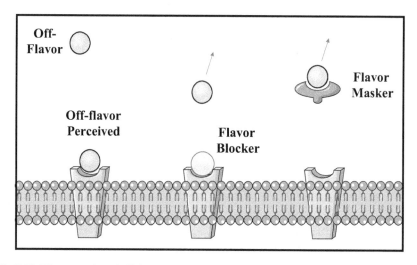

Fig. 2.12 The perception of off-flavors can be blocked or masked by using appropriate substances that bind to the flavor molecules or the taste receptors. These substances can be used to improve the flavor profiles of plant-based foods. (Images modified from Servier Medical Art licensed under CC BY 3.0 (https://creativecommons.org/licenses/by/3.0/))

2.5.3 pH Controllers

The pH of a plant-based product influences many of its physicochemical, biochemical, and sensory attributes, including its susceptibility to microbial growth, the rate of chemical reactions, ingredient solubility, ingredient interactions, texture, flavor, and appearance. For this reason, it is important to control the pH of the product. The pH can be adjusted by adding food-grade acids and bases, whereas the pH can be held at fixed levels by using appropriate buffering agents (Brady, 2013). As a specific example of the importance of pH controllers, plant-based creamers often destabilize when added to acidic coffee because the pH becomes close to the isoelectric point of the proteins (Chung et al., 2017a, 2017b). In this case, it may be important to maintain a relatively high pH and add buffering agents to increase the buffering capacity in the original product to inhibit this problem. Various kinds of food-grade acids, bases, and buffering agents can be used in plant-based foods to adjust and control their pH, such as hydrochloric acid, organic acids and their salts, carbonates, bicarbonates, hydrochlorides, phosphates, and polyphosphates (Kyriakopoulou et al., 2021; Lampila, 2013).

2.5.4 Crosslinking Agents

In semi-solid plant-based foods, such as meats, eggs, and cheeses, it may be necessary to increase the gel strength by adding crosslinking agents (Khalesi et al., 2020; McKerchar et al., 2019). For instance, enzymes like transglutaminase can be used to form covalent crosslinks between plant proteins, thereby increasing the mechanical strength of the product so that it more closely simulates the texture of animal-based products (Mattice & Marangoni, 2021). Indeed, this enzyme has been used to increase the texture of meat products for many years. Alternatively, cationic mineral ions, like calcium, can be added to aqueous solutions of anionic plant proteins to promote their crosslinking through electrostatic bridging effects (Zhao et al., 2017). In addition, acids or bases can be added to promote the physical crosslinking of proteins by decreasing their net electrical charge by bringing them close to their isoelectric point, thereby reducing the electrostatic repulsion between the protein molecules. As a result, the proteins aggregate with each other due to the van der Waals and hydrophobic attraction between them. This kind of physical attraction may be important in the formulation of plant-based yogurts and cheeses that are fabricated using a similar process as the conventional versions, *i.e.,* acidifying a colloidal dispersion containing proteins (Grygorczyk & Corredig, 2013).

2.5.5 Preservatives

Plant-based foods may also contain antimicrobials and antioxidants derived from botanical or microbial sources to increase their shelf-life, thereby reducing food waste, improving food quality, and enhancing food safety (Sha & Xiong, 2020). For instance, essential oils, plant extracts, tocopherols, carotenoids, spices, and herbs can all be used as natural antioxidants, whereas essential oils, curcumin, polyphenols, and nisin can be used as natural antimicrobials (Dominguez et al., 2021). It should be noted that botanical preservatives rarely work as well as synthetic ones, and so it is often necessary to use them at higher concentrations or in combination with each other. Moreover, it is important to take into account their partitioning and interactions within complex food matrices. For instance, hydrophobic preservatives may be solubilized within the non-polar interiors of lipid phases or charged preservatives may interact with oppositely charged proteins or polysaccharides, which can reduce their efficacy by decreasing the concentration available to interact with microbial membranes (Loeffler et al., 2014, 2020; Weiss et al., 2015).

2.5.6 Micronutrients and Nutraceuticals

It is important that plant-based foods have similar or better nutritional profiles than the animal-based counterparts they are designed to replace, otherwise there may be nutritional deficiencies associated with adopting a more plant-based diet (Chap. 5). For this reason, many food manufacturers fortify their plant-based products with micronutrients that might be lacking in a vegetarian or vegan diet, such as vitamin D, ω-3 fatty acids, vitamin B_{12}, calcium, iron, or zinc (Tuso et al., 2013). Having said this, there is little evidence that consuming a plant-based diet leads to health problems due to nutritional deficiencies in adult populations in regions where food is abundant, although it has been recommended to supplement the diet with vitamins B_{12} and D (McDougall & McDougall, 2013). In contrast, individuals consuming a predominantly plant-based diet may suffer diseases related to vitamin or mineral deficiency in regions in the world where foods are not abundant, *i.e.,* iron deficiency anemia is widespread in rural India where the diet is mainly cereal-based (Taneja et al., 2020). Moreover, there is also some concern that a vegetarian or vegan diet may adversely affect the development of infants due to a lack of adequate levels of vitamins and minerals required for healthy growth (Biesalski & Kalhoff, 2020). For this reason, it is important to fortify these products with bioavailable forms of essential nutrients that may be missing from a vegetarian or vegan diet. Plant-based foods may also be fortified with bioactive phytochemicals (nutraceuticals) that may improve human health by inhibiting certain diseases, such as carotenoids, curcumin, resveratrol, green tea polyphenols, or ginseng (Calabrese, 2021). Vitamins, minerals, and nutraceuticals often have to be encapsulated within well-designed colloidal delivery systems so that they can be successfully incorporated into food matrices in a stable form, as well as to ensure that they have a high bioavailability after ingestion (Dima et al., 2021; McClements, 2020). Alternatively, food matrix effects can be controlled to enhance the bioavailability of these bioactive substances (Nair & Augustine, 2018).

2.6 Ingredient Functionality

The ingredients used to formulate plant-based foods play a variety of roles in determining their physicochemical, sensorial, and gastrointestinal properties. For instance, they may provide desirable optical, textural, mouthfeel, fluid holding, flavor, and stability characteristics. In this section, an overview of some of the most important functional attributes of plant-derived ingredients is therefore given. The main focus will be on the three types of macronutrients (proteins, polysaccharides, and lipids) because these are the most important functional ingredients used to assemble plant-based foods.

2.6.1 Solubility

An important physicochemical attribute of many plant-based ingredients is their solubility in oil, water, or other solvents. In particular, the water-solubility of proteins and polysaccharides derived from plants plays an important role in determining the formation, stability, and quality of many plant-based foods. For instance, the ability to fully dissolve these functional proteins and polysaccharides in water often influences their ability to thicken solutions, form gels, and stabilize emulsions (Phillips & Williams, 2021).

The solubility of biopolymers is governed by their unique molecular features, such as their molar mass, surface hydrophobicity, and electrical characteristics (Guo et al., 2017). Typically, their solubility within aqueous solutions decreases as the number of non-polar groups on their surfaces increases because of a strengthening of the hydrophobic attraction between neighboring molecules. This phenomenon accounts for the relatively low water-solubility of hydrophobic proteins, like gliadin and zein (Davidov-Pardo et al., 2015), as well as the tendency for hydrophobic polysaccharides to self-associate when they are heated, as is the case with methylcellulose (BeMiller, 2019; Murray, 2009). Methylcellulose tends to form transparent solutions at low temperatures but cloudy gels at high temperatures due to the formation of hydrophobic crosslinks between the polysaccharide molecules (Arvidson et al., 2013; Spelzini et al., 2005). In addition, polysaccharides that are able to come close together and form strong intermolecular hydrogen bonds, such as linear cellulose molecules, also have low water solubilities (Guo et al., 2017).

Water-insoluble proteins like zein and gliadin have also been used to create structural features within plant-based meat products that mimic those found in real meat products, such as fibers or particulates (Mattice & Marangoni, 2020a). Similarly, water-insoluble polysaccharides like cellulose have been used as fat replacers, texture modifiers, or bulking agents in processed foods. In this case, the crystalline cellulose material is fragmented into microparticles or nanoparticles using chemical or mechanical means, which are then used as functional ingredients in foods (Duan et al., 2018; Khalil et al., 2014).

The ability of ingredients to dissolve in water also depends on their electrical characteristics. A solution of strongly charged proteins or polysaccharides often has a high water-solubility because the strong electrostatic repulsive forces generated by the molecules prevent them from coming close together (Curtis & Lue, 2006). The low water-solubility of many plant-based proteins around their isoelectric points is a result of them losing their electrical charge, so that van der Waals and hydrophobic attractive forces dominate the weak repulsive forces (Gehring et al., 2010). Moreover, in systems containing both anionic and cationic substances, the water-solubility can be reduced due to electrostatic attraction between the oppositely charged molecules. An example of this phenomenon is when a cationic protein is mixed with an anionic polysaccharide, which can lead to the formation of insoluble electrostatic complexes (Weiss et al., 2019). The formation of these complexes may be advantageous in some cases because they lead to desirable structural features, such as fibers or particles. Finally, the solubility of large molecules, like most plant-based biopolymers, often increases as their molecular weight decreases because the entropy of mixing becomes more important, thereby favoring their random organization within water (Curtis & Lue, 2006).

The solubility characteristics of biopolymers can be related to their molecular features using theoretical models by assuming they are polymers or colloidal particles dispersed in a solvent, and then calculating the various kinds of forces acting between them, *e.g.*, van der Waals, steric, electrostatic, hydrophobic, and entropic interactions (Curtis & Lue, 2006). The biopolymers tend to remain soluble when the repulsive forces dominate the attractive ones but become insoluble when the attractive forces dominate.

2.6.2 Molecular Binding Interactions

Many proteins and polysaccharides used to formulate plant-based foods have complex surface chemistries containing polar, non-polar, anionic, and/or cationic surface groups (Phillips & Williams, 2021). These surface groups may vary in their type, number, and spatial distribution along the biopolymer chains, which impacts their ability to interact with other molecules in their environment (Foegeding & Davis, 2011; Stephen et al., 2006). Biopolymers with exposed non-polar regions on their surfaces, such as many plant-based globular proteins, can bind non-polar molecules through attractive hydrophobic interactions. Biopolymers with exposed anionic or cationic regions can bind oppositely charged molecules in their environment as a result of attractive electrostatic interactions (Blackwood et al., 2000).

Knowledge of these types of interactions is often useful when designing plant-based food products. For instance, plant-based proteins may bind to aromatic flavor molecules in beverages, thereby altering the overall flavor profile (Guo et al., 2019; Wang & Arntfield, 2015). In some cases, this would be desirable because it reduces the level of off-flavors, but in other cases it may be undesirable because it alters the anticipated flavor profile. Mathematical models have been developed to describe the influence of the molecular characteristics of flavor molecules on their binding to proteins in foods and how this influences the overall flavor profile (Viry et al., 2018).

In plant-based milk and cream analogs, it is important to identify and understand any binding interactions that occur between the fat droplets/oil bodies and other ingredients in the surrounding aqueous phase, as this can impact the stability of the overall system. As an example, the binding of an oppositely charged polysaccharide to the surfaces of protein-coated oil droplets can promote their aggregation through a bridging flocculation mechanism (Dickinson, 2019). Conversely, the complete coating of protein-coated oil droplets with a layer of oppositely charged polysaccharide molecules can protect it from aggregation by increasing the steric and electrostatic repulsion between the droplets (Li & de Vries, 2018; Xu et al., 2020). Plant protein-coated oil droplets may also be influenced by binding to mineral ions in the aqueous phase. For instance, the binding of cationic calcium ions to the surfaces of anionic proteins or protein-coated oil droplets can reduce the electrostatic repulsion between them, thereby promoting their aggregation (Marquez et al., 2018). Thus, elucidating, understanding, and controlling the binding interactions in these kinds of products is critical.

2.6.3 Fluid-Holding Capacity

Semi-solid plant-based foods, such as meat, egg, cheese, and yogurt analogs, typically contain substantial amounts of fluids, such as pure liquids (*e.g.,* water or oil), solutions (*e.g.,* brine or sugar solutions), or dispersions (*e.g.,* emulsions or suspensions). The presence of these fluids influences the appearance, texture, flavor, and stability of these foods. For instance, the retention of fluids in meat-like products plays a critical role in determining their perceived texture and juiciness (Cornet et al., 2021), whereas the separation of fluids from yogurts ("syneresis") is visibly unappealing (Grasso et al., 2020). Consequently, it is often important to control the ability of plant-based foods to retain their fluids during storage and food preparation.

Biopolymers are often incorporated into plant-based foods, such as meat, egg, and yogurt analogs, to improve their water holding capacity (WHC), *i.e.,* their ability to retain water (or other aqueous-based solutions or dispersions) within the food structure (Cornet et al., 2021; Grasso et al., 2020). As highlighted earlier, the presence of these fluids plays a vital role in determining the texture and mouthfeel of some plant-based food products, such as the juiciness of meat analogs. Consequently, it is important to ensure that the water does not separate from the product. Biopolymers can promote water retention by forming a porous 3D-network that holds water through capillary and hydration forces (Blackwood et al., 2000; Stevenson et al., 2013). The amount of water molecules that bind directly to the surfaces of biopolymer molecules depends on their surface chemistry, as well as on the total surface area exposed. The strength of the capillary forces can be approximated by the Laplace pressure of a porous material (Stevenson et al., 2013):

$$\Delta P = 2\gamma \cos\theta / r \tag{2.1}$$

In this expression, ΔP represents the capillary pressure, γ represents the surface tension, θ is the water-material contact angle, and r is the pore radius. This equation shows that the capillary pressure increases as the pore size decreases and the surface tension increases. Physically, the WHC is a measure of a food's ability to retain water in the presence of an external force, such as gravity or an applied stress (*e.g.,* pressing or centrifugation). An increase in the number and/or decrease in the size of the pores in a porous material tends to lead to an increase in its WHC. Consequently, it is often useful to create a fine uniform porous structure in a food to avoid water loss.

Using a different approach, researchers have attempted to relate the WHC of polymeric porous foods (including meats and plants) to their structural and chemical properties using thermodynamic approaches (van der Sman, 2012, 2013; van der Sman et al., 2013). The mathematical theories resulting from this kind of analysis have highlighted that there are three important contributions that influence the swelling pressure and consequently the WHC of biopolymer-based foods: (*i*) biopolymer-water mixing; (*ii*) biopolymer-ion interactions; and (*iii*) elastic effects resulting from gel deformation. These models can be useful for designing foods with improved fluid holding properties by identifying the importance of key molecular and physicochemical properties (Cornet et al., 2021; van der Sman et al., 2013). Typically, the ability of a biopolymer matrix to absorb water and swell is reduced as the crosslinking density between the biopolymer molecules is increased because this increases the shear modulus of the gel network, thereby making it more difficult for expansion to occur (Cornet et al., 2021). At present, the relationship between the capillary (pore size/hydration) theories and the thermodynamic (mixing/ion/elastic) theories is unclear and requires further elucidation.

2.6.4 Thickening

Plant-based biopolymers (especially extended polysaccharides) are often used as functional ingredients in fluid or semi-solid plant-based food products to increase the viscosity of the aqueous phase, including milks, creams, fluid eggs, dressings, and sauces (Williams & Phillips, 2021). These thickening agents may be utilized to generate desirable textural or mouthfeel attributes, or to reduce the rate of gravitational separation of particulate matter (such as fat droplets, oil bodies, plant tissue fragments, protein aggregates, herbs, or spices). The ability of a thickening agent to enhance the viscosity of an aqueous solution can be described by its *thickening*

2.6 Ingredient Functionality

power, which is a measure of how much of the ingredient is required to produce a large increase in viscosity. A good thickening agent typically only requires a small amount to effectively increase the viscosity. The thickening power of a biopolymer depends on its volume ratio (R_V), which is the effective volume occupied by the biopolymer molecule when it is dissolved in solution (biopolymer molecule + entrapped water) divided by the volume occupied by the biopolymer chain alone (Bai et al., 2017). The higher the value of R_V, the greater the thickening power of the polysaccharide. In general, R_V increases as the molar mass increases, the degree of branching decreases, and the conformation of the molecule becomes more extended (Fig. 2.13). For this reason, stiff elongated polysaccharides (such as xanthan gum) have a much higher thickening power than compact polysaccharides (such as gum arabic) or proteins (such as pea or soy proteins).

In addition, the ability of a biopolymer to thicken an aqueous solution can be modeled using the following equation (Bai et al., 2017):

$$\eta = \eta_1 \left(1 - \frac{\phi_E}{0.57}\right)^{-2} \tag{2.2}$$

Here, η and η_1 represent the shear viscosities of the biopolymer solution and the solvent, ϕ_E is the *effective volume fraction* of the solution occupied by the hydrated biopolymer molecules (including any entrained water). Roughly, the value of ϕ_E is given by the following expression (Bai et al., 2017):

$$\phi_E = \frac{4}{3}\pi r_H^3 \left(\frac{cN_A}{M}\right) \tag{2.3}$$

Here, r_H represents the hydrodynamic radius of the biopolymer molecules in solution, c is the concentration of the biopolymers, N_A is Avogadro's number, and M is the molecular weight of the biopolymer molecules. Predictions made using these equations show that the thickening power of a biopolymer becomes stronger as its effective volume increases (Fig. 2.13).

To a first approximation, the thickening power (TP) is given by the following simple expression, which represents the reciprocal of the biopolymer concentration required to produce a large increase in viscosity (Grundy et al., 2018):

$$TP = \frac{r_H^3}{21M} \tag{2.4}$$

In this expression, the units of TP are wt%$^{-1}$ when the molecular weight of the biopolymer is given in kg mol^{-1}, and the hydrodynamic radius is given in nanometers. This expression indicates that the thickening power becomes stronger as the hydrodynamic radius of the biopolymer increases (at a fixed molecular weight). In other words, biopolymer molecules that entrap large quantities of water are the most effective at thickening solutions (Fig. 2.13).

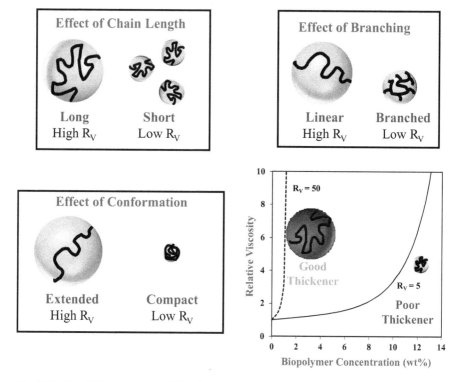

Fig. 2.13 The thickening power of biopolymers depends on their molecular characteristics, such as the volume ratio R_V. Typically, the thickening power increases with increasing molecular weight, increasing extension, and decreasing branching

Another important attribute of a thickening agent is how the viscosity of the solution changes as the applied shear rate is increased. Most biopolymer-based thickening agents exhibit shear-thinning behavior, *i.e.,* the viscosity decreases as the shear rate ($\dot{\gamma}$) increases. Typically, the shear viscosity has a constant value at low shear rates, then decreases at intermediate shear rates, and then reaches a constant value at high shear rates (Fig. 2.14). This kind of behavior can be described by the Cross model:

$$\eta = \eta_\infty + \frac{\eta_0 - \eta_\infty}{1 + (K\dot{\gamma})^{n-1}} \qquad (2.5)$$

Here, K is the Cross constant and n is the shear thinning index. For ideal fluids, where the viscosity does not depend on the applied shear rate, $n = 1$. For shear thinning fluids, $n < 1$, and for shear thickening fluids, $n > 1$. At intermediate shear stresses, the following expression can be used, which is known as the power-law model (Fig. 2.14):

2.6 Ingredient Functionality

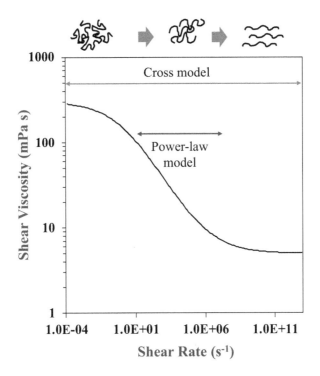

Fig. 2.14 Prediction of the change in shear viscosity with shear rate for a hydrocolloidal solution that exhibits shear thinning. The Cross model works over the entire shear rate range (which is often not experimentally accessible), while the power-law model works at intermediate shear rates

$$\eta = K\dot{\gamma}^{n-1} \tag{2.6}$$

Here, K is the consistency index and n is the power index. It should be noted that the values of K and n in the power-law model and Cross model need not have the same numerical values. Shear thinning is important because it affects the pourability of fluids, such as plant-based dips, sauces, and dressings. It also influences the perceived mouthfeel of food products. Consequently, it is important to ensure that the thickening agent used provides the required shear-thinning behavior. The molecular origin of shear thinning is the increasing disentanglement and alignment of the biopolymer chains as the shear rate increases, which leads to a reduction in energy dissipation due to friction. Substantial shear thinning is typically observed in concentrated biopolymer solutions where the molecules overlap with each other and may lead to a decrease in viscosity of several orders of magnitude.

It should also be noted that some semi-solid plant-based foods are required to have a yield stress. When a stress below this value is applied to them, they behave like elastic solids, but when the applied stress exceeds this value, they behave like viscous fluids (Rao, 2007). Products such as dips, sauces, and dressings should have a yield stress. This value should be high enough that the products do not collapse under their own weight after they have been applied to a food but low enough so that they flow from a container. The rheological properties of materials with a yield stress can be described by the following equation, which is known as the Herschel Bulkley model (Kaltsa et al., 2018):

$$\tau - \tau_0 = K\dot{\gamma}^n \tag{2.7}$$

In this case, τ_0 is the yield stress, K is the consistency index, and n is the power index. This equation only applies once the applied stress exceeds the yield stress. Then the material flows like a non-ideal fluid.

2.6.5 Gelling

Biopolymers, such as proteins and polysaccharides, are often used as gelling agents to provide desirable textural characteristics to semi-solid plant-based foods, such as meat, egg, cheese, and yogurt analogs. Many plant-based biopolymers can associate with each other through physical or chemical interactions under appropriate conditions, thereby forming a porous 3D network of biopolymer chains that traps water and other fluids inside (Phillips & Williams, 2021). The physicochemical attributes of these biopolymer hydrogels, including their optical, rheological, and fluid holding properties, as well as their sensitivity to environmental conditions (such as temperature, pH, or salt) are governed by the type and concentration of biopolymers used, as well as the nature of the crosslinks between them. Biopolymer gels can be formed that vary in their appearance (transparent, cloudy, or opaque), texture (soft/hard, rubbery/brittle, fracture properties), setting characteristics (heat-, cold-, ion-, pH-, or enzyme-set), and gastrointestinal fate (digestible/indigestible, fermentable/non-fermentable) (McClements, 2021). Consequently, it is important for formulators to select an appropriate biopolymer or a combination of biopolymers to obtain the desired characteristics in the final product, such as hardness, chewiness, or fracture during mastication. Some of the key factors influencing the nature of the gels formed from plant-based proteins and/or polysaccharides are highlighted in this section.

The overall gel strength (elastic modulus) of a biopolymer gel is determined by the type, concentration, and interactions of the biopolymer molecules within the hydrogel network (Vilgis, 2015). For relatively simple systems, theoretical models are available to link the molecular characteristics of biopolymer gels to their textural attributes (Gabriele et al., 2001; Rubinstein et al., 1996). Predictions made by these models indicate that the gel strength increases as the biopolymer concentration, crosslinking density, and bond strength increase (Cornet et al., 2021).

Many plant-based foods contain particles embedded within a hydrogel matrix, such as fat droplets or protein particles within plant-based meat, fish, egg, yogurt, or cheese analogs. The rheological behavior of these composite systems can also be described using mathematical models (Fu et al., 2008). Predictions made by these models indicate that the concentration, dimensions, and interactions of the embedded particles within the polymer network influence their textural attributes, such as gel strength, stiffness, toughness, and fracture properties. Knowledge of these theoretical models can be beneficial to food formulators because they help to identify the major factors that impact the textural attributes of semi-solid plant-based foods. Having said this, the compositional and structural complex nature of most plant-based foods limits most predictions to be only of qualitative character (rather than quantitative). However, a number of the most important attributes of biopolymer gelling agents that need to be considered when selecting them for plant-based food applications are briefly summarized here:

- *Final gel strength*: The final gel strength (elastic modulus) determines the hardness/softness of the gel, as well as the sensory perception of the gel during mastication (particularly the first bite).
- *Gelation temperature*: In some applications, the temperature (range) at which a gel forms or melts is important. For instance, a plant-based egg analog should form an opaque gel when it is heated above about 70 °C to mimic the cooking properties of real egg, whereas a plant-based gelatin analog should form a clear gel when it is cooled below about 35 °C.
- *Minimum concentration*: The lowest biopolymer concentration required to form a gel depends on the structure and interactions of the biopolymer molecules. A gel is only formed when the biopolymer molecules form a 3D network that extends throughout the entire volume of the material. Small compact spherical molecules like native globular proteins require much higher concentrations (often >10%) to form gels compared to large linear molecules like polysaccharides (often <1 wt%).
- *Fragmentation properties*: The stress or stain at which a gel breaks when a force is applied determines its brittleness or rubberiness. Moreover, the way that the gel fractures when it breaks is also important (especially during mastication), *e.g.*, the size and shape of the fragments formed. The fragmentation properties are particularly important in products like plant-based cheese and meat analogs.

Examples of different kinds of plant-based biopolymers and their gelling properties are highlighted in Table 2.5. In addition, examples of plant-based proteins that form gels are also included. Many plant-based proteins form gels when they are heated because thermal energy causes the globular proteins to unfold, thereby exposing non-polar surface groups that can associate with each other through hydrophobic attraction (Mezzenga & Fischer, 2013). Nevertheless, the pH and ionic strength must also be controlled to reduce the electrostatic repulsion between the protein molecules. If the electrostatic repulsion is too strong, then the proteins may not be able to come close enough together to aggregate. Once the unfolded proteins have aggregated through hydrophobic attraction, then disulfide bonds may form via sulfur-containing groups. Alternatively, plant proteins can be made to form gels by adding crosslinking enzymes such as transglutaminase, which binds neighboring proteins together through covalent bonds (McKerchar et al., 2019).

Table 2.5 Characteristics of gels formed using different kinds of animal- and plant-based biopolymers. Only the common gelling mechanisms of animal proteins are shown

Biopolymer type	Gelation mechanisms	Interactions	Gel properties
Animal proteins			
Gelatin	Cold-set reversible	Hydrogen bonds	Transparent
Egg	Heat-set irreversible	Hydrophobic and disulfide	Opaque
Whey protein	Heat-set irreversible	Hydrophobic and disulfide	Opaque
Casein	Acid-set irreversible	Van der Waals, hydrophobic, and ion bridges	Opaque
Plant proteins			
Globular proteins	Heat-set irreversible	Hydrophobic and disulfide	Opaque
	Enzyme-set irreversible	Covalent	Opaque
	Ion-set irreversible (Ca^{2+})	Salt bridges	Opaque
Plant polysaccharides			
Agar	Cold-set reversible	Hydrogen bonds	Transparent-turbid
Alginate	Ion-set irreversible (Ca^{2+})	Salt bridges	Transparent-turbid
Carrageeanan	Ion-set irreversible (Ca^{2+} or K^+)	Salt bridges	Transparent-turbid
Gellan gum	Cold set reversible	Hydrogen bonds and salt bridges	Transparent-turbid
Methyl cellulose	Heat-set reversible	Hydrophobic	Opaque
LM pectin	Ion-set irreversible (Ca^{2+})	Salt bridges	Transparent-turbid
HM pectin	Acid and sugar-set	Van der Waals, hydrophobic, and osmotic	Transparent-turbid
Starch	Heat-set irreversible	Hydrophobic	Turbid-opaque

Plant-based polysaccharides can also be made to form gels through various mechanisms (Table 2.4) (Williams & Phillips, 2021). Agar, which is derived from seaweed (red algae), can form clear gels through a thermo-reversible mechanism: it gels when cooled and melts when heated. Such seaweed-derived ingredients can be produced in a sustainable manner by carefully maintaining and harvesting them in coastal areas, which is a desirable attribute for many plant-based foods applications. Similarly, gellan gum, which is derived from microbial fermentation, can also form cold-set thermo-reversible gels. These cold-set polysaccharides exist as random coil molecules at high temperatures but undergo a coil-to-helix transition upon cooling. Any cationic ions in the system can then serve as salt bridges that link the anionic helixes on different molecules together. Alginate, which is obtained from brown seaweed, can form gels at ambient temperatures in the presence of cationic divalent mineral ions, especially calcium (Ca^{2+}) ions. These ions form salt bridges between the anionic carboxyl groups on neighboring alginate molecules. The strength of these gels can be modulated by controlling the calcium concentration, alginate concentration, and alginate type (*e.g.*, M-to-G block ratio, molecular weight). Carrageenan based ingredients, which are extracted from red seaweed, can also be used for their gelling properties. There are a number of different kinds of carrageenan that differ in their molecular structure and functional properties, with the iota and kappa types being the most widely used in the food industry as gelling agents. Iota carrageenan, which contains two sulfate groups per disaccharide unit, forms gels in the presence of calcium ions. In contrast, kappa carrageenan, which has only one sulfate group per disaccharide unit, forms gels in the presence of potassium ions. In addition, reversible heat-setting polysaccharides that form transparent gels have been developed as plant-based alternatives to gelatin (Jaswir et al., 2016). Like gelatin, these polysaccharides gel when they are cooled below a critical temperature but melt when they are heated above another (higher) critical temperature. The main challenges in this area are to obtain gelation and melting temperatures, as well as rheological and mouthfeel properties, that simulate those of real gelatin. Methylcellulose tends to gel when it is heated due to an increase in the hydrophobic attraction between the methyl groups on the molecules (Spelzini et al., 2005). This phenomenon can be useful in the formulation of plant-based fish and meat analogs where a semi-solid texture is required after cooking.

On a final note, there are also some other hydrophobic substances that can be used to structure oils, leading to the formation of oleogels (Puşcaş et al., 2020). These substances include waxes, phytosterols, monoglycerides, and modified celluloses (such as ethyl cellulose). Typically, an appropriate amount of oleogelator is dispersed within the oil phase and then the mixture is heated to melt it. The liquid mixture is then cooled down, which causes the oleogelator to crystallize and form a 3D network that provides mechanical strength to the oil phase. An oleogelator can be used to form a solid-like fat that does not contain high levels of saturated or trans-fatty acids, which may have health benefits.

2.6.6 Binders and Extenders

Biopolymers are often added to plant-based food products as binders, which are ingredients whose purpose is to hold the different components in the product together so it does not collapse (Williams & Phillips, 2021). They can also be used as extenders whose purpose is to add more texture to the product. Binders and extenders work through a variety of mechanisms, including fluid holding, thickening, gelling, and adhesion. Various kinds of plant proteins and polysaccharides can be used as binders or extenders, including soy, wheat, and pea proteins, as well as different types of starches. These ingredients are useful in plant-based foods where a semi-solid texture should be maintained, *e.g.*, in uncooked burgers, sausages, or nuggets. Commonly, proteins or polysaccharides are used as binders and extenders. For example, wheat gluten is used to bind extruded protein particles together. Polysaccharides such as pectin, guar gum, carrageenan, cellulose, and methylcellulose are also often employed as binders in meat analog products. The molecular characteristics and functionality of these ingredients have recently been reviewed elsewhere (Kyriakopoulou et al., 2021).

2.6.7 Emulsification

Emulsifiers are ingredients that can adsorb to oil-water interfaces and stabilize the fat droplets (McClements et al., 2017). They do this by forming a protective coating around the droplets that provides some mechanical rigidity and/or that generates repulsive forces that help to prevent them from aggregating with each other. The most common repulsive forces are electrostatic and steric repulsion, which depend on the surface charge and thickness of the adsorbed emulsifier layer (McClements, 2015). Plant-based ingredients that can act as emulsifiers are typically surface-active molecules that have some polar and some non-polar regions on their surfaces, or they are small colloidal particles that have appropriate wetting characteristics, *i.e.*, they are partly wetted by both oil and water (McClements & Gumus, 2016). Molecular plant-based emulsifiers include some proteins (*e.g.*, from pea, fava bean, lentil, and soy), polysaccharides (*e.g.*, modified starch and gum arabic), phospholipids (*e.g.*, from soy or sunflower), and saponins (*e.g.*, from quillaja or tea) (McClements & Gumus, 2016). Particle plant-based emulsifiers include nanoparticles fabricated from proteins and/or polysaccharides, such as zein, kafirin, and soy protein (Sarkar & Dickinson, 2020; Shi et al., 2020). An advantage of using particulate emulsifiers is that they tend to provide strong protection of the fat droplets against coalescence, which may be useful in applications where the droplets are in close proximity for long times (such as plant-based dressings and mayonnaise). Conversely, it is difficult to make small fat droplets using particulate emulsifiers, which would lead to rapid creaming in low viscosity products (like plant-based milks or creams).

Emulsifiers are often used to facilitate the formation and enhance the stability of the fat droplets in emulsified plant-based foods, such as milk analogs, cream analogs, dressings, and sauces (McClements & Gumus, 2016). In this case, a water phase containing a plant-based emulsifier is homogenized with an oil phase, which leads to the formation of small emulsifier-coated fat droplets. The size and charge on the fat droplets have to be carefully controlled to ensure that they are stable to gravitational separation and aggregation. Emulsified fats play a variety of roles in determining the desirable appearance, texture, mouthfeel, flavor profile, and stability of many plant-based foods (McClements et al., 2019; McClements & Grossmann, 2021b). For instance, the presence of lipid droplets makes a product look creamy or opaque due to light scattering, as well as increasing the viscosity and creamy mouthfeel by disturbing the fluid flow and lubricating the tongue. Moreover, the lipid droplets may act as a reservoir for lipophilic aroma molecules, which alters the flavor profile of fatty foods.

Emulsifiers may be used in plant-based meat analogs to help form simulated adipose tissue. In mammals, adipose tissue consists of large fatty cells (typically tens to hundreds of micrometers) embedded within a network of collagen fibers (Urrutia et al., 2018). This kind of structure can be simulated using plant-based ingredients by preparing a concentrated oil-in-water emulsion containing emulsifier-coated fat droplets, which may be embedded in a biopolymer gel network. These plant-based high internal phase emulsions (HIPEs) are optically opaque semi-solid materials that have physicochemical attributes that mimic some of those of adipose tissue.

2.6.8 Foaming

Foaming agents are ingredients that can adsorb to air-water interfaces and facilitate the formation and stabilization of gas bubbles (Amagliani & Schmitt, 2017; Narsimhan & Xiang, 2018). Like emulsifiers, they create a protective coating around the gas bubbles that improves their stability by introducing some mechanical rigidity and/or by generating repulsive forces that can inhibit their aggregation. Foaming agents are useful in products such as plant-based whipped creams, ice cream, and baked foods, which may contain gas bubbles to provide desirable textural and stability characteristics. A variety of molecular and particulate plant-based foaming agents have been identified that may be suited for this type of application (Amagliani & Schmitt, 2017; Narsimhan & Xiang, 2018). These materials are often plant-based proteins or protein nanoparticles, but the functional performance of these proteins is often improved by using them in conjunction with plant-based polysaccharides. In addition, in some cases, the formation of a fat crystal network around the air bubbles can be used to help to stabilize them. In conventional whipped cream or ice cream, the partially crystalline milk fat globules aggregate with each other due to partial coalescence, forming a semi-solid shell around the air bubbles that helps to prevent them from collapsing. It may therefore be important for the creation of plant-based analogs of these products to identify suitable sources of plant-based fat droplets that can carry out a similar function.

2.6.9 Melting and Crystallization

The melting and crystallization behavior of plant-derived lipids play a major role in determining the texture, stability, and mouthfeel of some plant-based foods. In particular, they may be important for creating plant-based meat, cheese, and ice cream analogs that mimic the desirable textural attributes of the conventional versions of these products. In this case, the solid fat content (SFC) *versus* temperature profile is important (Fig. 2.10), as well as the morphology, polymorphic state, and interactions of the fat crystals formed (Marangoni et al., 2012; Ramel et al., 2016). At a sufficiently high concentration, a 3D network of small, aggregated fat crystals is formed that provides semi-solid characteristics, such as a yield stress and elastic modulus. The gel strength of these fat crystal networks typically increases as the crystal concentration, crosslinking density, and bond strength increase. Ideally, a plant-based lipid phase should match the desirable thermal and textural characteristics of the animal-based one it is designed to replace. This often requires blending different kinds of lipids together to get an appropriate SFC – temperature profile. Commonly, high-melting plant-based lipids like coconut oil, cocoa butter or palm oil are used for this purpose. If this is done correctly, then it will lead to the required SFC – temperature profile, as well as to the formation of a network of fat crystals with the required size, shape, polymorphic form, interactions, and mechanical properties.

2.6.10 Nutrition

Some additives are used to improve the nutritional profile of plant-based foods (Kyriakopoulou et al., 2021; McClements & Grossmann, 2021b) (Chap. 5). Vitamins and minerals may be added to mimic those found in animal-based foods or to provide micronutrients that might be lacking from a plant-based diet, such as vitamins B_{12}, D, iron, calcium, and zinc. Plant proteins provide a source of essential amino acids to the diet. However, individual plant proteins are typically deficient in one or more essential amino acids, whereas animal proteins are not (Herreman et al., 2020). This is highlighted in Table 2.6, which compares the digestible indispensable amino acid scores (DIAAS) and essential amino acids lacking for different plant and animal sources. Lysine is the main limiting amino acid in cereal proteins, whereas methionine and cysteine are limiting in pulse proteins. It may therefore be possible to overcome this problem by using appropriate blends of cereal and pulse proteins in the same product, or by eating a variety of different sources of protein throughout the day (Herreman et al., 2020). Nevertheless, further research is required to understand how specific protein blends behave within the human gut in different plant-based food matrices, as their nutritional effects will depend on their amino acid profiles, as well as on the rate and extent of their digestion in the upper gastrointestinal tract (Reynaud et al., 2021). As mentioned earlier, lipid sources rich in ω-3 oils (such as flaxseed or algae oils) may also be used to fortify plant-based foods with essential fatty acids, thereby improving their nutritional profile.

2.6.11 Gastrointestinal Fate

An aspect of ingredient functionality that is often overlooked is their gastrointestinal fate, *i.e.*, how they behave within the human gut after ingestion (McClements, 2021). Plant-based ingredients behave differently during digestion compared to the animal-based ones they are designed to replace (Ogawa et al., 2018). In particular, there are differences in the location, rate, and extent of nutrient digestion and

2.6 Ingredient Functionality

Table 2.6 Comparison of the digestible indispensable amino acid scores (DIAAS) and essential amino acids lacking for different plant and animal sources

Protein source	DIAAS	Limiting amino acid
Corn	36	Lysine
Rice	47	Lysine
Wheat	48	Lysine
Hemp	54	Lysine
Fava bean	55	Methionine + Cysteine
Oat	57	Lysine
Rapeseed	67	Lysine
Lupin	68	Methionine + Cysteine
Pea	70	Methionine + Cysteine
Canola	72	Lysine
Mung bean	86	Leucine
Soy	91	Methionine + Cysteine
Potato	100	N/A
Gelatin	2	Tryptophan
Whey	85	Histidine
Casein	117	N/A
Milk	116	N/A
Egg	101	N/A
Pork	117	N/A
Chicken	108	N/A
Beef	112	N/A

Adapted from McClements and Grossmann (2021b) with permission.

absorption within the gastrointestinal tract (GIT), which influences the pharmacokinetic profiles of nutrients and their metabolites in the bloodstream. Moreover, the composition and digestion rate of a food impacts the hormonal and metabolic responses, which impact the susceptibility to diseases such as diabetes, obesity, and heart disease (Xie et al., 2020). Many animal products mainly consist of proteins and fats, with very few carbohydrates, including meat, fish, and eggs (Toldra, 2017). In contrast, many plant-based foods contain digestible and indigestible carbohydrates, such as sugars, starches, and dietary fibers, which behave differently inside the human gut (Mariotti, 2017). A recent *in vitro* study (INFOGEST) showed that there were appreciable differences in the digestibility profiles of different plant and animal proteins, *i.e.,* garden pea, grass pea, soybean lentil, casein, and whey proteins (Santos-Hernandez et al., 2020). In general, the soy proteins were less digested in the gastric and intestinal fluids than the other proteins.

Moreover, the ability of dietary fibers to be fermented by colonic bacteria in the large intestine (colon) also impacts the healthiness of plant-based foods (Wilson et al., 2020). The nutritional and health implications of switching from an omnivore to a more plant-based diet are obviously important and will be a critical area for future research (Hemler & Hu, 2019). Further information about the potential nutritional and health effects of plant-based foods and their ingredients are discussed in Chap. 5.

2.6.12 Miscellaneous Functions

In addition to the properties just described, plant-based ingredients may provide various other desirable properties that are important for mimicking the characteristics of traditional animal-based foods. For instance, biopolymers may help to suppress the formation of large ice crystals in plant-based frozen meals and ice creams, thereby improving their texture and mouthfeel. They may also inhibit moisture migration, which can help to avoid water loss due to syneresis in food products. This is especially important in plant-based yogurt analogs, where a visible layer of water on top of the product is commonly tried to avoid.

Other ingredients such as flavors, colors, pH regulators, and preservatives are also important for creating plant-based foods that accurately simulate those of animal origin. The type of flavors and colors depends on the precise nature of the product being simulated. A number of specific examples are given later in this book in the chapters on specific plant-based foods, such as meat, egg, and dairy analogs.

2.7 Ingredient Utilization

After an appropriate set of ingredients has been selected to prepare a plant-based food product, it is important to utilize and combine them in an appropriate manner. There are a number of factors that need to be considered when doing this. First, it is important that the ingredients are stored under appropriate conditions (such as temperature, humidity, light, and oxygen levels) to avoid them deteriorating prior to use. Second, it may be important to dissolve or disperse them in a certain medium prior to utilization. For instance, many hydrophilic ingredients (such as proteins and polysaccharides) may have to be dissolved or dispersed within an aqueous solution before being used. It is often important to control the pH, ionic composition, and temperature of the aqueous solution, as well as the duration and intensity of any stirring conditions used to disperse the ingredients. Many functional polysaccharides must be heated prior to use to ensure they undergo a helix-to-coil transition as this increases their solubility and functionality. Similarly, many hydrophobic ingredients (such as flavors, colors, or oil-soluble vitamins) may need to be dissolved within an appropriate oil prior to utilization. Again, this may require control of the temperature and stirring conditions to ensure good dispersion and dissolution. For instance, the oil phase may have to be heated to melt any crystalline materials prior to use. Third, the order of addition of different ingredients is often important. As an example, if a plant-based food contains emulsified fats, then it may be better to homogenize the oil, water, and emulsifier first to form the emulsion, and then add any particulate materials, as these could block the homogenizer. For chemically labile ingredients, it may be important to add them towards the end of the manufacturing process to avoid promoting their degradation. The order that ingredients are added can also affect their rate and extent of dissolution. In some cases, it may be better to hydrate ingredients before combining them together to avoid clumping effects. Fourth, the possibility of ingredient interactions that negatively impact the quality attributes of the final product should be considered. There are many kinds of these interactions that can possibly occur, and we only give a few examples here to highlight their importance. Mixing polyunsaturated lipids and iron together can promote lipid oxidation reactions because Fe^{2+} and Fe^{3+} are potent prooxidants. Mixing proteins and calcium ions together at pH values above the protein's isoelectric point can promote aggregation and precipitation because the cationic Ca^{2+} ions form bridges between the anionic protein molecules. Similarly, mixing cationic proteins and anionic polysaccharides together can lead to the formation of biopolymer aggregates that have a tendency to sediment. Finally, mixing hydrophobic flavors with globular proteins can decrease the perceived flavor intensity because the flavor molecules bind to hydrophobic patches on the surfaces of the protein. Consequently, it is important to understand the different kinds of ingredients that might be present in a plant-based food and how they might interact with each other. It is then possible to select a combination of ingredients that will give the desired final properties.

2.8 Minimally-Processed Ingredients

In this chapter, we have largely been focused on the utilization of relatively pure ingredients (such as proteins, polysaccharides, and lipids) that have been isolated from plant sources. The isolation and purification of these ingredients is often an energy-intensive and time-consuming process involving many different steps, which generates a number of sidestreams that may not be utilized as human food. There is therefore interest in developing "minimally-processed" ingredients for the application in these products. In this case, the original plant material is only processed to an extent that produces an ingredient that can provide the required functional attributes, rather than producing a highly purified ingredient. As a result, the sustainability and economic viability of the ingredient are improved. An example of this approach is the isolation of oil bodies from plant-based materials, rather than isolation of the pure oil and protein fractions (Iwanaga et al., 2007). Oil bodies are small colloidal particles that are naturally found in many plants (especially seeds). They consist of a triacylglycerol core surrounded by a layer of phospholipids with proteins embedded within them. These oil bodies can be utilized as a source of pre-emulsified oil, thereby reducing the need to purify the oil and protein, as well as to homogenize oils and proteins together to form an emulsion (which reduces costs and energy use). An added advantage of this approach is that it may reduce the number of ingredients that appear on a food label, which is often desirable to consumers. Consequently, there is a considerable amount of research being carried out to identify ingredients that can be produced using minimal processing approaches.

2.9 Conclusions and Future Directions

The formulation of high-quality plant-based foods requires the selection of the most appropriate combination of functional ingredients, which depends on many factors, such as regulatory status, cost, functional attributes, interactions, label friendliness, ease of use, reliability, supply, and allergenicity. One of the major factors holding back the development of the next generation of high-quality plant-based foods is the lack of an abundant and consistent supply of affordable plant-derived ingredients with the required functional attributes, particularly plant proteins. There is currently a lack of understanding of the range of functional attributes that specific plant proteins can have, as well as of the factors that impact these attributes. This makes it challenging to select the most appropriate ingredient for a particular application. Moreover, there are often large variations in the functional performance of plant protein ingredients (even from the same species) from supplier-to-supplier and from batch-to-batch. For instance, the solubility, emulsifying, foaming, binding, gelling or thickening properties of a plant protein may vary considerably between batches, which makes it challenging to formulate plant-based foods with the required quality attributes. These variations are often a result of changes in the native state, aggregation state, or impurities within a plant protein ingredient. In the future, it will therefore be important to create a wide range of plant-derived ingredients with consistent functional attributes. To achieve this goal, it will be important to develop agricultural crops that are specifically bred to contain proteins with the required functionality and that can be isolated without altering their desirable functional properties. It will also be necessary to develop isolation and purification procedures that lead to plant protein ingredients with consistent properties, such as denaturation state, aggregation state, and purity. Several small and large companies are already working in these areas, and more and more high-quality plant protein ingredients are becoming commercially available. Another challenge is that many of the novel sources of plant proteins that have been shown to have good functional performances cannot currently be produced at a large enough scale to make them commercially viable. Consequently, research is still required to scale up the agricultural and manufacturing processes required to produce them. Again, many companies are currently working in this area to scale-up their current processes. Finally, it will be important to establish a standardized analytical protocol to systematically characterize the functional properties of different plant proteins under different conditions (Table 2.7). The information obtained could then be included in an open-access database that could be used by food companies to select the most appropriate functional ingredient for specific applications.

Table 2.7 Proposed standardized methods for characterizing the functional performance of the plant proteins used to formulate plant-based foods

Protein ingredient properties	Functional characterization
Composition	*Proximate analysis* – Moisture (drying or Karl Fischer); protein (Kjeldahl or Dumas); lipids (Soxhlet); ash content (muffle furnace or atomic spectroscopy); carbohydrate (phenol-sulfuric acid or mass balance) *Fourier transform infrared (FTIR)* – Moisture, protein, lipids, ash, and carbohydrates after calibration with known standards
Molecular characteristics	*Protein type, molecular weight distribution, and aggregation* – SDS PAGE; non-denaturing electrophoresis; size exclusion chromatography; light scattering *Isoelectric point* – Isoelectric focusing, microelectrophoresis *Native state* – Differential scanning calorimetry, fluorescence spectroscopy *versus* temperature
Functional properties	*Solubility* – Incubation/centrifugation/quantification at differnet pH values; turbidity measurements *versus* pH *Emulsifying* – Interacial tension *versus* concentration; droplet size *versus* concentration under standardized homogenization conditions (pressure, passes); particle size *versus* pH, ionic strength, and temperature. *Foaming* – Foaming capacity and foam stability measured under standardized foaming conditions *Thickening* – Shear viscosity *versus* concentration *Gelling* – Minimum gelation concentration; shear modulus *versus* temperature; texture profile analysis; appearance; water holding capacity

References

Alba, K., & Kontogiorgos, V. (2021). Techniques for the chemical and physicochemical characterization of polysaccharides. In P. A. Williams & G. O. Phillips (Eds.), *Handbook of hydrocolloids* (pp. 27–74). Woodhead Publishing.

Amagliani, L., & Schmitt, C. (2017). Globular plant protein aggregates for stabilization of food foams and emulsions. *Trends in Food Science and Technology, 67*, 248–259.

Amici, E., Clark, A. H., Normand, V., & Johnson, N. B. (2001). Interpenetrating network formation in agarose-sodium gellan gel composites. *Carbohydrate Polymers, 46*(4), 383–391.

Arab-Tehrany, E., Jacquot, M., Gaiani, C., Imran, M., Desobry, S., & Linder, M. (2012). Beneficial effects and oxidative stability of omega-3 long-chain polyunsaturated fatty acids. *Trends in Food Science and Technology, 25*(1), 24–33.

Arvidson, S. A., Lott, J. R., McAllister, J. W., Zhang, J., Bates, F. S., Lodge, T. P., Sammler, R. L., Li, Y., & Brackhagen, M. (2013). Interplay of phase separation and Thermoreversible gelation in aqueous methylcellulose solutions. *Macromolecules, 46*(1), 300–309.

Bai, L., Liu, F. G., Xu, X. F., Huan, S. Q., Gu, J. Y., & McClements, D. J. (2017). Impact of polysaccharide molecular characteristics on viscosity enhancement and depletion flocculation. *Journal of Food Engineering, 207*, 35–45.

Bangratz, K., & Beller, M. L. (2020). Masking flavours for plant proteins. *Food Science and Technology, 34*(1), 29–31.

BeMiller, J. N. (2019). Cellulose and cellulose-based hydrocolloids.

Biesalski, H. K., & Kalhoff, H. (2020). Contra vegan diet during childhood growth and development – a commentary from the nutritional medicine perspective. *Aktuelle Ernahrungsmedizin, 45*(2), 104–113.

Blackwood, A. D., Salter, J., Dettmar, P. W., & Chaplin, M. F. (2000). Dietary fibre, physicochemical properties and their relationship to health. *The Journal of the Royal Society for the Promotion of Health, 120*(4), 242–247.

Brady, J. W. (2013). *Introductory food chemistry*. Cornell University Press.

Bychkov, A. L., Gavrilova, K. V., Bychkova, E. S., Akimenko, Z. A., Chernonosov, A. A., Kalambet, Y. A., & Lomovskii, O. I. (2019). Fractionation and hydrolysis of proteins of plant raw materials obtaining functional nutrition products. In S. Xin (Ed.), *3rd international conference on new material and chemical industry* (Vol. 479).

Calabrese, E. J. (2021). Hormesis mediates acquired resilience: Using plant-derived chemicals to enhance health. *Annual Review of Food Science and Technology, 12*(1), 355–381.

Celik, E., & Calik, P. (2012). Production of recombinant proteins by yeast cells. *Biotechnology Advances, 30*(5), 1108–1118.

Chandan, R. C., & Kilara, A. (2013). *Manufacturing yogurt and fermented milks* (2nd ed.). Wiley-Blackwell.

Christie, W. W., & Han, X. (2010). Lipid analysis: Isolation, *Separation, Identification and Lipidomic Analysis (Oily Press Lipid)* : Woodhead Publishing

Chung, C., Sher, A., Rousset, P., Decker, E. A., & McClements, D. J. (2017a). Formulation of food emulsions using natural emulsifiers: Utilization of quillaja saponin and soy lecithin to fabricate liquid coffee whiteners. *Journal of Food Engineering, 209*, 1–11.

Chung, C., Sher, A., Rousset, P., & McClements, D. J. (2017b). Use of natural emulsifiers in model coffee creamers: Physical properties of quillaja saponin-stabilized emulsions. *Food Hydrocolloids, 67*, 111–119.

Cornet, S. H. V., Snel, S. J. E., Lesschen, J., van der Goot, A. J., & van der Sman, R. G. M. (2021). Enhancing the water holding capacity of model meat analogues through marinade composition. *Journal of Food Engineering, 290*, 110283.

Curtis, R. A., & Lue, L. (2006). A molecular approach to bioseparations: Protein-protein and protein-salt interactions. *Chemical Engineering Science, 61*(3), 907–923.

Damodaran, S. (2021). Amino acids, peptides, and proteins. In S. Damodaran, K. L. Parkin, & O. R. Fennema (Eds.), *Fennema's food chemistry* (4th ed., pp. 235–356). CRC Press.

Davidov-Pardo, G., Joye, I. J., & McClements, D. J. (2015). Encapsulation of resveratrol in biopolymer particles produced using liquid antisolvent precipitation. Part 1: Preparation and characterization. *Food Hydrocolloids, 45*, 309–316.

Delcour, J. A., Joye, I. J., Pareyt, B., Wilderjans, E., Brijs, K., & Lagrain, B. (2012). Wheat gluten functionality as a quality determinant in cereal-based food products. *Annual Review of Food Science and Technology, 3*(1), 469–492.

Dickinson, E. (2019). Strategies to control and inhibit the flocculation of protein-stabilized oil-in-water emulsions. *Food Hydrocolloids, 96*, 209–223.

Dima, C., Assadpour, E., Dima, S., & Jafari, S. M. (2021). Nutraceutical nanodelivery; an insight into the bioaccessibility/bioavailability of different bioactive compounds loaded within nanocarriers. *Critical Reviews in Food Science and Nutrition, 61*(18), 3031–3065.

Dominguez, R., Pateiro, M., Munekata, P. E. S., McClements, D. J., & Lorenzo, J. M. (2021). Encapsulation of bioactive phytochemicals in plant-based matrices and application as additives in meat and meat products. *Molecules (Basel, Switzerland), 26*(13).

Duan, B., Huang, Y., Lu, A., & Zhang, L. N. (2018). Recent advances in chitin based materials constructed via physical methods. *Progress in Polymer Science, 82*, 1–33.

Farrell, H. M., Qi, P. X., Brown, E. M., Cooke, P. H., Tunick, M. H., Wickham, E. D., & Unruh, J. J. (2002). Molten globule structures in milk proteins: Implications for potential new structure-function relationships. *Journal of Dairy Science, 85*(3), 459–471.

Foegeding, E. A., & Davis, J. P. (2011). Food protein functionality: A comprehensive approach. *Food Hydrocolloids, 25*(8), 1853–1864.

Fox, P. F., Guinee, T. P., Cogan, T. M., & McSweeney, P. L. H. (2016). *Fundamentals of cheese science*. Springer.

References

Fu, S. Y., Feng, X. Q., Lauke, B., & Mai, Y. W. (2008). Effects of particle size, particle/matrix interface adhesion and particle loading on mechanical properties of particulate-polymer composites. *Composites Part B-Engineering, 39*(6), 933–961.

Gabriele, D., de Cindio, B., & D'Antona, P. (2001). A weak gel model for foods. *Rheologica Acta, 40*(2), 120–127.

Gehring, C. K., Gigliotti, J. C., Tou, J. C., Moritz, J. S., & Jaczynski, J. (2010). The biochemistry of isoelectric processing and nutritional quality of proteins and lipids recovered with this technique.

GFI. (2021). *State of the industry report: Plant-based meat, eggs, and dairy* (pp. 1–85). Good Food Institute.

Giuseppin, M. L. F. (2008). Native potato protein isolates. In W. I. P. Organization (Ed.), (Vol. WO2008069650A1)

Grabowska, K. J., Zhu, S. C., Dekkers, B. L., de Ruijter, N. C. A., Gieteling, J., & van der Goot, A. J. (2016). Shear-induced structuring as a tool to make anisotropic materials using soy protein concentrate. *Journal of Food Engineering, 188*, 77–86.

Grasso, N., Alonso-Miravalles, L., & O'Mahony, J. A. (2020). Composition, physicochemical and sensorial properties of commercial plant-based yogurts. *Food, 9*(3).

Grommers, H. E., & van der Krogt, D. A. (2009). Chapter 11 – potato starch: Production, modifications and uses. In J. BeMiller & R. Whistler (Eds.), *Starch* (3rd ed., pp. 511–539). Academic.

Grundy, M. M. L., McClements, D. J., Ballance, S., & Wilde, P. J. (2018). Influence of oat components on lipid digestion using an in vitro model: Impact of viscosity and depletion flocculation mechanism. *Food Hydrocolloids, 83*, 253–264.

Grygorczyk, A., & Corredig, M. (2013). Acid induced gelation of soymilk, comparison between gels prepared with lactic acid bacteria and glucono-delta-lactone. *Food Chemistry, 141*(3), 1716–1721.

Gunstone, F. (1996). *Fatty acid and lipid chemistry*. Blackie Scientific.

Guo, M. Q., Hu, X., Wang, C., & Ai, L. (2017). Polysaccharides: Structure and solubility. In Z. Xu (Ed.), *Solubility of polysaccharides* (pp. 1–17). IntechOpen.

Guo, J., He, Z. Y., Wu, S. F., Zeng, M. M., & Chen, J. (2019). Binding of aromatic compounds with soy protein isolate in an aqueous model: Effect of pH. *Journal of Food Biochemistry, 43*(10).

Hemler, E. C., & Hu, F. B. (2019). Plant-based diets for cardiovascular disease prevention: All plant foods are not created equal. *Current Atherosclerosis Reports, 21*(5).

Herreman, L., Nommensen, P., Pennings, B., & Laus, M. C. (2020). Comprehensive overview of the quality of plant- and animal-sourced proteins based on the digestible indispensable amino acid score. *Food Science & Nutrition, 8*(10), 5379–5391.

Holscher, H. D. (2017). Dietary fiber and prebiotics and the gastrointestinal microbiota. *Gut Microbes, 8*(2), 172–184.

Hu, F. B., Manson, J. E., & Willett, W. C. (2001). Types of dietary fat and risk of coronary heart disease: A critical review. *Journal of the American College of Nutrition, 20*(1), 5–19.

Huber, K. C., & BeMiller, J. N. (2021). Carbohydrates. In S. Damodaran, K. L. Parkin, & O. R. Fennema (Eds.), *Fennema's food chemistry* (4th ed., pp. 91–170). CRC Press.

Iwanaga, D., Gray, D. A., Fisk, I. D., Decker, E. A., Weiss, J., & McClements, D. J. (2007). Extraction and characterization of oil bodies from soy beans: A natural source of pre-emulsified soybean oil. *Journal of Agricultural and Food Chemistry, 55*(21), 8711–8716.

Jacobsen, C. (2015). Some strategies for the stabilization of long chain n-3 PUFA-enriched foods: A review. *European Journal of Lipid Science and Technology, 117*(11), 1853–1866.

Jacobsen, C., Sorensen, A. D. M., & Nielsen, N. S. (2013). Stabilization of omega-3 oils and enriched foods using antioxidants. In C. Jacobsen, N. S. Nielsen, A. F. Horn, & A. D. M. Sorensen (Eds.), *Food enrichment with Omega-3 fatty acids* (Vol. 252, pp. 130–149).

Jaswir, I., Alotaibi, A., Jamal, P., Octavianti, F., Lestari, W., Hendri, R., & Alkahtani, H. (2016). Optimization of extraction process of plant-based gelatin replacer. *International Food Research Journal, 23*(6), 2519–2524.

Johnson, L. A. (1999). *Process for producing improved soy protein concentrate from genetically-modified soybeans*. Iowa State University Research Foundation. Available at: https://patents.google.com/patent/US5936069/en?oq=soy+protein+concentrate (accessed on: August 29, 2021).

Kaltsa, O., Yanniotis, S., Polissiou, M., & mandala, I. (2018). Stability, physical properties and acceptance of salad dressings containing saffron (Crocus sativus) or pomegranate juice powder as affected by high shear (HS) and ultrasonication (US) process. *LWT- Food Science and Technology, 97*, 404–413.

Kessel, A., & Ben-Tal, N. (2018). *Introduction to proteins: Structure, function, and motion* (2nd ed.). CRC Press.

Khalesi, H., Lu, W., Nishinari, K., & Fang, Y. P. (2020). New insights into food hydrogels with reinforced mechanical properties: A review on innovative strategies. *Advances in Colloid and Interface Science, 285*.

Khalil, H., Davoudpour, Y., Islam, M. N., Mustapha, A., Sudesh, K., Dungani, R., & Jawaid, M. (2014). Production and modification of nanofibrillated cellulose using various mechanical processes: A review. *Carbohydrate Polymers, 99*, 649–665.

Konwinski, A. H. (1992). Process for making soy protein concentrate. In U. P. Office (Ed.). Solae LLC

Kornet, R., Veenemans, J., Venema, P., van der Goot, A. J., Meinders, M., Sagis, L., & van der Linden, E. (2021). Less is more: Limited fractionation yields stronger gels for pea proteins. *Food Hydrocolloids, 112*, 106285.

Kyriakopoulou, K., Keppler, J. K., & van der Goot, A. J. (2021). Functionality of ingredients and additives in plant-based meat analogues. *Food, 10*(3), 600.

Lampila, L. E. (2013). Applications and functions of food-grade phosphates. *Annals of the New York Academy of Sciences, 1301*, 37–44.

Li, X. F., & de Vries, R. (2018). Interfacial stabilization using complexes of plant proteins and polysaccharides. *Current Opinion in Food Science, 21*, 51–56.

Loeffler, M., McClements, D. J., McLandsborough, L., Terjung, N., Chang, Y., & Weiss, J. (2014). Electrostatic interactions of cationic lauric arginate with anionic polysaccharides affect antimicrobial activity against spoilage yeasts. *Journal of Applied Microbiology, 117*(1), 28–39.

Loeffler, M., Schwab, V., Terjung, N., Weiss, J., & McClements, D. J. (2020). Influence of protein type on the antimicrobial activity of LAE alone or in combination with Methylparaben. *Food, 9*(3).

Loveday, S. M. (2019). Food proteins: Technological, nutritional, and sustainability attributes of traditional and emerging proteins. In M. P. Doyle & D. J. McClements (Eds.), *Annual review of food science and technology, Vol 10* (pp. 311–339).

Loveday, S. M. (2020). Plant protein ingredients with food functionality potential. *Nutrition Bulletin, 45*(3), 321–327.

Ludwig, D. S., Willett, W. C., Volek, J. S., & Neuhouser, M. L. (2018). Dietary fat: From foe to friend? *Science, 362*(6416), 764–770.

Marangoni, A. G., Acevedo, N., Maleky, F., Co, E., Peyronel, F., Mazzanti, G., Quinn, B., & Pink, D. (2012). Structure and functionality of edible fats. *Soft Matter, 8*(5), 1275–1300.

Mariotti, F. (2017). *Vegetarian and plant-based diets in health and disease prevention*. Academic.

Marquez, A. L., Wagner, J. R., & Palazolo, G. G. (2018). Effects of calcium content and homogenization method on the microstructure, rheology, and stability of emulsions prepared with soybean flour dispersions. *European Journal of Lipid Science and Technology, 120*(7).

Mattice, K. D., & Marangoni, A. G. (2020a). Comparing methods to produce fibrous material from zein. *Food Research International, 128*, 108804.

Mattice, K. D., & Marangoni, A. G. (2020b). Evaluating the use of zein in structuring plant-based products. *Current Research in Food Science, 3*, 59–66.

Mattice, K. D., & Marangoni, A. G. (2021). Physical properties of zein networks treated with microbial transglutaminase. *Food Chemistry, 338*.

References

McClements, D. J. (2015). *Food emulsions: Principles, practice, and techniques* (2nd ed.). CRC Press.

McClements, D. J. (2020). Nano-enabled personalized nutrition: Developing multicomponent-bioactive colloidal delivery systems. *Advances in Colloid and Interface Science, 282*, 02211.

McClements, D. J. (2021). Food hydrocolloids: Application as functional ingredients to control lipid digestion and bioavailability. *Food Hydrocolloids, 111*.

McClements, D. J., & Decker, E. (2018). Interfacial antioxidants: A review of natural and synthetic emulsifiers and Coemulsifiers that can inhibit lipid oxidation. *Journal of Agricultural and Food Chemistry, 66*(1), 20–35.

McClements, D. J., & Grossmann, L. (2021a). A brief review of the science behind the design of healthy and sustainable plant-based foods. *Npj Science of Food, 5*(1).

McClements, D. J., & Grossmann, L. (2021b). The science of plant-based foods: Constructing next-generation meat, fish, milk, and egg analogs. *Comprehensive Reviews in Food Science and Food Safety, 20*(4), 4049–4100.

McClements, D. J., & Gumus, C. E. (2016). Natural emulsifiers – biosurfactants, phospholipids, biopolymers, and colloidal particles: Molecular and physicochemical basis of functional performance. *Advances in Colloid and Interface Science, 234*, 3–26.

McClements, D. J., Bai, L., & Chung, C. (2017). Recent advances in the utilization of natural emulsifiers to form and stabilize emulsions. In M. P. Doyle & T. R. Klaenhammer (Eds.), *Annual review of food science and technology, Vol 8* (pp. 205–236).

McClements, D. J., Newman, E., & McClements, I. F. (2019). Plant-based milks: A Review of the science underpinning their design, fabrication, and performance. *Comprehensive Reviews in Food Science and Food Safety, 18*(6), 2047–2067.

McClements, D. J., Decker, E. A., & Xiao, H. (2021). Lipids. In S. Damodaran, K. L. Parkin, & O. R. Fennema (Eds.), *Fennema's food chemistry* (4th ed., pp. 171–234). CRC Press.

McDougall, C., & McDougall, J. (2013). Plant-based diets are not nutritionally deficient. *The Permanente Journal, 17*(4), 93–93.

McKerchar, H. J., Clerens, S., Dobson, R. C. J., Dyer, J. M., Maes, E., & Gerrard, J. A. (2019). Protein-protein crosslinking in food: Proteomic characterisation methods, consequences and applications. *Trends in Food Science and Technology, 86*, 217–229.

Mezzenga, R., & Fischer, P. (2013). The self-assembly, aggregation and phase transitions of food protein systems in one, two and three dimensions. *Reports on Progress in Physics, 76*(4).

Mitchell, J. R., & Hill, S. E. (2021). Starch. In P. A. Williams & G. O. Phillips (Eds.), *Handbook of hydrocolloids* (pp. 239–272). Woodhead Publishing.

Murray, J. C. F. (2009). Cellulosics. In G. O. Phillips & P. A. Williams (Eds.), *Handbook of hydrocolloids* (2nd ed., pp. 710–723).

Nair, K. M., & Augustine, L. F. (2018). Food synergies for improving bioavailability of micronutrients from plant foods. *Food Chemistry, 238*, 180–185.

Nardelli, C. A. (1994). In C. P. Office (Ed.), *Separation of phytate from plant protein using ion exchange* (Vol. CA2143280A1). Abbott Laboratories.

Narsimhan, G., & Xiang, N. (2018). Role of proteins on formation, drainage, and stability of liquid food foams. In M. P. Doyle & T. R. Klaenhammer (Eds.), *Annual review of food science and technology, Vol 9* (pp. 45–63).

Nielsen, S. S. (2017). *Food analysis* (5th ed.). Springer.

Nogueira, M. S., Scolaro, B., Milne, G. L., & Castro, I. A. (2019). Oxidation products from omega-3 and omega-6 fatty acids during a simulated shelf life of edible oils. *LWT- Food Science and Technology, 101*, 113–122.

Ogawa, Y., Donlao, N., Thuengtung, S., Tian, J. H., Cai, Y. D., Reginio, F. C., Ketnawa, S., Yamamoto, N., & Tamura, M. (2018). Impact of food structure and cell matrix on digestibility of plant-based food. *Current Opinion in Food Science, 19*, 36–41.

Osborne, T. (1924). *The vegetable protein*. Longmans Green & Co.

Pelgrom, P. J. M., Vissers, A. M., Boom, R. M., & Schutyser, M. A. I. (2013). Dry fractionation for production of functional pea protein concentrates. *Food Research International, 53*(1), 232–239.

Phillips, G. O., & Williams, P. A. (2021). *Handbook of hydrocolloids* (3rd ed.). Woodhead Publishing.

Pojic, M., Misan, A., & Tiwari, B. (2018). Eco-innovative technologies for extraction of proteins for human consumption from renewable protein sources of plant origin. *Trends in Food Science and Technology, 75*, 93–104.

Puşcaş, A., Mureşan, V., Socaciu, C., & Muste, S. (2020). Oleogels in food: A review of current and potential applications. *Food, 9*(1).

Puski, G. (1987). In U. P. Office (Ed.), *Process for preparing low phytate soy protein isolate* (Vol. US4697004A). Bristol-Myers Company.

Puski, G., Jr GHH, & Talbott, R. D. (1987). *Process for preparing low phytate soy protein isolate*. US4697004A. Available at: https://patents.google.com/patent/US4697004/en?oq=soy+protein+isolate+production (accessed on: August 29, 2021).

Qian, C., Decker, E. A., Xiao, H., & McClements, D. J. (2012). Physical and chemical stability of beta-carotene-enriched nanoemulsions: Influence of pH, ionic strength, temperature, and emulsifier type. *Food Chemistry, 132*(3), 1221–1229.

Rackis, J. J., Sessa, D. J., & Honig, D. H. (1979). Flavor problems of vegetable food proteins. *Journal of the American Oil Chemists Society, 56*(3), 262–271.

Rajaram, S. (2014). Health benefits of plant-derived alpha-linolenic acid. *The American Journal of Clinical Nutrition, 100*(1), 443S–448S.

Ramel, P. R., Co, E. D., Acevedo, N. C., & Marangoni, A. G. (2016). Structure and functionality of nanostructured triacylglycerol crystal networks. *Progress in Lipid Research, 64*, 231–242.

Rani, H., Sharma, S., & Bala, M. (2021). Technologies for extraction of oil from oilseeds and other plant sources in retrospect and prospects: A review. *Journal of Food Process Engineering, 44*(2).

Rao, M. A. (2007). Introduction: Food rheology and structure. In G. V. Barbosa-Canovas (Ed.), *Rheology of fluid and semisolid foods: Principles and applications* (2nd ed., pp. 1–26). Springer.

Rasala, B. A., & Mayfield, S. P. (2015). Photosynthetic biomanufacturing in green algae; production of recombinant proteins for industrial, nutritional, and medical uses. *Photosynthesis Research, 123*(3), 227–239.

Ratnaningsih, E., Reynard, R., Khoiruddin, K., Wenten, I. G., & Boopathy, R. (2021). Recent advancements of UF-based separation for selective enrichment of proteins and bioactive peptides-a review. *Applied Sciences-Basel, 11*(3).

Ratnayake, W. S., Hoover, R., & Warkentin, T. (2002). Pea starch: Composition, structure and properties — a review. *Starch – Stärke, 54*(6), 217–234.

Ren, Y., Bai, Y., Zhang, Z., Cai, W., & Del Rio Flores, A. (2019). The preparation and structure analysis methods of natural polysaccharides of plants and fungi: A review of recent development. *Molecules (Basel, Switzerland), 24*(17), 3122.

Reynaud, Y., Buffiere, C., Cohade, B., Vauris, M., Liebermann, K., Hafnaoui, N., Lopez, M., Souchon, I., Dupont, D., & Remond, D. (2021). True ileal amino acid digestibility and digestible indispensable amino acid scores (DIAASs) of plant-based protein foods. *Food Chemistry, 338*.

Rhein-Knudsen, N., Ale, M. T., Ajalloueian, F., Yu, L. Y., & Meyer, A. S. (2017). Rheological properties of agar and carrageenan from Ghanaian red seaweeds. *Food Hydrocolloids, 63*, 50–58.

Rosenthal, A., Pyle, D. L., & Niranjan, K. (1996). Aqueous and enzymatic processes for edible oil extraction. *Enzyme and Microbial Technology, 19*(6), 402–420.

Rubinstein, M., Colby, R. H., Dobrynin, A. V., & Joanny, J. F. (1996). Elastic modulus and equilibrium swelling of polyelectrolyte gels. *Macromolecules, 29*(1), 398–406.

Saini, R. K., & Keum, Y. S. (2018). Omega-3 and omega-6 polyunsaturated fatty acids: Dietary sources, metabolism, and significance – a review. *Life Sciences, 203*, 255–267.

Salome, J. P. (2007). In U. P. Office (Ed.), *Process for extracting the components of pea flour*. Roquette Freres SA.

References

Santos-Hernandez, M., Alfieri, F., Gallo, V., Miralles, B., Masi, P., Romano, A., Ferranti, P., & Recio, I. (2020). Compared digestibility of plant protein isolates by using the INFOGEST digestion protocol. *Food Research International, 137*.

Sari, Y. W., Mulder, W. J., Sanders, J. P. M., & Bruins, M. E. (2015). Towards plant protein refinery: Review on protein extraction using alkali and potential enzymatic assistance. *Biotechnology Journal, 10*(8), 1138–1157.

Sarkar, A., & Dickinson, E. (2020). Sustainable food-grade Pickering emulsions stabilized by plant-based particles. *Current Opinion in Colloid & Interface Science, 49*, 69–81.

Schutyser, M. A. I., Pelgrom, P. J. M., van der Goot, A. J., & Boom, R. M. (2015). Dry fractionation for sustainable production of functional legume protein concentrates. *Trends in Food Science and Technology, 45*(2), 327–335.

Segal, K. I., & Green, B. E. (2017). In U. P. Office (Ed.), *Soy protein products of improved water-binding capacity*. Burcon Nutrascience MB Corp.

Sha, L., & Xiong, Y. L. L. (2020). Plant protein-based alternatives of reconstructed meat: Science, technology, and challenges. *Trends in Food Science and Technology, 102*, 51–61.

Shahidi, F., & Ambigaipalan, P. (2018). Omega-3 polyunsaturated fatty acids and their health benefits. In M. P. Doyle & T. R. Klaenhammer (Eds.), *Annual review of food science and technology, Vol 9* (pp. 345–381).

Sharan, S., Zanghelini, G., Zotzel, J., Bonerz, D., Aschoff, J., Saint-Eve, A., & Maillard, M. N. (2021). Fava bean (Vicia faba L.) for food applications: From seed to ingredient processing and its effect on functional properties, antinutritional factors, flavor, and color. *Comprehensive Reviews in Food Science and Food Safety, 20*(1), 401–428.

Shi, A. M., Feng, X. Y., Wang, Q., & Adhikari, B. (2020). Pickering and high internal phase Pickering emulsions stabilized by protein-based particles: A review of synthesis, application and prospective. *Food Hydrocolloids, 109*.

Singh, N. (2006). Process for manufacturing a soy protein concentrate having high isoflavone content. In U. P. Office (Ed.). Solae, LCC.

Singh, N. (2007). Bland tasting soy protein isolate and processes for making same. In W. I. P. Organization (Ed.). United States: Solae, LLC.

Spelzini, D., Rigatusso, R., Farruggia, B., & Pico, G. (2005). Thermal aggregation of methyl cellulose in aqueous solution: A thermodynamic study and protein partitioning behaviour. *Cellulose, 12*(3), 293–304.

Stephen, A. J., Phillips, G. O., & Williams, P. A. (2006). *Food polysaccharides and their applications* (2nd ed.). CRC Press.

Stevenson, C. D., Dykstra, M. J., & Lanier, T. C. (2013). Capillary pressure as related to water holding in polyacrylamide and chicken protein gels. *Journal of Food Science, 78*(2), C145–C151.

Taneja, D. K., Rai, S. K., & Yadav, K. (2020). Evaluation of promotion of iron-rich foods for the prevention of nutritional anemia in India. *Indian Journal of Public Health, 64*(3), 236–241.

Thomas, R. L. (2001). In U. P. Office (Ed.), *Soy proteins and methods for their production* (Vol. US6,313,273B1). Abbott Laboratories.

Toldra, F. (2017). *Lawrie's meat science* (8th ed.). Woodhead Publishing.

Tornberg, E. (2013). Engineering processes in meat products and how they influence their biophysical properties. *Meat Science, 95*(4), 871–878.

Tornberg, E., Andersson, K., Andersson, A., & Josell, A. (2000). The texture of comminuted meat products. *Food Australia, 52*(11), 519–524.

Tuso, P. J., Ismail, M. H., Ha, B. P., & Bartolotto, C. (2013). Nutritional update for physicians: Plant-based diets. *The Permanente Journal, 17*(2), 61–66.

Urrutia, O., Alfonso, L., & Mendizabal, J. A. (2018). Cellularity description of adipose depots in domesticated animals. In L. Szablewski (Ed.), *Adipose tissue* (pp. 1–15). InTechOpen.

USDA. (2021). FoodData Central In: USDA.

van der Sman, R. G. M. (2012). Thermodynamics of meat proteins. *Food Hydrocolloids, 27*(2), 529–535.

van der Sman, R. G. M. (2013). Modeling cooking of chicken meat in industrial tunnel ovens with the Flory-Rehner theory. *Meat Science, 95*(4), 940–957.

van der Sman, R. G. M., Paudel, E., Voda, A., & Khalloufi, S. (2013). Hydration properties of vegetable foods explained by Flory-Rehner theory. *Food Research International, 54*(1), 804–811.

Vilgis, T. A. (2015). Soft matter food physics-the physics of food and cooking. *Reports on Progress in Physics, 78*(12).

Viry, O., Boom, R., Avison, S., Pascu, M., & Bodnar, I. (2018). A predictive model for flavor partitioning and protein-flavor interactions in fat-free dairy protein solutions. *Food Research International, 109*, 52–58.

Waglay, A., Achouri, A., Karboune, S., Zareifard, M. R., & L'Hocine, L. (2019). Pilot plant extraction of potato proteins and their structural and functional properties. *LWT, 113*, 108275.

Wang, K., & Arntfield, S. D. (2015). Effect of salts and pH on selected ketone flavours binding to salt-extracted pea proteins: The role of non-covalent forces. *Food Research International, 77*, 1–9.

Watson, E. (2019). In conversation with Givaudan: How do you create 'meaty' flavors in plant-based meat? In *Food navigator*. https://www.foodnavigator-usa.com/Article/2019/2008/2022/In-conversation-with-Givaudan-How-do-you-create-meaty-flavors-in-plant-based-meat)

Weiss, J., Loeffler, M., & Terjung, N. (2015). The antimicrobial paradox: Why preservatives lose activity in foods. *Current Opinion in Food Science, 4*, 69–75.

Weiss, J., Salminen, H., Moll, P., & Schmitt, C. (2019). Use of molecular interactions and mesoscopic scale transitions to modulate protein-polysaccharide structures. *Advances in Colloid and Interface Science, 271*, 101987.

Williams, P. A., & Phillips, G. O. (2021). *Handbook of hydrocolloids* (3rd ed.). Woodhead Publishing.

Wilson, A. S., Koller, K. R., Ramaboli, M. C., Nesengani, L. T., Ocvirk, S., Chen, C. X., Flanagan, C. A., Sapp, F. R., Merritt, Z. T., Bhatti, F., Thomas, T. K., & O'Keefe, S. J. D. (2020). Diet and the human gut microbiome: An international review. *Digestive Diseases and Sciences, 65*(3), 723–740.

Xie, C., Jones, K. L., Rayner, C. K., & Wu, T. Z. (2020). Enteroendocrine hormone secretion and metabolic control: Importance of the region of the gut stimulation. *Pharmaceutics, 12*(9).

Xu, X. F., Sun, Q. J., & McClements, D. J. (2020). Effects of anionic polysaccharides on the digestion of fish oil-in-water emulsions stabilized by hydrolyzed rice glutelin. *Food Research International, 127*.

Zhang, D.-Q., Mu, T.-H., Sun, H.-N., Chen, J.-W., & Zhang, M. (2017). Comparative study of potato protein concentrates extracted using ammonium sulfate and isoelectric precipitation. *International Journal of Food Properties, 20*(9), 2113–2127.

Zhang, J. C., Liu, L., Liu, H. Z., Yoon, A., Rizvi, S. S. H., & Wang, Q. (2019). Changes in conformation and quality of vegetable protein during texturization process by extrusion. *Critical Reviews in Food Science and Nutrition, 59*(20), 3267–3280.

Zhang, R. J., Zhang, Z. P., & McClements, D. J. (2020). Nanoemulsions: An emerging platform for increasing the efficacy of nutraceuticals in foods. *Colloids and Surfaces. B, Biointerfaces, 194*.

Zhao, H. B., Wang, Y. S., Li, W. W., Qin, F., & Chen, J. (2017). Effects of oligosaccharides and soy soluble polysaccharide on the rheological and textural properties of calcium sulfate-induced soy protein gels. *Food and Bioprocess Technology, 10*(3), 556–567.

Chapter 3
Processes and Equipment to Create Plant-Based Foods

3.1 Introduction

The conversion of plant-derived ingredients into plant-based foods requires the design and implementation of a suitable food manufacturing *process*. This process consists of a series of *unit operations* assembled into a specific sequence to achieve the desired transformation. These unit operations are categorized according to their intended purpose (Berk, 2013). For instance, filtration and centrifugation are unit operations designed for the separation of particles or phases, whereas homogenization and milling are unit operations designed to reduce the size of particles or phases. Each unit operation typically involves the use of specific kinds of manufacturing *equipment*, such as mixers, extruders, heat exchangers, filters, centrifuges, homogenizers, or mills, which vary in their design and operating principles. Unit operations may also induce different kinds of molecular or physicochemical transformations in foods, such as mixing, phase separation, phase transitions, aggregation, and conformation changes. Food manufacturers must therefore select the most suitable unit operations, process, and manufacturing equipment to create a particular plant-based product, such as a milk, egg, meat, or seafood analog. This requires a good understanding of the behavior of plant-derived ingredients, as well as of the principles of food engineering and processing.

Most of the procedures used to manufacture plant-based foods have been used by the food industry for many years to create other kinds of food products. Nevertheless, there have been some recent innovations in this area, which have been specifically developed to create plant-based foods. Moreover, in the past, food manufacturers often focused on creating foods that were safe, delicious, affordable, shelf-stable, and convenient. But more recently there has been growing emphasis on also making them healthier and more sustainable, which requires innovations in processing.

In this chapter, the most important processes, unit operations, and equipment suitable for creating different kinds of plant-based foods are described. Our aim is to provide knowledge about how plant-based foods are usually produced, the equipment that is required, and how this equipment is normally operated.

3.2 Molecular Approaches to Structuring Plant-Based Ingredients

The creation of plant-based foods with specific physicochemical, textural, and sensorial attributes often involves controlling the structural organization of the ingredients they contain. This can sometimes be achieved using thermomechanical processing methods as described in Sect. 3.4. However, in some cases, it can also be achieved by controlling the molecular interactions between the different ingredients used to formulate plant-based foods, such as proteins, polysaccharides, or lipids. In this section, we describe several molecular and soft matter physics approaches that can be used to structure plant-based ingredients.

3.2.1 Biopolymer Phase Separation

The ingredients used to produce plant-based foods are often a mixture of biopolymers, especially proteins and polysaccharides. These ingredients may be intentionally combined, or they may naturally occur together. For instance, plant protein concentrates often contain >30% carbohydrates (Boye et al., 2010; Ingredion, 2020; Pelgrom et al., 2013). Thus, the interactions of proteins and polysaccharides need to be considered during processing. Protein-polysaccharide interactions can be utilized to create specific structures and physicochemical properties in plant-based foods. For instance, anisotropic fibrous structures, similar to those found in real meat, can be produced using mixtures of plant proteins and polysaccharides by controlling the interactions between them and superimposing shear forces (Dekkers et al., 2016). In this section, we present the underlying molecular and physicochemical mechanisms that can be used to structure plant-based ingredients to obtain desirable textural and other functional attributes.

Proteins and polysaccharides interact with each other through various kinds of molecular and colloidal interactions, as well as through various kinds of entropy effects (McClements, 2006; Tolstoguzov, 1991). For example, biopolymers with charged functional groups can exhibit repulsive or attractive electrostatic interactions, whereas those with non-polar groups exhibit attractive hydrophobic interactions. Moreover, configurational and mixing entropy influence the conformation and distribution of biopolymers in solution. These molecular characteristics influence the mixing behavior of biopolymers by altering the overall free energy of the system (Fang, 2021). As biopolymers exhibit different molecular characteristics, the mixing of different combinations of proteins and polysaccharides can result in different phenomena, including miscibility, association, or segregation (Fig. 3.1).

3.2 Molecular Approaches to Structuring Plant-Based Ingredients

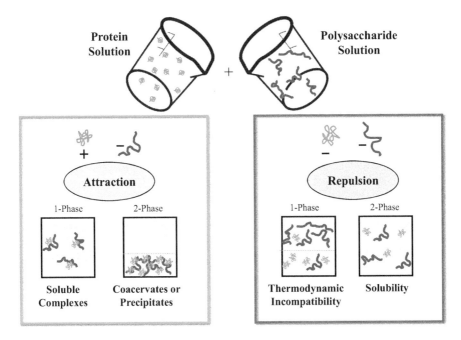

Fig. 3.1 A mixture of two plant-based biopolymers, typically a protein and polysaccharide, may form one phase or two phases depending on the sign and strength of the interactions between the biopolymer molecules. Phase separation may occur due to thermodynamic incompatibility or complexation. (Reprinted from (McClements & Grossmann, 2021) with permission of Elsevier)

Biopolymers tend to be fully miscible and form a single phase after mixing when the entropy of mixing effects are greater than the attractive or repulsive interactions acting between them. In many cases, however, biopolymers do interact with each other when combined in solution. Upon interacting, they can still be present as a single phase and appear soluble, even though there is a net attraction or repulsion between them. For instance, they may be present as soluble complexes or co-soluble individual polymers, respectively (Weiss et al., 2019).

Depending on the solution and environmental conditions, the biopolymers can also separate into distinct phases when there is a sufficiently strong attraction or repulsion between them. Associative separation occurs when the two biopolymers attract each other and form molecular complexes known as coacervates or precipitates. Coacervates have relatively loose open structures, whereas precipitates have relatively tight compact structures. In this case, the system separates into one phase that contains the biopolymer complexes and another phase that consists mainly of water (and any excess biopolymers not incorporated into the complexes). Segregative phase separation occurs when the two biopolymers repel each other, for example, because of opposite charges or excluded volume effects. In this case, the system also separates into two phases: one phase that is rich in proteins but depleted in polysaccharides, as well as another phase that is rich in polysaccharides but depleted in proteins (McClements & Grossmann, 2021). The segregative mechanism has been utilized to create meat-like structures in meat analogs, whereas the associative mechanism has been explored for its potential to improve the stability of milk analogs (Dekkers et al., 2018; Kyriakopoulou et al., 2018).

3.2.1.1 Segregative Phase Separation

The segregative regime typically occurs in biopolymer mixtures under conditions where both the biopolymer molecules carry a similar net charge, *e.g.*, an anionic polysaccharide and an anionic protein (pH > pI). Here, the pI is the isoelectric point of the protein, *i.e.*, the pH where it has a net-zero charge. Under segregative conditions, the protein and polysaccharide are thermodynamically incompatible because they both carry a negative charge and repel each other. Moreover, there are excluded volume effects associated with the fact that two biopolymer molecules cannot occupy the same space, which promotes phase separation at sufficiently high biopolymer concentrations. Depending on the nature of the system, the repulsion between the biopolymer chains leads to co-solubility (single-phase system) or phase separation (two-phase system). A single-phase system tends to form when the biopolymer concentrations are below some critical level, whereas a two-phase system tends to form when the biopolymer concentrations exceed this level. In the case of phase separation, the mixed biopolymer solution separates into different phases, with one phase being rich in proteins and the other phase being rich in polysaccharides (Weiss et al., 2019). This kind of system can be gently stirred to form a "water-in-water" emulsion that consists of one biopolymer-rich phase that is dispersed in the other biopolymer-rich phase.

This phenomenon can be utilized to form anisotropic structures by shearing and setting the phase-separated biopolymer mixture (Fig. 3.2). Commonly, a protein with a pI in the acidic range (pH 4–5) and an anionic polysaccharide are used for this purpose because these biopolymers are widely used in many foods and are available in large quantities. Additionally, the protein has a net negative charge under the pH conditions found in many plant-based meat formulations, *i.e.*, pH of 5–7 (De Marchi et al., 2021). Researchers who have utilized this phenomenon for

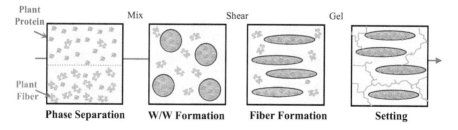

Fig. 3.2 Meat-like fibrous structures can be produced from plant proteins and polysaccharides using soft-matter physics approaches: (i) phase separation; (ii) mixing and water-in-water emulsion formation; (iii) shearing and fiber-formation; and (iv) gelling and setting. (Reprinted with permission of Elsevier from (McClements & Grossmann, 2021))

the formation of plant-based meats have commonly used soy protein isolate and pectin as model biopolymers to provide insights into the underlying mechanisms involved. It has also been demonstrated that the polysaccharides present in soybean protein concentrates are important to achieve anisotropic structures through this mechanism (Dekkers et al., 2016; Grabowska et al., 2016).

Once the biopolymers are mixed with water at sufficiently high concentrations (around 2–4% pectin and 40–42% soy protein), they separate into two distinct phases (Dekkers et al., 2016). Because the phases do not mix, there is an interfacial tension between them and they form a water-in-water emulsion upon shearing. The interfacial tension between the dispersed and continuous phases (polysaccharide-rich and protein-rich) is very low ($\approx \mu N/m$) because both phases are hydrophilic (Scholten et al., 2004). This means that the water droplets formed have a very low Laplace pressure (*i.e.*, the pressure difference between the inside and outside of the droplets) and can easily be deformed by external stresses. Moreover, the temperature is important because the viscosity decreases and the molecular flexibility increases with increasing temperature.

This phase separation mechanism is used to form fibrous meat-like structures by using the shear cell technology. In this device, the biopolymers, water, and any other ingredients are sheared and heated simultaneously in a rotating cone-in-cone cell (see *Shear Cell* section). As a result, the biopolymer mixture separates into a protein-rich phase and a polysaccharide-rich phase. The shearing induces an elongation of the 'droplets', which are subsequently embedded into the biopolymer matrix. For example, when soy proteins were used alone they formed a rubbery, layered structure during shearing and heating, whereas anisotropic structures were obtained when the soy proteins were mixed with 2.2% pectin (Dekkers et al., 2016). The shearing induced elongated filaments of pectin, oriented in the direction of the shear flow that were embedded in a denatured continuous phase of soy protein. Most likely, the pectin weakens protein-protein interactions by spatially separating the proteins, which fosters the formation of fibrous structures similar to the connective tissue that surrounds muscle fibers. However, the precise origin of the structure formation in these kinds of biopolymer blends is still under debate and new models are being developed to describe this process (Sandoval Murillo et al., 2019).

These studies indicate that the nature of the fibrous structures that can be formed by segregative phase separation depends on the selection of appropriate biopolymers, biopolymer concentrations, shearing conditions, and temperatures. In general, phase separation of mixed biopolymer solutions can be achieved at much lower temperatures (ambient conditions) when lower biopolymer concentrations are used.

Meat-like structures can also be formed using this approach without the need for a shear cell or extrusion device (Fig. 3.2). In this case, only simple mixing operations are required to form anisotropic structures in the biopolymer mixtures. First, two thermodynamically incompatible biopolymers (usually a protein and a polysaccharide) are dispersed in water at a sufficiently high concentration, which leads to spontaneous phase separation. Second, the biopolymer dispersion is gently mixed, which leads to the formation of a W/W emulsion. Third, the biopolymer mixture is sheared, which causes the dispersed phase droplets to become elongated. Finally, the dispersed and/or continuous phases are gelled, which sets the fibrous structure of the system. The final setting step can be achieved using various approaches depending on the nature of the biopolymers used, including heating, cooling, dehydrating, pH adjustment, salt addition, or adding gelling agents (Sect. 3.2.3).

In conclusion, controlling the attractive and repulsive forces between biopolymers has important consequences for the production of plant-based foods. In a following chapter, we discuss how this phase separation mechanism can be used to produce plant-based meat analogs using the shear cell device (Chap. 6).

3.2.1.2 Associative Phase Separation

Associative phase separation has been used to a lesser extent in plant-based food formulations, although it does have considerable potential for some kinds of products. This kind of phase separation is induced by promoting a sufficiently strong attractive interaction between the biopolymers, which is often an electrostatic attraction between them caused by opposite charges. The biopolymer molecules then interact with each other and form soluble complexes, coacervates, or precipitates depending on the precise nature of the interactions involved.

Proteins and anionic polysaccharides are often used for this purpose. The proteins need to be near or below their isoelectric point (pI) in order to carry positively charged groups on their surfaces that can interact with negatively charged residues on the polysaccharides. Theoretically, two proteins with different pIs could also be used to achieve associative phase separation, *e.g.*, by using a protein with an isoelectric point at a pH of 5 and another one with one at a pH of 8. However, there are only a few food proteins that have high pI values and are economically viable. Food-grade proteins that possess high pI values, such as lysozyme (from eggs) or lactoferrin (from milk), occur only at low concentrations. For example, lactoferrin ($pI > 8$) was reported to occur in bovine milk in concentrations ranging from 32 µg to 486 µg per mL (Cheng et al., 2008). Therefore, combinations of proteins and anionic polysaccharides are more frequently used to promote associative phase separation because these kinds of polysaccharides are already commonly used as functional ingredients in foods (Chap. 2).

At present, protein-polysaccharide complexes have mainly been explored for application in plant-based milks to improve the emulsifying performance of plant proteins. The main advantage of using protein-polysaccharide complexes rather than single proteins as emulsifiers is their ability to increase the resistance of oil-in-water emulsions to alterations in environmental conditions, such as pH, ionic strength, and temperature changes (Evans et al., 2013). By combining proteins and polysaccharides, the desirable functional attributes of both types of biopolymers can be exploited. Many plant proteins are surface-active because they are amphiphilic molecules with both hydrophilic and hydrophobic regions. As a result, they can adsorb to the interfaces of the oil droplets formed during homogenization. The resulting protein-coated oil droplets are mainly stabilized by electrostatic repulsion arising from the charged proteins at the oil-water interfaces. However, they may aggregate when the pH becomes too close to the p*I* of the proteins or when a sufficiently high amount of salt is added because this reduces the electrostatic repulsion. Moreover, they may also aggregate when they are heated above the thermal denaturation temperature (T_m) of the adsorbed proteins because this increases the hydrophobic attraction between them. Consequently, plant proteins are often good at forming emulsions but not at stabilizing them against environmental stresses. In contrast, many anionic polysaccharides are strongly hydrophilic molecules that are not surface active and cannot adsorb to oil droplet interfaces and form emulsions. However, they are highly effective at generating strong steric and electrostatic repulsion.

Electrostatic complexes assembled from proteins and anionic polysaccharides can adsorb to oil droplet interfaces and can therefore be used to form emulsions. Moreover, once they have adsorbed they can generate strong steric and electrostatic repulsive forces between the droplets, which increases their resistance to environmental stresses (like pH changes, salt addition, or heating). Traditionally, this approach was used to improve the stability of emulsions containing oil droplets coated by dairy proteins, such as casein or whey. For example, it has been shown that casein-coated oil droplets, which are normally destabilized when the pH is around the isoelectric point of the adsorbed casein (p*I* ~ 4.6), remained stable from pH 2 to 7 when dextran sulfate was added (Jourdain et al., 2008). This effect was attributed to the ability of the strongly anionic dextran sulfate to increase both the electrostatic and steric repulsion between the oil droplets. The same approach can also be used to improve the stability of oil droplets coated by plant proteins, which may be useful for creating plant-based milks that do not break down when added to, for example, hot acidic coffees. For instance, the addition of anionic pectin or xanthan gum to model rice milks was shown to increase the stability of the rice protein-coated oil droplets against aggregation (Xu et al., 2020).

In conclusion, controlling the attraction and repulsion between biopolymers, especially polysaccharides and proteins, is of importance during the production of many plant-based foods. Repulsive interactions are especially relevant for the production of anisotropic, meat-like structures, whereas associative interactions are a promising tool to stabilize plant-based milk analogs. However, there may also be numerous other applications of controlled phase separations in the design and development of plant-based foods.

3.2.2 Gelation

The formation of solid-like textures is important in several plant-based food products, including meat, seafood, cheese, and yogurt analogs. These textures can be achieved by the controlled gelation of plant-derived proteins and/or polysaccharides, which usually involves adjusting solution conditions so there is an attractive interaction between the biopolymer molecules. At a sufficiently high concentration, the biopolymer molecules associate with each other and form a 3D network that extends throughout the entire volume of the system, thereby generating some elastic properties. Gelation can be achieved through various mechanisms depending on the nature of the biopolymers in the system:

- *Heat gelation* – Many plant proteins form heat-set gels because of the unfolding and aggregation of the globular proteins when they are heated above their thermal denaturation temperatures. Some polysaccharides can also be used to form heat-set gels. For instance, starch granules form gels when heated because they absorb water and swell, whereas methylcellulose forms gels when heated because the strength of the hydrophobic attraction between the methyl groups increases with temperature.
- *Cold gelation* – Some biopolymers form cold-set gels, which is often the result of hydrogen bond formation or salt bridge formation between helical regions at low temperatures. Typically, the biopolymers are first heated to a temperature where they unfold and then they are cooled, which promotes their self-association. In the case of polysaccharides, this typically involves a transition from a random coil structure at high temperatures to a helical structure at low temperatures. The helical regions formed on the polysaccharide molecules then interact with each other through a mechanism that depends on the polysaccharide type. Some polysaccharides gel because of the formation of electrostatic salt bridges between anionic helices and cations (*e.g.*, carrageenan and potassium), whereas other polysaccharides gel because of hydrogen bonding between different helical regions (*e.g.*, agar). Cold-setting biopolymers are widely used in plant-based foods to replace gelatin.
- *Ionic gelation* – Mineral ions can be mixed with oppositely charged biopolymers to induce gelation through electrostatic screening or bridging mechanisms at ambient temperatures. For instance, cationic ions (like calcium) can be used to crosslink anionic biopolymers like pectin, alginate, carrageenan, and proteins above their isoelectric point. The characteristics of the gels formed depend on the type and concentration of the mineral ions used, as well as the properties of the biopolymers.
- *Enzymatic gelation* – Specific enzymes can also be used to crosslink biopolymers and form gels. For instance, transglutaminase and tyrosinase can be used to crosslink proteins (Grossmann et al., 2017), whereas laccase can be used to crosslink polysaccharides (beet pectin) (Jung & Wicker, 2012).
- *pH gelation* – Gelation can also be promoted by adjusting the pH of the aqueous solution surrounding the biopolymer molecules. The electrostatic interactions between biopolymers with ionizable functional groups (*e.g.*, -COO$^-$ or -NH$_3^+$) depend on the pH because this influences the degree of ionization of these groups. Some proteins form gels around their isoelectric point because the electrostatic repulsion between them decreases and so they can contact each other. This mechanism is particularly important in conventional yogurt and cheese production where the casein molecules lose their charge and aggregate with each other to form a 3D protein network. It may also be useful in the production of plant-based versions of these dairy products.

3.2.3 Phase Transitions

Controlled phase transitions, particularly solid-to-liquid (melting) or liquid-to-solid (crystallization), can be used to create desirable structures, textures, and other functional attributes in plant-based foods. The melting and crystallization of fat crystals play an important role in many conventional meat and dairy products, and so it is important to simulate this kind of behavior in plant-based alternatives. Coconut oil or cocoa butter are often used for this purpose because they melt around ambient and body temperatures, which mimics the melting behavior of animal fats (Chap. 2). Moreover, the melting and crystallization of water are important in plant-based ice creams, and so it is important to control it, especially the temperature range at which it occurs and the nature of the ice crystals formed.

3.3 Advanced Particle Technologies

Recently, there have been many innovations in the design, development, and application of advanced particle technologies to extend or enhance the performance of foods (Fig. 3.3). The composition, structure, and properties of these particles are manipulated to obtain specific functional attributes in foods, such as improved ingredient dispersibility, stability, or performance. Many of these advanced particle technologies can be applied in plant-based foods. In this section, we highlight several kinds of advanced particles that can be created entirely from plant-derived ingredients and highlight some of their potential applications in plant-based foods. A more detailed review of these advanced particle technologies can be found in the literature (Bai et al., 2021; Tan & McClements, 2021).

3.3.1 Types of Advanced Particles

3.3.1.1 Emulsions

Emulsion technology is one of the most versatile tools for creating novel structures and functionalities in foods. Emulsions are thermodynamically unstable colloidal dispersions consisting of two immiscible liquids, usually oil and water. In the simplest case, one of the immiscible liquids is dispersed in the other in the form of small emulsifier-coated droplets (droplet diameter, d, usually 100 nm to 100 μm). These kinds of emulsions can be classified as oil-in-water (O/W) or water-in-oil (W/O) depending on whether the oil or the water phase forms the droplets, respectively (Fig. 3.4). However, more structurally sophisticated emulsions can be designed, such as water-in-oil-in-water (W/O/W) or oil-in-water-in-oil (O/W/O) emulsions, which are both kinds classified as multiple emulsions (also known as

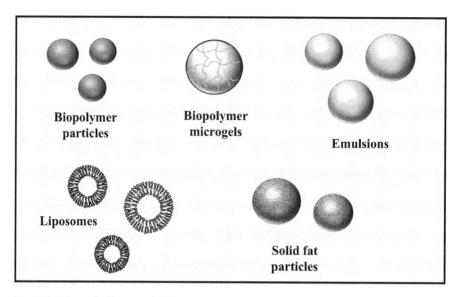

Fig. 3.3 Schematic diagram of different types of advanced particles that can be assembled from plant-based ingredients

double emulsions). Moreover, other kinds of advanced emulsions can be formulated, including nanoemulsions, multilayer emulsions, Pickering emulsions, and high internal phase emulsions (HIPEs), which may be of the O/W or W/O type (Fig. 3.4). Nanoemulsions are like emulsions, but they contain smaller droplets ($d < 100$ nm). Multilayer emulsions contain droplets coated by multiple layers of surface-active substances, often an initial emulsifier layer followed by one or more oppositely charged biopolymer layers, rather than just a single emulsifier layer. Pickering emulsions are stabilized by colloidal particles rather than surface-active molecules. HIPEs are emulsions containing relatively high droplet concentrations (usually >73%), which causes the droplets to be so closely packed together that the overall system has solid-like characteristics. Each type of emulsion has its own advantages and disadvantages for different applications.

Emulsions are usually formed using mechanical devices known as homogenizers, such as high-shear blenders, colloid mills, high-pressure valve homogenizers, sonicators, or microfluidizers. The type of device used, the operating conditions employed, and the nature of the starting material determine the type of emulsion produced, as well as the sizes of the droplets it contains. Typically, the droplet size decreases with increasing energy intensity and duration. Controlling the droplet size is often important for manipulating the optical, textural, stability, flavor, and gastrointestinal attributes of plant-based foods. Emulsions can be formulated entirely from plant-based functional ingredients, such as lipids and emulsifiers.

3.3 Advanced Particle Technologies

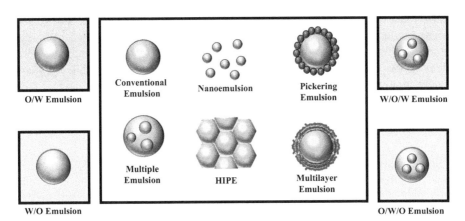

Fig. 3.4 Schematic diagram of various kinds of advanced emulsions that can be used in plant-based foods

3.3.1.2 Solid Fat Particles

Solid fat particles have structures like O/W emulsions, but the lipid phase is fully or partly solidified. Consequently, they consist of emulsifier-coated solid fat particles dispersed within water (Fig. 3.3). Solid fat particles can be fabricated using the same methods used to prepare emulsions, but homogenization should be performed at a temperature above the melting point of the lipid phase and then the emulsion created should be cooled to promote fat crystallization. These kinds of particles can also be formed using spray chilling methods, which involve spraying a hot lipid into a cold chamber. Solid fat particles are usually prepared from plant-based lipids that are crystalline at ambient temperature, *e.g.*, coconut oil, cocoa butter, palm oil, or some waxes. The crystalline nature of the lipid phase can increase the chemical stability of hydrophobic bioactive agents trapped inside the solid fat particles by reducing their interactions with prooxidants in their environment. Moreover, it can also be used to control the retention and release of encapsulated substances, such as colors or flavors. For instance, an active ingredient encapsulated within the solid fat particles could be released when they are heated above their melting point during cooking.

3.3.1.3 Liposomes

Liposomes are colloidal particles comprised of one or more concentric bilayers, with each bilayer consisting of phospholipid molecules aligned tail-to-tail (Fig. 3.3). They can be formulated from plant-based ingredients such as soybean or sunflower lecithin. Liposomes typically range in diameter from around 50 nm to 50 μm depending on their composition and fabrication method. They have both polar and non-polar regions inside them and so can be used to encapsulate both water- and oil-soluble functional ingredients. Several approaches can be used to form liposomes, with high-pressure valve homogenization and microfluidization being the most suitable for large-scale production. Some of the challenges to using liposomes in plant-based foods are that they are relatively expensive and too fragile for many applications.

3.3.1.4 Biopolymer Particles

Biopolymer particles are typically assembled from proteins and/or polysaccharides that are physically or chemically crosslinked and form a relatively dense particle interior (Fig. 3.3). These particles are usually spherical in shape and have diameters ranging from around 50 nm to 50 μm. The composition, dimensions, and surface properties of these particles can be varied by using different ingredients and processing methods to manufacture them, which enables their functional attributes to be tailored for specific applications. Several methods are available for fabricating biopolymer particles, with antisolvent precipitation and controlled denaturation-aggregation being the most common. For example, zein particles can be formed using antisolvent precipitation by dissolving the hydrophobic protein in a concentrated ethanol solution and then injecting it into water. Soy protein particles can be formed using the denaturation-aggregation method by heating a solution of the globular proteins above their thermal denaturation temperature under controlled pH and ionic strength conditions. This causes the protein molecules to unfold and aggregate with each other, thereby promoting the formation of small protein particles.

3.3.1.5 Biopolymer Microgels

Biopolymer microgels are also fabricated from proteins and/or polysaccharides, but they are usually bigger (100 nm to 1000 μm) and more porous than biopolymer particles. Indeed, they usually consist of 3D networks of physically or chemically crosslinked biopolymers that trap large quantities of water inside (Fig. 3.3). A variety of approaches can be used to prepare biopolymer microgels, with the most common ones being injection-gelation and coacervation methods. The first method involves injecting a solution of biopolymers (*e.g.*, alginate) into another solution containing a gelling agent (*e.g.*, calcium ions), which leads to the formation of microgels. The second approach usually involves preparing a solution containing a mixture of proteins and polysaccharides and then adjusting the pH conditions so that there is a net electrostatic attraction between them, which again leads to the formation of microgels. The size, shape, and stability of these microgels can be further controlled by applying shear and then crosslinking the biopolymers. The functional attributes of biopolymer microgels can be manipulated by controlling their composition, dimensions, shape, internal structure, and surface properties.

3.3.2 Applications of Advanced Particle Technologies

In this section, we provide a brief overview of some potential applications of advanced particle technologies in plant-based foods (Bai et al., 2021; Tan & McClements, 2021).

3.3.2.1 Water Dispersibility

Several of the functional ingredients used in plant-based foods are strongly hydrophobic substances that cannot be directly incorporated into an aqueous food matrix, including non-polar pigments, flavors, preservatives, vitamins, and nutraceuticals. The water-dispersibility of these hydrophobic substances can usually be improved by encapsulating them inside colloidal particles with hydrophilic exteriors and hydrophobic interiors, such as oil droplets, solid fat particles, liposomes, and some types of biopolymer particles.

3.3.2.2 Chemical Stability

Some functional ingredients used in plant-based foods are chemically labile and tend to degrade when exposed to certain environmental conditions (such as heat, light, and oxygen) or in the presence of other ingredients (such as transition metals or enzymes). For instance, omega-3 fatty acids, oil-soluble vitamins, curcumin, and carotenoids are susceptible to oxidation leading to the loss of valuable nutrients, the generation of off-flavors, and/or color fading. The chemical stability of these ingredients can sometimes be improved by encapsulating them within colloidal particles. The particles may isolate the functional ingredients from pro-oxidants in the surrounding aqueous phase or the particles may contain antioxidants that inhibit chemical degradation reactions. As an example, many plant proteins are natural antioxidants and so encapsulating labile functional ingredients within particles made from them may increase their oxidative stability.

3.3.2.3 Controlled Release

In some applications, it is desirable to control the release of functional ingredients to obtain specific effects in plant-based foods, such as a change in flavor or color during heating. The release profile of functional ingredients can often be controlled by trapping them inside colloidal particles. Typically, a particle is designed to retain the functional ingredient under one set of conditions but then release it under another set of conditions. The release profile, such as burst, sustained, or triggered, can be controlled by changing the properties of the particles, such as their composition, structure, and size. The rate of release of encapsulated substances can usually be controlled by changing the particle size or matrix viscosity: the larger the particle diameter or the higher the matrix viscosity the slower the release. Triggered release can be achieved by designing particles that disintegrate or undergo a phase change in response to some external trigger, such as pH, ionic strength, temperature, or enzyme activity. For instance, a solid fat particle may melt or a biopolymer microgel may undergo a gel-sol transition when heated above a particular temperature.

These phenomena may be useful for creating a color change in a plant-based food during cooking. For instance, a pigment could be trapped inside a solid fat particle, which then releases it when heated above the melting point of the lipid phase. As a result, the pigment interacts with other molecules in its environment, which promotes a color change. Moreover, the targeted release of active ingredients (such as vitamins or nutraceuticals) within the human gut may be controlled by encapsulating them in colloidal particles that break down in particular regions of the gastrointestinal tract due to a change in enzyme activity. For instance, starch-based particles mainly break down in the mouth and small intestine due to the action of amylases, protein-based particles mainly breakdown in the stomach and small intestine due to the action of proteases, lipid-based particles mainly breakdown in the small intestine due to the action of lipases, and dietary fiber-based particles may only breakdown after they reach the colon due to the action of enzymes released by colonic bacteria. Particles can also be designed that will release encapsulated active agents in response to changes in the pH and ionic strength of their environment. For instance, biopolymer microgels held together by electrostatic attraction may swell or disintegrate when the pH or ionic strength is changed due to alterations in the electrical charge of the biopolymer molecules or electrostatic screening effects. There is therefore great interest in developing colloidal particles that can be used to control the release of functional ingredients in plant-based foods.

3.3.2.4 Flavor Masking

Some functional ingredients used to formulate plant-based foods have bitter or astringent tastes, such as bioactive polyphenols or peptides. The unpleasant flavor of these ingredients may be masked by encapsulating them within particles that remain intact in the mouth but break down and release them in the stomach, small intestine, or colon after ingestion. For instance, polyphenols could be encapsulated in oil droplets, solid fat particles, protein particles, or liposomes, which should reduce their tendency to interact with the tongue in the mouth but then release them in the gastrointestinal tract.

3.3.2.5 Control of Macronutrient Digestion

The rate and extent of the digestion of macronutrients (lipids, starch, and proteins) within the gastrointestinal tract, and the subsequent absorption of their digestion products (fatty acids, monoacylglycerols, glucose, amino acids, and peptides) into the bloodstream, impacts their bioavailability, as well as the bodies hormonal and metabolic response to ingested foods (such as appetite/satiety and insulin levels). As a result, the digestion profile of macronutrients may impact human health and well-being. Macronutrient digestion can be manipulated by controlling the nature of any particles containing them. Rapid digestion can be achieved by converting the macronutrients into colloidal particles that have high surface areas (small diameters), such as lipid, starch, or protein nanoparticles. Conversely, slow digestion can be achieved by trapping macronutrients inside large indigestible particles with small pore sizes, such as biopolymer microgels comprised of dietary fibers.

3.3.2.6 Texture Modification

The incorporation of certain kinds of particles into plant-based foods may be utilized to alter their textural characteristics. The viscosity of fluid plant-based foods, such as milk analogs, can be increased by incorporating colloidal particles, such as emulsion droplets, biopolymer microgels, or biopolymer particles. The textural attributes of semi-solid plant-based foods, such as meat, seafood, egg, or cheese analogs, can be modified by introducing similar kinds of colloidal particles. In this case, the gel strength and fracture properties of the food depend on the interactions of the particles with the gel network, as well as their size and shape. Particles that behave as active fillers are attracted to the gel network and tend to strengthen the material, whereas particles that behave as inactive fillers are not attracted and tend to weaken it. The interactions between the particles and the gel network can be controlled by altering the types of molecules at their surfaces. For example, in emulsions, the type of emulsifier used can be varied to obtain different particle-network interactions. The incorporation of small particles into a gel network may also impact the pore size, which could alter the release characteristics and fluid holding properties of the system.

3.3.2.7 Modification of Optical Properties

The appearance of plant-based foods can also be controlled by incorporating colloidal particles into them because these particles scatter light waves. The degree of light scattering, and therefore the opacity or lightness of a food, depends on the size, concentration, and refractive index of the particles present. Typically, the intensity of light scattering by particles increases with increasing refractive index contrast, increasing particle concentration, and has a maximum value at a particle size of around a few hundred nanometers (where the dimensions of the particles are near those of light waves). Consequently, the optical properties of plant-based foods, such as their opacity, can be altered by incorporating particles with appropriate characteristics, such as oil droplets, solid fat particles, or protein particles.

3.3.2.8 Macronutrient Replacement

To produce healthier plant-based foods it may be desirable to reduce the levels of fat or starch they contain, since these are high-calorie nutrients that are often rapidly digested and absorbed into the lymphatic system and bloodstream, respectively. These nutrients are often present in the form of particles, such as oil droplets or starch granules, which provide desirable optical, textural, mouthfeel, and flavor characteristics to foods. Consequently, their removal can lead to an undesirable reduction in product quality. There has therefore been interest in developing fat or starch mimetics using healthier ingredients, such as proteins and dietary fibers. For instance, biopolymer particles or microgels assembled from proteins or dietary fibers may be used to replace oil droplets or starch granules, thereby reducing the calorie content of a food. Alternatively, multiple emulsions (W/O/W) may be used to replace conventional emulsions (O/W) in plant-based foods, as this would reduce the total fat content because some of the oil phase inside the fat droplets is replaced by water (Fig. 3.4).

In summary, advanced particle technologies have numerous potential applications in plant-based foods, but further research is still required to ensure that the particles used will perform as expected in real food applications, as well as being cost-effective and capable of large-scale production.

3.4 Mechanical Processing Methods

As mentioned earlier, a variety of different unit operations are required to convert plant-derived ingredients into plant-based foods. These unit operations usually involve applying mechanical forces and/or heat to transform the properties of plant-derived ingredients. In this section, we highlight some of the most common processing operations used for this purpose.

3.4.1 Size Reduction

A number of unit operations utilize specialized equipment to reduce the size of the particles in food materials, such as mills, grinders, or homogenizers. Mills and grinders are commonly used to reduce the particle size of solids or slurries. Mills are often employed in the production of plant-based protein powders and milks, whereas grinders are often used in the production of plant-based meats. Homogenizers are mainly used to convert bulk oil and water phases into emulsions or to reduce the size of the droplets in existing emulsions. In this section, we describe common types of equipment used for size reduction in plant-based food applications.

3.4.1.1 Milling

Dry milling devices, such as hammer mills, are commonly used for the grinding of seeds (Fig. 3.5). Hammer mills consist of a cylindrical container with an embedded rotating shaft that contains "hammers", which may have various shapes. The material enters the mill from the top and gets crushed by the impact of the rotating hammers. Additionally, the hammers cause the seeds to accelerate towards the chamber walls where they get crushed by the impact forces generated. The dimensions of the particles within the sample are reduced by these processes until a critical particle size is achieved and the particles exit the screen at the bottom of the device. Such milling devices are commonly used to crush seeds and then fractionate them based on their size using air classification, which results in the production of protein concentrates (Chap. 2).

A second mill used for dry grinding is the corundum stone mill, which is manufactured from a special type of rock. This mill is often used for milling nuts. For example, nuts can be converted into a paste containing fine particles by combining a colloid mill with a corundum stone mill (Fig. 3.6, **right**). Corundum stone mills consist of two round discs parallel to each other and the material is fed into the gap in between. The material is crushed by the action of at least one rotating disc, which transfers friction and pressure to the particles. For some mills, both discs rotate in opposite directions to increase the disruptive forces generated. The particle size obtained – often in the range of 40–150 µm – can be controlled by adjusting the gap between the discs, as well as by changing their rotational speeds.

Wet milling devices are also often utilized during the production of plant-based milks. They are used to break down the structure of the plant materials and produce a colloidal dispersion with an appropriate stability and mouthfeel. One of the main challenges during the production of milk analogs is the generation of off-flavors due to the activity of enzymes in the plant materials (Cosson et al., 2022). For instance, lipoxygenase catalyzes a reaction between oxygen and unsaturated fatty acids that leads to the formation of hydroperoxides (R-O-OH). The lipid hydroperoxides then decompose and form secondary oxidation products that are volatile and are responsible for an unpleasant beany and greeny note (Hayward et al., 2017). Other enzymes that are still active in the native seed, such as lipases and proteases, may also produce off-flavors and so need to be inactivated. Thus, most food processors try to inactivate these enzymes as early as possible in the process to avoid the formation of undesirable off-flavors. For example, prior to milling, soybean seeds are transferred into a hopper and hot water (>60 °C, typically 95 °C) is added at a water-to-seed ratio of about 4:1, which softens the plant material and reduces the enzyme activity (Prabhakaran & Perera, 2006). For oats, the grains are treated with saturated steam at 88–98 °C for several minutes (Head et al., 2011). However, some

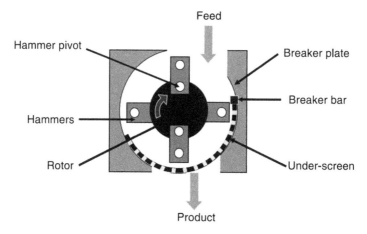

Fig. 3.5 Hammer mills are a type of impact mill often used during dry grinding of seeds to produce protein concentrates. The main working principle is the impact of the rotating hammer that crushes the seeds. (Reprinted with permission of Elsevier from (Shi et al., 2003))

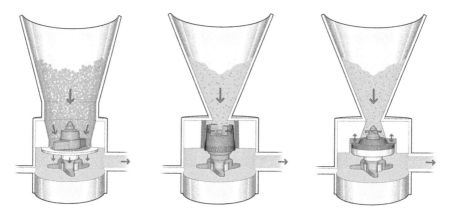

Fig. 3.6 Working principle of a perforated disc mill (left), toothed colloid mill (middle), and corundum stone mill (right). The particle size achieved after grinding tends to decrease from left to right. Perforated disc mills and colloid mills can process particles with dimensions below about 5 cm before grinding. Corundum stone mills are best suited for particle sizes below 1 cm. (The image was kindly provided by Fryma Koruma AG (Rheinfelden, Switzerland))

manufacturers do not use a thermal treatment during their plant-based milk production to maintain the native state of the proteins, which may be beneficial for further processing. In this case, the off-flavor notes can be removed by a deodorization step (Herrmann, 2009).

The soaking step requires the seeds to be milled by a wet-milling approach. Two different kinds of mills are often used for grinding seeds (such as soybeans and oats) using this technique: a disc mill and a colloid mill. In a disc mill, the (hot) seed-water mixture is passed through the device to break down the seeds into coarse particles (Prabhakaran & Perera, 2006). Often, perforated disc mills (Fig. 3.6, **left**) are used to crush the seeds (Herrmann, 2009). These mills consist of a perforated disc with a knife that is rotating close to the stationary disc. The material is fed into the mill from the top and then crushed by the action of the knife until the size reaches about 0.5–5 mm. These small particles then pass through the holes in the disc and are transported to the next mill: the colloid mill. The colloid mill crushes the coarse particles formed by the disc mill into a fine paste. Colloid mills are designed with a conical rotor that is shaped like a shortened traffic cone that is cut at the top (Fig. 3.6, **middle**) (McClements, 2015). The rotor sits in a static conical case and the material is forced between the case and the rotating conical rotor, which facilitates particle disruption. As with the common disc mill, the gap and the surface of the rotor can be adjusted to achieve the desired particle size (usually 100–500 µm). A rotor with a toothed design is often used to enhance the particle disruption efficiency. Colloid mills can have capacities of up to 40,000 kg/h (inline).

The result of these milling operations is a slurry that contains particles that vary in size and density. Some of these particles are relatively large ($d > 100$ µm) and dense ($\rho > 1000$ kg m^{-3}) and so would have a tendency to rapidly sediment if left in the final product (Lan et al., 2021). For this reason, they are often removed from the product before packaging (Li et al., 2012).

3.4.1.2 Grinding

The production of some plant-based foods (especially meat analogs) involves a unit operation where larger pieces of material are broken down into smaller ones. The equipment used to reduce the size of larger pieces of plant-foods (*i.e.*, >5 cm) is often similar to the ones used in meat and meat product manufacturing, such as grinders, vacuum filler-grinders, and bowl choppers. This type of equipment may be used to break down extruded plant-based products into smaller fragments, mix different ingredients, and form an emulsified final product. A short overview of the principles of this type of equipment is given here:

- *Grinders* consist of a pitched screw that pushes the material through a cutting system that consists of hole plates and knives. The material is fed from a hopper (up to 1000 L) by a feeding screw to the main screw that pushes the material through the cutting system. Grinders are used to break down solid materials with large dimensions (*e.g.*, extruded plant-based materials). The diameter of the hole plates decreases as the material proceeds through the device (Fig. 3.7b) and the material is progressively ground into smaller particles by the action of the rotating knives and hole plates. The final coarseness can be adjusted by selecting specific knife and hole-plate setups. Grinders are often used to produce plant-based mince and reduce the size of textured vegetable proteins.
- *Vacuum filler-grinders* are a versatile piece of equipment in the production of meat analogs. Conventional vacuum fillers are widely used to fill highly viscous materials into containers. They can also be modified to grind materials by adding

an inline grinding system to the machine. Such a set-up has been used to produce emulsified, coarse dry-cured sausages and minced meat (Irmscher et al., 2015). They consist of a vacuum pump that pulls a strong vacuum in a rotary vane pump (positive displacement pump) (Fig. 3.7c). The material is sucked into the cell compartments by the vacuum and the action of the auger in the hopper. As the compartments rotate, the material is pushed out at the opposite end through the inline grinder, which creates pressures that are 4–6 times higher than in a conventional grinding system (Weiss et al., 2010). This enables very fine grinding, because hole plates with smaller diameters can be employed. Vacuum filler-grinders have been used to produce plant-based mince that has been pre-chopped with a traditional grinder.

- *Bowl choppers* consist of a revolving bowl (up to 1000 L) and rotating knives that sit in the bowl and cut the material (Fig. 3.7a). The bowl can be heated and cooled to induce phase transitions or to cook the material. Moreover, by employing a lid and installing a vacuum pump, the bowl can be operated under vacuum. Bowl choppers are used to finely chop and emulsify materials and have been used in the production of plant-based sausages.

3.4.1.3 Homogenization

Homogenization is typically used to form an emulsion from separate oil and water phases or to reduce the size of the droplets in an existing emulsion. It can also be used to disrupt any biopolymer particles in a colloidal dispersion. Reducing the particle size of fluid products is desirable because it increases their resistance to

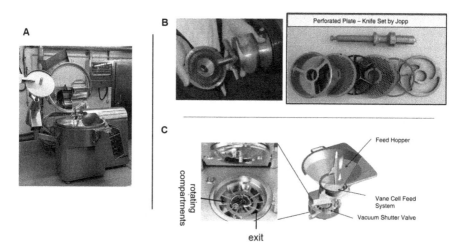

Fig. 3.7 Equipment used to grind and chop larger pieces of solid food materials. (**a**) = bowl chopper; (**b**) = grinder with hole-plate knife set-up; (**c**) = vacuum filler with vane cell feed system. (Reprinted with permission of Elsevier from (Weiss et al., 2010; Rust & Knipe, 2014))

3.4 Mechanical Processing Methods

gravitational separation and aggregation, as well as improves their mouthfeel. For these reasons, homogenization is commonly employed in the production of plant-based milks. It may also be used to emulsify hydrophobic additives (such as colors, flavors, vitamins, or oils) that need to be incorporated into aqueous-based matrices, such as those in plant-based egg, meat, or seafood. Homogenization can be carried out using various kinds of equipment, including high shear blenders, colloid mills, high-pressure valve homogenizers, sonicators, and microfluidizers. However, the high-pressure valve homogenizer is the most widely device used for this purpose and so will be the main focus of this section.

High-pressure homogenizers are capable of relatively high production rates 80,000 L h^{-1}, which is an advantage for large-scale manufacture of plant-based food products. Basically, they consist of an electrical motor, a piston, and a valve. The liquid to be homogenized is pulled into the device and then forced through the valve by the action of the piston. As the liquid moves through the valve, it is subjected to intense disruptive forces that break down and intermingle the separate phases or particles. Plant-based milks are typically homogenized at pressures of around 100–250 bar using a valve with a diameter of around 0.1 mm, which causes the fluids to be accelerated to speeds of up to 400 m/s (Fig. 3.8) (Bylund, 2015). The operating pressure can be varied by changing the gap width of the valve: the smaller the gap, the higher the pressure.

The high speeds of the fluids inside the homogenizer lead to intense shear and turbulent forces being generated, which help to break up the droplets. Moreover, a phenomenon known as cavitation is induced within the valve. Cavitation occurs because there is a transition in the pressure as particles start to move, which is described by the Bernoulli equation:

$$p_0 + \frac{1}{2}\rho v^2 + \rho gh = \text{const.} \tag{3.1}$$

Here, p_0 is the static pressure, ρ is the density, v is the fluid velocity, g is the gravitational constant, and h is the height. The first term in this equation is the static pressure, which acts because of the random motion of the molecules. The second term is the dynamic pressure, which is related to the kinetic energy of the fluid and acts in the longitudinal direction. The third term is the hydrostatic pressure.

When no external forces are applied particles move randomly in all directions because of their thermal motion, which results in the static pressure. During homogenization, the particles are forced to move in one direction through the valve as a result of the action of the piston, which causes the dynamic pressure to increase relative to the static pressure. If the particles move fast enough, the static pressure decreases below the boiling point of the liquid, which leads to the formation of small vapor bubbles in the liquid. After leaving the valve, the liquid slows down again because the diameter of the tube increases. Consequently, the dynamic pressure decreases and the static pressure increases. As a result, the vapor bubbles rapidly implode, which generates a hydraulic shock wave in the fluids that is strong enough to disrupt any oil droplets or biopolymer particles.

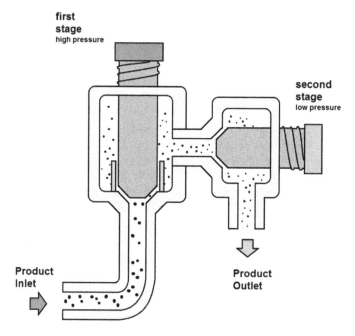

Fig. 3.8 Valve set up of a two-stage high-pressure homogenizer. The second stage disrupts possible aggregates that can arise from uncovered interfaces. (Adapted from (Comuzzo & Calligaris, 2019) under CC BY 4.0 (http://creativecommons.org/licenses/by/4.0/))

3.4.2 Separation and Fractionation Methods

Plant seeds contain numerous components that may or may not be wanted in the final product. For example, larger insoluble particles like hulls or cell wall fragments typically need to be removed before processing because they easily sediment (milk analogs), prevent structure formation (meat analogs), or cause a gritty mouthfeel (milk analogs). Therefore, separation and fractionation methods are required to selectively remove certain components during the manufacturing process. Various kinds of equipment can be utilized to achieve this goal.

3.4.2.1 Decanter Centrifuge

Decanters are used to accelerate liquid-liquid separation or liquid-solid separation based on the density differences of the different phases. They are commonly used to separate large particles from suspensions during the production of plant-based milks to ensure that the product can be homogenized and does not block the valve. Decanters can be used to separate particles with sizes greater than 10 μm, even when they are present at relatively high concentrations (9–60%) (Berk, 2013). They are capable of processing up to tens of thousands of liters per hour. The main

3.4 Mechanical Processing Methods

differences between a centrifuge and a decanter are that a decanter works continuously and that the suspension enters the equipment horizontally rather than vertically.

As the suspension is pumped into the decanter, it enters a conical rotating bowl with a screw inside, which is also rotating (Fig. 3.9). As the particles enter the bowl through inlet zones in the screw, they get separated through centrifugal forces based on their density. Particles with higher density are deposited on the drum walls, where they get compressed and discharged at the conical end of the device by the action of the screw, which rotates at a lower speed than the bowl. In contrast, the clarified liquid flows through the flights and gets discharged at the opposite end.

An important parameter in decanter processing is the feed handling capacity, which is described by the following equation that is based on the Stokes' settling velocity (Menesklou et al., 2021):

$$\dot{V} = \frac{4\pi^3 \Delta \rho}{9\eta} d_p^2 n_{rot}^2 L_c R_c^2 \quad (m^3 s^{-1}) \tag{3.2}$$

Here, $\dot{V}$ is the volume flow of the feed (m³ s⁻¹), $\Delta \rho$ is the density difference between the solid and liquid phase (kg m⁻³), η is the dynamic viscosity of the liquid, (Pa s) d_p is the smallest particle diameter (m), n_{rot} is the rotational speed of the drum (s⁻¹), R_c is the characteristic radius (m) and L_c is the characteristic length (m), which depends on the geometry of the decanter. It should be noted that this equation is somewhat limited because the Stokes' settling viscosity is not applicable for high volume fractions and the equation assumes only a 50% recovery (Menesklou et al., 2021). However, it still highlights the important parameters to be considered when selecting and operating a decanter: the density difference, the particle size, and the viscosity.

Another important parameter is the separation efficiency of compounds by the machine (Haller et al., 2021):

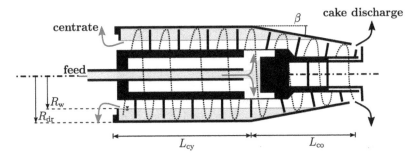

Fig. 3.9 Schematic overview of a decanter centrifuge for solid-liquid separation. (Adopted from (Menesklou et al., 2021) under CC BY 4.0 (https://creativecommons.org/licenses/by/4.0/))

$$\eta_{sep} = 1 - \frac{c_{centrate}}{c_{inlet}} \tag{3.3}$$

with V and c being the volume and concentration in the centrate (the liquid output) and in the inlet, respectively.

3.4.2.2 Hydrocyclones

Hydrocyclones are also utilized for liquid-solid separation during the production of some plant-based foods. Like decanters, they are also based on the utilization of centrifugal forces to separate the particles from the fluids. However, they commonly operate at lower solid contents, are suited to separate particles with sizes >10 μm, and the exiting solids have a higher residual water content. Hydrocyclones are commonly used to separate starches from slurries during protein extraction procedures (Chap. 2). In contrast to decanters, the particles are not accelerated by the action of a rotating element but by the centrifugal acceleration that results from pumping the dispersion tangentially into the cyclone. Thus, hydrocyclones do not require any moving parts. The design of a hydrocyclone with a conical shape at the bottom and a cylindrical shape at the entrance is shown schematically in Fig. 3.10.

As the suspension enters the cyclone at the top, it gets drawn into a rotational flow pattern, and the inertia of the particles forces them towards the wall. The inertial forces get higher as the particles move downward in the conical part of the cyclone because the particles move at a faster speed. As a result, progressively smaller particles are separated as the suspension moves through the hydrocyclone. The particles exhibit a radial force, which pushes them towards the outside of the cyclone towards the wall. By the impact of radial movement and gravity, the particles spiral to the apex of the cyclone and get discharged. Because the liquid descends closer to the center and the vortex finder is smaller than the apex, the liquids reverse their flow in the conical part and get redirected to the top by forming an inner vortex where they exit through the overflow.

3.4.2.3 Filtration

Decanter centrifuges and hydrocyclones employ centrifugal forces to separate particles in suspensions based on their density differences. In contrast, filtration techniques separate and concentrate particles based on their dimensions (and sometimes on their surface properties) using semipermeable membranes. Filtration techniques are classified into two main groups depending on their design: dead-end-filtration and crossflow-filtration. In dead-end-filtration, the particles accumulate on top of the membrane and form a filter cake because the flow of the feed is perpendicular to the membrane surface. In crossflow-filtration, a continuous fluid flow is applied parallel to the membrane surface, which minimizes the build-up of a thick filter

3.4 Mechanical Processing Methods

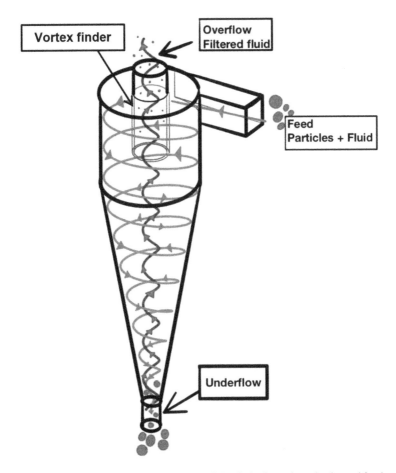

Fig. 3.10 Schematic overview of the working principle of a hydrocyclone that is used for the separation of solids from liquids. (Modified from (Durango-Cogollo et al., 2020) under CC BY 4.0 (http://creativecommons.org/licenses/by/4.0/))

cake. In the remainder of this section, we focus on crossflow-filtration because it is more commonly used for separation and concentration purposes.

The different crossflow-filtration techniques can be classified based on the type and size of particles they can separate: reversed osmosis; nanofiltration; ultrafiltration; and microfiltration (Fig. 3.11). In all of these techniques, the initial liquid is referred to as the *feed*, the liquid that passes through the membrane is the *permeate*, and the material that is retained and re-circulated is called the *retentate*. Depending on the application, either the permeate or the retentate may be the desired part to be recovered. Filtration equipment typically consists of a frame, a feed reservoir, a filtration module, tubes, feed pump(s), and retentate/permeate pumps (if required) (Fig. 3.12). The feed pump is commonly a centrifugal, piston, diaphragm, or a Mohno-type pump, depending on the device (Wagner, 2001).

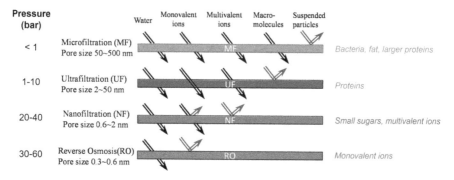

Fig. 3.11 Filtration techniques can be classified according to their pore size and the particles they can retain. (Modified from (Kotobuki et al., 2021) under CC BY 4.0 (http://creativecommons.org/licenses/by/4.0/))

Fig. 3.12 A typical ultrafiltration pilot plant setup. (Adapted from (Gienau et al., 2018) under CC BY 4.0 (http://creativecommons.org/licenses/by/4.0/))

The feed pump is responsible for establishing a positive pressure on the retentate side, which creates a pressure across the membrane that is the main driving force for the filtration process. The transmembrane pressure difference creates the flux through the membrane, which is given as liters of permeate per square meter of membrane surface area per hour ($l/m^2/h$). The flux is mainly influenced by the viscosity of the feed, the area of the membrane, and the resistance of it, as well as by the tendency for a fouling layer to develop during operation (Kessler, 2002). The transmembrane pressure is calculated using the following equation:

3.4 Mechanical Processing Methods

$$\Delta p_{TM} = \frac{p_1 + p_2}{2} - p_3 \, (\text{Pa}) \qquad (3.4)$$

Here, p_{TM} is the transmembrane pressure, p_1 is the feed stream's pressure at the inlet (high), p_2 is the concentrate stream pressure at the outlet (low), and p_3 is the permeate stream pressure at the outlet.

The transmembrane pressure induces the flow across the membrane and the process is carried out until a certain concentration factor is obtained (i.e., when the compound reaches a desired concentration). The concentration factor and the concentration of the compound in the retentate and permeate can be calculated using the following equations (PS Prozesstechnik, 2021):

$$X = \frac{V_0}{V_k} \qquad (3.5)$$

$$c_k = c_0 X^R \qquad (3.6)$$

$$\overline{cp} = c_0 \frac{X}{X-1}\left(1 - X^{R-1}\right) \qquad (3.7)$$

In these equations, R is the rejection factor, which reflects the amount of desired compounds not passing through the membrane:

$$R = 1 - \frac{c_p}{c_F} \qquad (3.8)$$

Here, X is the concentration factor, V_0 is the initial volume, V_k is the end volume, c_k is the end concentration in the retentate, $\overline{cp}$ is the concentration of collected permeate and c_p/c_f are the momentary permeate and feed concentration, respectively.

The core part of a membrane filtration system is the filtration modules, which are responsible for the separation process. Four different types of modules are commonly employed and the module selected for a particular application depends on the nature of the feed material and the filtration technique: plate and frame (ultrafiltration); tubular ceramic (microfiltration, ultrafiltration); spiral-wound (reverse osmosis, nanofiltration, ultrafiltration); and hollow-fiber (Bylund, 2015). The membranes used to carry out the separation must also be selected based on feed material type and operating conditions. For example, membranes fabricated from organic materials (e.g., cellulose acetate and polysulfone) are typically less resistant to extreme environmental conditions such as high temperatures or high/low pH compared to membranes fabricated from inorganic materials (e.g., aluminum oxide). An overview of the different membrane types is given here and shown schematically in Fig. 3.13:

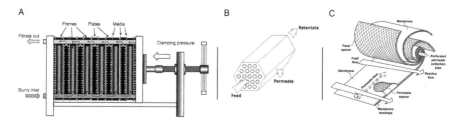

Fig. 3.13 Different membrane module systems are used to separate and concentrate particles. (**a**) = plate and frame; (**b**) = tubular ceramic, (**c**) = spiral-wound. (Adapted from (Abdul Latif et al., 2021; Gupta & Yan, 2016; Hakami et al., 2020) with permission from Elsevier and under CC BY 4.0 (https://creativecommons.org/licenses/by/4.0/))

- *Plate and frame modules* consist of stacks that are sandwiched together and hold the membranes. The feed flows between the membranes in specific flow patterns, similar to a plate heat exchanger.
- *Tubular ceramic modules* consist of filters manufactured from fine-grained inorganic ceramic materials. Typically, the module consists of several inner channels to increase the filtration area.
- *Spiral-wound modules* consist of tubular layers of thin sheets of membranes separated by a feed channel spacer that transports the feed. Layers of membranes and spacers (e.g., membrane-feed spacer-membrane-permeate spacer-membrane-feed spacer-membrane) are assembled to obtain a high throughput and a high filtration area.
- *Hollow fiber modules* often come as cartridges that contain several hollow fiber membranes consisting of polymers, such as cellulose acetate or polysulfone. A cartridge contains several fiber membranes that retain large particles but allow small particles and liquids to permeate into the internal fiber, which then get discharged as permeate.

Membrane filtration systems are commonly employed in the production of plant-based foods and ingredients. For example, hollow fiber ultrafiltration devices containing polysulfone membranes with a molecular weight cut-off of 10 kDa have been used to concentrate soy milk at a flux of 3.13 L/m^2/h (Giri & Mangaraj, 2014). This technique reduced the concentration of antinutritional compounds (20 kDa cut off) in the final product and proved useful for concentrating the soy milk before further processing, *e.g.*, for plant-based yogurt production.

3.5 Structure Formation Methods

Real meat and seafood products are semi-solid materials with fibrous internal structures comprised of numerous kinds of proteins (Chap. 6). The unique physicochemical and sensory attributes associated with these products are strongly influenced by their fibrous structure. A major challenge in the production of meat and seafood

3.5 Structure Formation Methods

analogs is therefore to mimic these kinds of fibrous structures using plant proteins (Grossmann & Weiss, 2021).[1] The majority of proteins found in plants are storage proteins, which are usually packaged into small dense particles (protein bodies) that are stored within specific locations in the plant cells until they get a signal to be released by the growing plant (Grossmann & Weiss, 2021). These storage proteins, therefore, play a very different role in nature than the muscle proteins in animals. The individual storage proteins tend to be globular proteins with approximately spherical structures that are a few nanometers in diameter (Fig. 3.14) (Glantz et al., 2010; Guo et al., 2012). Consequently, food processing techniques are required to transform these globular proteins into a 3D anisotropic (direction-dependent) fibrous network that resembles the texture of muscle fibers. These muscle fibers exhibit different responses when stresses are applied either perpendicular or parallel to their direction of orientation. Typically, the stresses needed to extend the fibers in a parallel direction are different from those required to extend them in a perpendicular direction (McClements et al., 2021).

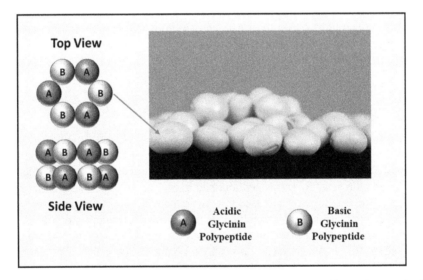

Fig. 3.14 The globular proteins in plant materials are often present as multimers that are physically and/or covalently linked together. The functionality of the proteins depends on their denaturation and aggregation state. Here, the proposed structure of the native form of the 11S glycinin molecule is shown, which consists of acidic (**a**) and basic (**b**) polypeptide chains. (Image of soybeans from CSIRO (Creative Commons 3.0). Reprinted from (McClements & Grossmann, 2021) with permission from Elsevier)

---

[1] The fibrous structures of meat and seafood products can be mimicked using some kinds of fungi (such as *Fusarium venenatum*) that produce protein-rich filamentous hyphae (Dai et al., 2021). These fungi have been used to produce meat and seafood analogs but they are not strictly plant-based.

Fibrous structures can be formed from globular plant proteins by promoting their unfolding and aggregation under carefully controlled processing conditions. Typically, the raw materials used to create plant-based meat analogs are powdered protein isolates or concentrates rather than whole plant seeds. This facilitates handling and storage of the raw material but also means that the powder needs to be rehydrated before use. Typically, the plant protein powders are therefore mixed with water to form a viscous slurry prior to further processing.

Two types of processing equipment are commercially used to produce meat analogs from plant proteins at present: extruders and shear cell devices. Both of these devices employ thermomechanical processing to create fibrous semi-solid structures from plant proteins. Extruders have been utilized for over a century, with the first patent being filed in 1869 (Bouvier & Campanella, 2014). Since then they have been used in all kinds of industries. In food production, extruders are employed for manufacturing numerous kinds of foods, including breakfast cereals, snacks, crispbread, and pasta. Extruders have also been used to produce meat analogs and meat extenders since the 1960s. The shear cell device was developed as a means of structuring proteins in the early 2000s by researchers at Wageningen University in the Netherlands (Manski et al., 2007). This device is currently being utilized by Rival Foods to create plant-based meat and seafood products. However, both have the same overall aim: unfold the globular plant proteins and assemble them into an anisotropic fibrous viscoelastic solid.

3.5.1 Extrusion

Extruders are currently the most common method of producing plant-based meat analogs. Depending on the intended application, food extruders are available with outputs as little as a few grams per hour and as high as several thousand kilograms per hour. The smaller units are used for research and development purposes, while the larger ones are used for commercial production. Extruders are widely used because they combine several unit operations into one device: mixing, heating, and structuring. Moreover, they can be operated continuously, which is a benefit for large-scale production. An extruder contains three main parts: an electrical motor (with power ratings of several hundred kW); one or two-segmented screws embedded within a temperature-controlled barrel; and a die through which the final material is extruded. Intermeshing twin-screw extruders are commonly used for the production of plant-based meats.

In this device, the temperature-controlled barrel is fitted with two screws that rotate in each other in the same direction (co-rotating). The co-rotating system enables a high mixing efficiency and outputs of up to 1000 kg h^{-1} with the current technology utilized for plant-based meat production. Gravimetric or volumetric feeders are used to dose the powdered protein into the feeding section of the extruder, while pumps are used to introduce water into the second section. Oils can be added at the beginning or end of the barrel section (Kendler et al., 2021). The

3.5 Structure Formation Methods

barrel and the screw are assembled into different sections to facilitate the processing and transformation of materials using different operating conditions. The screw elements are designed to fulfill different tasks during the extrusion process: conveying, mixing, kneading, compressing, and shearing. Some examples of commonly used screw elements are shown in Fig. 3.15.

The screw elements used in extrusion can be characterized according to their pitches (which is the length of one full rotation) and their main purpose. The screw and other elements inside an extruder fulfill different tasks depending on their design and diameter D (Maskan & Altan, 2012; Riaz, 2000):

- Elements with large pitches (≥1D) are typically used in the feeding zone. These screws have a large free volume and a high capacity for transporting.
- Elements with medium pitches (0.5–1.0D) are used in the kneading and melting sections.
- Elements with short pitches (0.25–0.75D) are employed in the metering section of the extruder and before the die to increase the pressure.
- Mixing elements are designed to disrupt the flow pattern, increase the filling level, and convert mechanical energy into heat by friction. Commonly used mixing elements are paddle blocks (Fig. 3.15, front left element) that are staggered in different geometries to achieve different mixing, shearing, and conveying actions.
- Cut flight screws are standard screws that are cut like gear wheels to enhance the leakage and distributive mixing.
- Reverse screws enhance the flow resistance and increase the pressure, which induces back mixing.

For meat analog production, a typical screw setup would consist of a screw with a length-to-diameter ratio (L/D) > 20 that is assembled of forward conveying elements with decreasing pitch and mixing elements in the first sections, followed by kneading, shearing, backward, and forward conveying elements in the middle section of the extruder barrel, and forward conveying elements in the final section before the die (Caporgno et al., 2020; Pietsch et al., 2017). In principle, the same

Fig. 3.15 Different elements are used to assemble the screw. Their function is described in more detail in the text. The screw elements shown are conveying elements with different pitches and kneading blocks with various thicknesses. (Images kindly provided by Coperion GmbH (Stuttgart, Germany))

screw design can be used for low- and high moisture extrusion as shown by Samard et al. (2019).

The temperatures in the different sections of an extruder are controlled by the heated barrel. In the feeding zone, the extruder is not heated and is maintained at room temperature to safely feed the material into the screw. In the following initial sections, the barrel is heated to temperatures below 100 °C to hydrate and mix the powders. In the middle sections, barrel temperatures of around 100 °C are used. In the final sections, barrel temperatures of 140–160 °C are employed to obtain the desired structures from the plant protein materials (Pietsch et al., 2019). As they pass through the different sections of the extruder, the raw materials (plant proteins and other ingredients) are mixed, sheared, hydrated, and heated. This causes the globular protein molecules to become denatured, which exposes reactive functional groups at their surfaces, such as non-polar and sulfhydryl groups. As a result, they have a tendency to associate with other proteins in the later parts of the extruder due to hydrophobic attraction and disulfide bond formation.

After the barrel section, the heated mass is fed into the die, which is mainly responsible for structuring the proteins into an anisotropic shape. The type of die used determines the type of end product created: low-moisture or high-moisture texturized vegetable protein (Fig. 3.16).

Low-moisture texturized vegetable protein products are created by extruding protein-rich powders with low water levels (<50%). These products are produced using a short shaping-die. The simplest geometry used would be a hole nozzle. In this die type, the proteins are structured by the pressure drop at the die exit, which is typically from around 20–100 bar to ambient pressure. As described earlier, the temperature in the final part of the barrel is around 140–160 °C and the product is

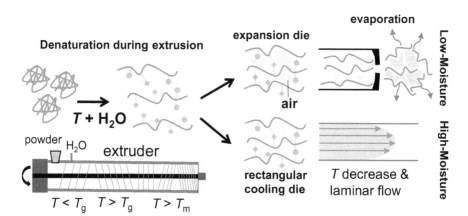

Fig. 3.16 Extrusion process of low- and high-moisture texturized vegetable protein. An extrusion process and equipment typically consists of motor, feeder, water pump, extruder barrel with screws, different elements such as conveying and kneading blocks, breaker plate (hole plate with small holes between the barrel and the die to separate the material and force it into a cylindrical shape), cooling die and die outlet. *Abbreviations*: T_g glass transition temperature, T_m denaturation temperature

3.5 Structure Formation Methods

under pressure because of the compressing action of the screw and the high temperatures. The material is hydrated and heated above the glass transition temperature, which results in the formation of a rubbery mass followed by a further increase in temperature above the denaturation temperature and the transformation into a flowable melt. When the material exits the die, there is a sudden drop in pressure and the hot protein dispersion expands through flash evaporation. The rapid evaporation of the water causes the protein molecules to become aligned into anisotropic structures. Being a flowable melt in the extruder, the structure is now solidified below the glass transition temperature of the material through the sudden drop in temperature and the removal of water to a content of around 20%, which allows new bonds to be formed between the protein molecules. These low-moisture texturized vegetable proteins are typically subjected to a drying step (final water content of around 7% and protein >50%) after the extruder to enhance their shelf life, which enables the product to be stored at room temperature. Thus, the finished product is a dry product that needs to be rehydrated before consumption.

In contrast, high-moisture texturized vegetable protein is obtained by extruding protein-rich powders with high water levels (around 50–70%). This is achieved by attaching a cooling die that prevents a rapid pressure drop at the exit of the die and consequently retains the water in the protein structure. This process is usually referred to as 'high-moisture extrusion' or 'wet extrusion' and typically involves pressures of <30 bar before the die (Pietsch et al., 2019). After the die, the textured protein is cut from the die and transported for further processing (such as freezing, shredding, marinating, or frying).

Commonly, the die has a rectangular slit shape that forms the protein melt and the protein mass is released from the die as a cuboid sheet (Fig. 3.17). For example, a cooling die with dimensions of 15 × 30 × 380 mm (H × W × L) was used in studies with screw diameters of 25.5 mm and a L/D of 29 (Pietsch et al., 2017). However, a disadvantage of this shape is the limited output because the dimensions are restricted by the cooling efficiency – which is directly related to structuring properties – of the die. Therefore, new die shapes are emerging that employ double cylindrical shapes to increase the surface area available for heat conduction. This type of die-shape allows outputs of up to 1000 kg/h (Fig. 3.17).

In both die shapes, the protein dispersion is forced into the die. Here, the pressure and temperature are reduced, which leads to the formation of anisotropic structures. The temperature of the material within the die is usually reduced using temperature-controlled water (20–80 °C). The precise nature of structure formation within the die during extrusion is still a matter of debate and many research groups are working on uncovering the mechanisms involved. It has been proposed that bonds in the protein are weakened in the extruder barrel at high temperatures (except for hydrophobic bonds that strengthen with increasing temperature) and are partly reformed in the cooling die (Cornet et al., 2021). Thus, the material is in a partially aggregated state when it exits the extruder and is pushed into the die where it flows in a laminar flow pattern. Once it cools down in the die, new bonds are formed. Because this happens while the material is flowing the bonds are first reformed near the cooled wall (Fig. 3.18).

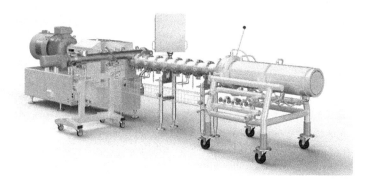

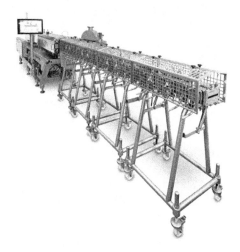

Fig. 3.17 For high-moisture plant-based meat production in an extruder, the material is transferred into the extruder for thermomechanical processing and pushed into the cooling die to cool down the material below 100 °C and induce anisotropic structure formation. *Top:* Cylindrical cooling die shape PolyCool1000 on a BCTF93 PolyTwin extruder (screw diameter 93 mm, maximum power 630 kW). In this setup, the material enters the extruder via the pre-conditioner from the left in which the plant protein powder and water are mixed. (Reprinted with permission of Bühler AG (Uzwil, Switzerland)). *Bottom:* Cuboid cooling die shape attached to a ZSK 43 twin-screw extrusion system. (Reprinted with permission from Coperion GmbH (Stuttgart, Germany))

The protein molecules closest to the die wall are cooled more rapidly, which causes them to form attractive interactions with their neighbors more quickly, thereby causing them to increase their viscosity and move more slowly. In addition, the wall shear stress slows them down in relation to the molecules in the middle. Thus, the material towards the middle section flows faster and cools down more slowly. As a result, an anisotropic structure is formed. This phenomenon is promoted when proteins precipitate at lower temperatures because it causes the

3.5 Structure Formation Methods

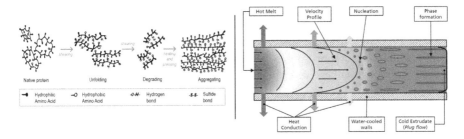

Fig. 3.18 Proposed mechanism of structure formation during protein processing in extruders. Most bonds are broken at high temperatures and re-formed at lower temperatures. The flow pattern and cooling in the cooling die result in the formation of anisotropic structures (right). (Adapted from (Zahari et al., 2020) under CC BY 4.0 (http://creativecommons.org/licenses/by/4.0/) and reprinted with permission of Elsevier from (Sandoval Murillo et al., 2019))

formation of protein-rich and protein-depleted regions inside the material, also known as spinodal phase separation (Sandoval Murillo et al., 2019). These regions are subsequently deformed by the shear stress superimposed by the flow pattern in the die (Fig. 3.18). Again, this occurs first close to the wall and continues to the middle of the cooling die. Thus, phase separations seem to play an essential part in the formation of such structures during thermomechanical processing since they have also been reported in shear cell processing (see next section) when polysaccharides and protein mixtures are present under segregative conditions (Sect. 3.2.1.1) (Cornet et al., 2021). Such phase separations interrupt the formation of protein-protein interactions, which fosters the formation of anisotropic structures. Recently, researchers have reported that there are no new bonds formed during the extrusion of meat analogs (Wittek et al., 2021). Most likely the type and total number of bonds do not change substantially, but their spatial location does. For example, the bonds may change from intramolecular to intermolecular bonds. Further research is clearly needed to fully uncover the molecular and physicochemical origins of the structure formation in meat analogs during extrusion.

An important process parameter that influences the efficiency of extrusion processing is the specific mechanical energy input (SME), which describes the energy input per unit mass (J kg^{-1}). This parameter is influenced by several parameters including process temperature, screw assembly, and die design. The SME is helpful to compare the effect of different treatment conditions on the raw materials or to design an extrusion process and to scale-up from pilot-plant to large-scale production. The SME can be calculated using the following expression (Pietsch et al., 2017):

$$\text{SME} = \frac{P - P_0}{\dot{m}} = \frac{\dfrac{n}{n_{\max}} \dfrac{M_d - M_{d,\text{empty}}}{100}}{\dot{m}} P_{\max} \quad \left(\text{kJ kg}^{-1}\right) \tag{3.9}$$

Here, n and n_{max} are the actual and maximum screw speed (s^{-1}), M_d and $M_{d,empty}$ are the actual and idle torque (Nm), $\dot{m}$ is the total mass flow (kg h^{-1}), and P_{max} is the maximum engine power (W).

Assuming no pressure drops, the average shear rate for conveying elements in an extruder can be calculated using the following equation (Vergnes, 2021):

$$\dot{\gamma} = \frac{2\pi NR}{60h} \quad (s^{-1}) \tag{3.10}$$

Here, $\dot{\gamma}$ is the average shear rate, N is the rotation speed (expressed in rpm), R is the screw radius, and h is the channel depth. In addition, to calculate the shear rate in the cooling die assuming no wall slip, the following equation can be used (Cornet et al., 2021):

$$\dot{\gamma}_{apparent} = \frac{6\dot{Q}_{net}}{wh2} \quad (s^{-1}) \tag{3.11}$$

In this equation $\dot{\gamma}_{apparent}$ is the apparent shear rate in the die, w is the width of the die, h is the height of the die, and $\dot{Q}_{net}$ is the volumetric flow rate.

Another important parameter in extrusion is the specific thermal energy (STE) input (Caporgno et al., 2020):

$$STE = c_p(T)dT \quad (kJ kg^{-1}) \tag{3.12}$$

Here, $c_p(T)$ is the temperature-dependent heat capacity of the material and dT is the incremental temperature difference. Phase transitions are not taken into account in this equation but they can be included by adding the effective enthalpy of fusion ΔH_m. This is especially important for formulations containing fats that melt or crystallize during the extrusion process.

On a final note, proteins from different plants have been successfully used to prepare plant-based meats using extrusion. For example, soy protein concentrate (Pietsch et al., 2019), soy protein isolate (Wittek et al., 2021), pea protein isolate (Beck et al., 2017; Ferawati et al., 2021), faba bean concentrate (Ferawati et al., 2021), lupin protein isolate/concentrate (Palanisamy et al., 2019), and others (Mosibo et al., 2020).

3.5.2 Shear Cell

The shear cell developed by Wageningen University in the Netherlands is being increasingly used to create meat analogs. Initially, this device was developed to better understand and control the structuring of proteins by shear forces. It is challenging to control, model, and understand the structuring of proteins in the barrel and die of an extruder. In contrast, the parameters in a shear cell are easier to control (such as shear rate, temperature, and pressure), which facilitates modeling and understanding of the structuring process. Another advantage of shear cells is that the meat analogs produced have dimensions more closely resembling those of whole-cuts of animal products, like beef or fish, especially in terms of their height. However, the shear cell also has some disadvantages. It has lower throughputs than extrusion because it is a batch process and the materials must be mixed before processing. In addition, longer residence times are required for shear cell processing (up to 20 min) than for extrusion processing (30 s to 3 min).

Typically, the raw materials used for creating meat analogs using the shear cell are similar to those used for extrusion. A powdered plant protein material (usually a concentrate or isolate) is mixed with water and a polysaccharide (if needed) before transferring the slurry into the cell at dry matter contents of around 45% (Dekkers et al., 2016). The material is then sheared at high temperatures to obtain the required structure (Cornet et al., 2021). Two types of shear cell geometries have been designed to create meat analogs: cone-in-cone and cylinder-in-cylinder.

The cone-in-cone design consists of an upper cone that fits into a lower cone-shaped cavity (Fig. 3.19). The two cones are typically made from stainless steel. The lower cone is heated and capable of rotating, while the upper cone is stationary. Initially, the material to be processed is placed into the lower cone. The upper cone is then lowered into this material and a defined gap is established between the two cones, which is where the protein blends are sheared. The shear cell is sealed to reduce water evaporation during the process.

The cylinder-in-cylinder design – also called the Couette design – was inspired by a coaxial cup-and-bob rotational rheometer measurement cell. The material to be processed is mixed and then placed in the lower cylinder ("cup") and the upper cylinder ("bob") is lowered into it. The material is sheared in the narrow gap between the two cylinders, whose dimensions are controlled (gap ~30 mm). This type of shear cell is designed with a steam-heated stationary outer cylinder with a lid and a heated inner cylinder that is rotated using a drive shaft (Krintiras et al., 2016). When the inner cylinder is rotated, the protein dispersion is sheared and heated.

Both types of shear cells are normally operated at high temperatures, similar to those used in extrusion processing. As an example, fibrous protein structures have been produced using the cone-in-cone system by shearing at 0 to 100 rpm at temperatures up to 140 °C for several minutes (Grabowska et al., 2016), whereas they have been produced using the cylinder-in-cylinder system by shearing at 20 rpm at 120 °C for 30 min (Krintiras et al., 2016).

Both designs rely on shearing the protein blend under defined conditions and the shear rate can be calculated according to the following equation for cone-in-cone designs, Newtonian fluids, and narrow gaps (Peighambardoust et al., 2004):

$$\dot{\gamma} = \frac{\omega}{\tan\theta} \quad (s^{-1})$$

(3.13)

Fig. 3.19 Cone-in-cone shear cell equipment for plant-based meat production. The production of plant-based meats in a shear cell involves pre-mixing and transferring the protein suspension into the cell for high-temperature processing (>100 °C) for 15–20 min at low pressures and constant shear (around $\dot{\gamma}$ = 40 s⁻¹). This contrasts with extrusion processing, which involves typically no pre-mixing and thermomechanical processing at high pressures and temperatures for a short time of 1–2 min (Cornet et al., 2021). (Reprinted from (Kyriakopoulou et al., 2019) with permission from Elsevier)

Here, ω is the rotor speed (s⁻¹) and θ is the angle between the cones (tan $\theta \approx \theta$ for small angles). The SME in this setup can be calculated using the following expression:

$$\text{SME} = \frac{\left(\int_{t=0}^{t} \omega M(t) dt\right)}{m} \quad \left(\text{kJ kg}^{-1}\right) \tag{3.14}$$

Here, m is the mass of material in the shear cell (kg) and M is the torque (Nm) at treatment time t. The shear rate for the cylinder-in-cylinder design can be calculated using the following expression (Krintiras et al., 2015):

$$\dot{\gamma} = \frac{2R_i \pi \text{rpm}}{60h} \quad \left(\text{s}^{-1}\right) \tag{3.15}$$

Here, R_i is the radius of the inner cylinder and h is the gap between the rotating and stationary cylinder. It should be noted, these equations have some limitations and work has also been published to calculate the shear rate for wider gaps for either Newtonian liquids (Michels et al., 2010) or non-Newtonian liquids (Krintiras et al., 2016). Nonetheless, the structuring mechanism remains the same for both designs.

In both setups, the proteins are denatured in the gap under the influence of heat, which is similar to extrusion processing. The velocity gradient that is generated by the shearing action of the cone or cylinder facilitates the formation of fibrous or layered structures. The obtained structures show a typical structured gradient based on the velocity profile (wall shear stress) that develops within the gap. For wider gaps, a non-linear 'J-curve' is obtained due to the shearing profile between the inner and outer walls (Krintiras et al., 2016). The successful formation of fibrous structures also depends on the ingredients used. Thermodynamic incompatibility between different proteins, or between proteins and polysaccharides (Sect. 3.2.1.1) is thought to be the main driver for the formation of anisotropic structures during shear cell processing by the formation of water-in-water emulsions that are deformed under shear and fixed by the action of heat (Cornet et al., 2021).

In conclusion, the shear cell technology is a promising means of producing batch amounts of structured plant proteins. Researchers have already used it to successfully structure soy protein isolate with pectin (Dekkers et al., 2016), soy protein isolate and wheat gluten (Krintiras et al., 2015), soy protein concentrate (Grabowska et al., 2016), pea protein isolate, wheat gluten (Schreuders et al., 2019), and starch/zein blends (Habeych et al., 2008). It is especially useful for producing meat analogs with dimensions similar to those of whole meat cuts.

3.5.3 Additive Manufacturing

Meat and seafood analogs can also be produced using additive manufacturing, which is also known as 3D printing. In this process, a plant-based food is assembled by extruding edible inks from one or more moving nozzles that add layer upon layer of material at specific locations based on a 3D digital template. Different edible inks can be used to simulate different parts of the food. For instance, a protein-rich red ink could be used to print the 'muscle fibers', while a lipid-rich white ink could be used to print the 'adipose tissue'. These different parts of the food can be printed sequentially using a single nozzle (*i.e.*, first one part, then the other) or simultaneously using multiple nozzles (with each nozzle containing a different edible ink).

In general, several 3D printing technologies are available for additive manufacturing purposes (Le-Bail et al., 2020). However, the most common technology used for creating structures that resemble meat and seafood is '3D extrusion' (Le-Bail et al., 2020; Dick et al., 2019). This process should not be confused with the extrusion process discussed earlier. 3D printing involves pushing a material through a nozzle, which is somewhat similar to extrusion processing. However, 3D printers do not usually contain elements that shear and heat the samples during the manufacturing process. Moreover, they tend to be much smaller than extruders. Finally, 3D printers create structures and shapes using a layer-by-layer additive manufacturing approach, whereas extruders create them using a specific die design.

The design of a typical 3D printer based on extrusion is shown in Fig. 3.20. This device typically consists of a few major parts:

- A digital processing unit that converts the 3D digital dimensions of the material stored in the computer into a movement of the nozzles and/or platform.
- Servo-motors that move the nozzles and/or platform along the X-, Y- and Z-axis.
- A (heated) barrel with a nozzle head as an outlet that contains the material
- A piston device to push the material out of the nozzle
- A moving platform to deposit the material on

The 3D printing device shown in this diagram utilizes a piston to force the edible ink out of the nozzle. It consists of a syringe and a step-motor that drives the piston. Alternatively, an edible ink can be forced out of the syringe using screw-based extrusion. This design consists of a screw that sits inside a cartridge. The material is loaded into the cartridge and the edible ink is deposited onto the platform when the screw turns. Pneumatic dispensing units that use air pressure to force the edible ink out of the cartridge are also available (Derakhshanfar et al., 2018). However, this pneumatic technique is more suitable for printing fairly fluid materials, because it

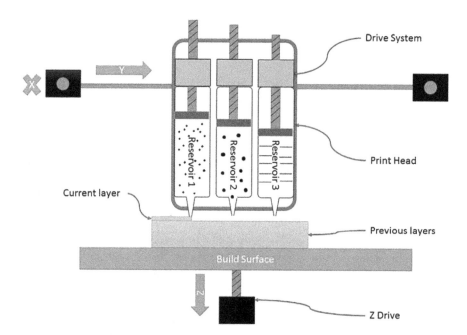

Fig. 3.20 A 3D printer consists of one or more nozzles that are driven by motors above a platform. In this case, an extrusion-based system is shown that pushes the material out of the nozzles using a piston and then deposits it on a platform using a layer-by-layer process. (Reprinted from (Lipton et al., 2015) with permission from Elsevier)

3.5 Structure Formation Methods

cannot generate the forces required to extrude solid or semi-solid materials (Sun et al., 2018). For this reason, it is less useful for printing plant-based meats because they often use semi-solid edible inks. During 3D extrusion, the material can be extruded at room temperature or the cartridge can be heated to liquefy the material before exiting the nozzle. The material may then undergo a liquid-to-solid transition on the platform.

The materials used for 3D extrusion printing need to fulfill certain properties that are important for their (*i*) extrudability, and (*ii*) buildability. Extrudability depends on the material properties linked to the deposition process, such as flow through the nozzle and extrusion force. Buildability depends on the material properties linked to the ability of the edible ink to form a semi-solid or solid structure after printing onto the platform, such as yield stress, gelling temperature, or gelling time (Wilms et al., 2021). Mathematical models have been developed to describe the printability of certain kinds of food materials. For instance, the Herschel-Bulkley model can be used to describe the flow behavior of edible inks that exhibit plastic properties. This model is often used to describe the material flow behavior during 3D printing:

$$\tau = \tau_y + K\dot{\gamma}^n \quad (\text{Pa}) \tag{3.16}$$

Here, τ is the shear stress (Pa), τ_y is the yield stress, K is the consistency index (Pa s^n), $\dot{\gamma}$ is the shear rate (s^{-1}), and n is the flow index. For $n < 1$ the fluid is shear-thinning, whereas for $n > 1$ the fluid is shear-thickening. This type of edible ink behaves like a solid below the yield stress and does not flow but behaves like a fluid above the yield stress that does flow. An example of a food material that exhibits this kind of behavior is mashed potatoes. When printing using this kind of edible ink, a stress must be applied to the material in the syringe that exceeds the yield stress and causes it to be extruded through the nozzle. However, once the material is printed onto the platform, the applied stress (due to gravity), is below the yield stress, which causes the material to become solid and retain its shape.

Another helpful equation is the Hagen-Poiseuille equation that describes the pressure drop of a Newtonian fluid in a laminar flow through a cylinder, which is useful for describing the pressure required to force an edible ink through a syringe:

$$\Delta P = \frac{8\eta L Q}{\pi R^4} \quad (\text{Pa}) \tag{3.17}$$

Here, ΔP is the pressure drop (Pa), η is the apparent viscosity (Pa s), L is the specific length (m), Q is the volumetric flow rate (m^3 s^{-1}), and R is the radius of the cylinder (m). Because this equation is only valid for Newtonian liquids, the equation has been modified to account for the nonlinear behavior of non-ideal fluids (Wilms et al., 2021):

$$\Delta P = \frac{2KL}{R}\left(\frac{3n+1}{n}\right) n \left(\frac{Q}{\pi R^3}\right)^n \quad (\text{Pa}) \tag{3.18}$$

Here, K is the consistency index and n is the flow index. This equation also has some limitations because it ignores wall slip, clogging, extrusion instabilities, and other factors, which have been discussed by Wilms et al. (2021).

Despite their limitations, these equations provide valuable insights into the major factors influencing 3D printing, including material properties, operating conditions, and the nozzle geometry. Table 3.1 provides an overview of the different parameters that influence the predictions made using these equations. In general, extrudability is favored at low viscosities but buildability is favored at high yield stresses. The material should be able to continuously flow out of the nozzle but then maintain its structure after it is deposited on the platform, which involves careful control of the viscosity and the yield stress. A compromise between different material properties needs to be established when developing an edible ink that exhibits plastic properties.

Edible inks can also be created that rely on other mechanisms to form a semi-solid structure after printing. For instance, a material that undergoes a liquid-to-solid transition can be printed by controlling the temperature of the syringe and the platform. The syringe can be heated above the transition temperature of the material. In this way, the edible ink is fluid when it is extruded. In contrast, the platform can be maintained below the transition temperature so the edible ink solidifies after it is printed. Food ingredients that exhibit this kind of behavior include solid fats (like coconut oil and cocoa butter) and cold-set hydrocolloids (like agar). Alternatively, a gelling hydrocolloid (like alginate) and a gelling agent (like calcium) can be co-extruded through a co-axial syringe to promote gelation of the hydrocolloid on the platform (Sun et al., 2018).

3D printing has already been used to create plant-based meats that resemble real meats. The technology is able to simultaneously print more than one material by utilizing several nozzles or by co-axial extrusion (synchronous extrusion of two materials from an inner and an outer nozzle). In this way, a fat and a protein phase can be printed to resemble a real meat cut with intramuscular fat in it. Several global start-up companies are developing and commercializing 3D printing technologies for this purpose, including Novameat (Barcelona, Spain) and Redefine Meat (Rehovot, Israel). Moreover, numerous academic studies have highlighted the potential of using 3D printing for producing whole-cut meat alternatives.

Chen et al. (2021) studied the printability and buildability of mixtures of textured and non-textured soy proteins combined with different hydrocolloids. The printer consisted of a syringe with a 22 mm diameter that was equipped with a 0.8 mm nozzle. The material was ejected from the nozzle by the action of a piston and the nozzle moved at a speed of 20 mm/s at room temperature. The 'steak' produced using this process had dimensions of 60 mm × 30 mm × 8 mm. Interestingly, the samples containing hydrocolloids showed superior printability and buildability. The formulations prepared from non-textured soy protein had low water-binding and low yield stresses, which prevented successful structure build-up. The addition of xanthan considerably improved printability and buildability. Moreover, the meat analogs produced maintained their shape and texture after frying. These effects can be related to the high water binding capacity of xanthan gum, which was also shown to synergistically interact with soy proteins to increase the yield stress (Sánchez

3.5 Structure Formation Methods

Table 3.1 3D printing extrusion toolbox that describes the impact of material properties and processing conditions on the rheological properties and critical shear rate, as well as on extrusion pressure and extrusion instabilities

	Dependent variables					Independent variables	Dependent variables	
	Product rheology[a]			Critical shear rates				
Independent variables	Consistency factor[b] (K)	Flow index[c] (n)	Yield stress (τ_y)	Upper boundary (DST)	Lower boundary (LPM)		Extrusion pressure (ΔP)	Extrusion instabilities
Material properties						Product rheology		
Liquid phase[a]						Apparent viscosity[k] (η)	↑	
Consistency factor[b] (K)	↑	n/a	n/a	↓	↓	Flow index (n)	↑	
Flow index[c] (n)	n/a	↑	n/a	↓	↓	Trouton ratio[l]	↑	
Yield stress (τ_y)	n/a	n/a	↑			Processing conditions		
Suspended particles						Volumetric flow rate (Q)	↑	↑
Particle volume fraction (φ_m)	↑	↓	↑	↓	↑	Temperature (T)	↓	↓
Particle size ($D_{,4,3}$)	↓[d]		↓	↓[e]	↑	Nozzle design		
Width particle size distribution	↓	↑	↓	↑		Length (L)	↑	↓
Particle shape anisotropy[f] (r_p)	↑	↓	↑	↓		Radius (R)	↓	↑
Suspension						Entry angle (θ)	→[m]	↑
Interparticle forces[g]	↑	↓	↑[h]	→[i]				
Processing conditions								
Temperature (T)	↓		↓	↑	↑			
Time and shear (during flow)[j] (t)	↓		↓	↑	↑			
Time (after deposition)[i] (t)	↑		↑	n/a	n/a			

(continued)

Table 3.1 (continued)

Arrows indicate an increase or decrease, if the independent variable is increased. *n/a* no relationship, *DST* discontinuous shear thickening, *LPM* liquid phase migration. Adapted and modified from (Wilms et al., 2021)

[a] Described as a Herschel Bulkley fluid ($\tau = \tau_y + K\gamma^n$); [b] Including shear and extensional contributions; *c* Assuming a flow index <1, i.e. a shear-thinning fluid; [d] Especially true for colloidal particles, where the interparticle forces become more relevant; [e] Only when the particle size is significantly smaller than the dimensions of the nozzle, or geometrical constraints can become important (Cheyne et al., 2005); [f] Related to the aspect ratio, an increasing value means further away from the optimum of $r_p \sim 0.5$ (Gan et al., 2004); [g] Both attractive and repulsive interactions; [h] Only attractive interactions; [i] Theoretically only deflocculated suspensions show DST, as flocculated suspensions already have a high viscosity (Barnes, 1989); [j] Assuming thixotropic behavior. In the case of rheopectic behavior, the relationships are of the opposite nature. [k] Shear stress divided by the instantaneous shear rate; [l] Ratio between extensional viscosity and shear viscosity, higher Trouton values result in an increased pressure loss during convergent flow; [m] Competition between shear and extensional flow, possibly having a local minimum at intermediate entry angles of 30°-45° (Ansari et al., 2010; Ardakani et al., 2013)

et al., 1995). The authors also showed that the 3D printing conditions had to be optimized to produce meat analogs that closely simulated the textural attributes (hardness, gumminess, and chewiness) of real meat using an edible ink consisting of textured soy protein and xanthan.

Another study used a coaxial extrusion nozzle to 3D print a meat alternative (Ko et al., 2021). Two different edible inks were prepared to simulate the different parts of meat products. The first edible ink consisted of 17% soy protein isolate, 17% potato starch, 1% $CaCl_2$, 1% KCl, 0.5% xanthan, and 63.5% demineralized water. This ink was used to produce the protein-rich "muscle fiber" regions in the meat analog. The second edible ink consisted of mixtures of carrageenan, glucomannan, and alginate and was used to produce the "connective tissue" regions. The printer used had dimensions with an inner nozzle diameter of 1.0 mm and a nozzle speed of 20 mm/s, and an infill level of 70%. A rectilinear infill pattern at an extrusion speed of 0.03 mL/min was used. The nozzle was designed to extrude the fiber solution from the inner nozzle, while the protein solution was pumped from the surrounding outer nozzle, which also contained calcium and potassium ions to induce the crosslinking of alginate and carrageenan. This resulted in a considerable increase in gel strength in the first few minutes after depositing both materials onto the platform by the crosslinking action of the mineral ions with the hydrocolloids. Optimum results were obtained for mixtures of 1.0% alginate, 1.5% carrageenan, and 1.5% glucomannan that were co-extruded with the protein onto the platform. The printed meat analogs exhibited extensive fibrousness with a higher tensile strength but lower hardness values compared to a beef control sample. Moreover, samples with hydrocolloids had a lower cooking loss and transversal as well as longitudinal shrinkage compared to beef. This was related to the high-water retention capacity of the co-extruded hydrocolloids employed that increased the water retention and contributed to the development of the fibrous structure in conjunction with the proteins. Thus, these results indicate that a 3D printing technology that utilizes co-extrusion could be used to create meat-like structures.

3.5 Structure Formation Methods

As mentioned before, one advantage of 3D printing is that different edible inks can be extruded, which is useful when creating a plant-based meat with fatty regions that mimic marbled meat. The challenge is to create a fat-mimetic that has textural properties that allow for good printability and buildability. Liquid oils can be printed but are not able to build structures, whereas solid fats are difficult to print because they have very high yield stresses (Gonzalez-Gutierrez & Scanlon, 2018). This problem can sometimes be overcome by controlling the solid fat content of the system, which can be achieved by controlling the fatty acid composition and printing/deposition temperatures. Alternatively, other kinds of materials can be utilized to formulate edible inks. For instance, emulsion gels have been used as edible inks to create the fatty regions in meat analogs. The 20% oil-in-water emulsions used for this purpose consisted of lecithin-coated oil droplets dispersed within a continuous phase gelled using a mixture of potato starch (5–25%) and inulin (40%) (Wen et al., 2021). These fat mimetics were extruded through a nozzle (diameter 1.1 mm) at room temperature. The potato starch increased the shear modulus, printability, and hardness of the fat mimetics but decreased their meltability at 160 °C when used at high concentrations. Low starch concentrations resulted in low buildability because the network was too weak, whereas high starch concentration resulted in a solid network that exhibited poor meltability. However, animal fat also retains some of its structure under melting conditions because of the connective tissue network that stabilizes the lipid network at high temperatures. Thus, it is important to find the optimum hydrocolloid concentration to obtain the desired processing parameters and final product texture. This study concluded that edible inks that exhibited comparable meltability to pork and beef fat could be fabricated from emulsions prepared with either soybean oil or coconut oil when emulsified and mixed with 15% potato starch.

In conclusion, the development of 3D printing for the production of plant-based foods is still in its infancy but it does have potential for selected applications in the future. In particular, it could become a viable process for the de-centralized production of plant-based meat analogs. For instance, 3D printing could be used to create meat analogs on a small scale in a restaurant or home. Nevertheless, there are still several hurdles that need to be overcome before 3D printing finds widespread application. First, the printing speed is relatively slow and it can take several hours to produce a single whole-cut meat analog. The printing speed can be increased by increasing the flow rate of the edible ink and the movement of the nozzle and platform but this often decreases printing accuracy and product quality (Le-Bail et al., 2020). Nevertheless, improvements in 3D extrusion technology have enabled the production of several kilograms per hour of plant-based meats (Redefine Meat, 2020). Moreover, it is possible to run several printers simultaneously, which is being explored by some plant-based food companies (*e.g.*, revo-foods.com). Therefore, this technology might be suitable for de-centralized production of meat analogs where time constraints are less relevant than in large-scale industrial production. Another concern is related to food safety. As the printing process is commonly carried out at room temperature and is rather slow, microbial growth needs to be controlled. Thus, the instrument must be carefully cleaned between each printing run to avoid microbial contamination. Finally, standardized edible inks need to be developed that have a reliable functional performance and a high nutritional value. There is still a lack of published information on the nutritional quality of edible inks suitable for the production of plant-based meats.

3.6 Thermal Processing Methods

Thermal processing methods may be used for a variety of reasons, including deactivating undesirable enzymes (blanching), inactivating spoilage or pathogenic microorganisms (pasteurization/sterilization), or promoting thermal transformations in specific ingredients (cooking). In this section, we provide a brief overview of the methods used for the thermal processing of plant-based foods.

3.6.1 Blanching

Vegetables and fruits are often blanched to deactivate enzymes, decrease microbial contamination, and remove air. Blanching is commonly carried out to preserve food quality during storage (usually dry, cold, or frozen storage) until further processing. The process usually involves rapid heating to a specified temperature, holding at this temperature for a specific time, and then rapid cooling to a final temperature. For example, peas are blanched by heating them in water at 80 °C for 2 min and then cooling them in an ice bath, which leads to a 90% reduction of lipoxygenase activity, thereby reducing the development of off-flavors (Gökmen et al., 2005). Blanching is therefore important for maintaining the desirable sensory attributes of seeds after harvesting. For this reason, this process is commonly used during the production of plant-based foods and ingredients, such as milk analogs, meat analogs, and protein powders.

Blanching may be carried out by bringing the food into contact with either hot water or saturated steam. Typically, steam blanching leads to lower nutrient loss because less water is available to solubilize them. Different designs are available for both of these methods. Hot water blanchers expose the foods to temperatures ranging from 70 to 100 °C before cooling and dewatering them (Fellows, 2017). A reel blancher is a hot water blancher that consists of a rotating cylindrical mesh drum that is partly immersed in hot water and slowly moves the food forward (Fig. 3.21). These kinds of blanchers are often used to blanch peas (Featherstone, 2016). Pipe, conveyor, and rotary drum blanchers are other kinds of hot water blanchers that are also used for this purpose (Fellows, 2017). Steam blanchers often consist of a conveyer belt that transports the food through a long tunnel (< 20 m) filled with saturated steam (Fig. 3.21). Conveyer belt blanching processes often utilize multistage countercurrent cooling to reduce energy costs and increase sustainability. In these

3.6 Thermal Processing Methods

 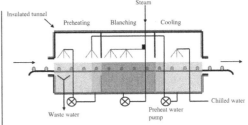

Fig. 3.21 Blanching is carried out using either hot water or steam to inactivate enzymes, microbes, and other compounds that may decrease the quality and safety of foods during storage. Shown here is a reel blancher that employs hot water (left) and a steam blancher with countercurrent cooling (right). (Reprinted with permission of Elsevier from (Fellows, 2017))

processes, water used to cool the exiting product is recirculated to the beginning of the blanching process to pre-heat the entering product.

3.6.2 Inactivation of Antinutrients

Many plants contain antinutrients that are substances that interfere with the digestion and absorption of nutrients (Table 3.2). The presence of these antinutrients can be attributed to evolutionary pressures that made the plants less desirable to eat or that prevented the digestion of the seeds in the gastrointestinal tract so they could survive and germinate after excretion, thereby facilitating their distribution to new locations. However, the presence of antinutrients in plant-based foods may cause nutritional problems by interfering with the normal digestion process.

For these reasons, food processors try to remove or inactivate antinutrients. In this section, we mainly discuss the removal or inactivation of antinutrients in the two most common plant-based food types: plant-based meats and milks.

High-temperature extrusion has been reported to reduce antinutritional compounds such as inositol hexaphosphate (IP6), tannins, and enzyme inhibitors (Cotacallapa-Sucapuca et al., 2021). However, some residual antinutrient activity is still often observed, even after these high-temperature treatments. For plant-based milks, the production conditions are often not sufficient to reduce the activity or concentration of certain antinutritional substances. However, some of the antinutritional compounds (such as phytic acid, saponins, tannins, and lectins) are fully or partially removed using the separation procedures discussed earlier (*e.g.*, decanting or centrifuging) (Swallah et al., 2021). Nonetheless, trypsin inhibitors are still typically present in raw or only mildly processed plant-milks, such as soy milk (Chen et al., 2014). For example, blanching soybeans at 85 °C for 90 s only reduced trypsin inhibitor activity by 42% (Yuan et al., 2008). Trypsin inhibitors are proteins that reduce the catalytic activity of pancreatic enzymes like trypsin and chymotrypsin, thereby inhibiting the hydrolysis and absorption of proteins in the digestive tract. As

Table 3.2 Many plant-based ingredients contain antinutritional factors. A selection of some of the most important ones is highlighted in this table

Source	Type	Amount
Legumes: Soy, lentils, chickpeas, peanuts, beans	Phytic acid Saponins Cyanide Tannins Trypsin inhibitor Oxalates	386–714 mg/100 g 106–170 mg/100 g 2–200 mg/100 g 1.8–18 mg/g 6.7 mg/100 g 8 mg/kg
Grains: Wheat, barley, rye, oat, millet, corn, spelt, kamut, sorghum	Phytic acid Oxalates	50–74 mg/g 35–270 mg/100 g
Pseudo-grains: Quinoa, amaranth, wheat, buckwheat, teff	Phytic acid Lectins Saponins Goitrogens	0.5–7.3 g/100 g 0.04–2.14 ppm
Nuts: Almonds, hazelnut, cashew, pignolias, pistachio, Brazil nuts, walnuts, macadamia, *etc.*	Phytic acid Lectins Oxalates	150–9400 mg/100 g 37–144 µg/g 40–490 mg/100 g
Seeds: Sesame, flaxseed, poppy seed, sunflower, pumpkin	Phytic acid Alpha-amylase inhibitor Cyanide	1–10.7 g/100 g 0.251 mg/mL 140–370 ppm
Tubers: Carrot, sweet potato, Jerusalem artichoke, manioc (or tapioca), yam	Oxalates Tannins Phytates	0.4–2.3 mg/100 g 4.18–6.72 mg/100 g 0.06–0.08 mg/100 g
Nightshades: Potato, tomato, eggplant, pepper	Phytic acid Tannins Saponins Cyanide	0.82–4.48 mg/100 g 0.19 mg/100 g 0.16–0.25 mg/100 g 1.6–10.5 mg/100 g

Modified from Popova and Mihaylova (2019)

a result, they can lead to overproduction of trypsin by the pancreas, which can have adverse health effects (Gilani et al., 2012). An individual ultra-high temperature treatment step (>135 °C for several seconds) is therefore often used during the production of plant-based milks to deactivate trypsin inhibitors.

Direct or indirect-heat treatment can be used for this purpose. Steam injection is commonly used because it can achieve a fast heating cycle (Fig. 3.22). In this operation, steam (5–8 bar; 160–175 °C) is directly injected into the fluid food using steam injection heads at one or more locations, followed by a short holding tube, which facilitates rapid heating due to the exothermic transition that occurs when steam is condensed back to liquid water. The fluid food is held for several seconds at the

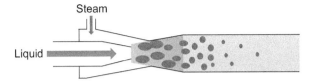

Fig. 3.22 Steam injection is used for UHT treatment to ensure fast heating. After the nozzle, the liquid is held for a specific time and then flash cooled using a vacuum to remove the surplus liquid

desired temperature, and then flash cooled in a vacuum chamber that removes the added water. The fast heating and cooling ensures minimal degradation of valuable nutrients but inactivates most of the trypsin inhibitors.

Typical processing conditions for the inactivation of trypsin inhibitors in plant-milks have been described previously (Prabhakaran & Perera, 2006; Yuan et al., 2008). Prabhakaran and Perera (2006) reported that heating soy milk at 120 °C for 80 s inactivated approximately 80% of the investigated trypsin inhibitor, thereby enhancing the digestibility of the final product. Other researchers reported that indirect ultra-high temperature processing resulted in a 77% decrease in trypsin inhibitor activity after heating at 140 °C for 4 s. Moreover, combining blanching for 2 min at 80 °C with an ultra-high temperature treatment resulted in a reduction of trypsin inhibitor activity of up to 89%, depending on the time-temperature combination used (Yuan et al., 2008). Thus, different processing options are available to reduce the content and activity of antinutritional compounds, and the most appropriate conditions need to be carefully selected based on the raw materials used.

3.6.3 Pasteurization and Ultra-High Temperature (UHT) Treatments

The aim of a thermal treatment is to produce a safe and shelf-stable food. There are two main factors that impact the time and temperature profile selected to thermally process a food. First, the treatment should inactivate any viable pathogens, which is usually achieved at temperatures below 100 °C (Montville et al., 2012). If the food matrix contains conditions that still allow microbial growth (such as a low salt content and pH > 4.6), then these products need to be stored in a refrigerated area because pasteurization does not inactivate all enzymes – which can cause off-flavors – and does not inactivate the dormant forms of microorganisms (spores). Spores are more resistant to high temperatures than living microorganisms and will start to grow once the environmental conditions allow. Spores and enzymes are inactivated when foods are heated to sufficiently high temperatures. Consequently, when viable microorganisms, spores, and enzymes are known to be a problem in a particular food it can be heated to a higher temperature to inactivate them, which is known as sterilization. Compared to sterilization, ultra-high temperature (UHT) processing is a less severe thermal treatment that does not inactivate all enzymes but all relevant spores. In practice, even sterilized foods are not 100% sterile but can be considered to be "commercially" sterile, *i.e.*, free of microorganisms that could grow under the conditions found in the product (Richardson, 2001). Sterilized products can be stored in closed containers at room temperatures for several months and are safe to consume for years (although their quality attributes may deteriorate).

In the remainder of this section, we will focus on methods used to pasteurize and sterilize fluid foods, like plant-based milk and eggs. Plant-based solid foods can be transformed into shelf-stable products using processes such as steaming, cooking, autoclaving, or extrusion.

Most plant-based milks currently on the market are subjected to ultra-high temperature processing. This is mainly to ensure the deactivation of antinutrients and because the turnover of plant-based milks is still relatively low, which means that they may remain on the supermarket shelves for extended periods. UHT processing employs temperatures of around 140–145 °C for 5–10 sec (Prabhakaran & Perera, 2006). This ensures the inactivation of spores while still preserving heat-sensitive compounds (like vitamins) more effectively than sterilization, which involves heating the food at 121 °C in an autoclave for several minutes.

Direct UHT treatment involves the injection or infusion of steam directly into the liquid food followed by flash evaporation to cool it down and remove the added water. The amount of steam needed per kg of plant-based milk in this process can be calculated as (Kessler, 2002):

$$\frac{\dot{m}_S}{\dot{m}_L} = \frac{c_{L,F}T_F - c_{L,I}T_I}{h_F - c_{W,F}T_F} \quad (\text{kg}/\text{kg}) \tag{3.19}$$

Here, $\dot{m}_S$ is the mass flow of the steam, $\dot{m}_L$ is the mass flow of the liquid (plant-based milk), h_S is the enthalpy of the steam at a given pressure, c is the heat capacity, T is the temperature, and the subscripts, *L*, *W*, *F*, and *I* refer to liquid, water, final, and inlet, respectively. This process is quite energy-intensive compared to indirect UHT processing and therefore some manufacturers choose the indirect method.

During indirect UHT treatment, the product is never in direct contact with the heating medium, *e.g.*, steam or hot water. Typically, plate-heat exchangers or tubular heat exchangers are used for this operation. Figure 3.23 shows the layout of a typical plate heat exchanger.

A plate heat exchanger is able to process around 30,000 l/h and consists of several thin metal plates screwed together. The pack of stainless-steel plates is clamped within a frame and is commonly divided into different sections such as pre-heating and final heating sections. The gap between the thin plates is typically 3–6 mm, which ensures a high surface-to-volume ratio and thus rapid heating rates (Kessler, 2002). During operation, the heating medium is pumped through one side of the plate, while the plant-based milk is pumped through the opposite side. The heat is transferred through the metal wall from the heating medium to the milk by conduction and convection. Typically, hot water that is heated by steam is used as a heating medium. The heated water enters the plate-heat exchanger and flows in a

3.6 Thermal Processing Methods

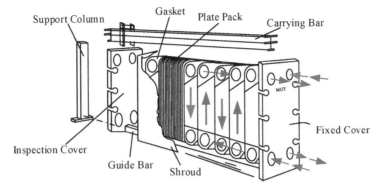

(a) Components of a plate heat exchanger

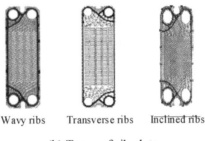

(b) Types of rib plates

Fig. 3.23 A plate heat exchanger can be used to carry out indirect UHT treatment of fluid foods. The food is heated and cooled in a countercurrent flow pattern. (Modified from (Phila et al., 2019) under CC BY 4.0 (http://creativecommons.org/licenses/by/4.0/))

countercurrent direction (constant dT_M) on the opposite side of the fluid and heats it to the desired temperature. The fluid flows through a short holding tube to ensure the necessary holding time and is inserted back into the plate heat exchanger, where it gets cooled by the incoming milk and cooling water to the desired storage temperature (~4 °C). The following equation can be used to calculate the area needed for the heating process in plate heat-exchangers:

$$A = \frac{\dot{Q}}{kT_m} = \frac{\dot{m}c_p \Delta T}{kT_m} \quad (m^2) \tag{3.20}$$

Here, A is the required heat transfer area (m²), $\dot{Q}$ is the heat flow (kJ s^{-1}), k is the heat transfer coefficient (W m^{-2} K^{-1}), T_m is the logarithmic mean temperature difference (°C), $\dot{m}$ is the mass flow rate (kg s^{-1}), c_p is the specific heat capacity (kJ kg^{-1} °C^{-1}), and ΔT is the temperature change (°C) of the product in the heat exchanger. The heat transfer coefficient k can be calculated or is provided by the equipment manufacturer. The logarithmic mean temperature can be calculated using the following equation:

$$T_m = \frac{(T_{i2} - T_{o1}To1) - (T_{o2} - T_{i1})}{\ln\frac{(T_{i2} - T_{o1}To1)}{(T_{o2} - T_{i1})}} \quad (°C) \quad (3.21)$$

Here, T_i and T_o are the inlet and outlet temperatures of the two fluids (*e.g.*, hot water and plant-based milk), and the subscripts 1 and 2 refer to the cool and hot liquid, respectively.

Another concept is the tubular heat exchanger, which employs the shell and tube (tube-in-tube) principle. It is basically one or more tubes placed in a second tube (Fig. 3.24). The heating medium flows through the outer tube and heats the liquid that flows in a countercurrent direction through the inner tubes. The diameter of these tubes is greater than the gaps between the plates in a plate heat exchanger, which means that tubular heat exchangers are usually more suitable for processing liquid foods containing particles or fibers.

Fig. 3.24 Tubular heat exchangers have a single (**a**) or multiple (**b**) tube-in-tube design. The liquid to be processed (*e.g.*, a plant-based milk) flows through the inner tube, whereas the heating or cooling medium circulates through the surrounding outer tube. (Reprinted with permission of Elsevier from (Fellows, 2017))

3.7 Fermentation Methods

Fermentation processes are used to produce some important plant-based foods, including yogurt and cheese analogs. In this section, we briefly review enzymatic and microbial approaches for fermenting plant-derived ingredients.

3.7.1 Enzymatic Fermentation

Some plant-based milks are fermented by enzymes or microorganisms to break down residual starches and improve their texture and flavor profile (thickness, mouthfeel, and sweetness). For instance, oat-based beverages are treated with specialized enzymes to hydrolyze the starch, which decreases their viscosity, increases their sweetness by releasing glucose, and leads to a smoother mouthfeel by removing large particles (Deswal et al., 2014). Similar approaches may be useful for other cereal-based milk analogs that contain relatively high levels of starch compared to protein, such as rice milk. In pulse-based plant-based milks, the starches are commonly not degraded but are separated by centrifugal forces because the raw materials are rich in proteins rather than starches. The enzymes used to break down starches are glucosidases that hydrolyze glycosidic bonds at random locations within the starch molecule (α-amylase), or at the terminal reducing end to produce maltose (β-amylase) or glucose (α-glucosidase) (Gong et al., 2020). The viscosity, mouthfeel, and sweetness of plant-based milks can therefore be controlled by regulating the type and amount of enzymes used to treat the raw materials, as well as the incubation time and temperature.

Different types of reactors can be used to carry out enzymatic conversion processes, with some examples shown in Fig. 3.25. The most simple design is a stirred batch reactor. The enzymes can be directly added to the tank or they can be solubilized in another smaller tank under controlled atmosphere conditions and then added to the reactor. The reaction is then performed under optimum environmental conditions (T, pH, I) until the desired amount of substrate is converted. Other reactor types include immobilized enzyme reactors that minimize the loss of enzymes by retaining the enzymes within the reactor until they lose their activity. Another technique is to utilize membrane filtration to recover the enzymes from the medium and reuse them in a fresh batch.

3.7.2 Microbial Fermentation

Microbial fermentation can be carried out to reduce the content of antinutritional compounds and to decrease the presence of off-flavors. Fermentation is also often used to create yogurt or cheese analogs from plant-based milk products.

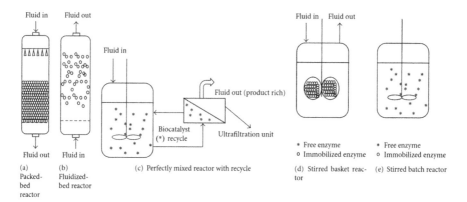

Fig. 3.25 Overview of some important reactor types used for enzymatic conversion processes. The stirred batch reactor is the most common one used (e). (Adapted from (Fernandes, 2010) under CC BY 3.0 (https://creativecommons.org/licenses/by/3.0/))

Common species used for the fermentation of plant-based foods include *Lactobacillus delbrueckii* subsp. *bulgaricus*, *Streptococcus thermophilus*, *Lactobacillus acidophilus*, *Bifidobacterium infantis*, and others (Tangyu et al., 2019). Most of these species are also used to ferment dairy-based milk products but they have been shown to also enhance the quality of plant-based milks. For example, fermenting plant-based milks has been shown to improve their flavor by decreasing beany off-notes, such as n-hexanal and n-hexanol, as well as lowering the content of antinutritional compounds, such as phytic acid and trypsin inhibitors, thereby increasing calcium bioavailability (Tangyu et al., 2019).

Fermentation is commonly carried out in heated stainless-steel tanks at optimum temperatures for the microorganisms (e.g., mesophilic bacteria – optimal growth temperatures of 20–30 °C, thermophilic bacteria – optimal growth temperatures of 40–45 °C) for several hours before the plant-based milk is cooled down again and packed under aseptic conditions. Typically, the starter cultures are obtained from a specialized company in a freeze-dried or frozen form. Companies often offer highly concentrated starter cultures that can directly be used but some food companies prefer to produce their own concentrated starter cultures that they can employ for production. To achieve this, they have to propagate an initial commercial culture and gradually increase its volume until a final production culture is obtained (Bylund, 2015). During the whole propagation process, it is critically important to prevent contamination of the equipment and media. The propagation process involves a series of steps: heat treatment of the medium → cooling to inoculation temperature → inoculation → incubation → cooling of the finished culture → storage of the culture (Bylund, 2015). The starter culture can then be used directly or it can be frozen or freeze-dried for later use.

Prior to adding the starter culture for propagation, a food product that is selected as a medium (such as a milk analog) is usually heat-treated to inactivate any microorganisms or bacteriophages it contains, as they could interfere with the

3.7 Fermentation Methods

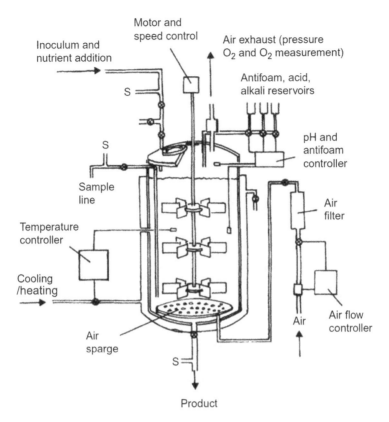

Fig. 3.26 Typical bioreactor setup. Depending on the propagation conditions for the respective starter culture this setup might differ in terms of gas supply (if any), heating, and cooling medium as well as stirrer and gas exhaust. (Reprinted with permission of Elsevier from (Fellows, 2017))

fermentation process. This process can be carried out in the fermentation reactor (Fig. 3.26), *e.g.*, by heating the product at 90 °C for 45 min using a double-wall reactor. Subsequently, the commercial starter culture is transferred into the food product under hygienic conditions. The microorganisms start to propagate at optimum conditions until a desired cell density is obtained. The fermentation process can then be stopped using an appropriate method, *e.g.*, cooling, heating, or pH changes.

3.8 Process Design Examples: Soy, Oat, and Nut Milk

In this chapter, we have discussed several unit operations and equipment that can be used to produce plant-based foods. In this last section, we highlight how these individual elements can be put together into a process using plant-based milks as an example.

As discussed in Chap. 8, plant-based milk analogs can be produced using tissue disruption methods or homogenization methods. In this section, we focus on the tissue disruption methods because they are currently the most widely used. The production of milk analogs using these methods involves the conversion of solid plant materials (such as seeds and grains) into colloidal fluids. The unit operations used for this purpose are fairly similar for different kinds of starting materials: grinding, fermentation, separation, standardization, heat treatment, and homogenization. The overall process is designed to produce a high-quality, reliable, safe, and shelf-stable product. In particular, the final product should have a consistent appearance, texture, mouthfeel, flavor profile, shelf-life, and nutrient composition.

In the remainder of this section, we describe the production of three common plant-based milks: soy, nut, and oat milks. The raw materials used to produce these milk analogs vary in their compositions: soybeans and nuts are rich in proteins and lipids, whereas oats are rich in starch. The processes described should only be considered as representative examples. In practice, the processes used commercially to create milk analogs depend on the nature of the starting material, as well as the preferences of the food manufacturer. An overview of process adaptions can be found in Aydar et al. (2020).

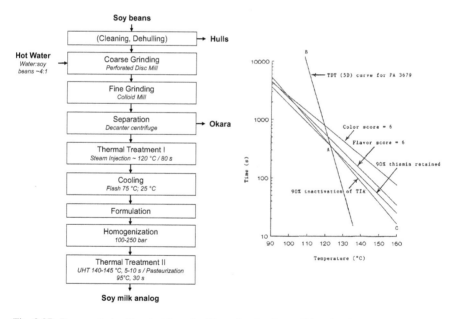

Fig. 3.27 Process design for plant-based milk production (soymilk) and a time-temperature diagram for a 5-log reduction in spores (TDT) and 90% trypsin inhibitor reduction (TIA). Auxiliary equipment and streams include plate-heat exchangers (hot water production from steam), vacuum pumps (flash evaporation), and ice water (cooling liquid). (Adapted from (Prabhakaran & Perera, 2006; Poliseli-Scopel et al., 2012) and from (Kwok et al., 2002) with Copyright 2002 American Chemical Society)

Figure 3.27 shows the process for soymilk production. An additional microbial fermentation step could be added to this process to enhance the flavor profile and decrease the antinutritional compounds. In the process shown, the soymilk is subjected to two thermal treatments: the first one is designed to inactivate trypsin inhibitors, while the second one is designed to inactivate microorganisms (including spores). It should be noted that soymilks can also be produced using a single heat treatment, *e.g.*, 143 °C for 60 s (Yuan et al., 2008). The UHT processes are often designed to give a 9 log reduction in spore counts, but regulations vary from country to country, *e.g.*, a 12 log reduction of *Clostridium botulinum* spores is required in some countries (Bylund, 2015). Processing windows for a 5D log reduction of *Clostridium sporogenes* are shown in Fig. 3.27.

Differences in the compositions of the raw materials used to produce soymilk (protein- and lipid-rich) and oat milk (starch-rich) lead to differences in the processing operations used to manufacture them. An important goal in soymilk production is to solubilize and disperse the proteins and oil bodies, whereas in oat milk production it is to hydrolyze the starch granules (Peterson, 2011). As discussed earlier, the starch granules in oat milks are often converted into sugars using enzymes, which improves the stability, texture, mouthfeel, and flavor of the final product. Moreover, it may be necessary to incorporate emulsified oils into the formulation because oats are typically low in lipids. The starch granules may be removed from raw plant materials that contain significant amounts of soluble protein (*e.g.*, peas) using separation processes like hydrocyclones or decanters. The overall process designs for manufacturing oat- and nut-based milk analogs are shown in Fig. 3.28.

The overall processing equipment required to produce milk analogs from oats and soybeans is quite similar. In both cases, the raw materials are dispersed, milled, centrifuged, heat-treated, and homogenized. More details about the processes used to create plant-based milks are given in Chap. 8.

3.9 Conclusions and Future Directions

The manufacture of plant-based foods often involves the use of unit operations and equipment already widely employed in the food industry, such as mixers, mills, grinders, homogenizers, extruders, decanters, centrifuges, filters, and heat-exchangers. This is advantageous because it means that the equipment and knowledge required to create these products are already well-established. In some cases, however, innovative equipment is required to create plant-based food products that accurately simulate the unique characteristics of some types of animal-based food, such as meat and seafood. For instance, the shear cell technology has been specifically developed to create fibrous structures from proteins, which can be useful in creating meat and seafood analogs. In addition, 3D printing is a versatile technology that can be used to create complex food matrices from plant-derived ingredients,

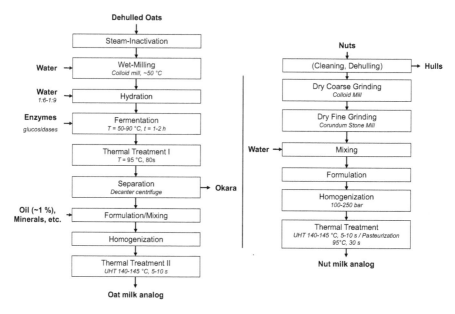

Fig. 3.28 Process layout of plant-based milk production: oat and nut milk. (Adapted and modified from (Alfa Laval, 2021))

such as proteins, polysaccharides, and lipids. Another important innovation will be to combine an understanding of the molecular/physicochemical behavior of plant-derived ingredients with processing technologies to more accurately simulate the structure and behavior of animal-based products.

References

Abdul Latif, A. A., Lau, K. K., Low, S. C., & Azeem, B. (2021). Multicomponent spiral wound membrane separation model for CO2 removal from natural gas. *Membranes, 11*(9), 654. https://doi.org/10.3390/membranes11090654

Alfa Laval, A. B., (Lund, Sweden). (2021). *Plant-based drink production.* Retrieved from: https://www.alfalaval.com/industries/food-dairy-beverage/beverage-processing/plant-based-drink-production/. Accessed Oct 2021.

Ansari, M., Alabbas, A., Hatzikiriakos, S. G., & Mitsoulis, E. (2010). Entry flow of polyethylene melts in tapered dies. *International Polymer Processing, 25*(4), 287–296. https://doi.org/10.3139/217.2360

Ardakani, H. A., Mitsoulis, E., & Hatzikiriakos, S. G. (2013). A simple improved mathematical model for polytetrafluoroethylene (PTFE) paste extrusion. *Chemical Engineering Science, 89*, 216–222. https://doi.org/10.1016/j.ces.2012.11.040

Aydar, E. F., Tutuncu, S., & Ozcelik, B. (2020). Plant-based milk substitutes: Bioactive compounds, conventional and novel processes, bioavailability studies, and health effects. *Journal of Functional Foods, 70*, 103975. https://doi.org/10.1016/j.jff.2020.103975

Bai, L., Huan, S., Rojas, O. J., & McClements, D. J. (2021). Recent innovations in emulsion science and technology for food applications. *Journal of Agricultural and Food Chemistry, 69*(32), 8944–8963. https://doi.org/10.1021/acs.jafc.1c01877

Barnes, H. A. (1989). Shear-thickening ("Dilatancy") in suspensions of nonaggregating solid particles dispersed in Newtonian liquids. *Journal of Rheology, 33*(2), 329–366. https://doi.org/10.1122/1.550017

Beck, S. M., Knoerzer, K., & Arcot, J. (2017). Effect of low moisture extrusion on a pea protein isolate's expansion, solubility, molecular weight distribution and secondary structure as determined by Fourier Transform Infrared Spectroscopy (FTIR). *Journal of Food Engineering, 214*, 166–174. https://doi.org/10.1016/j.jfoodeng.2017.06.037

Berk, Z. (2013). *Food process engineering and technology* (2nd ed.). Academic.

Bouvier, J.-M., & Campanella, O. H. (2014). *Extrusion processing technology: Food and non-food biomaterials* (1st ed.). Wiley-Blackwell.

Boye, J. I., Aksay, S., Roufik, S., Ribéreau, S., Mondor, M., Farnworth, E., & Rajamohamed, S. H. (2010). Comparison of the functional properties of pea, chickpea and lentil protein concentrates processed using ultrafiltration and isoelectric precipitation techniques. *Food Research International, 43*(2), 537–546. https://doi.org/10.1016/j.foodres.2009.07.021

Bylund, G. (2015). *Dairy processing handbook* (2nd ed.). Tetra Pak.

Caporgno, M. P., Böcker, L., Müssner, C., Stirnemann, E., Haberkorn, I., Adelmann, H., ... Mathys, A. (2020). Extruded meat analogues based on yellow, heterotrophically cultivated Auxenochlorella protothecoides microalgae. *Innovative Food Science & Emerging Technologies, 59*, 102275. https://doi.org/10.1016/j.ifset.2019.102275

Chen, Y., Xu, Z., Zhang, C., Kong, X., & Hua, Y. (2014). Heat-induced inactivation mechanisms of Kunitz trypsin inhibitor and Bowman-Birk inhibitor in soymilk processing. *Food Chemistry, 154*, 108–116. https://doi.org/10.1016/j.foodchem.2013.12.092

Chen, Y., Zhang, M., & Bhandari, B. (2021). 3D printing of steak-like foods based on textured soybean protein. *Food, 10*(9), 2011. https://doi.org/10.3390/foods10092011

Cheng, J. B., Wang, J. Q., Bu, D. P., Liu, G. L., Zhang, C. G., Wei, H. Y., ... Wang, J. Z. (2008). Factors affecting the Lactoferrin concentration in bovine Milk. *Journal of Dairy Science, 91*(3), 970–976. https://doi.org/10.3168/jds.2007-0689

Cheyne, A., Barnes, J., & Wilson, D. I. (2005). Extrusion behaviour of cohesive potato starch pastes: I. Rheological characterisation. *Journal of Food Engineering, 66*(1), 1–12. https://doi.org/10.1016/j.jfoodeng.2004.02.028

Comuzzo, P., & Calligaris, S. (2019). Potential applications of high pressure homogenization in winemaking: A review. *Beverages, 5*(3), 56. https://doi.org/10.3390/beverages5030056

Cornet, S. H. V., Snel, S. J. E., Schreuders, F. K. G., van der Sman, R. G. M., Beyrer, M., & van der Goot, A. J. (2021). Thermo-mechanical processing of plant proteins using shear cell and high-moisture extrusion cooking. *Critical Reviews in Food Science and Nutrition, 0*(0), 1–18. https://doi.org/10.1080/10408398.2020.1864618

Cosson, A., Oliveira Correia, L., Descamps, N., Saint-Eve, A., & Souchon, I. (2022). Identification and characterization of the main peptides in pea protein isolates using ultra high-performance liquid chromatography coupled with mass spectrometry and bioinformatics tools. *Food Chemistry, 367*, 130747. https://doi.org/10.1016/j.foodchem.2021.130747

Cotacallapa-Sucapuca, M., Vega, E. N., Maieves, H. A., Berrios, J. D. J., Morales, P., Fernández-Ruiz, V., & Cámara, M. (2021). Extrusion process as an alternative to improve pulses products consumption. A review. *Foods, 10*(5), 1096. https://doi.org/10.3390/foods10051096

Dai, X., Sharma, M., & Chen, J. (Eds.). (2021). *Fungi in sustainable food production*. Springer. https://doi.org/10.1007/978-3-030-64406-2

De Marchi, M., Costa, A., Pozza, M., Goi, A., & Manuelian, C. L. (2021). Detailed characterization of plant-based burgers. *Scientific Reports, 11*(1), 2049. https://doi.org/10.1038/s41598-021-81684-9

Dekkers, B. L., Nikiforidis, C. V., & van der Goot, A. J. (2016). Shear-induced fibrous structure formation from a pectin/SPI blend. *Innovative Food Science & Emerging Technologies, 36*, 193–200. https://doi.org/10.1016/j.ifset.2016.07.003

Dekkers, B. L., Boom, R. M., & van der Goot, A. J. (2018). Structuring processes for meat analogues. *Trends in Food Science & Technology, 81*, 25–36. https://doi.org/10.1016/j.tifs.2018.08.011

Derakhshanfar, S., Mbeleck, R., Xu, K., Zhang, X., Zhong, W., & Xing, M. (2018). 3D bioprinting for biomedical devices and tissue engineering: A review of recent trends and advances. *Bioactive Materials, 3*(2), 144–156. https://doi.org/10.1016/j.bioactmat.2017.11.008

Deswal, A., Deora, N. S., & Mishra, H. N. (2014). Optimization of enzymatic production process of Oat milk using response surface methodology. *Food and Bioprocess Technology, 7*(2), 610–618. https://doi.org/10.1007/s11947-013-1144-2

Dick, A., Bhandari, B., & Prakash, S. (2019). 3D printing of meat. *Meat Science, 153*, 35–44. https://doi.org/10.1016/j.meatsci.2019.03.005

Durango-Cogollo, M., Garcia-Bravo, J., Newell, B., & Gonzalez-Mancera, A. (2020). CFD modeling of Hydrocyclones—A study of efficiency of hydrodynamic reservoirs. *Fluids, 5*(3), 118. https://doi.org/10.3390/fluids5030118

Evans, M., Ratcliffe, I., & Williams, P. A. (2013). Emulsion stabilisation using polysaccharide–protein complexes. *Current Opinion in Colloid & Interface Science, 18*(4), 272–282. https://doi.org/10.1016/j.cocis.2013.04.004

Fang, Y. (2021). Chapter 5 – Mixed hydrocolloid systems. In G. O. Phillips & P. A. Williams (Eds.), *Handbook of hydrocolloids* (3rd ed., pp. 125–155). Woodhead Publishing. https://doi.org/10.1016/B978-0-12-820104-6.00018-8

Featherstone, S. (Ed.). (2016). 1 – Canning of vegetables. In *A complete course in canning and related processes (fourteenth edition)* (pp. 3–84). Woodhead Publishing https://doi.org/10.1016/B978-0-85709-679-1.00001-5.

Fellows, P. (2017). *Food processing technology: Principles and practice*. Woodhead Publishing.

Ferawati, F., Zahari, I., Barman, M., Hefni, M., Ahlström, C., Witthöft, C., & Östbring, K. (2021). High-moisture Meat analogues produced from yellow pea and Faba bean protein isolates/concentrate: Effect of raw material composition and extrusion operating parameters on texture properties. *Food, 10*(4), 843. https://doi.org/10.3390/foods10040843

Fernandes, P. (2010). Enzymes in food processing: A condensed overview on strategies for better biocatalysts. *Enzyme Research, 2010*, 20201208031233.

Gan, M., Gopinathan, N., Jia, X., & Williams, R. A. (2004). Predicting packing characteristics of particles of arbitrary shapes. *Kona Powder and Particle Journal, 22*, 82–93. https://doi.org/10.14356/kona.2004012

Gienau, T., Brüß, U., Kraume, M., & Rosenberger, S. (2018). Nutrient recovery from anaerobic sludge by membrane filtration: Pilot tests at a 2.5 MWe biogas plant. *International Journal of Recycling of Organic Waste in Agriculture, 7*(4), 325–334. https://doi.org/10.1007/s40093-018-0218-6

Gilani, G. S., Xiao, C. W., & Cockell, K. A. (2012). Impact of antinutritional factors in food proteins on the digestibility of protein and the bioavailability of amino acids and on protein quality. *British Journal of Nutrition, 108*(S2), S315–S332. https://doi.org/10.1017/S0007114512002371

Giri, S. K., & Mangaraj, S. (2014). Soymilk concentration by ultrafiltration: Effects of pore size and transmembrane pressure on filtration performance. *International Journal of Food Science & Technology, 49*(3), 666–672. https://doi.org/10.1111/ijfs.12348

Glantz, M., Devold, T. G., Vegarud, G. E., Lindmark Månsson, H., Stålhammar, H., & Paulsson, M. (2010). Importance of casein micelle size and milk composition for milk gelation. *Journal of Dairy Science, 93*(4), 1444–1451. https://doi.org/10.3168/jds.2009-2856

Gökmen, V., Savaş Bahçeci, K., Serpen, A., & Acar, J. (2005). Study of lipoxygenase and peroxidase as blanching indicator enzymes in peas: Change of enzyme activity, ascorbic acid and chlorophylls during frozen storage. *LWT – Food Science and Technology, 38*(8), 903–908. https://doi.org/10.1016/j.lwt.2004.06.018

References

Gong, L., Feng, D., Wang, T., Ren, Y., Liu, Y., & Wang, J. (2020). Inhibitors of α-amylase and α-glucosidase: Potential linkage for whole cereal foods on prevention of hyperglycemia. *Food Science & Nutrition, 8*(12), 6320–6337. https://doi.org/10.1002/fsn3.1987

Gonzalez-Gutierrez, J., & Scanlon, M. G. (2018). Chapter 5 – Rheology and mechanical properties of fats. In A. G. Marangoni (Ed.), *Structure-function analysis of edible fats* (2nd ed., pp. 119–168). AOCS Press. https://doi.org/10.1016/B978-0-12-814041-3.00005-8

Grabowska, K. J., Zhu, S., Dekkers, B. L., de Ruijter, N. C. A., Gieteling, J., & van der Goot, A. J. (2016). Shear-induced structuring as a tool to make anisotropic materials using soy protein concentrate. *Journal of Food Engineering, 188*, 77–86. https://doi.org/10.1016/j.jfoodeng.2016.05.010

Grossmann, L., & Weiss, J. (2021). Alternative protein sources as technofunctional food ingredients. *Annual Review of Food Science and Technology, 12*(1), 93–117. https://doi.org/10.1146/annurev-food-062520-093642

Grossmann, L., Wefers, D., Bunzel, M., Weiss, J., & Zeeb, B. (2017). Accessibility of transglutaminase to induce protein crosslinking in gelled food matrices – Influence of network structure. *LWT – Food Science and Technology, 75*, 271–278. https://doi.org/10.1016/j.lwt.2016.09.005

Guo, J., Yang, X.-Q., He, X.-T., Wu, N.-N., Wang, J.-M., Gu, W., & Zhang, Y.-Y. (2012). Limited aggregation behavior of β-Conglycinin and its terminating effect on Glycinin aggregation during heating at pH 7.0. *Journal of Agricultural and Food Chemistry, 60*(14), 3782–3791. https://doi.org/10.1021/jf300409y

Gupta, A., & Yan, D. (Eds.). (2016). Chapter 15 – Solid liquid separation – Filtration. In *Mineral processing design and operations* (2nd ed., pp. 507–561). Elsevier. https://doi.org/10.1016/B978-0-444-63589-1.00015-0.

Habeych, E., Dekkers, B., van der Goot, A. J., & Boom, R. (2008). Starch–zein blends formed by shear flow. *Chemical Engineering Science, 63*(21), 5229–5238. https://doi.org/10.1016/j.ces.2008.07.008

Hakami, M. W., Alkhudhiri, A., Al-Batty, S., Zacharof, M.-P., Maddy, J., & Hilal, N. (2020). Ceramic microfiltration membranes in wastewater treatment: Filtration behavior, fouling and prevention. *Membranes, 10*(9), 248. https://doi.org/10.3390/membranes10090248

Haller, N., Greßlinger, A. S., & Kulozik, U. (2021). Separation of aggregated β-lactoglobulin with optimised yield in a decanter centrifuge. *International Dairy Journal, 114*, 104918. https://doi.org/10.1016/j.idairyj.2020.104918

Hayward, S., Cilliers, T., & Swart, P. (2017). Lipoxygenases: From isolation to application. *Comprehensive Reviews in Food Science and Food Safety, 16*(1), 199–211. https://doi.org/10.1111/1541-4337.12239

Head, D., Cenkowski, S., Arntfield, S., & Henderson, K. (2011). Storage stability of oat groats processed commercially and with superheated steam. *LWT – Food Science and Technology, 44*(1), 261–268. https://doi.org/10.1016/j.lwt.2010.05.022

Herrmann, M. (2009). *Method for producing a soy milk*. Patent number: US20090317533A1. Retrieved from: https://patents.google.com/patent/US20090317533A1/en?oq=soybean+milk+perforated+disc+mill. Accessed Aug 2021.

Ingredion. (2020). *Ingredion VITESSENCETM Pulse 1550 Protein*. Retrieved from: https://www.ingredion.com/content/dam/ingredion/technical-documents/na/VITESSENCE%20Pulse%201550%20%20%2037403E00%20%20%20Nutritional.pdf. Accessed Oct 2021.

Irmscher, S. B., Rühl, S., Herrmann, K., Gibis, M., Kohlus, R., & Weiss, J. (2015). Determination of process-structure relationship in the manufacturing of Meat batter using vane pump-grinder systems. *Food and Bioprocess Technology, 8*(7), 1512–1523. https://doi.org/10.1007/s11947-015-1514-z

Jourdain, L., Leser, M. E., Schmitt, C., Michel, M., & Dickinson, E. (2008). Stability of emulsions containing sodium caseinate and dextran sulfate: Relationship to complexation in solution. *Food Hydrocolloids, 22*(4), 647–659. https://doi.org/10.1016/j.foodhyd.2007.01.007

Jung, J., & Wicker, L. (2012). Laccase mediated conjugation of sugar beet pectin and the effect on emulsion stability. *Food Hydrocolloids, 28*(1), 168–173. https://doi.org/10.1016/j.foodhyd.2011.12.021

Kendler, C., Duchardt, A., Karbstein, H. P., & Emin, M. A. (2021). Effect of oil content and oil addition point on the extrusion processing of wheat gluten-based Meat analogues. *Food, 10*(4), 697. https://doi.org/10.3390/foods10040697

Kessler, H. G. (2002). *Food and bio process engineering – Dairy technology*. Kessler, N. Verlag A. Kessler.

Ko, H. J., Wen, Y., Choi, J. H., Park, B. R., Kim, H. W., & Park, H. J. (2021). Meat analog production through artificial muscle fiber insertion using coaxial nozzle-assisted three-dimensional food printing. *Food Hydrocolloids, 120*, 106898. https://doi.org/10.1016/j.foodhyd.2021.106898

Kotobuki, M., Gu, Q., Zhang, L., & Wang, J. (2021). Ceramic-polymer composite membranes for water and wastewater treatment: Bridging the big gap between ceramics and polymers. *Molecules, 26*(11), 3331. https://doi.org/10.3390/molecules26113331

Krintiras, G. A., Göbel, J., van der Goot, A. J., & Stefanidis, G. D. (2015). Production of structured soy-based meat analogues using simple shear and heat in a Couette cell. *Journal of Food Engineering, 160*, 34–41. https://doi.org/10.1016/j.jfoodeng.2015.02.015

Krintiras, G. A., Gadea Diaz, J., van der Goot, A. J., Stankiewicz, A. I., & Stefanidis, G. D. (2016). On the use of the Couette cell technology for large scale production of textured soy-based meat replacers. *Journal of Food Engineering, 169*, 205–213. https://doi.org/10.1016/j.jfoodeng.2015.08.021

Kwok, K.-C., Liang, H.-H., & Niranjan, K. (2002). Optimizing conditions for thermal processes of Soy Milk. *Journal of Agricultural and Food Chemistry, 50*(17), 4834–4838. https://doi.org/10.1021/jf020182b

Kyriakopoulou, K., Dekkers, B. L., & van der Goot, A. J. (2018). Plant-based Meat analogues. In *Sustainable meat production and processing* (pp. 103–126). https://doi.org/10.1016/B978-0-12-814874-7.00006-7.

Kyriakopoulou, K., Dekkers, B., & van der Goot, A. J. (2019). Chapter 6 – Plant-based Meat analogues. In C. M. Galanakis (Ed.), *Sustainable meat production and processing* (pp. 103–126). Academic. https://doi.org/10.1016/B978-0-12-814874-7.00006-7

Lan, Q., Lin, Z., Dong, H., Wu, D., Lin, D., Qin, W., ... Zhang, Q. (2021). Influence of okara with varying particle sizes on the gelling, rheological, and microstructural properties of glucono-δ-lactone-induced tofu. *Journal of Food Science and Technology, 58*(2), 520–531. https://doi.org/10.1007/s13197-020-04563-7

Le-Bail, A., Maniglia, B. C., & Le-Bail, P. (2020). Recent advances and future perspective in additive manufacturing of foods based on 3D printing. *Current Opinion in Food Science, 35*, 54–64. https://doi.org/10.1016/j.cofs.2020.01.009

Li, B., Qiao, M., & Lu, F. (2012). Composition, nutrition, and utilization of Okara (soybean residue). *Food Reviews International, 28*(3), 231–252. https://doi.org/10.1080/87559129.2011.595023

Lipton, J. I., Cutler, M., Nigl, F., Cohen, D., & Lipson, H. (2015). Additive manufacturing for the food industry. *Trends in Food Science & Technology, 43*(1), 114–123. https://doi.org/10.1016/j.tifs.2015.02.004

Manski, J. M., van der Goot, A. J., & Boom, R. M. (2007). Formation of fibrous materials from dense calcium Caseinate dispersions. *Biomacromolecules, 8*(4), 1271–1279. https://doi.org/10.1021/bm061008p

Maskan, M., & Altan, A. (Eds.). (2012). *Advances in food extrusion technology*. CRC Press. https://doi.org/10.1201/b11286

McClements, D. J. (2006). Non-covalent interactions between proteins and polysaccharides. *Biotechnology Advances, 24*(6), 621–625. https://doi.org/10.1016/j.biotechadv.2006.07.003

McClements, D. J. (2015). *Food emulsions: Principles, practices, and techniques, Third Edition*. CRC Press.

McClements, D. J., & Grossmann, L. (2021). The science of plant-based foods: Constructing next-generation meat, fish, milk, and egg analogs. *Comprehensive Reviews in Food Science and Food Safety, 20*(4), 1–52. https://doi.org/10.1111/1541-4337.12771

McClements, D. J., Weiss, J., Kinchla, A. J., Nolden, A. A., & Grossmann, L. (2021). Methods for testing the quality attributes of plant-based foods: Meat- and processed-meat analogs. *Food, 10*(2), 260. https://doi.org/10.3390/foods10020260

Menesklou, P., Sinn, T., Nirschl, H., & Gleiss, M. (2021). Scale-up of decanter centrifuges for the particle separation and mechanical dewatering in the minerals processing industry by means of a numerical process model. *Minerals, 11*(2), 229. https://doi.org/10.3390/min11020229

Michels, M. H. A., van der Goot, A. J., Norsker, N.-H., & Wijffels, R. H. (2010). Effects of shear stress on the microalgae Chaetoceros muelleri. *Bioprocess and Biosystems Engineering, 33*(8), 921–927. https://doi.org/10.1007/s00449-010-0415-9

Montville, T. J., Matthews, K. R., & Kniel, K. E. (2012). *Food microbiology: An introduction* (3rd ed.). ASM Press.

Mosibo, O. K., Ferrentino, G., Alam, M. R., Morozova, K., & Scampicchio, M. (2020). Extrusion cooking of protein-based products: Potentials and challenges. Critical Reviews in Food Science and Nutrition, 0(0), 1–35. https://doi.org/10.1080/10408398.2020.1854674.

Palanisamy, M., Franke, K., Berger, R. G., Heinz, V., & Töpfl, S. (2019). High moisture extrusion of lupin protein: Influence of extrusion parameters on extruder responses and product properties. *Journal of the Science of Food and Agriculture, 99*(5), 2175–2185. https://doi.org/10.1002/jsfa.9410

Peighambardoust, S. H., van der Goot, A. J., Hamer, R. J., & Boom, R. M. (2004). A new method to study simple shear processing of wheat gluten-starch mixtures. *Cereal Chemistry, 81*(6), 714–721. https://doi.org/10.1094/CCHEM.2004.81.6.714

Pelgrom, P. J. M., Vissers, A. M., Boom, R. M., & Schutyser, M. A. I. (2013). Dry fractionation for production of functional pea protein concentrates. *Food Research International, 53*(1), 232–239. https://doi.org/10.1016/j.foodres.2013.05.004

Peterson, D. M. (2011). Chapter 8 – Storage proteins. In F. H. Webster & P. J. Wood (Eds.), *Oats (second edition)* (pp. 123–142). AACC International Press. https://doi.org/10.1016/B978-1-891127-64-9.50013-0

Phila, A., Thianpong, C., & Eiamsa-ard, S. (2019). Influence of geometric parameters of alternate axis twisted baffles on the local heat transfer distribution and pressure drop in a Rectangular Channel using a transient liquid crystal technique. *Energies, 12*(12), 2341. https://doi.org/10.3390/en12122341

Pietsch, V. L., Emin, M. A., & Schuchmann, H. P. (2017). Process conditions influencing wheat gluten polymerization during high moisture extrusion of meat analog products. *Journal of Food Engineering, 198*, 28–35. https://doi.org/10.1016/j.jfoodeng.2016.10.027

Pietsch, V. L., Bühler, J. M., Karbstein, H. P., & Emin, M. A. (2019). High moisture extrusion of Soy protein concentrate: Influence of thermomechanical treatment on protein-protein interactions and rheological properties. *Journal of Food Engineering, 251*, 11–18. https://doi.org/10.1016/j.jfoodeng.2019.01.001

Poliseli-Scopel, F. H., Hernández-Herrero, M., Guamis, B., & Ferragut, V. (2012). Comparison of ultra high pressure homogenization and conventional thermal treatments on the microbiological, physical and chemical quality of soymilk. *LWT – Food Science and Technology, 46*(1), 42–48. https://doi.org/10.1016/j.lwt.2011.11.004

Popova, A., & Mihaylova, D. (2019). Antinutrients in plant-based foods: A review. *The Open Biotechnology Journal, 13*(1). https://doi.org/10.2174/1874070701913010068

Prabhakaran, M. P., & Perera, C. O. (2006). Effect of extraction methods and UHT treatment conditions on the level of isoflavones during soymilk manufacture. *Food Chemistry, 99*(2), 231–237. https://doi.org/10.1016/j.foodchem.2005.06.055

PS Prozesstechnik. (2021). *Pressure driven membrane processes calculations*. Retrieved from: https://www.ps-prozesstechnik.com/images/membrane/calculations_membrane_processes.pdf. Accessed Oct 2021.

Redefine Meat. (2020). *What the meat? The Israeli invention printing plant based protein.*

Riaz, M. N. (2000). *Extruders in food applications* (1st ed.). Taylor & Francis.

Richardson, P. (2001). *Thermal technologies in food processing.* Taylor & Francis.

Rust, R. E., & Knipe, C. L. (2014). Processing equipment | mixing and cutting equipment. In M. Dikeman & C. Devine (Eds.), *Encyclopedia of meat sciences* (2nd ed., pp. 126–130). Academic. https://doi.org/10.1016/B978-0-12-384731-7.00224-5

Samard, S., Gu, B.-Y., & Ryu, G.-H. (2019). Effects of extrusion types, screw speed and addition of wheat gluten on physicochemical characteristics and cooking stability of meat analogues. *Journal of the Science of Food and Agriculture, 99*(11), 4922–4931. https://doi.org/10.1002/jsfa.9722

Sánchez, V. E., Bartholomai, G. B., & Pilosof, A. M. R. (1995). Rheological properties of food gums as related to their water binding capacity and to soy protein interaction. *LWT – Food Science and Technology, 28*(4), 380–385. https://doi.org/10.1016/0023-6438(95)90021-7

Sandoval Murillo, J. L., Osen, R., Hiermaier, S., & Ganzenmüller, G. (2019). Towards understanding the mechanism of fibrous texture formation during high-moisture extrusion of meat substitutes. *Journal of Food Engineering, 242*, 8–20. https://doi.org/10.1016/j.jfoodeng.2018.08.009

Scholten, E., Visser, J. E., Sagis, L. M. C., & van der Linden, E. (2004). Ultralow interfacial tensions in an aqueous phase-separated gelatin/dextran and gelatin/gum Arabic system: A comparison. *Langmuir, 20*(6), 2292–2297. https://doi.org/10.1021/la0351919

Schreuders, F. K. G., Dekkers, B. L., Bodnár, I., Erni, P., Boom, R. M., & van der Goot, A. J. (2019). Comparing structuring potential of pea and soy protein with gluten for meat analogue preparation. *Journal of Food Engineering, 261*, 32–39. https://doi.org/10.1016/j.jfoodeng.2019.04.022

Shi, F., Kojovic, T., Esterle, J. S., & David, D. (2003). An energy-based model for swing hammer mills. *International Journal of Mineral Processing, 71*(1), 147–166. https://doi.org/10.1016/S0301-7516(03)00035-8

Sun, J., Zhou, W., Yan, L., Huang, D., & Lin, L. (2018). Extrusion-based food printing for digitalized food design and nutrition control. *Journal of Food Engineering, 220*, 1–11. https://doi.org/10.1016/j.jfoodeng.2017.02.028

Swallah, M. S., Fan, H., Wang, S., Yu, H., & Piao, C. (2021). Prebiotic impacts of soybean residue (Okara) on Eubiosis/Dysbiosis condition of the gut and the possible effects on liver and kidney functions. *Molecules, 26*(2), 326. https://doi.org/10.3390/molecules26020326

Tan, C., & McClements, D. J. (2021). Application of advanced emulsion technology in the food industry: A review and critical evaluation. *Food, 10*(4), 812. https://doi.org/10.3390/foods10040812

Tangyu, M., Muller, J., Bolten, C. J., & Wittmann, C. (2019). Fermentation of plant-based milk alternatives for improved flavour and nutritional value. *Applied Microbiology and Biotechnology, 103*(23), 9263–9275. https://doi.org/10.1007/s00253-019-10175-9

Tolstoguzov, V. B. (1991). Functional properties of food proteins and role of protein-polysaccharide interaction. *Food Hydrocolloids, 4*(6), 429–468. https://doi.org/10.1016/S0268-005X(09)80196-3

Vergnes, B. (2021). Average shear rates in the screw elements of a Corotating twin-screw extruder. *Polymers, 13*(2), 304. https://doi.org/10.3390/polym13020304

Wagner, J. (2001). *Membrane filtration handbook practical tips and hints* (2nd ed.). Osmonics, Inc..

Weiss, J., Gibis, M., Schuh, V., & Salminen, H. (2010). Advances in ingredient and processing systems for meat and meat products. *Meat Science, 86*(1), 196–213. https://doi.org/10.1016/j.meatsci.2010.05.008

Weiss, J., Salminen, H., Moll, P., & Schmitt, C. (2019). Use of molecular interactions and mesoscopic scale transitions to modulate protein-polysaccharide structures. *Advances in Colloid and Interface Science, 271*, 101987. https://doi.org/10.1016/j.cis.2019.07.008

Wen, Y., Che, Q. T., Kim, H. W., & Park, H. J. (2021). Potato starch altered the rheological, printing, and melting properties of 3D-printable fat analogs based on inulin emulsion-filled gels. *Carbohydrate Polymers, 269*, 118285. https://doi.org/10.1016/j.carbpol.2021.118285

References

Wilms, P., Daffner, K., Kern, C., Gras, S. L., Schutyser, M. A. I., & Kohlus, R. (2021). Formulation engineering of food systems for 3D-printing applications – A review. *Food Research International, 148*, 110585. https://doi.org/10.1016/j.foodres.2021.110585

Wittek, P., Zeiler, N., Karbstein, H. P., & Emin, M. A. (2021). High moisture extrusion of Soy protein: Investigations on the formation of anisotropic product structure. *Food, 10*(1), 102. https://doi.org/10.3390/foods10010102

Xu, X., Sun, Q., & McClements, D. J. (2020). Effects of anionic polysaccharides on the digestion of fish oil-in-water emulsions stabilized by hydrolyzed rice glutelin. *Food Research International, 127*, 108768. https://doi.org/10.1016/j.foodres.2019.108768

Yuan, S., Chang, S. K. C., Liu, Z., & Xu, B. (2008). Elimination of trypsin inhibitor activity and beany flavor in Soy milk by consecutive blanching and ultrahigh-temperature (UHT) processing. *Journal of Agricultural and Food Chemistry, 56*(17), 7957–7963. https://doi.org/10.1021/jf801039h

Zahari, I., Ferawati, F., Helstad, A., Ahlström, C., Östbring, K., Rayner, M., & Purhagen, J. K. (2020). Development of high-moisture meat analogues with hemp and soy protein using extrusion cooking. *Food, 9*(6), 772. https://doi.org/10.3390/foods9060772

Chapter 4
Physicochemical and Sensory Properties of Plant-Based Foods

4.1 Introduction

The physicochemical properties of plant-based foods, such as their optical, rheological, fluid holding, retention/release, and stability attributes, impact their processing, storage, preparation, consumption, digestion and other quality attributes. Consequently, it is important to understand the major factors contributing to the physicochemical attributes of these products. Plant-based foods exhibit a broad spectrum of properties, ranging from low viscosity fluids (milk analogs), to high viscosity fluids (heavy cream or mayonnaise analogs), to viscoelastic solids (meat, fish, egg, or cheese analogs). These products are compositionally and structurally complex materials containing many different ingredients that interact with each other through a variety of molecular and colloidal interactions (McClements & Grossmann, 2021). In this chapter, we present the physicochemical principles underlying the optical, rheological, retention/release, and stability properties of plant-based foods. The most important functional constituents in most of these products are biopolymers (such as proteins and polysaccharides) and colloids (such as fat droplets, fat crystals, protein aggregates, starch granules, air bubbles or ice crystals). For this reason, we give special emphasis to the presentation of mathematical models that can be used to describe and predict the behavior of biopolymer and colloidal materials. We also discuss the sensory attributes of plant-based foods in this chapter because they are partly determined by their physicochemical properties, as well as having a major impact on consumer acceptance and liking. More information about the physicochemical and sensory properties of specific plant-based foods is given in the chapters on meat, seafood, egg and dairy analogs. An improved understanding of the fundamental factors impacting the physicochemical properties of plant-based foods will help food manufacturers to create higher quality products.

4.2 Appearance

The look of a plant-based food product is usually the first sensory impression that a consumer receives and uses to decide on its acceptability and desirability. Consequently, it is important for food manufacturers to create plant-based products with appearances that meet consumer expectations for the animal-based product they are designed to replace. For instance, a milk analog should have a creamy white appearance, a scrambled egg analog should have a creamy yellow appearance, and a grilled beef steak analog should have a dark brown crust and regions of pinky-brown (lean) and whitish-yellow (adipose tissue) inside. In this section, we describe the physicochemical principles that can be used to understand and predict the optical properties of plant-based food products.

4.2.1 Factors Affecting Appearance

In general, the overall appearance of plant-based foods is governed by a number of factors related to their interactions with light waves and the human eye (Hutchings, 1999). As mentioned earlier, many plant based foods consist of colloidal particles and/or polymers dispersed within an aqueous medium and so their optical properties can be described by mathematical models originally developed for colloidal materials (McClements, 2002a).

4.2.1.1 Spatial Uniformity

Depending on the product, plant-based foods may be expected to have either a uniform or non-uniform appearance. For instance, a milk analog is expected to have a uniform milky appearance throughout, whereas a burger analog is expected to have a dark brown crust and a light brown interior that contains visibly discernable pieces. The resolution of the human eye is around 200 µm, which means that it cannot discern objects smaller than this size (Hutchings, 1999). Consequently, for a product to appear uniform it should have heterogeneities below this size, whereas to appear non-uniform it should have heterogeneities above this size. Moreover, for a product that is meant to have a non-uniform appearance it is important that the food manufacturer creates regions that have the correct size, shape, spatial distribution, color, and opacity. The heterogeneity of a product is often controlled by adding structural components with the required dimensions, or by controlling the processing conditions to create these structural components during the manufacturing process. Alternatively, they may be formed by selectively coloring specific regions of the food after it has been formed. It may also be important to ensure that regions with different appearances remain intact during the storage and processing of the product. One challenge to achieving this goal may be the diffusion of pigments from one region to another. Consequently, appropriate pigments and delivery systems to prevent this problem need to be chosen. For instance, a pigment with a very low water-solubility may be utilized to prevent the diffusion of the colors from one region to another in foods with a continuous aqueous phase. Alternatively, the colors may be trapped within colloidal particles that cannot easily move.

In some products, the spatial uniformity of a plant-based food is expected to change during cooking. For instance, a burger analog may start off with a pinkish appearance but be expected to become brownish after cooking with a dark brown exterior and a light brown or pinkish interior. In this case, it is important to include ingredients within the formulation that provide these color changes during cooking. This type of color change is often provided by the Maillard reaction that occurs between proteins and reducing sugars, which is accelerated by the high temperatures and moisture loss (until a critical water loss is reached) that occurs at the surfaces of foods during cooking. Alternatively, it may be achieved by using ingredients that undergo chemical degradation reactions during heating (Chap. 2).

4.2.1.2 Transmission and Reflection

When light waves encounter the surface of a plant-based food transmittance and reflectance may occur (also scattering and absorption may occur, see later) (Fig. 4.1). For a homogeneous transparent material, the fraction of light reflected (R) from the surface is given by:

$$R = \left(\frac{m-1}{m+1}\right)^2 \tag{4.1}$$

Here, the relative refractive index (m) is the refractive index of the material (n_2) divided by the refractive index of the medium through the light wave is originally

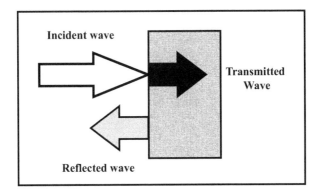

Fig. 4.1 When a light wave encounters an object, it can be partly transmitted and partly reflected by an amount that depends on the refractive indices of the different media

traveling through (n_1): $m = n_2/n_1$. Most plant-based foods predominantly consist of water ($n_2 = 1.33$), proteins ($n_2 = 1.50$), carbohydrates ($n_2 = 1.50$), and fat ($n_2 = 1.43$), whereas the medium that light waves travel through is usually air ($n_1 = 1.00$). Consequently, one would only expect about 2–4% of the light waves to be reflected from the surface of the material with the rest being transmitted into the material. In practice, the fraction of reflected light waves may be much higher than this value because of scattering of light waves by irregularities inside a plant-based food, such as fibers, particles, or droplets (see later).

4.2.1.3 Surface Gloss

The surface of a plant-based food product may be expected to be glossy (like raw meat) or matt (like cooked meat). Consequently, it is important for food manufacturers to control the surface glossiness of their products. The glossiness of a material is mainly governed by its surface roughness, *i.e.*, the dimensions of any surface irregularities relative to the dimensions of the wavelength of light (Arino et al., 2005). Surfaces with irregularities smaller than a few micrometers usually appear glossy because the light waves are reflected in a specular fashion, whereas surfaces with irregularities around and above this size appear matt because they scatter or reflect light waves in all directions leading to more diffuse reflection (Fig. 4.2). The surfaces of raw meat are fairly smooth because they are wet, while the surfaces of cooked meat are fairly rough because they are dehydrated during cooking, which leads to the formation of surface irregularities. Consequently, it is important to mimic this kind of behavior when trying to develop plant-based meat or seafood analogs that change from glossy to matt during cooking. This may be achieved by including structural heterogeneities within a formulation that generate a surface

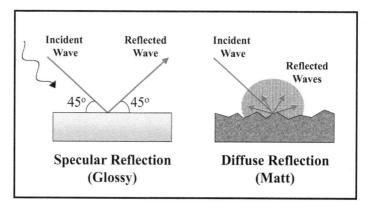

Fig. 4.2 An object may appear glossy or matt depending on its surface roughness compared to the wavelength of the incident light wave

4.2 Appearance

roughness of around a few hundred nanometers to a few micrometers when the surface of the product is dried as a result of cooking.

4.2.1.4 Selective Absorption

The color of a food, such as its redness (raw meat), brownness (cooked meat), yellowness (whole egg), or orangeness (cheddar cheese), depends on the selective absorption of certain wavelengths of light (Hutchings, 1999). White light consists of a 'mixture' of different colors with wavelengths that extend from around 380 to 750 nm: violet (380–450 nm), blue (450–495 nm), green (495–570 nm), yellow (570–590 nm), orange (590–620 nm) and red (620–750 nm). Foods contain certain types of molecules (chromophores) that have outer shell electrons capable of undergoing a transition to a higher energy level by absorbing photons in the visible region of the electromagnetic spectrum (Table 4.1). Different chromophores have different electronic structures, which causes them to selectively absorb light waves in different regions of the electromagnetic spectrum, thereby leading to different colors. For instance, a substance that appears red contains chromophores that absorb light waves in the wavelength range corresponding to violet to orange so that only red light waves are reflected backwards. As an example, the reflectance spectra of light from two model foods (oil-in-water emulsions) with and without a red dye are shown in Fig. 4.3. In the absence of the dye, the light is strongly reflected back at all wavelengths, which leads to a white appearance. In contrast, in the presence of the red dye, light is selectively absorbed by the model food from 380 to 620 nm (violet to orange) but not from 620–750 nm (red). As a result, only the wavelengths of light corresponding to red are reflected backwards leading to a reddish appearance. The reflectance spectrum, and therefore color, of a material depends on the type and concentration of chromophores present, as well as by light scattering effects (see next section). Typically, an increase in light scattering leads to a reduction in the intensity of the perceived color, since the light waves cannot penetrate as far into the

Table 4.1 Photon energies and wavelengths of different colored light waves

Color	Photon energy (eV)	Wavelength (nm)
Violet	2.75–3.26	380–450
Blue	2.50–2.75	450–495
Green	2.17–2.50	495–570
Yellow	2.10–2.17	570–590
Orange	2.00–2.10	590–620
Red	1.65–2.00	620–750

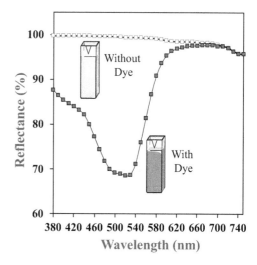

Fig. 4.3 Reflectance spectra of a model food (oil-in-water emulsions) in the absence and presence of a red dye

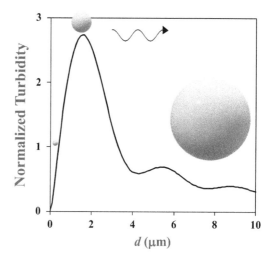

Fig. 4.4 Impact of particle dimensions on the turbidity of a colloidal dispersion. The strongest light scattering occurs when the dimensions of the particles are a few micrometers big because this is close to the wavelength of light

sample, which means that there is less absorption. Consequently, it is important to control the nature of the chromophores present in a plant-based food, as well as the concentration, size, and refractive index of any heterogeneities that scatter light, including fat droplets, protein aggregates, or air bubbles.

4.2.1.5 Scattering

Scattering occurs when a light wave encounters a heterogeneity (such as a fat droplet, oil body, air bubble, biopolymer fiber, biopolymer particle, or plant-tissue fragment) that has a different refractive index than the surrounding matrix (usually water). As a result, the light waves are redirected into directions that are different to that of the original wave, thereby leading to a scattering pattern (Bohren & Huffman, 1998; Hergert & Wriedt, 2012). The fraction of light scattered, as well as the angular dependence of the scattered wave, depends on the size, shape, refractive index, and concentration of the heterogeneities (McClements, 2002a). When the dimensions of the objects are relatively small compared to the wavelength of light ($d < < \lambda$), the scattering is weak and most of the light is scattered evenly in all directions (Hutchings, 1999). For intermediate-sized particles ($d \approx \lambda$), the scattering is relatively strong and the scattering pattern has a complex form. For relatively large particles ($d > > \lambda$), the scattering becomes weaker again and most of the light is scattered in the forward direction. Sufficiently large particles ($d > 200$ µm) can actually be discerned as individual objects by the human eye. The scattering of light waves by particles can be described by a series of equations known as the Mie theory (Hergert & Wriedt, 2012). A measure of the influence of particle size on the magnitude of the light scattering can be obtained by plotting the normalized turbidity *versus* particle diameter for a colloidal suspension containing spherical particles (Fig. 4.4). When the particles are much smaller than the wavelength of light ($d < 40$ nm) the system appears transparent. As the particle size increases, the turbidity increases until it reaches a maximum value around a few micrometers ($d \approx 1.6$ µm), after which it falls again. The calculations highlight the importance of the size of the particles in plant-based foods on their optical characteristics.

The optical properties of colloidal dispersions are also strongly determined by particle concentration (Fig. 4.5). In highly dilute systems, a light wave only encounters one particle as it travels through a material and then exits, which is known as single scattering. In more concentrated systems, a light wave is scattered by one particle and then encounters another particle and is scattered again before exiting the material, which is known as multiple scattering. At sufficiently high particle concentrations, a light wave is scattered by so many different particles that it can be

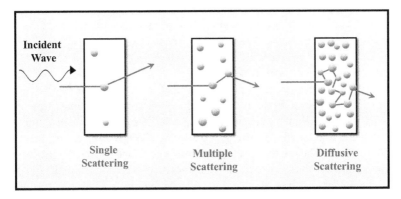

Fig. 4.5 Impact of particle concentration on the scattering of light waves by a colloidal dispersion. Single scattering occurs in dilute systems where the light waves are only scattered by a single particle, whereas multiple scattering occurs in more concentrated systems. Diffusive scattering occurs is very concentrated systems

considered to travel through the material *via* a diffusion-like process, which is known as diffuse scattering. Mathematical models are available to accurately describe the scattering patterns of dilute colloidal dispersions (Mie theory), which are useful for predicting the impact of their composition and structure on their optical properties. Mathematical models are also available for more concentrated systems, but these are usually much more complex and less accurate.

The scattering of light by colloidal dispersions influences their turbidity, opacity, cloudiness and lightness, thereby influencing their overall appearance. In plant-based milks, the scattering is mainly from spherical particles like fat droplets or oil bodies, although irregular shaped plant-tissue fragments may also contribute to the scattering. In plant-based meat and fish products, the scattering may be due to the fat droplets used to mimic adipose tissue, as well as the protein fibers used to mimic the muscle fibers in meat (Purslow et al., 2020).

4.2.2 Modeling and Prediction of Appearance

Ideally, it would be advantageous to have a mathematical model to predict the appearance of plant-based foods from the knowledge of their structures and compositions. In practice, this is challenging because of the structural and compositional complexity of most real food products. Nevertheless, some insights can be obtained using simple model systems that include some of the important features of more complex real systems. For instance, a plant-based food can be considered to consist of a suspension of spherical or cylindrical particles dispersed in a continuous matrix, such as fat droplets in water (milk analog) or protein fibers in water (muscle analog). The particles and/or surrounding matrix may contain chromophores that selectively absorb light waves and therefore provide color. The overall appearance of the system depends on the concentration, size, and refractive index of the particles, as well as the concentration and absorption spectra of the chromophores (McClements, 2002a, b). As an example, we consider modeling the optical properties of a colloidal dispersion containing spherical particles dispersed in a continuous matrix. Analysis of this simple model provides valuable insights into the major factors influencing the optical properties of plant-based foods.

Initially, it is useful to consider the physical processes occurring when white light waves encounter a material containing colloidal particles and chromophores (Fig. 4.6). After the light wave encounters the surface of the material, part of it is transmitted and the remainder is reflected (Berns, 2000; Bohren & Huffman, 1998; Hutchings, 1999). The relative fractions of transmitted and reflected light depend on the microstructure and composition of the system. The transmitted light travels through the material and encounters the colloidal particles where it is scattered. As mentioned earlier, the scattering pattern depends on the size, shape, and refractive index of the particles (Bohren & Huffman, 1998; Hergert & Wriedt, 2012). The

4.2 Appearance

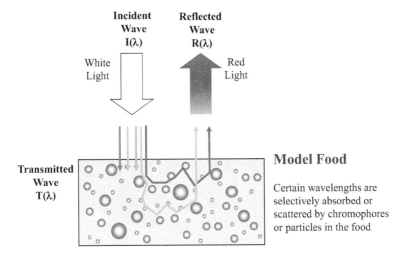

Fig. 4.6 Many plant-based foods can be considered to consist of a dispersion of colloidal particles in an aqueous medium. Their optical properties depend on the reflection, transmission, scattering, and absorption of the light waves incident upon their surfaces

scattered waves may then encounter one or more other colloidal particles and be scattered again (multiple or diffusive scattering). The presence of chromophores within the colloidal particles and/or surrounding medium leads to selective absorption of the light waves, which is the origin of the color of the material. The intensity and hue of the color depend on the concentration and type of chromophores present, which is modulated by scattering effects. The overall optical properties of the material are therefore governed by a combination of light scattering and absorption effects. The opacity of the material is mainly determined by scattering effects, whereas the color is mainly determined by absorption effects.

The color of plant-based foods can be objectively described using tristimulus color coordinates, which can conveniently be measured using instrumental methods such as UV-visible spectrophotometers or colorimeters. One of the most commonly used is the $L^*a^*b^*$ coordinate system (Fig. 4.7), which is discussed in more detail later. The use of instrumental tristimulus color coordinates overcomes many of the challenges that humans have in objectively describing color. In the remainder of this section, a brief overview of a theory that has been developed to relate the composition and structure of colloidal foods to their tristimulus color coordinates is given, which is based on the light scattering theory (McClements, 2002a, b). It is assumed that a plant-based food can be treated as a colloidal dispersion containing a single type of spherical particles dispersed within a continuous matrix, but more complex and realistic theories could be derived to more accurately describe real systems. This theory can be programmed on a personal computer and then used to predict the impact of structural characteristics (such as particle size, refractive index, and concentration) and chromophore characteristics (such as absorption spectrum and concentration) on the appearance of plant-based foods, which could facilitate the design

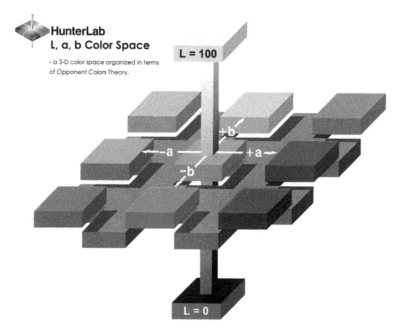

Fig. 4.7 The optical properties of plant-based foods can be described by the tristimulus color coordinates, like the L^*, a^*, b^* color space shown here. (Diagram kindly provided by Ken Wendt (HunterLab) with permission to use)

of high quality products. Because most plant-based foods are optically opaque only reflectance measurements are considered. The various steps used to relate the particle and chromophore characteristics of the model plant-based food to its tristimulus coordinates are shown in Fig. 4.8. Only a brief overview of this model is given here because it has been described in detail elsewhere (McClements, 2002b).

4.2.2.1 Calculation of Scattering Characteristics of Particles

The first piece of information required to predict the optical properties of colloidal dispersions is the scattering characteristics of the individual particles. In particular, the scattering efficiency (Q_s) and asymmetry factor (g) of the particles need to be known across the visible region of the electromagnetic spectrum, *i.e.*, from 380 to 750 nm (Fig. 4.8). The scattering efficiency provides a measure of the fraction of light waves scattered by the particles, while the asymmetry factor describes the angular dependence of the scattered waves (Kerker, 1969). These parameters are typically calculated as a function of a dimensionless size parameter (x) using a mathematical model known as the Mie theory by using the size and relative refractive index of the particles (Hergert & Wriedt, 2012):

4.2 Appearance

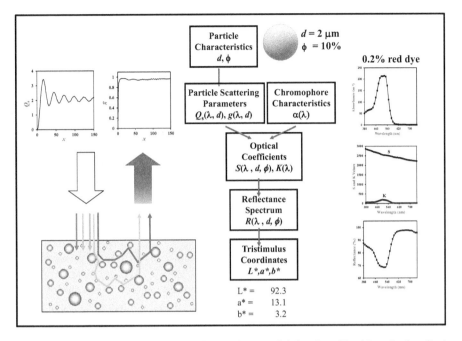

Fig. 4.8 The optical properties of colloidal dispersions (model plant-based foods) can be described using light scattering theory, which relates particle and chromophore characteristics to $L^*a^*b^*$ values

$$x = \pi d n_1 / \lambda \qquad (4.2)$$

Here, d is the diameter of the colloidal particles, n_1 is the refractive index of the surrounding medium, and λ is the wavelength of light. The values of Q_s and g can be calculated as a function of x by programming the Mie theory model using a suitable software program or by using an on-line Mie theory calculator. The dependences of Q_s and g on x for a colloidal dispersion consisting of fat droplets dispersed in water are shown in Fig. 4.8. These predictions show that the extent and direction of light scattering depend on the size of the colloidal particles compared to the wavelength of light. As a result, they indicate that the optical properties of plant-based foods are influenced by the dimensions of any structural heterogeneities they contain, such as fat droplets, plant-tissue fragments, biopolymer particles, or fibers. The Mie theory is only suitable for calculating the scattering characteristics of isolated colloidal particles in dilute systems, and so its predictions are not accurate for aggregated or concentrated dispersions (Kerker, 1969). In this case, more advanced mathematical models are required to model the scattering characteristics of the particles.

For plant-based foods, the fat droplets in plant-based milks, eggs, or meat analogs can usually be modelled as spheres. However, the fibrous structures in plant-based meat or seafood analogs should be modelled as cylinders. As a consequence, different mathematical equations are required to calculate the dependence of the Q_s

and *g* values on *x* for fibers, which depend on the refractive index, orientation, and thickness of the fibers (Bohren & Huffman, 1998). Nevertheless, similar general trends are observed as for spheres.

4.2.2.2 Determination of Absorption Spectra of Chromophores

The second piece of information required to predict the optical properties of a colloidal dispersion is the absorption characteristics of any chromophores present (Fig. 4.8). The wavelength-dependence of the absorption coefficient ($\alpha(\lambda)$) of the chromophores in a colloidal dispersion is usually measured using an UV-visible spectrophotometer. Typically, the chromophores are first dissolved within a solvent that represents the phase they would normally be located in a plant-based food, such as oil for hydrophobic pigments or water for hydrophilic ones (using the same solvent as a blank). The absorbance of the chromophore solution is then measured from 380 to 750 nm and the absorption coefficient is calculated as a function of wavelength: $\alpha(\lambda) = 2.303 \times A(\lambda)/L$, where $A(\lambda)$ is the measured absorption at a wavelength of λ and L is the length of the cuvette used to contain the solution.

Many plant-based foods are multiphasic systems that contain chromophores in different phases. In the case of a simple colloidal dispersion, the chromophores may be present in the dispersed and continuous phases, which are typically oil and water in many plant-based foods. The overall absorption spectrum of the colloidal dispersion can then be predicted using the following expression: $\alpha(\lambda) = \phi \times \alpha_O(\lambda) + (1-\phi) \times \alpha_W(\lambda)$. Here, ϕ is the volume fraction of the oil phase and the subscripts O and W refer to the oil and water phases, respectively. In the case of oil-soluble and water-soluble chromophores, it may therefore be necessary to measure the absorption spectrum of the oil and water phases separately.

4.2.2.3 Calculation of Spectral Reflectance of Colloidal Dispersion

The third stage in predicting the color of a colloidal dispersion is to calculate the spectral reflectance, $R(\lambda)$, which is the change in the reflectance with wavelength in the visible region of the electromagnetic spectrum. The reflectance depends on the scattering of light waves by the colloidal particles, as well as on the absorption of light waves by the chromophores. Most plant-based foods can be considered to be fairly concentrated colloidal dispersions that are optically opaque because they contain numerous particles that scatter light. As a result, their optical properties are usually characterized by reflectance (rather than transmission) measurements. The spectral reflectance is the fraction of light waves reflected (or back-scattered) from the surface of a material as a function of the wavelength. The reflectance of concentrated colloidal dispersions can be related to their scattering and absorption characteristics using a mathematical model known as the Kubelka-Munk theory (Kotrum, 1969):

4.2 Appearance

$$R = 1 + \frac{K}{S} - \sqrt{\frac{K}{S}\left[\frac{K}{S} + 2\right]}$$

(4.3)

In this expression, K and S are the absorption and scattering coefficients, respectively. The R, K and S values change with wavelength because the absorption and scattering of light by the chromophores and particles in the colloidal dispersion are both wavelength-dependent. Calculations of these values for a colloidal dispersion containing a red dye are shown in Fig. 4.8. The Kubelka-Munk theory was derived from a mathematical analysis of the propagation of light waves through a medium that absorbs and scatters light (Mudgett & Richards, 1971). Expressions have been derived to relate the K and S coefficients of a colloidal dispersion to the absorption spectrum of the chromophores and the scattering properties of the particles:

$$K = 2\alpha$$

(4.4)

$$S = \frac{3}{16}\pi d^2 Q_s [1-g] - \frac{1}{4}\alpha$$

(4.5)

Predictions of the reflectance spectrum of a colloidal dispersion with and without red dye were shown in Fig. 4.3. Without dye, the reflectance remains relatively high and constant across the whole of the visible wavelength range, which leads to a whitish appearance. With dye, the reflectance spectrum has a trough corresponding to the peak seen in the absorbance spectrum of the dye (Fig. 4.8). Typically, as the dye concentration is raised, the trough becomes deeper because a greater fraction of the light waves is absorbed by the dye, so less is reflected.

In practice, the simple equations described above have to be modified to take into account other factors, such as particle-particle interactions and the fact that a sample may be kept in a container (rather than directly exposed to light waves) and so the interaction of the light waves with the container walls have to be accounted for (McClements, 2002b).

4.2.2.4 Calculation of Tristimulus Color Coordinates from Reflectance Spectra

The final step is to calculate the tristimulus color coordinates of a colloidal dispersion from the reflectance spectrum (Fig. 4.8). Equations have been developed that allow one to calculate the L^*, a^*, and b^* values of a material from its reflectance spectrum (McClements, 2002a; Wyszecki & Stiles, 2000). These equations depend on three factors: (*i*) the reflectance spectra, $R(\lambda)$, of the material being tested (which depends on its structure and composition); (*ii*) the spectral distribution, $S(\lambda)$, of the standard illuminant used to irradiate the material, which is a measure of how the light intensity changes with wavelength; and (*iii*) the standardized response functions of the human eye, $x(\lambda)$, $y(\lambda)$ and $z(\lambda)$, which are a measure of how the sensitivity of the different light-sensitive receptors in the eye (rods and cones) change with wavelength. The equations used to calculate the L^*, a^*, and b^* values from the reflectance spectrum have been presented elsewhere (McClements, 2002b). In practice, it may be necessary to take into account that there may be different kinds of structural entities (such as fat droplets, plant tissue fragments, and protein aggregates) that have different concentrations, sizes, shapes, and refractive indices, since all of these particles may contribute to the overall scattering pattern.

The availability of mathematical theories to predict the color of plant-based foods based on their composition and structure allows food formulators to use computer models to estimate the relative importance of different factors on the overall appearance of the final product. For example, the importance of particle type and size on the appearance of the product can be estimated, as well as the effects of mixing different kinds of chromophores together. This may facilitate the design and creation of plant-based foods that have appearances that more closely match those of the animal-based products they are designed to replace.

4.2.3 Major Factors Impacting the Appearance of Plant-Based Foods

In this section, we give a brief overview of the major factors impacting the appearance of plant-based foods. We use theoretical calculations or experimental measurements made on simple model systems (colloidal dispersions consisting of fat droplets and chromophores dispersed in water) to highlight the importance of these factors.

4.2.3.1 Chromophore Type and Concentration

The overall color of a plant-based food depends on the type and concentration of the various chromophores it contains because this determines the fraction of light waves selectively absorbed and reflected at different wavelengths. As an example, the impact of chromophore concentration on the calculated reflectance spectra and tristimulus color coordinates of a colloidal dispersion containing different amounts of red dye is shown in Fig. 4.9. As expected, the depth of the trough in the reflectance spectra increases as the dye concentration increases because of an increase in the fraction of light waves absorbed in the non-red regions of the electromagnetic spectrum. Consequently, there is an increase in the magnitude of the positive a^* value (redness), as well as a smaller increase in the magnitude of the positive b^* value (yellowness). At the same time, there is a reduction in the L^* value (lightness) of the colloidal dispersion because less light is reflected back from its surface when it contains more chromophores that absorb light.

The impact of chromophore type on the measured reflectance spectra and calculated tristimulus coordinates of colloidal dispersions containing different kinds of

4.2 Appearance

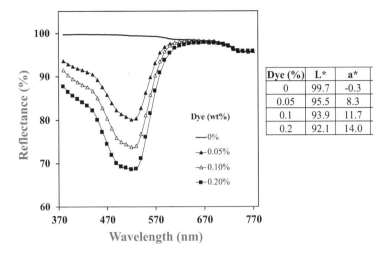

Fig. 4.9 Theoretical predictions of the influence of red dye concentration on the reflectance spectra and $L^*a^*b^*$ values of colloidal dispersions (model plant-based foods)

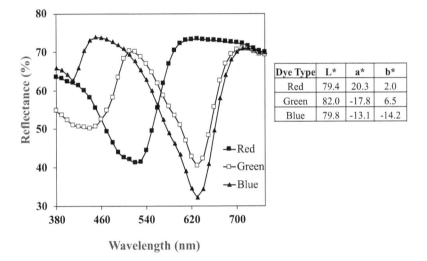

Fig. 4.10 Theoretical predictions of the influence of different kinds of dye on the reflectance spectra and $L^*a^*b^*$ values of colloidal dispersions (model plant-based foods)

food dyes (red, green, and blue) is shown in Fig. 4.10. As expected, the wavelengths where most of the light is selectively reflected back from the colloidal dispersions depends on the type of dye they contain: 430–510 nm (violet/blue) and 700–750 nm (red) for the blue dye; 500–540 nm (green) and 700–750 nm (red) for the green dye; and 590–750 nm (red) for the red dye. The calculated L^*, a^*, b^* values are consistent with these colors: moderately negative a^* (green) and negative b^* (blue) for the blue dye; moderately negative a^* (green) and slightly positive b^* (yellow) for the

Table 4.2 Examples of different kinds of pigments used to formulate plant-based foods

Pigment	Color	Notes
Leghemoglobin	Red to brown	Changes from red to brown when cooked. Isolated from the roots of soybeans or produced by microbial fermentation. Applications in meat analogs, especially plant-based burgers.
Beet juice extracts	Bright dark red	Heat stable (remains red during cooking). Good for simulating the "rare" center of meat products after cooking. Application in meat analogs.
Carmine	Pink red to orange	Water soluble. Color depends on pH. Good heat and light stability. Isolated from cochineal insects.
Annatto	Yellow to reddish-orange	Oil soluble (but water-dispersible forms available). Carotenoid derived from seeds of annatto tree. Used in dairy analogs.
β-carotene	Dark Orange	Oil soluble (but water-dispersible forms available). Carotenoid derived from carrots and other plants. Used in dairy and egg analogs.
Sienna fruit juice	Brown	Water soluble. Heat, light and acid-stable. Used in meat and dairy analogs.
Lycopene	Red	Oil soluble (but water-dispersible forms available). Carotenoid derived from tomatoes.
Capsanthins	Orange to orange-red	Oil soluble (but water-dispersible forms available). Carotenoids (capsanthins) derived from red peppers. Heat, water and pH stable. Used in meat, seafood, and dressings.
Curcumin	Bright yellow	Oil soluble (but water-dispersible forms available). Isolated from turmeric. Stable to heat, light and acidic-to-neutral conditions but degrades under basic conditions. Used in meat, seafood, egg, and dairy products.
Betanin	Bright bluish-red to bluish-violet	Water soluble. Isolated from red beet juice. Color depends on pH. Stable to heat, light and acidic-to-neutral conditions but degrades under basic conditions. Used in meat products.
Beet juice	Red	Water soluble. Isolated from sugar beet. Used in meat analogs.
Paprika	Orange	Isolated from red peppers. Used in meat and egg analogs.
Safflower	Light yellow	Isolated from safflower (Carthamus). Stable to heat and light. Used as substitute for saffron.
Caramel	Golden brown	Produced by heating carbohydrates in the presence of acids, alkalis or salts.
Titanium dioxide	White	Inorganic particles that scatter light strongly. Stable to cooking. Some concerns about potential toxicity. Used in cheese analogs.

The authors thank Zachary Henderson (Sensient) for valuable discussions about plant-based colors.

green dye; and highly positive a^* (red) and slightly positive b^* (yellow) for the red dye (Fig. 4.10). It would have been difficult for a human to specify the precise colors of the different samples in words, but this can be achieved using instrumental methods.

These results highlight the importance of controlling the type and concentration of chromophores in a plant-based food when trying to obtain a desired color. When creating a plant-based analog of a traditional animal-based food it is important to match the reflectance spectra and color coordinates as closely as possible. This typically requires controlling the type and concentration of different chromophores present in the system. This can be done experimentally or theoretically. Experimentally, different pigments could be combined together in different ratios and then the tristimulus color coordinates of the end product could be measured. The ratios could then be adjusted until the color of the plant-based food matches that of the animal-based one. Alternatively, the mathematical model described in Sect. 4.2.2. could be used to predict the optimum combination of pigments required to achieve the desired color.

The type of chromophores used in a specific application depends on the color required in the final product, as well as changes in color that might be desirable during food preparation or cooking (Table 4.2). For instance, for the color of a beef analog is expected to change from red to brown during cooking, while that of a chicken analog is expected to change from pinkish to whitish. In other products, the color may be required to be heat-stable, *i.e.,* its color does not change during cooking. For example, meatball analogs are expected to have a brownish color before and after cooking, whereas hot dog analogs are expected to maintain their reddish color. It is also important to consider any colors introduced by the other ingredients used to formulate the product, such as the plant proteins used to form a solid-like matrix. These ingredients can vary greatly in color, *e.g.,* whitish (some purified protein powders), brownish (some mushroom flours), or even reddish, yellowish, or greenish (some seaweed proteins). The base color provided by these ingredients must therefore be accounted for when trying to match the reflectance spectra and tristimulus color coordinates of plant-based foods to animal-based ones.

Another important aspect of colors (especially natural ones) is that they must remain stable in a product during food production, distribution, and preparation, or at least have a controlled instability (*e.g.,* change from red to brown during cooking). Consequently, the impact of temperature, light, pH, oxygen, ingredient interactions, and other factors must be considered when selecting an appropriate color for a particular application.

Many ingredient suppliers are now developing natural pigments for specific utilization in plant-based products, such as Sensient, Givaudan, and Biocon.

4.2.3.2 Particle Size and Concentration

The size and concentration of the different kinds of particles within a plant-based food also have a pronounced impact on its appearance. In particular, they will scatter the light waves, making the product look cloudy or opaque, which is an important attribute of many plant-based foods, including meat, seafood, egg, and dairy analogs. The particles present may be fat droplets, oil bodies, protein particles, polysaccharide particles, protein fibers, polysaccharide fibers, plant-tissue fragments, or mineral particles (like TiO_2). In this section, we examine the impact of particle size and concentration on the optical properties of model colloidal dispersions containing fat droplets and a red food dye.

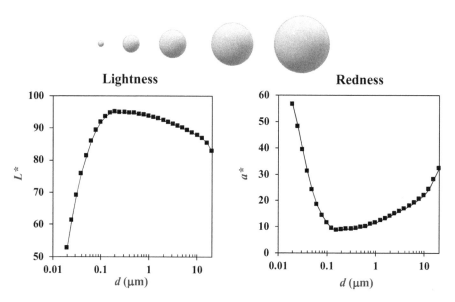

Fig. 4.11 Theoretical predictions of the influence of fat droplet size on the lightness (L^*) and redness (a^*) of colloidal dispersions (model plant-based foods) assuming a constant droplet concentration

The predicted changes in the lightness (L^*) and redness (a^*) of the colloidal dispersions with particle size are shown in Fig. 4.11. The lightness of the material increases as the particle size increases from around 10 to 200 nm, remains relatively high from around 200 to 800 nm, and then decreases when it is increased further. This effect can be attributed to changes in the light scattering efficiency of the particles as their size changes. Typically, the light scattering efficiency has a maximum value at a particle size that is around the wavelength of light (380–750 nm). The redness of the material followed the opposite trend. It decreased as the particle size increased from around 10 to 200 nm, remained relatively high from around 200 to 800 nm, and then decreased when it was increased further. This effect can be attributed to the fact that the light waves can penetrate further into the sample when light scattering is relatively weak, so that more selective absorption occurs.

The dependence of the lightness (L^*) and redness (a^*) of model colloidal dispersions on particle concentration are shown in Fig. 4.12. The lightness increases steeply when the particle concentration is raised from around 0 to 5% but then only increases slightly when it is raised further. The increase in lightness can be attributed to an increase in the fraction of the light waves scattered by the particles. At sufficiently high particle concentrations, the light waves are scattered by so many particles that the majority of them are reflected back. The redness of the colloidal dispersion decreases with increasing particle concentration, which can be attributed to the fact that more of the light waves are scattered backwards and so do not penetrate into the interior of the material where they can be selectively absorbed. These phenomena have important consequences for designing plant-based foods. For

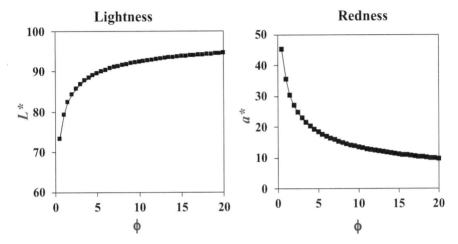

Fig. 4.12 Theoretical predictions of the influence of fat droplet concentration on the lightness (L^*) and redness (a^*) of colloidal dispersions (model plant-based foods). The photograph shows measurements of the influence of droplet concentration on emulsions with the same dye concentration

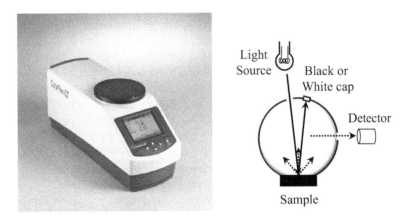

Fig. 4.13 The optical properties of plant-based foods can be measured using an instrumental colorimeter that determines the tristimulus color coordinates (L^*, a^*, b^*) by reflecting a light wave from their surface. (Images kindly provided by Ken Wendt (HunterLab, Reston VA, USA)

instance, changing the concentration of particles in a product will influence both its lightness and color intensity.

4.2.3.3 Refractive Index Contrast

The refractive indices of the particles in a plant-based food product also has an important impact on their appearance. The greater the contrast in refractive index between the particles and surrounding food matrix, the stronger the light scattering. The refractive index contrast can be taken to be the ratio of the refractive index of the particles divided by that of the surrounding matrix ($m = n_2/n_1$). At a fixed particle concentration and size, the lightness and color fading increase as the refractive index contrast increases. Thus, titanium dioxide particles ($n_2 = 2.5$) are much more effective at increasing the lightness of foods than fat particles ($n_2 = 1.43$). Consequently, they can be used at much lower concentrations to produce the same effect, which is one of the reasons they have been commonly used as lightening agents in foods. However, due to consumer and regulator concerns about the possible toxicity of this type of inorganic particle, there has been interest in identifying alternatives, such as other forms of mineral-, protein-, or polysaccharide-based nanoparticles that scatter light strongly (but not usually as strong as TiO_2).

4.2.4 Measurement of the Appearance of Plant-Based Foods

Plant-based foods tend to be optically opaque and so their appearances are normally determined using light waves that are reflected or scattered from their surfaces (rather than transmitted through them). A variety of instruments are available for characterizing their optical properties and the most suitable needs to be selected based on the particular application and depending on the type of information required, as well as the nature of the food being tested.

The overall spatial uniformity of plant-based foods can be characterized using digital photography combined with image analysis (Chmiel & Slowinski, 2013; Russ, 2012). Typically, a digital image of the food is acquired under standardized lighting conditions. An image analysis program (like ImageJ from NIH) is then used to quantify the number, area, dimensions, shape, and color of any heterogeneities. These methods can be used to compare the spatial uniformity of plant-based foods with animal-based ones so as to match their overall appearances.

The glossiness of a food material can be measured using specialized instruments known as glossmeters (Leloup et al., 2014). These instruments direct a light wave at the surface of a material at a particular incident angle and then measure the intensity of light that is reflected at the same angle (Fig. 4.2). For a highly glossy surface, the intensity of the detected light wave is relatively high because the majority of light is reflected at the same angle as the angle of incidence of the original light wave (specular reflectance). In contrast, for a matt surface, the intensity of the detected light wave is decreased because a high fraction of the light is scattered in all directions (diffuse reflectance).

The color and lightness of foods is usually quantified in terms of tristimulus color coordinates, such as the CIELAB system proposed by the Commission International de l'Eclairage (CIE) (Hutchings, 1999). This system specifies the optical properties of a material using three parameters, such as the L^*, a^* and b^* system. As mentioned earlier, L^* is the lightness, which ranges from 0 (black) to 100 (white); a^* is the redness/greenness, which ranges from strongly positive (red) to strongly negative (green); and b^* is the yellowness/blueness, which ranges from strongly positive (yellow) to strongly negative (blue) (Fig. 4.7). A major advantage of the CIELAB system is that it allows one to accurately quantify the color of a material using just three parameters. In contrast, it is often difficult for humans to accurately describe the color of materials, *e.g.,* a food that is red may have many different shades of red that are difficult to articulate, but easy to quantify using the CIELAB system. When developing a plant-based food, it is initially important to quantify the color coordinates ($L^*a^*b^*$) of the animal-based food that it is designed to mimic. In particular, it is important to establish the range of these values that lead to an end product that consumers find desirable (the target "color space"). The color and lightness of a plant-based food can then be designed to match those of the animal-based one by adding one or more pigments that absorb light and/or by controlling the size and concentration of any structural heterogeneities that scatter light (such as fat droplets, oil bodies, fibers, or particles).

The tristimulus coordinates of plant-based foods are usually measured using colorimeters or UV-visible spectrophotometers (McClements, 2015). Colorimeters direct a beam of white light at the surface of a plant-based food and then measure the intensity of the light that is reflected back as a function of wavelength under standardized conditions, such as light source and measurement cell design (Fig. 4.13). Typically, the instrument is first calibrated using standardized white and black plates, and then the sample is analyzed. The instrument software then calculates the tristimulus color coordinates ($L^*a^*b^*$). When using these instruments, it is important to select an appropriate standardized light source (such as one that represents daylight, incandescent light, or fluorescent light), as well as a standardized observer angle (*e.g.,* 2° or 10°). This allows comparisons to be made between different samples under similar conditions. A UV-visible spectrophotometer is typically used in combination with an integrating sphere attachment that allows the reflectance *versus* wavelength spectrum to be measured. This spectrum can then be converted into tristimulus color coordinates using an appropriate mathematical model (McClements, 2002a, b).

Table 4.3 Measured $L^*a^*b^*$ values of a number of animal- and plant-based food products

Product	L^*	a^*	b^*	Reference
Cow's milk (skim)	81.7	−4.8	4.1	McClements et al. (2019)
Cow's milk (full fat)	86.1	−2.1	7.8	McClements et al. (2019)
Almond milk	71.4	3.3	16.0	Zheng et al. (2021)
Oat milk	67.8	4.2	13.8	Zheng et al. (2021)
Soy milk	73.1	12.1	2.1	Durazzo et al. (2015)
Beef burger (raw)	48.1	16.1	19.7	Our laboratory
Plant-based burger (raw)	38.0	21.3	20.9	Our laboratory
Beef burger (cooked)	33.7	8.0	16.0	Our laboratory
Plant-based burger (cooked)	25.7	9.1	9.4	Our laboratory
Real scallops (raw)	65.8	−1.6	8.8	Our laboratory
Plant-based scallops (raw)	55.5	7.5	22.8	Our laboratory
Real scallops (cooked)	51.8	5.9	26.9	Our laboratory
Plant-based scallops (cooked)	64.6	4.6	22.9	Our laboratory
Hens whole egg (uncooked)	77	+0.6	+45	Our laboratory
Hens whole egg (cooked)	77	−3	+21	Li et al. (2018)
Hens whole egg (cooked)	87	−4	+28	Kassis et al. (2010)
Plant-based egg (uncooked)	71	+6	+53	Our laboratory
Cheddar cheese	56	+6	+28	Our laboratory
Plant-based cheddar cheese	46	+22	+42	Our laboratory
Coconut yogurt	62.3	−1.8	4.3	Grasso et al. (2020)
Soy yogurt	64.2	−2.8	9.7	Grasso et al. (2020)
Almond yogurt	64.2	−1.0	6.9	Grasso et al. (2020)
Dairy yogurt	66.6	−3.5	6.6	Grasso et al. (2020)
Egg-based mayonnaise	79.5	+7.7	+32.5	Huang et al. (2016)
Egg-based mayonnaise	73.4	+7.1	+35.5	Alu'datt et al. (2017)
Plant-based mayonnaise	74.4	+5.3	+26.5	Alu'datt et al. (2017)
Egg-based salad dressing	77.8	+0.94	+32.6	Song and McClements (2021)
Plant-based salad dressing	79.8	−10.8	+45.3	Kaltsa et al. (2018)

4.2.5 Color Attributes of Plant-Based Foods

Instrumental colorimeters have been widely used to measure the tristimulus color coordinates ($L^*a^*b^*$) of plant- and animal-based foods (Table 4.3). This knowledge is essential for matching the color of plant-based foods to the animal-based foods they are designed to replace. The plant-based product can be reformulated by adding different types and amounts of chromophores or scatterers so as to mimic the required color characteristics. These measurements show that each category of plant-based foods has different color requirements, and therefore requires different ingredients to obtain the desired appearance. As mentioned earlier, in some applications it is also important that the color changes in a specific way when the product is cooked, *e.g.*, burger analogs may be expected to change from pink to brown during cooking. In this case, it is important to match the color coordinates of the product before, during, and after cooking.

4.3 Texture

The textural attributes of plant-based foods play an important role in determining their processing, quality attributes, shelf life, and sensory properties (McClements et al., 2021). Moreover, these attributes should usually be designed to mimic as closely as possible those of the animal-based food they are designed to replace. Consequently, it is important to understand and control the textures of plant-based foods. In general, the texture of foods is typically characterized in terms of their rheological properties, *i.e.*, how they mechanically respond (flow or deform) when a well-defined stress is applied to them. Plant-based foods exhibit a broad range of rheological characteristics, including low viscosity fluids (like milk analogs), high viscosity fluids (like cream or sauce analogs), weak gels (like yogurt analogs), or strong gels (like meat or seafood analogs). In this section, we present various mathematical models that can be used to describe the rheological properties of plant-based foods, as well as give some insights into the major factors impacting these properties. Again, it is assumed that these foods can be represented as simple colloidal dispersions (fluids) or polymer or particle gels (solids) so that relatively simple models can be used to describe their behavior. Even so, these models still lead to valuable insights into the major factors impacting the textural attributes of these systems. The analytical instruments that can be used to measure the rheological properties of these systems are also briefly described.

4.3.1 Fluids

4.3.1.1 Definition and Description of Shear Viscosity

A number of plant-based foods are predominantly fluids, such as milk, cream, uncooked egg, sauce, and dressing analogs. These products are typically characterized by their shear viscosity (η), which is determined from the slope of a plot of shear stress *versus* shear rate. For an ideal (Newtonian) fluid, the shear stress (τ) is proportional to the shear rate ($\dot{\gamma}$) and so their rheological behavior can be described by the following equation:

$$\tau = \eta \dot{\gamma} \tag{4.6}$$

The shear stress has units of Pa whereas the shear rate has units of s^{-1} and so the units of the shear viscosity are Pa s. Many plant-based foods exhibit non-ideal behavior, which means the stress is not proportional to the shear rate (Fig. 4.14). Commonly, the shear viscosity decreases with increasing shear rate, which is referred to as shear thinning. This kind of behavior is exhibited by many milk and egg analogs due to the disruption of weak structures by the applied shear forces, such as entangled or aggregated polymers or particles (Chaps. 6 and 7). In some cases, the shear viscosity increases with increasing shear rate, which is referred to

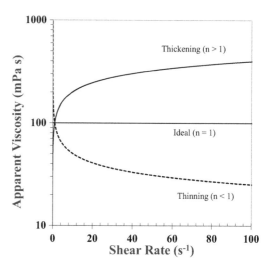

Fig. 4.14 Many fluid plant-based foods exhibit non-ideal behavior, such as shear thickening or shear thinning

as shear thickening (but this is much less common). This kind of behavior may occur when the application of the shear forces promotes aggregation of the polymers or particles in the sheared system. In general, the rheological properties of shear-dependent non-ideal fluids can be described by the following equation, which is referred to as the Cross model (McClements, 2015):

$$\eta = \eta_\infty + \frac{\eta_0 - \eta_\infty}{1 + (K\dot{\gamma})^{1-n}} \tag{4.7}$$

Here, η_0 and η_∞ are the apparent shear viscosities at very low and high shear rates, K is the Cross constant and n is the power index. The power index provides information about the non-ideal behavior of the fluid: for an ideal fluid $n = 1$; for a shear thinning fluid $n < 1$; and for a shear thickening fluid $n > 1$ (Fig. 4.14). Thus, the rheology of this type of fluid can be described by four parameters: η_0, η_∞, K and n. In aggregated systems, the value of K depends on the strength of the forces holding the structures together. A plot of the apparent viscosity *versus* shear rate predicted using the Cross model for a shear thinning fluid is shown in Fig. 4.15. However, in many cases, viscosity measurements can also be performed over an intermediate shear rate range, in which the apparent shear viscosity of the samples can also be modeled using a simple power law model (Hunter, 1994):

$$\eta = K(\dot{\gamma})^{n-1} \tag{4.8}$$

In this equation, the constants K and n are referred to as the consistency index and power index, respectively. In this case, the rheological properties of the fluid can be described in terms of two parameters: K and n. These parameters can be obtained by plotting $\log(\eta)$ *versus* $\log(\dot{\gamma})$, with the intercept being $\log(K)$ and the slope being

Fig. 4.15 The shear thinning behavior of fluid plant-based foods can be predicted by the Cross model. The range of the power-law model is also shown

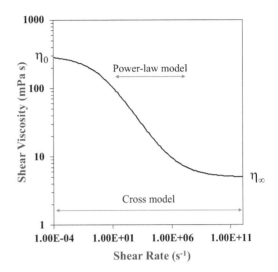

($n-1$). It should be noted that the power-law model should only be utilized once it has been shown that the log(η) *versus* log($\dot{\gamma}$) relationship is linear over the experimental conditions used.

The viscosity of non-ideal fluids may also depend on the length of time that the shear stress is applied. In some cases, the apparent shear viscosity decreases with increasing time (thixotropic), whereas in other cases it increases (rheopectic). Time-dependent shear thinning behavior tends to occur when the structures in a sample are progressively disrupted by the applied shear stress, *e.g.*, weakly aggregated particles or polymers. In contrast time-dependent shear thickening tends to occur when the applied shear stress promotes aggregation of particles or polymers.

4.3.1.2 Major Factors Impacting Viscosity

Most fluid plant-based foods can be considered to be colloidal dispersions containing particles or polymers dispersed in water. Consequently, their shear viscosities can be related to their composition and structure using the equations developed to model the rheological properties of colloidal dispersions (McClements, 2015). The shear viscosities of colloidal dispersions typically increase with increasing particle or polymer concentration, with the extent of the increase depending on the nature of the particles or polymers. For dilute systems containing non-interacting rigid spherical particles, the shear viscosity can be modeled using the Einstein equation:

$$\eta = \eta_1(1 + 2.5\phi) \qquad (4.9)$$

Here, η and η_1 are the shear viscosities of the colloidal dispersion and the aqueous phase, respectively, and ϕ is the volume fraction of the colloidal particles. This

Fig. 4.16 The viscosity of colloidal dispersions can be predicted using the Einstein equation (low concentrations) or effective medium theory (low to high concentrations)

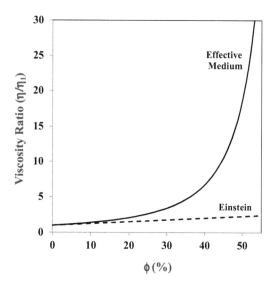

expression is typically applicable up to particle concentrations of about 5%, provided the particles meet the assumptions used to derive the equation. In the case of more concentrated colloidal dispersions, the shear viscosity is increased above the value predicted by the Einstein equation due to interactions between the particles. In this case, the shear viscosity can be modeled using a semi-empirical effective medium theory (Genovese et al., 2007; McClements, 2015):

$$\eta = \eta_1 \left(1 - \frac{\phi}{\phi_c}\right)^{-2}$$

(4.10)

In this equation, ϕ_c is a critical packing parameter (≈ 0.65) that is taken to be the volume fraction where the colloidal particles become so closely packed together that the overall system gains some solid-like features.

The dependence of the shear viscosity on particle concentration predicted using these equations is shown in Fig. 4.16. At low concentrations (<5%), both equations give similar values, but at higher concentrations the effective medium theory gives a much higher viscosity than the Einstein equation. For the effective medium theory, the shear viscosity increases relatively slowly from around 0 to 20% particles, but then increases much more steeply. In particular, at around 40 to 50% of particles in the system it increases dramatically because the particles become closely packed together i.e., ϕ approaches ϕ_c. This phenomenon is the reason that milk (<5% fat) is a low viscosity fluid while heavy cream (around 40% fat) is a high viscosity fluid. It also explains why mayonnaise (around 70% fat) is a semi-solid, even though the oil and water phases it is comprised of are both low viscosity fluids.

For colloidal dispersions containing polymers, the volume fraction of the particles (ϕ) should be replaced by the effective volume fraction of the polymers (ϕ_{eff}):

$\phi_{eff} = R_V\phi$, where R_V is the volume ratio of the polymer molecule. The volume ratio is the total volume occupied by a polymer molecule in solution (polymer chain plus entrained solvent) divided by the volume occupied by the polymer chain alone. This expression therefore takes into account that part of the volume occupied by the polymer molecules in solution is actually solvent. To a first approximation, the volume ratio of polymers can be described by the following expression:

$$R_V = \frac{4\pi r_H^3 \rho N_A}{3M} \qquad (4.11)$$

Here, r_H is the hydrodynamic radius of the polymer molecules, ρ is the density of the polymer chain, N_A is Avogadro's number, and M is the polymer molecular weight. For a fixed molecular weight, R_V increases with increasing hydrodynamic radius, which means that more extended molecules are more effective at increasing the viscosity than more compact ones. For molecules with the same conformation, R_V increases with increasing molar mass. This is because, r_H and M are not independent variables. For instance, the hydrodynamic radius is proportional to the molar mass for polymers with rigid rod structures ($r_H \propto M$), to the square root of the molar mass for polymers with random coil structures ($r_H \propto M^{1/2}$), and to the cube root of the molar mass for polymers with globular structures ($r_H \propto M^{1/3}$), thus R_V is proportional to M^2, $M^{3/2}$ or M for these three types of polymers, respectively (McClements, 2000). Plots of the effect of the volume ratio of polymers on the viscosity of aqueous solutions are shown in Fig. 4.17. These predictions show that compact globular proteins (R_V near unity) do not cause a large increase in viscosity until they reach high concentrations (>30 g/100 mL) but extended polysaccharides ($R_V \gg 1$) can

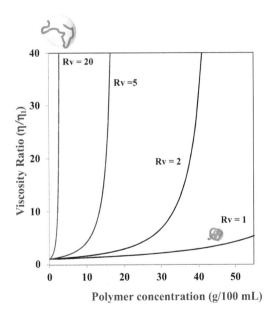

Fig. 4.17 The shear viscosity of polymer solutions increases with increasing polymer concentration, with the extent of the increase being greater for polymers with more extended structures

increase the viscosity at very low concentrations (<1 g/100 mL). This phenomenon is important for selecting plant-based ingredients for application in different kinds of products. For example, for a plant-based milk it may be important to have a high protein content without causing a large increase in viscosity, and so a compact globular protein should be used, such as a soy or pea protein. However, for a plant-based dressing it may be important to use a thickening agent that can greatly increase the viscosity of the aqueous phase, and so a polysaccharide with an extended structure should be used, such as guar, locust bean, or xanthan gum.

4.3.1.3 Rheological Characterization of Fluids

In research and development laboratories, the shear viscosity of fluid plant-based foods is usually measured using rotational viscometers or dynamic shear rheometers capable of accurately quantifying the fundamental rheological properties of the test materials (Fig. 4.18). In contrast, in quality assurance laboratories, the viscosity is often measured using cheaper versions of these instruments, as well as faster and simpler empirical methods, such as gravity flow tests or line spread measurements (Garcia et al., 2018; Rao, 2013). In this section, we only focus on the use of instrumental viscometers and rheometers because these provide reliable information that can be compared between laboratories. When using such instruments, the test sample is carefully placed into/onto a temperature-controlled measurement cell and then incubated at the required temperature for a specific time. A number of different measurement cell designs can be used depending on the nature of the sample and the amount of test material available, with the cone-and-plate and cup-and-bob types being the most commonly utilized (Fig. 4.18). For the cone-and-plate cell, the test sample is placed in the narrow gap between the cone and plate. For the

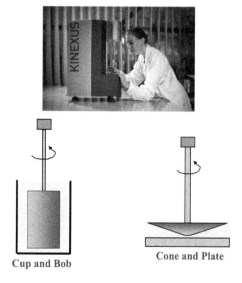

Fig. 4.18 The shear viscosity of fluid plant-based foods is usually measured using an instrumental rheometer or viscometer. The most common measurement cells are shown here. (Image of dynamic shear rheometer kindly supplied by Philip Rolfe from Netzsch (Selb, Germany))

cup-and-bob cell, the test sample is poured into the cup and then the bob is lowered into it. The pretreatment of the sample, as well as the time it spends in the measurement cell before analysis, often need to be standardized to obtain reliable results that can be compared between samples. A known shear stress is then applied to the plate or the bob and the resulting shear rate is determined (or *vice versa*) by the instrument. The change in shear stress with shear rate is then recorded and stored in the instrument's software and the apparent shear viscosity is calculated as a function of shear rate (or shear stress) from the slope of this curve (McClements, 2015). A suitable mathematical model, such as the Cross or power law model, can then be fitted to the measured data and the appropriate rheological parameters can be determined, such as the consistency and power law indices. In some cases, more sophisticated instruments are used to provide data that is more closely correlated to the mouthfeel of fluid foods, such as tribometers, which measure the influence of the fluid on the friction between two surfaces moving past each other (Prakash et al., 2013; Sarkar et al., 2021).

The shear viscosities of plant-based food products range from relatively low (milk analogs) to relatively high (egg analogs) (Table 4.4). Many of these products exhibit shear thinning behavior, but the power indices may vary considerably from relatively low (<0.1) to relatively high (1.0). The viscosity of fluid plant-based foods is usually determined by the presence of hydrocolloids or colloidal particles, such as polysaccharides, proteins, plant-tissue fragments, oil bodies, and fat droplets. More detailed information about the viscosity of specific fluid plant-based foods is given in the chapters on milk and egg analogs.

4.3.2 Solids

A number of plant-based foods can be considered to be soft solids that exhibit both elastic and viscous properties, such as meat, seafood, cooked egg, mayonnaise, yogurt, and cheese analogs. The textural properties of these materials depend on the

Table 4.4 Shear viscosity values for selected fluid animal-based and plant-based foods that have been described using a power-law model

Product type	Viscosity at 10 s^{-1} (mPa•s)	Consistency index K (Pa s^n)	Flow index (n)	Reference
Cow's milk	2.2–2.6	–	1.00	Jeske et al. (2017)
Almond milk	4.6–26.3	–	0.82–0.56	Jeske et al. (2017)
Oat milk	6.8	–	0.89	Jeske et al. (2017)
Soy milk	2.6–7.6	–	1.00–0.90	Jeske et al. (2017)
Nut milk	216	0.422	0.71	Silva et al. (2020)
Whole hen's egg	28	–	1.00	Panaite et al. (2019)
Salad dressing	2400	15	0.21	Briggs and Steffe (1997)

type, concentration, and interactions of the structural components they contain, especially colloidal particles and polymers. The formation of a 3D network of aggregated biopolymers (usually proteins) is the most important factor contributing to the rheology of many soft solid animal-based foods, such as the casein networks found in yogurt and cheese, the globular protein networks found in cooked eggs, and the fibrous muscle protein networks found in meat and seafood. For this reason, biopolymers (proteins and/or polysaccharides) are often used to simulate these network structures in plant-based foods so as to obtain similar textural attributes (McClements et al., 2021). However, other constituents within animal-based foods also contribute to their desirable textures, such as the adipose tissue in meat and seafood, the fat globules in dairy products, the lipoproteins in egg, the ice crystals and air bubbles in ice cream, or the fat crystals in butter. Consequently, structurally similar components are often used to mimic the textural attributes produced by these structural elements in plant-based foods, such as fat droplets, air bubbles, and fat crystals. Moreover, the textural attributes of solid foods depend on the nature of the physical and/or chemical interactions holding the different structural components together. The most important physical interactions are hydrogen bonds, hydrophobic attraction, van der Waals attraction, and salt bridging (McClements, 2015). The sign, strength, and range of these interactions depend on the solution and environmental conditions, such as pH, ionic strength, and temperature, which therefore alter the textural attributes of the overall system. In addition, covalent bonds may be formed between structural components through chemical or enzymatic reactions, such as disulfide bonds between proteins, which tend to be stronger and more robust than physical interactions. Controlling the number and type of interactions between the different structural components in soft solid plant-based foods is therefore important for controlling their textural attributes.

Solid plant-based foods are often characterized by their elastic modulus and fracture properties (Rao, 2013). The elastic modulus is a measure of the resistance of a material to deformation when a force is applied: a harder material requires a larger force to deform it by a certain amount than a softer material. The fracture properties of a food depend on the force required to break its structure (fracture stress), as well as how far the food is deformed before it breaks (fracture strain). For some solid plant-based foods other rheological parameters are required, such as their yield stress. The rheological attributes of solid plant-based foods influence their quality and sensory attributes, such as their hardness, softness, brittleness, pliability, spoonability, spreadability, cutting, and eating properties. In general, solid foods exhibit a broad range of rheological attributes, but these can usually be described using only a few relatively simple concepts and mathematical models (Tadros, 2010).

4.3.2.1 Ideal Solids

The simplest model of a solid to consider is for an ideal elastic material. For this type of material, the deformation is proportional to the applied force, which is known as Hooke's law. Moreover, the material instantaneously adopts and

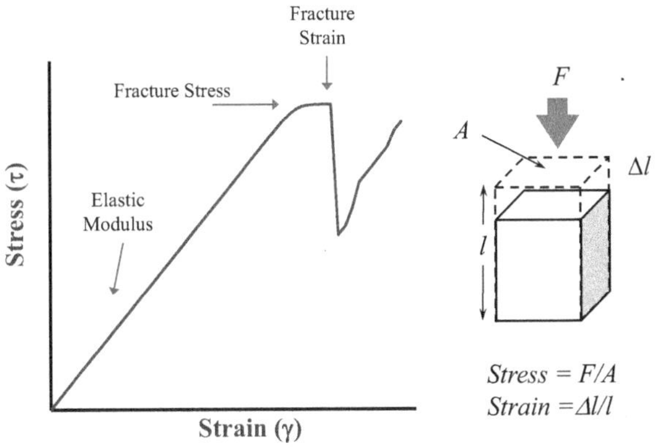

4.3 Texture

Fig. 4.19 The properties of solid plant-based foods is usually determined by measuring the stress versus strain profile using a mechanical rheometer. The elastic modulus, breaking stress and breaking strain can be determined from this profile

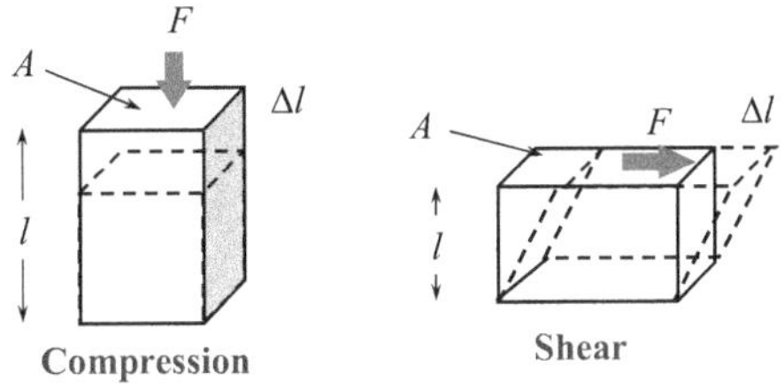

Fig. 4.20 The properties of solid plant-based foods are usually determined using either compression or shear testing methods

maintains its new dimensions when the force is applied and instantaneously returns to its original dimensions when the force is removed. In other words, all of the energy stored in the material when it is compressed is released when it is decompressed. As a result, there is no flow of the material. When testing solid foods, the stress (τ) is usually plotted as a function of the strain (γ), since this allows fundamental information about the rheological properties of the material to be determined (Fig. 4.19). Hooke's law can be expressed as follows:

$$\tau = E \times \gamma \quad (4.12)$$

Here, E is the elastic modulus, which is a fundamental property that characterizes the hardness of the material being tested. The stress is the force per unit surface area ($\tau = F/A$), while the strain is the fractional deformation of the material: $\gamma = \Delta l/l$, where F, A, Δl and l are the force, surface area, length change, and original length of the material (Fig. 4.19). The stress can be applied to the test material in various ways, with compression and shear stresses being the most commonly parameters used to test solid foods. Compression stresses are applied perpendicular to the surface of the material, whereas shear stresses are applied parallel to the surface (Fig. 4.20). The elastic modulus used in the above equation depends on the nature of the applied stress: Young's modulus (Y) is used for compression tests, whereas the shear modulus (G) is used for shear tests. It should be noted that the stress and strain can be defined in other ways depending on how the surface area and length of the material are expressed, which affects the calculated modulus (Walstra, 2003).

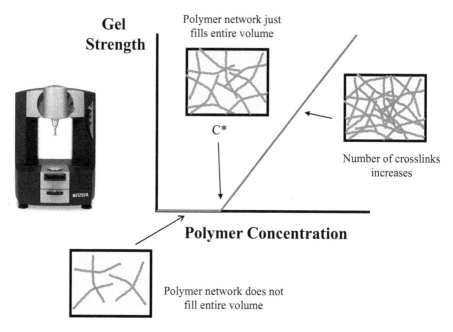

Fig. 4.21 The rheological properties of biopolymer and colloidal systems depends on the concentration of the structural components present. A gel will only form once the concentration exceeds a critical value (C^*) where a 3D network that extends throughout the system will form

For instance, the original surface area and length can be used in the calculations or the actual values during the deformation.

The rheological behavior of solid materials depends on the number, orientation, and strength of the intermolecular forces acting between the structural components (usually polymers or particles), as well as the concentration, morphology, and arrangement of these structural components (Walstra, 2003). As an example, consider the effect of polymer concentration on the elastic modulus of an aqueous solution containing polymer molecules that can form crosslinks with each other (Fig. 4.21). At relatively low concentrations, the system remains fluid because the polymer network does not extend throughout the entire volume of the system. Once the concentration exceeds a critical value (C^*), the polymer molecules form a 3D network that just extends throughout the entire volume of the system, leading to some elastic properties. As the polymer concentration is exceeded further, the shear modulus increases as the number of crosslinks in the system increases.

When a stress is applied to the surface of a solid material, the bonds between the structural components are deformed leading to compression of the whole material. The applied energy is then stored in the bonds. The number and strength of the bonds determine the resistance of the material to deformation and therefore its elastic modulus. Various kinds of mathematical models have been developed to describe the rheological properties of soft solids, which are usually based on an analysis of the forces acting between the different structural components within model polymer or colloidal systems. Different kinds of theoretical models are required to describe the textural attributes of different kinds of plant-based foods depending on the nature of the structural elements and interactions involved. For instance, a theory used to relate the texture of a mayonnaise-like product to the properties of the closely packed fat droplets it contains is different from one that relates the texture of a meat-like product to the packing and interactions of the fibers it contains. If a suitable mathematical model can be identified for a particular plant-based food, then it can provide valuable insights into the major factors influencing the texture of these systems (Cao & Mezzenga, 2020).

As an example of the utility of theoretical models, we consider an equation that has been derived to relate the rheology of a fibrous material to the properties of the semi-flexible filaments it contains (Broedersz & MacKintosh, 2014):

$$G \approx \frac{6\rho\kappa}{k_B T l_C^3} \tag{4.13}$$

Here, G is the shear modulus, ρ is the filament length density, κ is the filament bending rigidity, l_C is the spacing between crosslinks, k_B is Boltzmann's constant and T is the absolute temperature. This equation predicts that the shear modulus should increase as the concentration of the filaments increases (through ρ), the stiffness of the filaments increases (through κ), and the crosslinking density of the filaments increases (through l_C). In this case, the overall shear modulus is mainly a balance between entropy effects (the tendency for the filaments to be able to adopt as many different configurations as possible) and bending energy effects (the resistance of the filaments to deformation). This equation may be suitable for describing the behavior of fibrous networks, such as those found in meat or seafood products, as well as their plant-based analogs. As mentioned earlier, mathematical models must be identified that are appropriate for the particular food one is considering. For instance, if inclusions (such as fat droplets or starch granules) are embedded within a biopolymer matrix, then theories that take into account these inclusions should be used (Gravelle et al., 2015, 2019; Gravelle & Marangoni, 2021; Khalesi et al., 2021). In this case, the size, shape, and concentration of the inclusions should be considered, as well as their interactions with the surrounding biopolymer matrix. The following model has been developed to account for the influence of rigid spherical filler particles on the elastic modulus of a polymer matrix (Gravelle et al., 2015):

$$E_C = E_m \left(1 + \frac{15(1-v_m)(M-1)\phi_f}{(8-10v_m)M + 7 - 5v_m - (8-10v_m)(M-1)\phi_f} \right) \tag{4.14}$$

Here, $M = E_f/E_m$, and E_C, E_m, and E_f are the elastic moduli of the particle-filled composite, matrix, and filler, respectively. Also, v_m is the Poisson ratio and ϕ_f is the volume fraction of the particles embedded in the polymer matrix. In this model, it is assumed that the particles interact strongly with the surrounding polymer matrix (active fillers). In the case where the particles do not interact strongly (inactive fillers) then other models are required (Dickinson, 2012). In general, these models show that the elastic properties of foods that contain particles can be modulated by changing their concentration, size, shape, rheology, and surface properties, which provides food manufacturers with strategies they can use to control the textural attributes of their plant-based products.

4.3.2.2 Non-ideal Solids

Most solid plant-based foods are complex materials that can only be described by Hooke's law over a very limited range of conditions (*e.g.,* low applied stresses and strains) or not at all. For example, they may become fractured and therefore not return to the original shape after the force is removed, or they may exhibit both elastic and viscous properties, either simultaneously (viscoelastic materials) and/or sequentially (plastic materials). In this case, more complex mathematical models are needed to describe their rheological behavior.

Irreversible Deformation and Fracture Typically, a solid-like material only behaves like a Hookean solid in a range where the stress is proportional to the strain when relatively small deformations (<1%) of the material occur. At higher deformations, the bonds between different structural components may be disrupted, which prevents the material from returning to its original shape after the stress is removed. Understanding the properties of foods at large deformations is important for many practical applications, such as the cutting, spreading, or mastication (Rao, 2013; van Vliet, 2013; Walstra, 2003). At strains only slightly higher than the region where Hooke's law applies (around 1%), the stress is no longer proportional to the strain, and so an *apparent* elastic modulus is used to characterize the rheological properties. This value is calculated from the slope of a plot of stress *versus* strain at a specified stress or strain value. In this case, it is important to specify the stress or strain at which the apparent elastic modulus of the test material was measured. Under these conditions, the material may still return back to its original dimensions after the force is removed, even though it does not obey Hooke's law. However, once a particular stress or strain is exceeded, the material may no longer return back to its original shape once the stress is removed because it breaks or flows.

A material that breaks at low strains is referred to as brittle, whereas one that breaks at high strains is referred to as pliable. The fracture properties of solid materials are important for determining their physicochemical and sensory properties, such as their resistance to mechanical stresses during production, storage and transport, as well as their behavior in the mouth during mastication. Consequently, it is usually important to ensure that the fracture properties of plant-based foods closely match those of the animal-based foods they are designed to mimic. The critical stress at which a material first breaks when a force is applied is called the *fracture stress* (τ_{Fr}), whereas the critical strain at which this fracture occurs is called the *fracture strain* (γ_{Fr}) (Fig. 4.19). Quantifying these properties is therefore important when designing plant-based foods that simulate the textural attributes of animal-based

4.3 Texture

ones. Solids that have a tendency to flow under certain conditions are referred to as plastic or viscoelastic materials according to the nature of their flow properties, which are discussed in more detail later.

Food materials typically fracture or flow when the forces holding their structural elements together (typically biopolymers and/or colloidal particles) are exceeded (Walstra, 2003). Fracture typically begins at locations within a material where the bonds are relatively weak, such as cracks or dislocations. Consequently, it may be important to engineer plant-based materials that contain a specific number of these discontinuities.

Ideal Plastics: Some important plant-based foods have plastic-like behaviors, which means that they behave like solids below a critical applied stress (the yield stress) but as fluids above this stress, such as plant-based yogurts, cream cheeses, mayonnaise, and spreads. An ideal plastic material can be described by the following equations when a shear stress is applied:

$$\tau = G\gamma \quad \text{(for } \tau < \tau_Y\text{)} \tag{4.15}$$

$$\tau - \tau_Y = \eta \dot{\gamma} \quad \text{(for } \tau \geq \tau_Y\text{)} \tag{4.16}$$

Here, G is the shear modulus, τ_Y is the yield stress, and η is the plastic viscosity. These equations show that the effective stress is proportional to the strain below the yield stress (like a solid) but proportional to the strain rate above the yield stress (like a fluid). The rheological attributes of an ideal plastic are shown schematically in Fig. 4.22. For this kind of material, it is important that the elastic modulus, yield stress, and plastic viscosity of a plant-based food match those of the animal-based one it is designed to replace. As discussed later, various kinds of analytical instrument can be used to quantify the rheological properties of plastic materials, with shear rheometers being the most commonly used.

Plant-based foods that exhibit plastic-like properties are typically comprised of networks of interacting biopolymers or colloidal particles within a fluid medium. Plant-based yogurts or cream cheeses consist of a network of aggregated protein

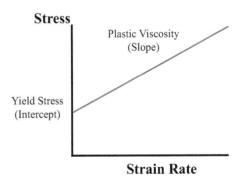

Fig. 4.22 The rheological properties of ideal plastic-like materials can be described by a yield stress and plastic viscosity

molecules dispersed in water, whereas plant-based spreads consist of a network of aggregated fat crystals dispersed in oil. When the applied stress is below the yield stress, the bonds between the structural entities are deformed but not disrupted, which leads to solid-like behavior. However, once the applied stresses exceed the yield stress, the bonds are disrupted and the structural entities move past each other, leading to fluid-like behavior. This kind of behavior is important for determining the pourability of dressings and mayonnaise, the spoonability of yogurt, as well as the spreadability of products like cheese cream.

Non-Ideal Plastics Plant-based foods in this category exhibit plastic-like characteristics because they behave like solids at low stresses and like fluids at high stresses, but they do not strictly follow ideal plastic behavior. For instance, below the yield stress they may exhibit some flow, whereas above the yield stress they may exhibit non-ideal fluid behavior such as shear thinning or thickening. As a result, the yield stress may not be a sharply defined value that is easy to measure. Instead, a small amount of flow may occur at low applied stresses (rather than no flow). This phenomenon tends to occur in materials where there is a gradual disruption of the network structure over a range of applied stresses, rather than a sudden breakdown at a particular applied stress. The non-ideal fluid-like behavior observed above the yield stress can be described by the following equation, known as the Herschel-Bulkley Model:

$$\tau - \tau_Y = K\dot{\gamma}^n \quad \left(\text{for } \tau \geq \tau_Y\right) \tag{4.17}$$

In this case, K and n are the plastic consistency index and plastic flow index, respectively. The values of K and n determined using this model are different from those that would be obtained by fitting the power law model to the same experimental data. Some representative values for these parameters for animal- and plant-based

Table 4.5 Rheological properties of selected animal- and plant-based foods that exhibit non-ideal plastic behavior that have been described by the Herschel-Bulkley Model

	Yield Stress (Pa)	Consistency index K (Pa s^n)	Flow index (n)	Reference
Coconut yogurt	30.4	1.34	0.87	Grasso et al. (2020)
Almond yogurt	28.4	6.45	0.37	Grasso et al. (2020)
Soy yogurt	27.2	3.52	0.45	Grasso et al. (2020)
Dairy yogurt	11.7	2.50	0.55	Grasso et al. (2020)
Whole hen's egg	0.20	0.030	0.97	Panaite et al. (2019)
Whole hen's egg	0.009	0.013	0.97	Our laboratory
PB whole egg	9.7	0.11	0.95	Our laboratory
Salad dressing	47	16.3	0.52	Hernandez et al. (2008)
Mayonnaise analog	81.4	82.6	0.21	Our laboratory

The authors thank Hualu Zhou (UMASS) for the measurements made in our laboratory.

products are included in Table 4.5. For plant-based foods exhibiting this kind of rheological behavior it is important to match the yield stress, consistency index, and flow index of the animal-based food they are designed to replace.

Viscoelastic Materials Many plant-based foods exhibit both solid- and liquid-like behavior simultaneously, rather than behaving as pure liquids, solids, or plastics (Rao, 2013; van Vliet, 2013; Walstra, 2003). For an ideal solid, a material is instantaneously deformed by a specific amount when a stress is applied to it but then instantaneously returns to its original shape once the stress is removed. In this case, all the mechanical energy used to compress the material is stored within the bonds between the structural elements and then released when the force is removed. For an ideal liquid, a material flows at a constant shear rate once a specific stress is applied. In this case, the applied mechanical energy is all converted into heat due to friction arising from the movement of different regions of the fluid past each other and it is therefore lost. For an ideal plastic, a material behaves like a solid below the yield stress (being deformed but not flowing) but like a fluid above the yield stress (flowing). In contrast, a viscoelastic material acts like both a solid and a liquid at the same time. When a stress is applied the material is both deformed and flows so that part of the energy is stored within the material and part of it is lost as heat. As a result, when a stress is applied to a *viscoelastic* material it does not instantaneously adopt its new dimensions nor does it instantaneously return to its original dimensions when the stress is removed (Fig. 4.23). For a viscoelastic solid when a stress is applied, the dimensions of a material change at a finite rate until they reach a constant value but when the stress is removed the dimensions return back to the original values at a finite rate. An example of this kind of behavior would be for a plant-based cheese at relatively low applied stresses. For a viscoelastic liquid, the material continues to change its shape as long as a stress is applied but once the stress is removed it only partially recovers its original shape and becomes partially but permanently deformed. An example of this kind of behavior would be a plant-based cream cheese or mayonnaise.

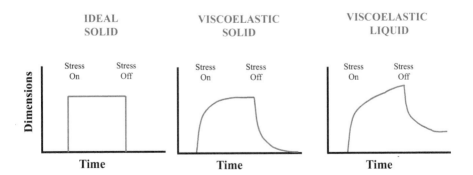

Fig. 4.23 Many plant-based foods exhibit viscoelastic behavior where they behave both as solids and fluids at the same time

The textural attributes of viscoelastic materials are usually characterized using dynamic shear rheology measurements that measure a complex shear modulus (G^*), which is composed of elastic (solid) and viscous (fluid) components:

$$G^* = G' + iG'' \qquad (4.18)$$

Here, G' is the storage modulus that represents the solid-like behavior and G'' is the loss modulus that represents the liquid-like behavior. Typically, the viscoelastic properties of a material are determined by applying an oscillating shear stress to the sample in the form of a sine wave using a dynamic shear rheometer. However, tests can also be carried out in other ways, *e.g.,* by applying an oscillating compression stress as in dynamic mechanical analysis (DMA).

Here, we consider the characterization of the rheological properties of a viscoelastic plant-based food using a dynamic shear rheometer, since this method is the most widely used (Rao, 2013; van Vliet, 2013). A sinusoidal shear stress is applied to the surface of the food material and the sinusoidal shear strain is measured (or *vice versa*). The maximum amplitude (τ_0) and angular frequency (ω) of the applied shear stress can be controlled. The measured strain wave has the same frequency as the applied stress wave, but the phase may be different as a result of relaxation mechanisms within the material that cause some of the mechanical energy to be lost due to viscous dissipation. For an ideal solid, the phase angle is zero ($\delta = 0°$), whereas for an ideal liquid it is 90° (Fig. 4.24). For a viscoelastic material, it is

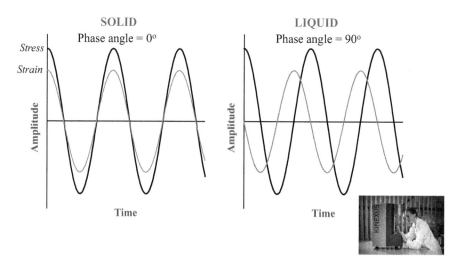

Fig. 4.24 The rheology of viscoelastic plant-based foods is usually measured by applying a sinusoidal shear stress and measuring the sinusoidal shear strain

somewhere in between these values, decreasing as the material becomes more solid-like. Typically, a material is assumed to be gelled when the phase angle falls below 45° (although this depends on the frequency used).

4.3.2.3 Rheological Characterization of Solids

A variety of analytical techniques are available for measuring the textural attributes of plant-based solids, including empirical, imitative, and fundamental methods (Rao, 2013). Empirical methods typically involve applying a force to a material and measuring its deformation using simple devices, such as a penetrometer. Imitative methods are designed to mimic some process that is an important feature of the food product, such as cutting, slicing, or chewing. The main advantages of these methods are that they are inexpensive, rapid and simple to use, while the main disadvantages are that they do not provide information about the fundamental rheological properties of the material being tested. Instead, the results depend on the piece of equipment used, which makes it difficult to compare data from one laboratory to another. In contrast, fundamental methods provide information about the intrinsic rheological properties of the material being tested and can measure changes in these properties when conditions such as the temperature, time, frequency, and shear stress are altered, which provides a more detailed understanding of their textural attributes. However, the instruments required to carry out these measurements tend to be relatively expensive to purchase and require comprehensive training of users.

In this section, we focus on the two fundamental methods that are commonly used to characterize the rheological properties of plant-based solids: compression testing (texture profile analysis) and shear testing (dynamic shear rheometry).

Compression Testing Typically, a test sample is placed on a flat plate and is then compressed/decompressed using another plate that can move downwards and upwards at a fixed speed (Fig. 4.25). The instrument has sensors that can measure the force acting on one of the plates, as well as the distance between the plates. As a result, the force *versus* distance (or time) profile can be measured as the material is compressed and then decompressed. Using measurements of the initial height and surface area of the sample, the stress *versus* strain profile can be calculated by the instrument software. Typically, a researcher has to specify the speed and maximum amount of compression the sample experiences, *e.g.*, 1 mm/min until 50% compression is reached. Information about the rheological properties of the material can then be obtained from the stress-strain curves, including parameters such as the elastic modulus, yield stress, fracture stress, and fracture strain depending on the nature of the material being tested (Fig. 4.19).

One of the most widely used compression methods is called Texture Profile Analysis (TPA), which involves compressing/decompressing a test material twice and recording the stress-strain profiles throughout this process (Fig. 4.25). Several

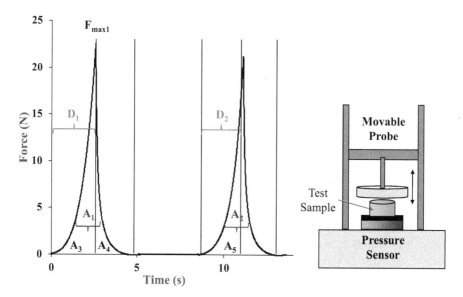

Fig. 4.25 The rheology of solid foods can be characterized by carrying out texture profile analysis (TPA), which involves measuring the force-time profile when the food is compressed/decompressed twice

parameters that can be related to important textural attributes of a food can then be obtained from these profiles (Table 4.6):

Shear Testing The rheological properties of soft solid foods are also often characterized using analytical instruments called dynamic shear rheometers (Fig. 4.18). These instruments are able to provide information about the dynamic rheological properties of viscoelastic materials, such as complex shear modulus (G' and G''). These properties are usually measured as a function of shear stress, frequency, time, or temperature depending on the nature of the information required. Experiments are typically carried out by placing the sample in an appropriate measurement cell, which is typically either a cup-and-bob or cone-and-plate arrangement, and then allowing it to reach the required temperature before the test is started.

Typically, dynamic shear rheology measurements should be performed within the linear viscoelastic region (LVR) of the test material, where the strain is low enough to prevent disruption of the material during the test. Consequently, an initial experiment should usually be carried out to establish the LVR by measuring the change in shear modulus with strain: the shear modulus is typically relatively constant at low strains but then falls steeply once a particular strain is exceeded due to the disruption of the material (Fig. 4.26).

Commercial instruments usually report the storage and loss modulus (G' and G'') of a material. Typically, these parameters are measured as a function of different variables, such as the applied shear stress, frequency, temperature, or time. Measurements are made as a function of shear stress/strain to ascertain the LVR and

4.3 Texture

Table 4.6 Textural characteristics that can be obtained from texture profile analysis (TPA) of foods

Parameter	Physical Meaning	Calculation
Hardness	A measure of the resistance of the material to compression	F_{max1}
Fracturability	A measure of the force required to first fracture a material	F_{frac1}
Cohesiveness	A measure of how well a material maintains its texture after the first deformation	A_2/A_1
Springiness	A measure of how well the material is able to spring back to its original shape after it is deformed and then allowed to rest for a specified time	D_2/D_1
Resilience	A measure of how well a material regains its original textural properties after a compression.	A_4/A_3
Gumminess (semi-solids)	A measure of the cohesiveness and stickiness of semi-solid foods.	Hardness × cohesiveness
Chewiness (solids)	A measure of the energy required to chew solid foods.	Hardness × cohesiveness × springiness

These parameters are based on those defined by the Texture Technologies Corporation (texturetechnologies.com). Note, that one must be careful to only select those parameters that are suitable for the particular material being tested.
https://texturetechnologies.com/resources/texture-profile-analysis#tpa-measurements
Key: Hardness, F_{max1} = Maximum force measured during the first compression; Fracturability, F_{frac1} = Force where the first peak is observed during the first compression (not always seen); Cohesiveness, A_2/A_1 = area under second peak divided by area under first peak; Springiness, D_2/D_1 = ratio of the distances from the start of compression until the maximum is reached for peak 2 and peak 1; Resilience, A_4/A_3 = is the area under the upstroke divided by the area under the downstroke during the first compression (see Fig. 4.25)

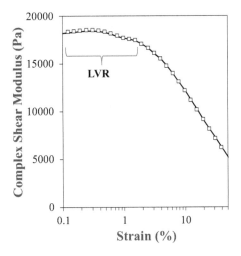

Fig. 4.26 Typically, it is important to measure the rheological properties of solid plant-based foods at low strains that fall within the linear viscoelastic region (LVR)

breaking stress/strain, whereas they are made as a function of frequency to provide information about the nature of the rheological properties of a material. Viscoelastic materials have characteristic relaxation times that correspond to the times required for the reorganization of certain structural elements to occur when a stress is applied. If a structural element has sufficient time to reorganize during the period that the stress is applied, then it can flow and exhibits fluid-like properties. Conversely, if a structural element does not have sufficient time to reorganize, then it cannot flow and exhibits solid-like properties. As the frequency of the applied shear stress increases, the time of each compression-expansion cycle decreases. Consequently, a material should become more solid-like as the frequency is increased. By measuring the shear modulus as a function of frequency it is therefore possible to obtain useful information about the characteristic relaxation times associated with the rearrangement of its structural elements, which may provide an insight into the microstructure and interactions in the system (Broedersz & MacKintosh, 2014).

Dynamic shear rheology measurements can be made as a function of time to determine the rate at which the rheological properties of a material change during processing or storage. For instance, a gelling agent may be added into a food matrix and then the change in shear modulus can be measured over time to determine how rapidly the material sets. Measurements can also be made when a material is heated or cooled at a controlled rate to identify critical temperatures where a material undergoes changes in its rheological properties. These changes may occur due to a variety of different physicochemical phenomena, such as the unfolding and aggregation of globular proteins, helix-coil transitions of some polysaccharides and proteins, or melting-crystallization of fats. For some plant-based foods, it is important to mimic the temperature-dependance of the rheological behavior of the corresponding animal-based foods, such as meat, seafood, or eggs. This is especially important when simulating the cooking properties of animal-based foods. Consequently, the temperature-dependence of the rheological properties of plant-based foods should be measured and matched to those made on the corresponding animal-based ones. As an example, the change in the complex shear modulus of an egg white protein solution with temperature is shown in Fig. 4.27 as it was heated and cooled at a controlled rate. When developing a plant-based version of an egg product it may therefore be important to simulate this kind of thermal behavior.

In some cases, the data from dynamic shear rheology measurements are reported as the magnitude of the complex modulus (G^*) and phase angle (δ), which can be calculated from the storage and loss modulus:

$$G^* = \sqrt{G'^2 + G''^2} \tag{4.19}$$

$$\delta = \tan^{-1}\left(\frac{G''}{G'}\right) \tag{4.20}$$

As mentioned earlier, the phase angle provides useful information about the viscoelastic characteristics of a material: $\delta = 0°$ for an ideal solid; $\delta = 90°$ for an ideal

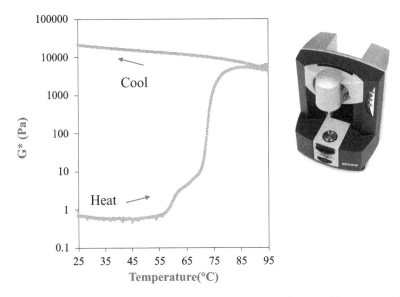

Fig. 4.27 Change in complex shear modulus with temperature for egg white measured by a dynamic shear rheometer during heating and cooling. Data kindly supplied by Hualu Zhou and Giang Vu (UMASS). The image of the rheometer (Kinexus) used to make the measurements was kindly provided by Netzsch (Selb, Germany)

liquid; and $0 < \delta < 90°$ for a viscoelastic material (Fig. 4.24). Thus, the smaller the phase angle of a material, the more solid-like it is. The gel point of a material is typically defined as the point where the phase angle first falls below 45° but this value depends on the frequency of the applied stress used and so this value should be specified when reporting gel points.

On a final note, dynamic shear rheometers are powerful tools for providing information about the rheology of plant-based foods but they are relatively expensive to purchase, and users often require comprehensive training to ensure that measurements are properly performed, analyzed, and presented.

4.3.3 Practical Considerations

In this section, we highlight some of the issues that should be considered when carrying out rheological analyses on plant-based foods. First, the properties of all the test samples should be kept as similar as possible (such as their size and shape), which means that they should be prepared in a consistent and carefully controlled manner. For instance, for meat-like products, it may be important to prepare cylindrical samples with the same height and diameter, and then store them under the same conditions (time, temperature, and humidity) before analysis. Second, it is important to account for any factors that may affect the accuracy of the results, such as "slip" at the boundary between a test sample and the measurement cell, gravitational separation of particles within the measurement cell, or the presence of particles that are too large relative to the gap in the measurement cell. Third, plant-based foods should not be exposed to excessive mechanical forces before carrying out the rheological measurements because this can lead to structural changes and water loss, which can alter the response of the material to the applied shear stress. Fourth, it may be important to prevent evaporation of moisture from the sample during the test (especially when heating for prolonged periods), which can be achieved by using specially designed covers or by placing a thin layer of mineral oil on any exposed surfaces.

4.4 Stability

The stability of a plant-based food can be defined as its ability to resist changes in its properties over time, which depends on various physical, chemical, and biological processes (McClements et al., 2021). The physical stability of a food is determined by its ability to resist changes in the spatial location of the different ingredients over time, *e.g.,* due to phase separation, aggregation, gravitational separation, or mass transport processes. The chemical stability is governed by the rate of various chemical reactions that lead to undesirable changes in the type of molecules present in the food, such as oxidation, reduction, or hydrolysis reactions that can lead to loss of nutrients, the formation of off flavors, or the degradation of desirable colors or flavors. The biological stability of foods depends on the growth of spoilage or pathogenic microorganisms, such as bacteria, yeasts, or molds, that can lead to undesirable changes in product quality or to foodborne illnesses. The resistance of a specific food product to these different processes determines its shelf-life. Consequently, it is important to carefully establish the relative importance of the various physical, chemical, and biological processes that occur in plant-based foods during processing, storage, and utilization. This knowledge can then be used to develop effective strategies to inhibit undesirable changes in product properties, such as controlling storage conditions, packaging, preservatives, or food matrix design.

In this section, we briefly highlight the physicochemical origin of a number of instability mechanisms that occur in various plant-based foods (Fig. 4.28). In practice, each food matrix is unique and so it is important to empirically establish the most important mechanisms for each particular product.

4.4 Stability

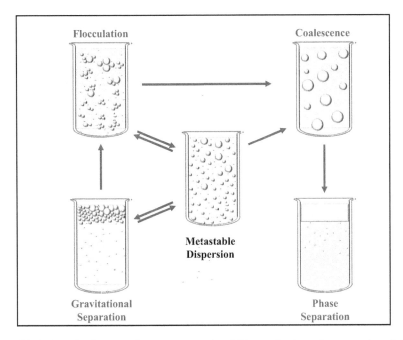

Fig. 4.28 Schematic diagram of most common instability mechanisms that occur in colloidal delivery systems: gravitational separation, flocculation, coalescence, Ostwald ripening and phase inversion

4.4.1 Gravitational Separation

Plant-based foods contain different kinds of particles that may cream (rise upwards) or sediment (fall downwards) depending on their density (McClements, 2015). Particles that are less dense than water (such as air bubbles, oil bodies, or fat droplets) have a tendency to move upwards, whereas particles that are denser than water (such as protein aggregates, starch granules, or plant cell fragments) have a tendency to move downwards (McClements, 2015). The creaming or sedimentation of the particles in a product is typically unwanted because it leads to undesirable changes in the appearance, such as a cream layer at the top or a sediment layer at the bottom. The rate at which gravitational separation occurs can be described using Stokes law, which was originally derived by considering the balance of gravitational and viscous forces acting on a rigid spherical particle as it moved through an ideal fluid:

$$v = -\frac{gd^2(\rho_2 - \rho_1)}{18\eta_1} \tag{4.21}$$

Here, v is the velocity that the particle moves upwards, d is the diameter of the particles, ρ_1 is the density of the surrounding fluid, ρ_2 is the density of the particles, g is the gravitational constant, and η_1 is the viscosity of the surrounding fluid. Stokes' law predicts that the rate of gravitational separation increases as the density contrast increases, the particle size increases, or the viscosity of the surrounding fluid decreases. The direction the particles move is governed by the sign of v, which depends on the relative densities of the particles and surrounding liquid: when v is positive ($\rho_2 < \rho_1$) the particles move upwards but when it is negative ($\rho_2 > \rho_1$) they move downwards. The impact of particle size on the creaming velocity of model fat droplets and plant-tissue fragments suspended in aqueous solutions with different viscosities was predicted using Stokes' law (Table 4.7). Typically, a creaming velocity above about 1 mm per day will lead to fairly rapid phase separation of a colloidal system. These calculations show that it is important to have relatively small particles in products with low viscosities, such as plant-based milks, so as to avoid excessive creaming or sedimentation during storage. This equation also highlights the fact that gravitational separation is not expected to be important in all plant-based foods. It is only important in fluids that have relatively low viscosities, such as plant-based milks or fluid eggs (prior to cooking). In contrast, the matrices

Table 4.7 Calculations of the impact of particle size on the creaming velocity of fat droplets (ρ_2 = 930 kg m^{-3}) and plant-tissue fragments (ρ_2 = 1350 kg m^{-3}) suspended in aqueous solutions (ρ_1 = 1050 kg m^{-3}) with different viscosities (1-500 mPa s) using Stokes' Law

Fat globules Diameter (μm)	Creaming velocity (mm/day) Viscosity (mPa s)					
	1	5	10	50	100	500
0.1	0.1	0.0	0.0	0.0	0.0	0.0
0.2	0.2	0.0	0.0	0.0	0.0	0.0
0.5	1.4	0.3	0.1	0.0	0.0	0.0
1	5.8	1.2	0.6	0.1	0.1	0.0
2	23.0	4.6	2.3	0.5	0.2	0.0
5	144	28.8	14.4	2.9	1.4	0.3
10	576	115	57.6	11.5	5.8	1.2
20	2304	461	230	46.1	23.0	4.6
50	14,400	2880	1440	288	144	28.8
Plant-fragments Diameter (μm)	Creaming velocity (mm/day) Viscosity (mPa s)					
	1	5	10	50	100	500
0.1	−0.1	0.0	0.0	0.0	0.0	0.0
0.2	−0.6	−0.1	−0.1	0.0	0.0	0.0
0.5	−3.6	−0.7	−0.4	−0.1	0.0	0.0
1	−14.4	−2.9	−1.4	−0.3	−0.1	0.0
2	−57.6	−11.5	−5.8	−1.2	−0.6	−0.1
5	−360	−72	−36.0	−7.2	−3.6	−0.7
10	−1440	−288	−144	−28.8	−14.4	−2.9
20	−5760	−1152	−576	−115	−57.6	−11.5
50	−36,000	−7200	−3600	−720	−360	−72.0

surrounding the particles in plant-based meat, seafood, and cooked egg products are typically gelled, so that gravitational separation is not an issue because the particles cannot move.

Stokes' law highlights various approaches that can be employed to inhibit gravitational separation in colloidal plant-based foods:

- *Decrease particle dimensions*: The rate of creaming or sedimentation can be reduced by decreasing the particle size. This might be achieved by using a mechanical homogenization device, such as a high-shear mixer, high-pressure valve homogenizer, microfluidizer, or sonicator to reduce the dimensions of the particles in the system. Alternatively, the particle size may be reduced by chemical (acid/alkali) or enzymatic treatment of the product. In addition, it may be important to prevent the particles from aggregating during storage, as this would increase their dimensions and therefore lead to faster separation. This can often be achieved by controlling the sign, magnitude, and range of the colloidal interactions acting between the particles, which is usually carried out by selecting appropriate emulsifiers and/or controlling solution conditions (such as pH, ionic strength, and temperature). Strategies for preventing particle aggregation are discussed in the following section.
- *Increase viscosity*: The rate of creaming or sedimentation can be reduced by increasing the viscosity of the fluids surrounding the particles. This can be achieved by adding thickening or gelling agents, such as hydrocolloids like xanthan, guar, or locust bean gum. Nevertheless, the type and concentration of hydrocolloids added have to be controlled to avoid causing depletion or bridging flocculation of the particles, which would accelerate gravitational separation (McClements, 2015). Moreover, the final viscosity should be consistent with the quality attributes expected for the final product. For instance, a plant-based milk should not be too viscous or consumers will find it undesirable.
- *Decrease the density contrast*: In principle, gravitational separation can be inhibited by reducing the density contrast ($\Delta\rho = \rho_2 - \rho_1$) between the particles and the surrounding fluid (McClements, 2015). In practice, this is difficult to achieve because plant-based oils typically exhibit a limited range of densities (between about 910 and 930 kg m^{-3}). Similarly, plant tissue fragments have relatively high densities that are difficult to control (around 1500 kg m^{-3}). For some plant-based foods (like milk analogs) it may be possible to increase the density of the fat droplets by wrapping thick layers of dense biopolymers around them. For cow's milk, the density of the milk fat globules can be altered by changing the temperature so as to change their solid fat content. Typically, the density of the fat globules increases as the solid fat content increases, which reduces the density contrast between the oil and water phases, thereby reducing the tendency for creaming to occur. However, one must be careful that partial coalescence does not occur, otherwise this will promote fat globule aggregation, which can alter the stability and texture of the final product (Fredrick et al., 2010). A similar approach may be used to control the density contrast for plant-based products,

i.e., use a fat phase that partially crystallizes so as to increase the density of the fat droplets.

4.4.2 Particle Aggregation

The shelf-life of fluid plant-based foods may be reduced because of aggregation of the colloidal particles they contain during storage, such as the fat droplets or plant tissue fragments in milk analogs (McClements, 2015; McClements et al., 2021). The aggregation of the particles in these products may reduce their quality attributes due to a number of effects: (*i*) large clumps may be formed that are visible to the eye; (*ii*) the increase in particle size may accelerate creaming or sedimentation leading to visible layering within the product; (*iii*) aggregation may lead to the formation of a 3D network that thickens or gels the product; and (*iv*) the presence of large particles may have an undesirable impact on the mouthfeel of the product. Consequently, it is usually desirable to avoid aggregation of the particles during storage. In more solid plant-based foods, the aggregation of the polymers or particles may increase the gel strength, which may be either desirable or undesirable depending on the product.

The tendency for particles or polymers to aggregate with each other depends on a delicate balance of attractive and repulsive forces acting amongst them (McClements, 2015). The major forms of attractive forces are van der Waals,

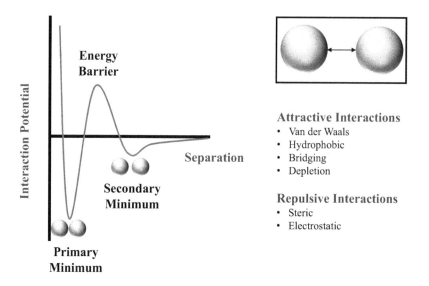

Fig. 4.29 The colloidal interactions between two particles can be modeled by calculating the interaction potential versus particle separation. The overall interaction depends on the combination of attractive and repulsive interactions, which is system specific. For electrostatically stabilized systems, the profile often has a primary minimum, secondary minimum, and an energy barrier

4.4 Stability

hydrogen bonding, hydrophobic, and salt bridge interactions, whereas the main repulsive ones are steric and electrostatic interactions (Fig. 4.29). Particles or polymers tend to stick together when the attractive forces dominate, but not when the repulsive forces dominate (McClements, 2015). In colloidal plant-based foods, such as milk, cream, or fluid egg analogs it is important to prevent the particles (fat droplets or plant tissue fragments) from aggregating with each other. There are a number of ways of achieving this (Fig. 4.30):

- *Reduce the particle size*: Typically, the strength of attractive and repulsive interactions decreases as the size of the particles decreases. As a result, the overall interactions become weaker relative to the thermal energy of the system (k_BT). Consequently, reducing the particle size can reduce the tendency for aggregation to occur in systems where there is a net attraction between the particles.
- *Increase steric stabilization*: The presence of polymeric materials at the surface of colloidal particles, such as polysaccharides or proteins, generates a short-range but strong steric repulsive force. This strong repulsion arises when polymer chains adsorbed to different particle surfaces overlap with each other as the two particles come close together, which leads to an unfavorable reduction in their configurational entropy. Typically, the effectiveness of steric stabilization increases as the thickness and density of the adsorbed layer of polymer molecules increases. Colloidal particles that are stabilized by this mechanism are usu-

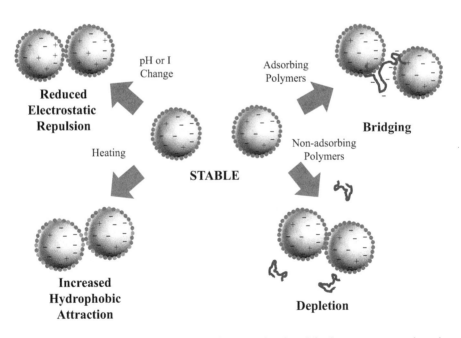

Fig. 4.30 The emulsifier-coated fat droplets in some plant-based foods may aggregate through numerous mechanisms, including bridging flocculation, depletion flocculation, reduced electrostatic repulsion, or increased hydrophobic attraction

ally more resistant to changes in pH and ionic strength than those stabilized by electrostatic repulsion.
- *Increase electrostatic stabilization*: The presence of charged groups on the surfaces of colloidal particles leads to an electrostatic repulsion between them that opposes their tendency to aggregate. The magnitude of this electrostatic repulsion increases as the number of charged groups per unit surface area of the particle increases. The number of ionized groups on the surfaces of the colloidal particles in plant-based foods often depends on the pH of the surrounding aqueous phase. This is because they have proteins, polysaccharides, or phospholipids at their surfaces, whose electrical characteristics are pH dependent. In particular, carboxyl and amino groups are protonated at low pH values ($-COOH$ and $-NH_3^+$) but are non-protonated at high pH values ($-COO^-$ and $-NH_2$). The ionic composition, particularly the type and concentration of counter ions, of the surrounding aqueous phase also influences the magnitude of the electrostatic interactions. An increase in ionic strength causes the magnitude and range of the electrostatic repulsion to decrease as a result of electrostatic screening effects, *i.e.*, the propensity for counter-ions to accumulate around oppositely charged surface groups on colloidal particles. Moreover, multivalent counter-ions may bind to the particle surfaces and change their net charge or act as salt bridges between neighboring charged particles, thereby promoting their aggregation.
- *Reduce hydrophobic attraction*: Certain types of colloidal particles in plant-based foods have non-polar groups on their surfaces that are exposed to water, which is thermodynamically unfavorable because of the hydrophobic effect. For instance, fat droplets coated by globular plant proteins (such as lentil, pea, or soy protein) may have some exposed non-polar groups on their surfaces, especially after they are heated above the thermal denaturation temperature of the proteins. Moreover, some plant proteins are inherently hydrophobic because of the nature of the amino acids in the polypeptide chains, such as zein and gliadin. As a result, there is a strong long-range hydrophobic attraction between the non-polar regions on the surfaces of different colloidal particles, which is often enough to promote particle aggregation. For this reason, it may be necessary to reduce the strength of the hydrophobic attraction by decreasing the number of non-polar surface groups exposed to water. This might be achieved by ensuring that globular proteins are not heated above their thermal denaturation temperatures or by adding surfactants or amphiphilic polymers that adsorb to the non-polar patches and cover them.
- *Reduce depletion attraction*: Plant-based colloidal dispersions may have appreciable levels of non-adsorbed biopolymers within the aqueous phase surrounding them. These biopolymer molecules are excluded from a region around each colloidal particle that is approximately equal to the radius of hydration of the biopolymers in solution. This phenomenon arises because the center of the biopolymer molecules cannot get closer to the surface of the particle than this distance. As a result, there is a biopolymer concentration gradient within the

4.4 Stability

system, which generates an osmotic stress that tends to push the particles together. The strength of this depletion attraction is influenced by biopolymer molecular weight and conformation and tends to increase as the number of non-adsorbed biopolymer rises. In this case, particle aggregation can be prevented by making sure that the concentration of non-adsorbed biopolymers in the aqueous phase is not enough to generate a strong depletion attraction.

- *Reduce bridging effects*: Some fluid plant-based foods contain biopolymer molecules that are attracted to the surfaces of the colloidal particles (such as fat droplets or oil bodies). As a result, a single biopolymer molecule may become attached to the surfaces of two or more colloidal particles, thereby causing them to become flocculated. Electrostatic interactions between charged biopolymers and oppositely charged colloidal particles are the most common example of this kind of instability mechanism. Bridging effects can therefore be reduced by designing the system so that it does not contain biopolymer molecules that are strongly attracted to the surfaces of the colloidal particles.

Colloidal particles may aggregate with each other through a number of mechanisms depending on the system properties and the nature of the interactions involved (Fig. 4.28). Flocculation occurs when two or more particles stick together, but they do not fuse together into a single particle. In contrast, coalescence occurs when two or more particles fuse together into a single larger particle. In the case of flocculation, the nature of the aggregates formed, such as their size, shape, and robustness, depends on the magnitude of the attractive forces acting between them. Flocs held together by relatively weak forces tend to be more compact and are easier to disrupt by dilution or shearing. Conversely, flocs held together by relatively strong forces tend to have more open structures and are more resistant to dilution and shearing. These differences in floc properties can have a pronounced influence on the rheology and stability of colloidal dispersions. Flocculation can lead to a pronounced increase in the shear viscosity of a product and may also lead to faster gravitational separation (in dilute systems) because of the increase in particle size. In concentrated systems, however, flocculation may inhibit gravitational separation because a 3D particle network is formed that prevents the particles from moving and leads to semi-solid properties. In some systems, the particles may also aggregate through other mechanisms. For instance, colloidal dispersions containing partly crystallized fat droplets may undergo partial coalescence, where a fat crystal in one droplet penetrates into the liquid region in another droplet, thereby forming a link between them. This can lead to the formation of large clusters that increase the viscosity and can eventually lead to phase separation. Partial coalescence is particularly important in dairy products, like ice cream, whipped cream, and butter, since it leads to the formation of desirable textures and stabilities. Consequently, it may be important to mimic this behavior in plant-based dairy analogs, which is discussed in the chapters on milk and dairy analog products.

4.4.3 Phase Separation

In some plant-based foods, instability may occur due to the separation of different phases within the product. For instance, the oil may separate from emulsified plant-based foods like meats, seafood, and dressings (oiling off) or the water may separate from hydrogels like yogurts (syneresis) and meats during heating (cooking loss). Moreover, different biopolymer phases within a product may separate from each other, such as protein-rich and polysaccharide-rich phases.

- *Oiling off*: Many plant-based foods contain regions of oil dispersed in an aqueous matrix, for example milk analogs contain fat droplets or oil bodies that simulate the fat globules in cow's milk, egg analogs contain fat droplets that simulate the lipoproteins in hen's eggs, and meat analogs contain fat domains that simulate the adipose tissue in muscle foods. Oil and water are immiscible because of the hydrophobic effect and therefore have a tendency to separate over time or in response to certain environmental changes (such as heating or shearing). In many cases, the separation of the oil is undesirable because it leads to adverse changes in product quality, such as an alteration in the appearance, texture, or mouthfeel. In some cases, however, it may be desirable, such as to facilitate the frying of plant-based meats or seafood.
- *Syneresis*: Many plant-based foods contain water trapped within a biopolymer matrix through hydration or capillary forces. This water may separate over time or when the environmental conditions are changed, leading to undesirable changes in product appearance and texture. A common example of this phenomenon is the layer of water that forms at the top of some yogurts during storage. This problem can be avoided by controlling the properties of the 3D network formed by the biopolymer molecules, such as the pore size and surface chemistry. This may require the addition of other ingredients to the product, such as hydrocolloids like xanthan, guar, or locust bean gum.
- *Cooking loss*: The amount of fluids lost from a plant-based food during cooking has an important impact on the quality attributes of the final product. In particular, the appearance, texture, and mouthfeel (such as juiciness) of plant-based meat and seafood products often depends on the degree of cooking loss. The fluids lost are mainly comprised of water but may also contain other components, such as fat droplets, salts, soluble proteins, and carbohydrates. The composition and structure of a plant-based food may therefore have to be controlled so that a consistent well-defined amount of cooking loss occurs during food preparation. This requires an understanding of the factors influencing the loss of fluids during cooking. Cooking loss may occur due to water evaporation, squeezing out of water molecules from within porous biopolymer structures due to a change in their structure caused by alterations in the molecular interactions (such as protein unfolding and aggregation), or other factors.
- *Biopolymer phase separation*: Some plant-based foods contain biopolymer domains that have different compositions, *e.g.*, a protein-rich and a polysaccharide-rich domain. Indeed, controlled biopolymer phase separation is being utilized to

create meat-like fibrous structures within meat analogs (McClements et al., 2021). In this case, it is important to form phase separated regions that have the appropriate structures during the food manufacturing process and then to ensure that they do not break down during storage, transport, and food preparation. This can often be controlled by gelling one or both of the biopolymer domains, although this will affect the textural attributes of the food product. The gels used may also have to be designed to resist environmental stresses, such as changes in temperature, ionic strength, pH, or mechanical forces.

4.4.4 Chemical Degradation

The quality attributes of plant-based foods may deteriorate during food manufacturing, storage, and preparation due to chemical reactions that alter the nature of the molecules present. There are many different kinds of chemical reactions that can occur, which depend on the type of ingredients present, as well as the environmental conditions that the food is exposed to, such as light, oxygen, heating, pH changes, and mineral ions. For each type of plant-based food it is therefore important to identify any possible detrimental chemical reactions that may occur. A few examples are given here:

- *Oxidation*: Some of the ingredients used to formulate plant-based foods are prone to oxidation, which leads to undesirable changes in product quality (Jacobsen, 2015; Jacobsen et al., 2013). For instance, the oxidation of lipids leads to the generation of reaction products that have unpleasant aromas and tastes (rancidity), as well as may exhibit toxicity. Foods containing polyunsaturated lipids, such as omega-3 fatty acids (like algal or flaxseed oils) are particularly prone to this problem. Other types of hydrophobic functional ingredients may also undergo oxidation within foods during storage or processing, such as carotenoids, which can lead to a reduction in their color intensity and a loss of their beneficial biological activities. Proteins are also susceptible to oxidation, which can lead to a reduction in their functional attributes, such as their ability to act as emulsifying, foaming, or gelling agents. Oxidation reactions can be inhibited by controlling environmental conditions, food matrix effects, or using additives. For instance, the rate of lipid oxidation can be slowed by reducing the exposure of a food to light, oxygen, or heat, by adding antioxidants or chelating agents, or by forming physical barriers that limit the ability of pro-oxidants to come into contact with labile ingredients. It should be noted that in some cases a limited amount of oxidation may be desirable, since it leads to the generation of aroma molecules that provide a desirable flavor profile.
- *Hydrolysis*: A number of the functional ingredients used to formulate plant-based foods are susceptible to hydrolysis, *i.e.,* the covalent bond between two atoms is broken due to the addition of a water molecule. For instance, pectin, which is a polysaccharide that can be used to formulate plant-based foods, is susceptible to

hydrolysis by acids and enzymes (Fraeye et al., 2007; Garna et al., 2006). As a result, its functional attributes may be reduced, such as its ability to thicken solutions, form gels, or create certain structures. Other kinds of biopolymer-based ingredients may also be hydrolyzed under some conditions found in plant-based foods during or after processing (Aida et al., 2010; Karlsson & Singh, 1999). Lipids are also susceptible to hydrolysis reactions during storage or processing as a result of chemical or enzymatic reactions (Swapnil & Arpana, 2019). For example, free fatty acids may be released from triacylglycerols and phospholipids, which can reduce food quality. An understanding of the susceptibility of different ingredients to hydrolysis and the factors that affect it is therefore important.

- *Crosslinking*: A number of covalent crosslinking reactions may occur between the ingredients in plant-based foods that alter their physicochemical and sensory properties. For instance, disulfide bonds may form between proteins containing sulfhydryl groups, especially when they are heated around neutral pH values (Nagy, 2013). The formation of these bonds may increase the gel strength of the system, which may be either desirable or undesirable depending on the nature of the product.
- *Maillard reaction*: The Maillard reaction is a nonenzymatic browning reaction that occurs between amino acids (proteins) and reducing sugars, especially at high temperatures, and leads to the formation of a complex mixture of end products that contribute to the color and flavor of cooked foods (Aljahdali & Carbonero, 2019; Lund & Ray, 2017). In particular, it is largely responsible for the characteristic dark brown color of the surfaces of animal-based cooked foods, like grilled, fried, or baked meat and seafood. The rate of the reaction increases at high temperatures and intermediate moisture contents, which is why the exteriors of the foods become dark brown (higher temperatures/lower moisture contents) while the interiors do not (lower temperatures/higher moisture contents). Indeed, the reaction usually starts to occur rapidly when the temperature exceeds about 140 to 165 °C. The extent of the Maillard reaction, and therefore the color and flavor of the final product, depend on the initial composition and structure of the food matrix, as well as the cooking conditions used (such as time and temperature). Consequently, it is important to optimize food formulation and cooking conditions to obtain plant-based foods that mimic the desirable appearances and flavors of cooked foods, such as meats or seafoods. It should be noted, however, that potentially toxic compounds (such as acrylamide) can be formed as end products if the reaction is carried out for too long or at too high a temperature (Aljahdali & Carbonero, 2019). This problem can be controlled by ensuring that the food is not overcooked or by adding certain types of additives that inhibit the Maillard reaction.
- *Caramelization*: Caramelization is a nonenzymatic browning reaction that occurs due to the thermal decomposition of sugars at sufficiently high temperatures (105–180 °C depending on sugar type). This reaction may therefore be important in cooked plant-based foods that contain high sugar levels.

In general, it is important to identify the different kinds of chemical reactions that can occur within a particular plant-based food formulation and to understand the key factors that impact their rate and reaction path (such as temperature, oxygen, light, pH, and other ingredients). The rate, extent, and direction of the reaction can then be controlled by manipulating food composition or food preparation procedures. For instance, to obtain a desirable brown crust on the surface of a plant-based burger the manufacturer may have to carry out experiments to establish the optimum cooking temperature and time, which can then be included in the cooking instructions printed on the package of the commercial product.

4.4.5 Microbial Contamination

Like all foods, the quality and safety of plant-based foods may deteriorate throughout their shelf life due to contamination with spoilage or pathogenic organisms. Consequently, it is important that the manufacturer takes appropriate steps to prevent or reduce microbial contamination. This involves maintaining a sanitary food production and distribution system, using processing operations that effectively deactivate microorganisms (such as thermal processing), using suitable packaging materials, controlling the storage conditions, and adding preservatives (such as antimicrobials). In general, it is important for food manufacturers to identify all potential sources of microbial contamination and develop a robust system to prevent, remove, or deactivate them. This varies considerably from food to food and is beyond the scope of this book.

4.4.6 Quantification of Stability

Changes in the stability of plant-based foods during storage or in response to alterations in their environment can be monitored using various kinds of analytical instruments and testing methods, which will be discussed in detail for each type of major plant-based food category in later chapters. In this section, we provide a general overview of the approaches that can be utilized.

The visible separation of foods due to gravitational separation (creaming or sedimentation) can be monitored by taking digital photographs of test samples kept in clear containers during storage. The rate and degree of separation can be quantified by measuring the heights of the boundaries between different layers, such as the cream at the top or the sediment at the bottom. However, it is sometimes challenging to unambiguously discern the location of these boundaries. Consequently, gravitational separation is often monitored using more sophisticated analytical instruments. Some of the most commonly used are based on laser profiling, nuclear

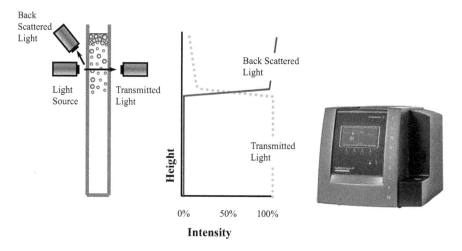

Fig. 4.31 The stability of fluid plant-based foods to sedimentation or creaming can be conveniently monitored using instruments that measure the change in transmitted and backscattered light with sample height over time. (Image of instrument kindly provided by Formulaction (Toulouse, France))

magnetic resonance (NMR) imaging, and X-ray tomography (McClements, 2015). For instance, the fraction of light reflected from the surface of a food material can be measured as a function of its height using a laser beam that can be moved upwards and downwards (Fig. 4.31). The reflectance typically increases as the particle concentration increases because more back scattering of the incident light occurs. Consequently, the creaming or sedimentation of particles can be monitored by measuring the change in reflectance with sample height over time. NMR imaging and X-ray tomography can provide detailed 3D images of the distribution of different components within a food (such as water and fat) and so can be used to monitor gravitational separation processes.

Changes in the aggregation state of the colloidal particles in plant-based foods, such as the fat droplets, oil bodies, or plant tissue fragments in fluid milk and egg analogs, are typically monitored using light scattering or microscopy methods (Fig. 4.32). Dynamic light scattering can be used to assess aggregation when the particles are relatively small (around 10 nm to 10 µm), whereas static light scattering can be used when the particles are relatively large (200 nm to 1000 µm). The most common forms of microscopy used to provide information about the aggregation state of particles are optical, confocal fluorescence, and electron microscopy. Phase separation processes can also be monitored using microscopy methods, as well as NMR imaging, multispectral imaging, and X-ray tomography methods.

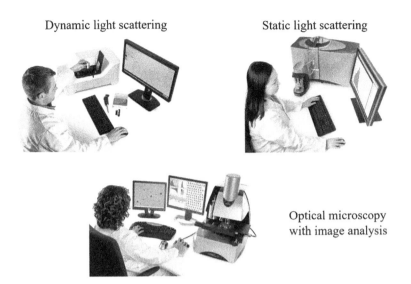

Fig. 4.32 Examples of instruments that can be used to characterize the particle size, morphology, and aggregation in colloidal plant-based foods. (Images kindly provided by Malvern Panalytical (Malvern, UK))

4.5 Fluid Holding and Cookability Properties

The ability of many solid plant-based foods, including meat, seafood, cheese, yogurt, and eggs, to retain or release fluids during their manufacture, distribution, and preparation plays a major role in determining their overall quality attributes. The fluids that are of concern in plant-based foods may be simple liquids like oil or water, solutions like salt or sugar solutions, or dispersions like emulsions or suspensions. The retention or release of these fluids has an impact on the way the foods look, feel, and taste, as well as on their shelf life. As an example, the formation of an aqueous layer on top of a yogurt is not appealing to consumers (Grasso et al., 2020). Moreover, the tenderness and juiciness of meat and seafood products depend on the amount of water they retain during cooking (Cornet et al., 2021). The release of fluids from meat or seafood analogs during cooking also impacts the sizzling sounds they make, which is an important sensory component of their desirable quality attributes. As a result, controlling the retention or release of fluids from plant-based foods is critical to ensure they have the required physicochemical properties, sensory attributes, and functional performance.

One of the most important quality attributes related to fluid holding properties of solid foods is the water holding capacity (WHC), which is a measure of the ability of the food to retain water (plus any dissolved solutes) when an external stress is applied, such as a normal force, centrifugaal force, or during a heat treatment (Cornet et al., 2021; Grasso et al., 2020). The concept of WHC is often ill-defined and has been defined differently by different authors. One common definition of WHC that can be used for meat and its plant-based analogs is given by the following expression (Gaviria et al., 2021):

$$\text{WHC} = 100 \times \frac{m_R}{m_T} \qquad (4.22)$$

Here, m_R is the mass of water retained in the system after some external force (such as centrifugation) is applied and m_T is the total mass of water present in the original system. The ability of solid foods to retain water is typically a result of the existence of a 3D network of entangled and/or crosslinked biopolymers (usually proteins or polysaccharides). These 3D networks hold water through three main mechanisms: (*i*) water-biopolymer mixing effects; (*ii*) ion effects (such as salts); (*iii*) network elastic deformation effects (Cornet et al., 2021; van der Sman et al., 2013). The water-biopolymer mixing effects depend on any enthalpy (*e.g.,* molecular interactions) and/or entropy (*e.g.,* mixing or configuration entropy) changes that occur when the water and biopolymer molecules are combined. As a result, it depends on the nature of the interactions between the water and biopolymer molecules, such as hydrogen bonding, hydrophobic interactions, and electrostatic interactions, as well as the surface area of the biopolymer network. The ion effects are due to differences in the concentration of the ions (such as mineral ions) inside and outside of the biopolymer network, which results in an osmotic pressure. This concentration gradient may arise due to preferential attraction of counter-ions to the surfaces of oppositely charged biopolymers in the gel network, such as Na^+ ions to anionic surface groups or Cl^- ions to cationic surface groups. The network elastic deformation effects are due to the mechanical resistance of the biopolymer network to compression or extension when an external force is applied. Mathematical models have been developed that can be used to describe the water holding properties of plant-based foods consisting of biopolymer gel networks (Cornet et al., 2021). The ability of a biopolymer network to hold water is usually increased when the number of crosslinks between the biopolymer molecules is raised, since this increases the gel strength, thereby making it more mechanically resistant to compression (Cornet et al., 2021). Thus, it may be possible to inhibit syneresis from plant-based yogurts by having stronger gel networks that are more resistant to compression, provided this does not cause adverse effects on other desirable physicochemical or sensory attributes of the product. The ability of a biopolymer network to hold water also depends on the pore size and it increases with decreasing pore size, since there is then a larger surface area of biopolymers to interact with the water and smaller pores lead to stronger capillary forces that hold the water into the gel network.

4.6 Partitioning, Retention and Release Properties

The location of the different constituents within plant-based foods, as well as the speed at which they move from one location to another, play an important role in determining their quality attributes. The chemical reactivity of some substances, such as natural colors, flavors, and bioactives, depends on whether they are surrounded by oil or water (Choi et al., 2009; Kharat et al., 2017). Typically, chemical degradation reactions occur faster when a substance is dissolved in water than when it is dissolved in oil. The flavor profile of plant-based foods depends on the type, concentration, and timing that different aroma and taste molecules reach the receptors in the human nose and mouth. The ability of a plant-based meat or fish product to change color during cooking may be due to an interaction between two components that are kept separate in the raw product but are made to come together and react with each other during the cooking process. It is therefore important to control the partitioning, retention, and release of the different functional ingredients in plant-based foods. This requires knowledge of the major factors impacting these parameters. The purpose of this chapter is therefore to highlight some of the most important factors influencing the partitioning, retention, and release of molecules in food matrices.

4.6.1 Partitioning Phenomena

Many plant-based foods are multiphase materials containing two or more different phases, with the most common ones being oil and water. In this section, we therefore consider some of the major factors impacting the partitioning of functional molecules in this type of multiphase material. More detailed information about partitioning phenomena can be found in earlier publications (McClements, 2014).

4.6.1.1 Equilibrium Partitioning Coefficients

To a first approximation, the distribution of a substance between the oil and water phases can be quantified by its oil-water equilibrium partition coefficient:

$$K_{OW} = \frac{c_O}{c_W} \qquad (4.23)$$

Here, c_O and c_W are the concentrations of the substance in the oil and water phases, respectively. The value of the partition coefficient depends on the relative affinity of the substance for the oil and water phases. Non-polar substances have a higher affinity for oil and therefore tend to preferentially accumulate in the oil phase ($K_{OW} > 1$), whereas polar substances have a higher affinity for water and therefore tend to preferentially accumulate in the water phase ($K_{OW} < 1$). Typically, the stronger the hydrophobicity of a substance, the higher its oil-water partition coefficient.

4.6.1.2 Partitioning of Substances in Multiphase Systems

It is often important to establish the amount of a particular substance in the different phases in a multiphase system, as this impacts its stability and functionality. To a first approximation, the fraction of the substance in the oil phase (Φ_O) of a food matrix containing both oil and water can be established using the following expression:

$$\Phi_O = \frac{\phi_O K_{OW}}{1 - \phi_O (1 - K_{OW})} \tag{4.24}$$

This equation shows that the fraction of the substance present in the oil phase increases as the oil phase volume fraction (ϕ_O) in the food matrix increases, as well as the oil-water partition coefficient increases (K_{OW}) (Fig. 4.33). This expression can be used to predict the location of substances in multiphase plant-based foods with different compositions, *e.g.*, oil contents. It should be noted, this equation assumes that the concentration of the substance is relatively small and below the saturation limit.

4.6.1.3 Partitioning of Flavors into the Headspace

The flavor profile of a plant-based food depends on the partitioning of volatile flavor molecules between the food and the gas phase above it (Fig. 4.33), as the volatile molecules must reach the aroma sensors within the nose (McClements, 2005). The headspace concentration of flavors depends on the volatility of the flavor molecules, as well as the composition of the food matrix. The partitioning of volatile substances between a food matrix containing oil and water phases, and a gas phase above it requires the definition of two other partition coefficients:

$$K_{GW} = c_G / c_W \text{ and } K_{GO} = c_G / c_O \tag{4.25}$$

Here, K_{GW} and K_{GO} are the gas-water or gas-oil partition coefficients, which describe the distribution of a volatile substance between the gas and water phases and between the gas and oil phases, respectively. More volatile substances have higher K_{GW} values and are therefore more likely to be in the headspace. The overall, partition coefficient between the gas phase and an emulsion is then given by (McClements, 2015):

4.6 Partitioning, Retention and Release Properties

$$K_{GE} = \left(\frac{\phi_O}{K_{GO}} + \frac{(1-\phi_O)}{K_{GW}} \right)^{-1} \tag{4.26}$$

These expressions can then be rearranged to develop an equation that relates the headspace concentration of the flavor molecules to the overall composition of the multiphase food matrix:

$$\Phi_G = \left(1 + \frac{V_E}{V_G} \left[\frac{\phi_O K_{OW}}{K_{AW}} + \frac{(1-\phi_O)}{K_{AW}} \right] \right)^{-1} \tag{4.27}$$

Here, Φ_G is the mass fraction of the flavor molecules in the gas phase. This is important because, flavor partitioning in plant-based foods has an important impact on their sensory attributes and so it is important to understand the main factors impacting the amount of flavor in the headspace. The above equation can be used to predict the impact of the fat content of plant-based foods on the fraction of flavor molecules in their headspace and therefore their flavor intensity (Fig. 4.33). These predictions show that the concentration of non-polar flavors ($K_{OW} > 1$) in the headspace decreases as the fat content increases, whereas the concentration of polar flavors ($K_{OW} < 1$) has the opposite effect. In reality, various other factors have to be taken into account, such as the binding of flavor molecules to proteins, polysaccharides, micelles, or other food ingredients, which reduces their headspace concentration and therefore the flavor intensity (McClements, 2015). In addition, the flavor release kinetics from the food matrix must also be considered.

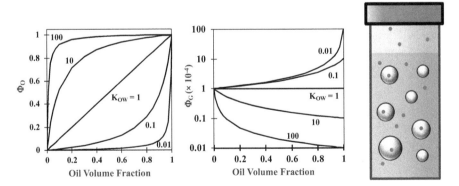

Fig. 4.33 Calculated impact of oil concentration and partition coefficient on the concentration of a substance in the oil and gas phases of an oil-water mixture in contact with a fixed volume of gas

4.6.2 Retention and Release Processes

In some plant-based foods, it is advantageous to have a particular ingredient retained in one environment during storage (*e.g.,* inside fat droplets) but then be released into another environment (*e.g.,* into the headspace above the food) when the conditions are changed (such as cooking). This is especially important for flavors that should be released during food preparation and consumption but it may also be important for designing various kinds of special effects into foods. For instance, it may be possible to separate two chemically reactive ingredients from each other during storage by locating them within different phases in the food matrix but then having them come into contact during cooking, which leads to a chemical reaction that gives a desirable change in color or texture. Ingredients may be retained within a certain phase due to equilibrium or non-equilibrium effects. Equilibrium effects are based on the partitioning phenomena discussed in the previous section, such as the tendency for a non-polar ingredient to be preferentially located within the oil phase. Non-equilibrium effects are based on trapping an ingredient within a particular phase through kinetic effects, *e.g.,* by creating a physical barrier that inhibits its release or by creating a solidified phase that retards its movement.

An ingredient may be released from a particular phase within a food matrix in response to changes in the environmental conditions, such as dilution, heating, or mechanical stresses. The rate at which the ingredient is released often plays an important role in determining the physicochemical and sensory attributes of foods. Consequently, it is important to be able to predict the release kinetics and to understand the major factors influencing the rate and extent of release. In general, the release kinetics depend on the physicochemical and structural attributes of the system, such as ingredient type and concentration, the dimensions and rheological properties of the different phases, and the application of shear forces. The release of substances from a particular phase may be a result of a variety of processes, such as simple diffusion, erosion, swelling, or disintegration (McClements, 2014). Here, we highlight the importance of mathematical modeling of these processes by considering the release of a substance from spherical particles embedded in a fluid matrix due to simple diffusion. This process can be described by the following equation, which is known as the Crank model (Lian et al., 2004):

$$\frac{M(t)}{M_\infty} = 1 - \exp\left[-\frac{4.8D\pi^2}{K_{OW}d^2}t\right]$$

(4.28)

Here, $M(t)$ is the mass of the ingredient that is released from inside the particles at time t, M_∞ is the mass of the ingredient that is released after an infinite time, D is the translational diffusion coefficient of the ingredient inside the particles, and d is the particle diameter. This equation is useful for predicting the impact of factors such as particle size, partition coefficient, and viscosity (inversely related to the diffusion coefficient) on the time taken for a substance to be released. As an example, the impact of particle size on the release kinetics of a hydrophobic ingredient

(K_{OW} = 1000) from fat droplets suspended in water is predicted using this equation (Fig. 4.33). Initially, there is a rapid release of the hydrophobic ingredient and then there is a more gradual release at longer times. The ingredient is released more rapidly as the particle size decreases because the molecules have a shorter distance to diffuse out of the particles. These calculations show that the release kinetics can be manipulated by controlling the dimensions of the particles. To a first approximation, the time taken for half of a substance to be released from a particle is given by the following equation (Lian et al., 2004):

$$t_{1/2} = \frac{d^2 K_{OW}}{68 D} \tag{4.29}$$

The influence of particle size and partition coefficient on $t_{1/2}$ for ingredients initially contained in fat droplets dispersed in water is shown in Table 4.8. These predictions show that the half time is very short for small fat droplets. For instance, for strongly hydrophobic ingredients (K_{OW} = 1000) it takes less than 1 second for half of the ingredient to be released from fat droplets with a diameter of 5 μm. For less hydrophobic ingredients, this time is even shorter. Indeed, the half time decreases as the ingredient becomes less hydrophobic (*i.e.*, K_{OW} decreases). These kinds of equations are therefore useful for designing food structures to control the release of flavors or other active ingredients from different regions in plant-based foods (Fig. 4.34).

In general, more complex equations are required to predict the release of flavor molecules into the headspace above foods during food preparation and consumption, which take into account their movement through the particles and surrounding matrix and then into the nose. A number of these equations have been reviewed elsewhere (McClements, 2015).

Table 4.8 Predictions of the influence of particle diameter and oil-water partition coefficient on the time taken for half of the molecules to move out of spherical oil droplets dispersed in water (Crank model)

d (μm)	K_{OW} = 1	K_{OW} = 10	K_{OW} = 100	K_{OW} = 1000
	Release time: $t_{1/2}$ (s)			
0.1	3.7E-07	3.7E-06	3.7E-05	3.7E-04
0.2	1.5E-06	1.5E-05	1.5E-04	1.5E-03
0.5	9.1E-06	9.1E-05	9.1E-04	9.1E-03
1	3.7E-05	3.7E-04	3.7E-03	3.7E-02
2	1.5E-04	1.5E-03	1.5E-02	1.5E-01
5	9.1E-04	9.1E-03	9.1E-02	9.1E-01
10	3.7E-03	3.7E-02	3.7E-01	3.7E+00
20	1.5E-02	1.5E-01	1.5E+00	1.5E+01
50	9.1E-02	9.1E-01	9.1E+00	9.1E+01
100	3.7E-01	3.7E+00	3.7E+01	3.7E+02
200	1.5E+00	1.5E+01	1.5E+02	1.5E+03
500	9.1E+00	9.1E+01	9.1E+02	9.1E+03
1000	3.7E+01	3.7E+02	3.7E+03	3.7E+04

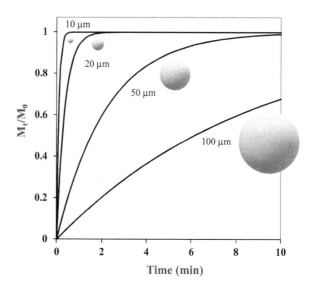

Fig. 4.34 Influence of particle size on the release kinetics of a hydrophobic ingredient ($K_{ow} = 1000$) from inside colloidal particles (fat droplets) suspended in water

4.7 Oral Processing and Sensory Attributes

As discussed in the first chapter, the sensory attributes of plant-based foods (like other foods) are usually the most important factor determining their desirability and acceptability to consumers. In this section, we therefore consider the methods that have been developed to understand the behavior of plant-based foods during the eating process, as well as to quantify the sensory attributes of these foods.

4.7.1 Oral Processing

The traditional compression and shear testing methods discussed in Sect. 4.3 are useful for providing information about the textural attributes of plant-based foods prior to consumption. In some cases, they can measure parameters that can be related to the behavior of foods in the mouth, such as the creaminess of fluid foods (shear viscosity) or the resistance of solid foods to breaking during the first bite (hardness). However, they are unable to provide information about the complex behavior of foods throughout the mastication process. Fluid foods may coat the surfaces of the oral cavity and reduce the friction between the tongue and palate, thereby acting as lubricants (Sarkar et al., 2021). These effects are related to the perceived wateriness, creaminess, or astringency of foods. Solid foods are mixed with saliva and progressively disrupted by the mechanical actions of the teeth and jaws (biting, grinding, and chewing), which leads to changes in their structure, physicochemical properties, and sensory perception over time (Panda et al., 2020). There has therefore been a great interest in studying the behavior of foods within the mouth, which is a discipline known as oral processing (Chen, 2015; Wang & Chen, 2017). One of the main goals of this discipline is to reduce the gap that currently exists between the results of conventional instrumental testing of food texture (Sect. 4.3) and the data obtained in sensory trials. There are some important potential benefits from establishing robust correlations between oral processing tests and sensory analysis. For instance, many oral processing methods do not involve the use of human panelists to taste foods, which can save time and money, as well as providing more quantitative data that can be easily be compared between different products and formulations.

A number of different analytical instruments have been developed to monitor the oral processing of foods (Panda et al., 2020; Sarkar et al., 2021; Wang et al., 2017). Tribology instruments that measure the friction that occurs when two soft surfaces move over each other (designed to simulate the tongue and palate) provide valuable information about the ability of foods to lubricate the mouth (Sarkar et al., 2021). These instruments are most suitable for characterizing the properties of fluid samples and are therefore useful for analyzing liquid foods directly or solid foods after they have been converted into a liquid by chewing. Some researchers have developed mechanical rheometers that are designed to simulate the movements of the human jaw, the secretion of saliva, and the formation and swallowing of the bolus (Panda et al., 2020). These instruments often contain a set of artificial teeth to break down the food, as well as sensors to measure changes in the properties of the food during simulated mastication. Some oral processing devices also involve the use of human subjects and are primarily utilized to better understand the behavior of foods during mastication. For instance, researchers have developed instruments that can track jaw movements and muscle activity during mastication by attaching small magnets to a person's teeth and/or by using video cameras (Çakır et al., 2012; Laguna et al., 2016; Wilson et al., 2016). These instruments can provide information about the number, frequency, duration, and force of the chews that a person uses to masticate a particular kind of food product (Çakır et al., 2012), which can then be related to the composition, structure, and properties of the food (Wagoner et al., 2016).

These oral processing instruments may be useful for the design of plant-based foods that better simulate the behavior of animal-based ones during mastication, thereby better matching their desirable sensory attributes. For instance, they can help understand how material properties (such as fiber length, thickness, and hardness) are linked to oral properties (such as number of chews and bite strength) and sensory attributes (such as perceived firmness, chewiness, and juiciness).

4.7.2 Sensory Evaluation

Ultimately, the quality attributes of any newly developed plant-based food, such as its appearance, texture, mouthfeel, and flavor, should be assessed by human beings who actually consume and rate them (Civille & Carr, 2015; Lawless & Heymann,

2010; Stone & Sidel, 2020). This is usually carried out using an appropriate number of sensory panelists or consumers under carefully controlled conditions so as to provide reliable and meaningful results. The people involved base their judgements on one or more sensory attributes of the test products such as their appearance, texture, or flavor. Plant-based foods are usually designed to accurately simulate the sensory attributes of animal-based ones. Consequently, the goal of sensory evaluation is often to establish the similarities or differences between these two types of products. In general, sensory studies are broadly classified as discrimination tests (detection of differences between products), descriptive tests (ranking of perceived intensities of specific attributes), and affective tests (liking of products) (Lawless & Heymann, 2010). All of these test methods can be used for testing plant-based foods, although descriptive and affective tests are the most commonly used (McClements et al., 2021).

Discrimination tests: These tests are used to establish whether there is a detectable difference in the specified sensory attributes of two or more products. These tests can be performed in various ways. For instance, in a *duo-trio* test a person is given three samples (AXY), one that is known (A) and two that are unknown (X and Y). The person then has to say whether X or Y is the same as A. In a *triangle* test, a person is given three unknown samples (XXY) and then they have to say which one is the odd one out. In an ABX test, a person is given two knowns (A and B) and one unknown (X) and has to match the unknown to one of the knowns. As a specific example, in an ABX test, a person could be given a real sausage (A) and a plant-based sausage (B) and then they are given an unknown sausage (X) and have to decide whether it is real or plant-based.

Descriptive tests: In a descriptive sensory test, panelists are asked to rate given attributes ("descriptors") of foods on a pre-defined scale. In some cases, the panelists may be asked to identify the most appropriate attributes to describe a product by themselves. As an example, meat analogs have been rated according to attributes such as 'fibrousness', 'firmness/hardness', 'juiciness', 'elasticity', 'brittleness', 'earthy', 'chicken', 'crumbly', 'moist', 'tenderness', 'taste', 'flavor' and 'smell' (Lin et al., 2002; Savadkoohi et al., 2014; Grahl et al., 2018; Palanisamy et al., 2018; Stephan et al., 2018; Chiang et al., 2019; Taylor et al., 2020). The intensities of each of these attributes are rated on a scale. A major challenge in descriptive tests is to define the most appropriate attributes and to establish the most appropriate intensity scale for each attribute. As a result, ratings can vary considerably across individuals tasting the same food. To overcome this problem, it is usually advised that descriptive sensory evaluations include between around 8 to 12 panelists and that these panelists undergo training before carrying out the sensory analysis so as to calibrate the panel with the defined attributes (Savadkoohi et al., 2014). Intensity scales usually range from 1 to 9, with each of the numbers being linked to a specific descriptor that increases in intensity, such as 1 = soft, 5 = firm, and 9 = hard. If possible, the trials should be carried out in testing booths under standardized conditions and the panelists should be given water between each product to reduce carryover effects (Chiang et al., 2019).

Affective tests: These tests are used to assess the liking of specified product attributes by consumers, as well as the overall acceptance of the product. In this method, non-trained consumers are usually asked to rate their liking of a food. Consequently, affective tests are a valuable tool to receive feedback about the desirability of a product, which will affect its potential success on the market. Typically, a 9-point hedonic scale ranging from 'dislike extremely' to 'like extremely' is used to assess the liking of a product or some of its specific attributes (Wichchukit & O'Mahony, 2015). However, other types of scale are also available that can be used where appropriate (Lawless et al., 2010). Affective tests have been used to assess the acceptance of various kinds of plant-based foods, including meat analogs formulated from pea, wheat, peanut, chick pea, soy protein, and mycoprotein (Rehrah et al., 2009; Kim et al., 2011; Savadkoohi et al., 2014; Yuliarti et al., 2021). However, different scales and testing procedures are often employed by different researchers, which makes it difficult to compare the results of different studies. This highlights the need to develop standardized procedures to characterize the sensory properties of plant-based foods.

For sensory analysis, it is important to recruit a sufficiently large number of people to participate to ensure the results have enough statistical power. These people may be trained (panelists) or untrained (consumers) depending on the type of test that is being carried out. Typically, more detailed insights can be obtained using trained panelists but it is much more time consuming and expensive. For this reason, a much smaller number of participants is typically used for trained panelists than for untrained consumers. For instance, around 8 to 12 trained panelists are typically used for descriptive tests, while around 60 to 120 untrained consumers are used for descriptive and affective tests. It should also be recognized that the results obtained by trained panelists do not always accurately reflect the preferences of normal consumers. It is also important to use appropriate controls for the material being tested, *e.g.,* by comparing a plant-based product (such as a chicken nugget analog) with the animal-based one it is designed to replace (such as a real chicken nugget). During sensory analysis, the products should be presented in a randomized sampling order, they should all have similar sizes, shapes and distributions on the plate, the experiments should be carried out under controlled environmental conditions (such as lighting and temperature), and the samples should be presented blind with a three-digit code to help reduce bias.

A number of researchers have compared the textural and sensory properties of plant-based foods. As an example, instrumental and sensory analysis was used to characterize the textural attributes of a meat analog fortified with *Arthrospira platensis* (Spirulina) (Grahl et al., 2018). Soy protein-based meat analogs were produced using high moisture extrusion and the impact of spirulina content, extrusion temperature, screw speed, and moisture content on their properties was determined. The meat analogs were analyzed by a trained sensory panel that partly developed the descriptors, by texture profile analysis, and by cutting force tests. The panelists assessed the meat analogs with various descriptors, including different smell, color, texture, and taste attributes. For example, 'brittle' was used as a descriptor for texture whereas 'umami' was used as a descriptor of aftertaste. An interesting finding of this study was that the incorporation of up to 50% spirulina still resulted in the formation of a fibrous texture during extrusion at low moisture contents (57%), while the cutting force and hardness of the meat analogs produced were not significantly altered by the presence of the spirulina. However, the intensity of the odor, flavor, aftertaste, and color increased at higher spirulina contents, which is most likely because of the strong intrinsic taste and intense color of spirulina. Additionally, the texture became less elastic, less fibrous, and softer with increasing spirulina content. Based on their instrumental texture and sensory tests, the researchers concluded that it was feasible to incorporate spirulina into soy-based meat analogs up to a certain level.

4.8 Conclusions

Plant-based foods are compositionally and structurally complex materials that exhibit a broad range of physicochemical properties, ranging from low viscosity fluids (milk analogs) to hard solids (frozen meat analogs). The physicochemical and functional properties of these foods determine their processing, distribution, preparation, sensory attributes, and digestion. The design and formulation of high quality plant-based foods therefore require a good understanding of the factors that impact their physicochemical properties. There has already been some progress in understanding the physicochemical properties of certain categories of plant-based foods but much work is still needed. In particular, there is a need to develop a much deeper fundamental understanding of the relationship between the composition and structure of these foods on one hand, and their cookability, appearance, stability, texture, mouthfeel, flavor, and gastrointestinal fate on the other. Advances in this area could be made by treating them as complex colloidal-polymer materials and then identifying or developing appropriate mathematical or computational models to describe their properties. Once established, these models could be used to identify the most important factors contributing to the desirable attributes of plant-based foods, as well as to design products with improved performance. An important part of this work will be to better understand how the molecular and physicochemical properties of plant-based foods impact their interaction with the human body during mastication and digestion. More details about the physicochemical properties of specific kinds of plant-based foods (such as meat, seafood, eggs, and dairy products) are given in later chapters.

References

Aida, T. M., Yamagata, T., Watanabe, M., & Smith, R. L. (2010). Depolymerization of sodium alginate under hydrothermal conditions. *Carbohydrate Polymers, 80*(1), 296–302.

Aljahdali, N., & Carbonero, F. (2019). Impact of Maillard reaction products on nutrition and health: Current knowledge and need to understand their fate in the human digestive system. *Critical Reviews in Food Science and Nutrition, 59*(3), 474–487.

References

Alu'datt, M. H., Rababah, T., Alhamad, M. N., Ereifej, K., Gammoh, S., Kubow, S., & Tawalbeh, D. (2017). Preparation of mayonnaise from extracted plant protein isolates of chickpea, broad bean and lupin flour: Chemical, physiochemical, nutritional and therapeutic properties. *Journal of Food Science and Technology-Mysore, 54*(6), 1395–1405.

Arino, I., Kleist, U., Mattsson, L., & Rigdahl, M. (2005). On the relation between surface texture and gloss of injection-molded pigmented plastics. *Polymer Engineering and Science, 45*(10), 1343–1356.

Berns, R. S. (2000). *Billmeyer and Saltzman's principles of color technology* (3rd ed.). Wiley-Interscience.

Bohren, C. F., & Huffman, D. R. (1998). *Absorption and scattering of light by small particles.* Wiley-VCH.

Briggs, J. L., & Steffe, J. F. (1997). Using Brookfield data and the Mitschka method to evaluate power law foods. *Journal of Texture Studies, 28*(5), 517–522.

Broedersz, C. P., & MacKintosh, F. C. (2014). Modeling semiflexible polymer networks. *Reviews of Modern Physics, 86*(3), 995–1036.

Çakır, E., Vinyard, C. J., Essick, G., Daubert, C. R., Drake, M., & Foegeding, E. A. (2012). Interrelations among physical characteristics, sensory perception and oral processing of protein-based soft-solid structures. *Food Hydrocolloids, 29*(1), 234–245. https://doi.org/10.1016/j.foodhyd.2012.02.006

Cao, Y. P., & Mezzenga, R. (2020). Design principles of food gels. *Nature Food, 1*(2), 106–118.

Chen, J. S. (2015). Food oral processing: Mechanisms and implications of food oral destruction. *Trends in Food Science & Technology, 45*(2), 222–228.

Chiang, J. H., Loveday, S. M., Hardacre, A. K., & Parker, M. E. (2019). Effects of soy protein to wheat gluten ratio on the physicochemical properties of extruded meat analogues. *Food Structure, 19*, 100102. https://doi.org/10.1016/j.foostr.2018.11.002

Chmiel, M., & Slowinski, M. (2013). Application of video image analysis in meat technology. *Medycyna Weterynaryjna-Veterinary Medicine-Science and Practice, 69*(11), 670–673.

Choi, S. J., Decker, E. A., Henson, L., Popplewell, L. M., & McClements, D. J. (2009). Stability of citral in oil-in-water emulsions prepared with medium-chain triacylglycerols and triacetin. *Journal of Agricultural and Food Chemistry, 57*(23), 11349–11353.

Civille, G. V., & Carr, B. T. (2015). *Sensory evaluation techniques* (5th ed.). CRC Press.

Cornet, S. H. V., Snel, S. J. E., Lesschen, J., van der Goot, A. J., & van der Sman, R. G. M. (2021). Enhancing the water holding capacity of model meat analogues through marinade composition. *Journal of Food Engineering, 290*.

Dickinson, E. (2012). Emulsion gels: The structuring of soft solids with protein-stabilized oil droplets. *Food Hydrocolloids, 28*(1), 224–241.

Durazzo, A., Gabrielli, P., & Manzi, P. (2015). Qualitative study of functional groups and antioxidant properties of soy-based beverages compared to cow milk. *Antioxidants, 4*(3), 523–532.

Fraeye, I., De Roeck, A., Duvetter, T., Verlent, I., Hendrickx, M., & Van Loey, A. (2007). Influence of pectin properties and processing conditions on thermal pectin degradation. *Food Chemistry, 105*(2), 555–563.

Fredrick, E., Walstra, P., & Dewettinck, K. (2010). Factors governing partial coalescence in oil-in-water emulsions. *Advances in Colloid and Interface Science, 153*(1–2), 30–42.

Garcia, J. M., Chambers, E., & Cook, K. (2018). Visualizing the consistency of thickened liquids with simple tools: Implications for clinical practice. *American Journal of Speech-Language Pathology, 27*(1), 270–277.

Garna, H., Mabon, N., Nott, K., Wathelet, B., & Paquot, M. (2006). Kinetic of the hydrolysis of pectin galacturonic acid chains and quantification by ionic chromatography. *Food Chemistry, 96*(3), 477–484.

Gaviria, L. M., Ospina-E, J. C., & Munoz, D. A. (2021). Phenomenological-based semiphysical model to predict the water holding capacity of processed meats in the mixing process. *Journal of Food Process Engineering*.

Genovese, D. B., Lozano, J. E., & Rao, M. A. (2007). The rheology of colloidal and noncolloidal food dispersions. *Journal of Food Science, 72*(2), R11–R20.

Grahl, S., Palanisamy, M., Strack, M., Meier-Dinkel, L., Toepfl, S., & Morlein, D. (2018). Towards more sustainable meat alternatives: How technical parameters affect the sensory properties of extrusion products derived from soy and algae. *Journal of Cleaner Production, 198*, 962–971.

Grasso, N., Alonso-Miravalles, L., & O'Mahony, J. A. (2020). Composition, physicochemical and sensorial properties of commercial plant-based yogurts. *Food, 9*(3).

Gravelle, A. J., Barbut, S., & Marangoni, A. G. (2015). Influence of particle size and interfacial interactions on the physical and mechanical properties of particle-filled myofibrillar protein gels. *RSC Advances, 5*(75), 60723–60735.

Gravelle, A. J., & Marangoni, A. G. (2021). Effect of matrix architecture on the elastic behavior of an emulsion-filled polymer gel. *Food Hydrocolloids, 119*.

Gravelle, A. J., Nicholson, R. A., Barbut, S., & Marangoni, A. G. (2019). Considerations for readdressing theoretical descriptions of particle-reinforced composite food gels. *Food Research International, 122*, 209–221.

Hergert, W., & Wriedt, T. (2012). *The Mie theory: Basics and applications*. Springer.

Hernandez, M. J., Dolz, J., Delegido, J., Cabeza, C., & Dolz, M. (2008). Thixotropic behavior of salad dressings stabilized with modified starch, pectin, and gellan gum. Influence of temperature. *Journal of Dispersion Science and Technology, 29*(2), 213–219.

Huang, L. Y., Wang, T., Han, Z. P., Meng, Y. L., & Lu, X. M. (2016). Effect of egg yolk freezing on properties of mayonnaise. *Food Hydrocolloids, 56*, 311–317.

Hunter, R. J. (1994). *Introduction to modern colloid science*. Oxford University Press.

Hutchings, J. B. (1999). *Food color and appearance* (2nd ed.). Springer.

Jacobsen, C. (2015). Some strategies for the stabilization of long chain n-3 PUFA-enriched foods: A review. *European Journal of Lipid Science and Technology, 117*(11), 1853–1866.

Jacobsen, C., Horn, A. F., & Nielsen, N. S. (2013). Enrichment of emulsified foods with omega-3 fatty acids. In C. Jacobsen, N. S. Nielsen, A. F. Horn, & A. D. M. Sorensen (Eds.), *Food enrichment with Omega-3 fatty acids* (Vol. 252, pp. 336–352).

Jeske, S., Zannini, E., & Arendt, E. K. (2017). Evaluation of physicochemical and glycaemic properties of commercial plant-based milk substitutes. *Plant Foods for Human Nutrition, 72*(1), 26–33.

Kaltsa, O., Yanniotis, S., Polissiou, M., & Mandala, I. (2018). Stability, physical properties and acceptance of salad dressings containing saffron (Crocus sativus) or pomegranate juice powder as affected by high shear (HS) and ultrasonication (US) process. *Lwt-Food Science and Technology, 97*, 404–413.

Karlsson, A., & Singh, S. K. (1999). Acid hydrolysis of sulphated polysaccharides. Desulphation and the effect on molecular mass. *Carbohydrate Polymers, 38*(1), 7–15.

Kassis, N., Drake, S. R., Beamer, S. K., Matak, K. E., & Jaczynski, J. (2010). Development of nutraceutical egg products with omega-3-rich oils. *Lwt-Food Science and Technology, 43*(5), 777–783.

Kerker, M. (1969). *The scattering of light and other electromagnetic radiation*. Academic.

Khalesi, H., Lu, W., Nishinari, K., & Fang, Y. P. (2021). Fundamentals of composites containing fibrous materials and hydrogels: A review on design and development for food applications. *Food Chemistry, 364*.

Kharat, M., Du, Z. Y., Zhang, G. D., & McClements, D. J. (2017). Physical and chemical stability of curcumin in aqueous solutions and emulsions: Impact of pH, temperature, and molecular environment. *Journal of Agricultural and Food Chemistry, 65*(8), 1525–1532.

Kim, K. Cho, B., Lee, I. Lee, H., Kwon, S., et al. (2011). Bioproduction of mushroom mycelium of Agaricus bisporus by commercial submerged fermentation for the production of meat analogue. *Journal of the Science of Food and Agriculture, 1*(9), 561–1568. https://doi.org/10.1002/jsfa.4348

Kotrum, G. (1969). *Reflectance spectroscopy: Principles, methods, applications*. Springer.

Laguna, L., Barrowclough, R. A., Chen, J., & Sarkar, A. (2016). New approach to food difficulty perception: Food structure, food oral processing and individual's physical strength. *Journal of Texture Studies, 47*(5), 413–422. https://doi.org/10.1111/jtxs.12190

Lawless, H. T., & Heymann, H. (2010). *Sensory evaluation of food: Principles and practices* (2nd ed.). Springer.

Lawless, H. T., Popper, R., & Kroll, B. J. (2010). A comparison of the labeled magnitude (LAM) scale, an 11-point category scale and the traditional 9-point hedonic scale. *Food Quality and Preference, 21*(1), 4–12. https://doi.org/10.1016/j.foodqual.2009.06.009

Leloup, F. B., Obein, G., Pointer, M. R., & Hanselaer, P. (2014). Toward the soft metrology of surface gloss: A review. *Color Research and Application, 39*(6), 559–570.

Li, J. H., Wang, C. Y., Zhang, M. Q., Zhai, Y. H., Zhou, B., Su, Y. J., & Yang, Y. J. (2018). Effects of selected phosphate salts on gelling properties and water state of whole egg gel. *Food Hydrocolloids, 77*, 1–7.

Lian, G. P., Malone, M. E., Homan, J. E., & Norton, I. T. (2004). A mathematical model of volatile release in mouth from the dispersion of gelled emulsion particles. *Journal of Controlled Release, 98*(1), 139–155.

Lin, S., Huff, H. E., & Hsieh, F. (2002). Extrusion process parameters, sensory characteristics, and structural properties of a high moisture soy protein meat analog. *Journal of Food Science, 67*(3), 1066–1072. https://doi.org/10.1111/j.1365-2621.2002.tb09454.x

Lund, M. N., & Ray, C. A. (2017). Control of Maillard reactions in foods: Strategies and chemical mechanisms. *Journal of Agricultural and Food Chemistry, 65*(23), 4537–4552.

McClements, D. J. (2000). Comments on viscosity enhancement and depletion flocculation by polysaccharides. *Food Hydrocolloids, 14*(2), 173–177.

McClements, D. J. (2002a). Colloidal basis of emulsion color. *Current Opinion in Colloid & Interface Science, 7*(5–6), 451–455.

McClements, D. J. (2002b). Theoretical prediction of emulsion color. *Advances in Colloid and Interface Science, 97*(1–3), 63–89.

McClements, D. J. (2005). *Food emulsions: Principles, practice, and techniques* (2nd ed.). CRC Press.

McClements, D. J. (2014). *Nanoparticle- and microparticle-based delivery systems*. CRC Press.

McClements, D. J. (2015). *Food emulsions: Principles, practice, and techniques* (2nd ed.). CRC Press.

McClements, D. J., & Grossmann, L. (2021). The science of plant-based foods: Constructing next-generation meat, fish, milk, and egg analogs. *Comprehensive Reviews in Food Science and Food Safety, 20*(4), 4049–4100.

McClements, D. J., Newman, E., & McClements, I. F. (2019). Plant-based milks: A review of the science underpinning their design, fabrication, and performance. *Comprehensive Reviews in Food Science and Food Safety, 18*(6), 2047–2067.

McClements, D. J., Weiss, J., Kinchla, A. J., Nolden, A. A., & Grossmann, L. (2021). Methods for testing the quality attributes of plant-based foods: Meat- and processed-meat analogs. *Food, 10*(2).

Mudgett, P. S., & Richards, L. W. (1971). Multiple scattering calculations for technology. *Applied Optics, 10*(7), 1485.

Nagy, P. (2013). Kinetics and mechanisms of thiol-disulfide exchange covering direct substitution and thiol oxidation-mediated pathways. *Antioxidants & Redox Signaling, 18*(13), 1623–1641.

Palanisamy, M., Töpfl, S., Aganovic, K., & Berger, R. G. (2018). Influence of iota carrageenan addition on the properties of soya protein meat analogues. *LWT, 87*, 546–552. https://doi.org/10.1016/j.lwt.2017.09.029

Panaite, T. D., Mironeasa, S., Iuga, M., & Vlaicu, P. A. (2019). Liquid egg products characterization during storage as a response of novel phyto-additives added in hens diet. *Emirates Journal of Food and Agriculture, 31*(4), 304–314.

Panda, S., Chen, J. S., & Benjamin, O. (2020). Development of model mouth for food oral processing studies: Present challenges and scopes. *Innovative Food Science & Emerging Technologies, 66*.

Prakash, S., Tan, D. D. Y., & Chen, J. S. (2013). Applications of tribology in studying food oral processing and texture perception. *Food Research International, 54*(2), 1627–1635.

Purslow, P. P., Warner, R. D., Clarke, F. M., & Hughes, J. M. (2020). Variations in meat colour due to factors other than myoglobin chemistry; a synthesis of recent findings (invited review). *Meat Science, 159*.

Rao, M. A. (2013). *Rheology of fluid, semisolid, and solid foods: Principles and applications* (3rd ed.). Springer.

Rehrah, D., Ahmedna, M., Goktepe, I., & Yu, J. (2009). Extrusion parameters and consumer acceptability of a peanut-based meat analogue. *International Journal of Food Science & Technology, 44*(10), 2075–2084. https://doi.org/10.1111/j.1365-2621.2009.02035.x

Russ, J. C. (2012). Image analysis of food microstructure. In D. W. Sun (Ed.), *Computer vision technology in the food and beverage industries* (pp. 233–252).

Sarkar, A., Soltanahmadi, S., Chen, J. S., & Stokes, J. R. (2021). Oral tribology: Providing insight into oral processing of food colloids. *Food Hydrocolloids, 117*.

Savadkoohi, S., Hoogenkamp, H., Shamsi, K., & Farahnaky, A. (2014). Color, sensory and textural attributes of beef frankfurter, beef ham and meat-free sausage containing tomato pomace. *Meat Science, 97*(4), 410–418.

Silva, K., Machado, A., Cardoso, C., Silva, F., & Freitas, F. (2020). Rheological behavior of plant-based beverages. *Food Science and Technology, 40*, 258–263.

Song, H. Y., & McClements, D. J. (2021). *Nano-enabled-fortification of salad dressings with curcumin: Impact of nanoemulsion-based delivery systems on physicochemical properties* (Vol. 145). Lwt-Food Science and Technology.

Stephan, A., Ahlborn, J., Zaju,l M., & Zorn, H. (2018). Edible mushroom mycelia of Pleurotus sapidus as novel protein sources in a vegan boiled sausage analog system: functionality and sensory tests in comparison to commercial proteins and meat sausages. *Eur Food Res Technol, 244*(5), 913–924. https://doi.org/10.1007/s00217-017-3012-1

Stone, H., & Sidel, J. L. (2020). *Sensory evaluation practices* (5th ed.). Elsevier Academic Press.

Swapnil, S. J., & Arpana, H. J. (2019). Applications of lipases. *Research Journal of Biotechnology, 14*(11), 130–138.

Tadros, T. F. (2010). *Rheology of dispersions: Principles and applications.* Wiley-VCH.

Taylor, J., Ahmed, I. A. M., Al-Juhaimi, F. Y., & Bekhit, A. E-D. A. (2020). Consumers' Perceptions and Sensory Properties of Beef Patty Analogues. *Foods, 9*(1), 63. https://doi.org/10.3390/foods9010063

van der Sman, R. G. M., Paudel, E., Voda, A., & Khalloufi, S. (2013). Hydration properties of vegetable foods explained by Flory-Rehner theory. *Food Research International, 54*(1), 804–811.

van Vliet, T. (2013). *Rheology and fracture mechanics of foods.* CRC Press.

Wagoner, T. B., Luck, P. J., Foegeding, E. A. (2016). Caramel as a model system for evaluating the roles of mechanical properties and oral processing on sensory perception of texture. *Journal of Food Science, 81*(3), S736–S744. https://doi.org/10.1111/1750-3841.13237

Walstra, P. (2003). *Physical chemistry of foods.* Marcel Decker.

Wang, X. M., & Chen, J. S. (2017). Food oral processing: Recent developments and challenges. *Current Opinion in Colloid & Interface Science, 28*, 22–30.

Wang, Q. L., Jin, G. F., Wang, N., Guo, X., Jin, Y. G., & Ma, M. H. (2017). Lipolysis and oxidation of lipids during egg storage at different temperatures. *Czech Journal of Food Sciences, 35*(3), 229–235. https://doi.org/10.17221/174/2016-cjfs

Wichchukit, S., & O'Mahony, M. (2015). The 9-point hedonic scale and hedonic ranking in food science: some reappraisals and alternatives. *Journal of the Science of Food and Agriculture, 95*(11), 2167–2178. https://doi.org/10.1002/jsfa.6993

Wilson, A., Luck, P., Woods, C., Foegeding, E. A., & Morgenstern, M. (2016). Comparison of jaw tracking by single video camera with 3D electromagnetic system. *Journal of Food Engineering, 190*, 22–33. https://doi.org/10.1016/j.jfoodeng.2016.06.008

Wyszecki, G., & Stiles, W. S. (2000). *Color science: Concepts and methods, quantitative data and formulae.* Wiley-Interscience.

Yuliarti, O., Kiat Kovis, T. J., & Yi, N. J. (2021). Structuring the meat analogue by using plant-based derived composites. *Journal of Food Engineering, 288*, 110138. https://doi.org/10.1016/j.jfoodeng.2020.110138

Zheng, B. J., Zhou, H. L., & McClements, D. J. (2021). *Nutraceutical-fortified plant-based milk analogs: Bioaccessibility of curcumin-loaded almond, cashew, coconut, and oat milks* (p. 147). LWT-Food Science and Technology.

Chapter 5
Nutritional and Health Aspects

5.1 Introduction

Many consumers believe that eating a plant-based diet will improve their health but this depends on the nature of the foods consumed (Hemler & Hu, 2019). A plant-based diet mainly consisting of plant-based burgers, sausages, and nuggets consumed with refined grains, fried potatoes, snacks, sweets, and sugary beverages is unlikely to be healthy. In contrast, a plant-based diet mainly comprised of fruits, vegetables, legumes, whole grains, and nuts is likely to be much healthier. Consequently, it is important that a plant-based diet contains foods that are considered to be healthy, which usually means that they contain an appropriate macronutrient balance (carbohydrates, proteins, and fats), contain sufficient quantities of bioavailable micronutrients (vitamins, minerals, and nutraceuticals), contain high levels of dietary fibers, and are not digested too rapidly in the human gut. In addition, the impact of plant-based foods on satiety and satiation (the feelings of fullness during or after a meal), as well as metabolism (*e.g.,* insulin response), is also important, as this may affect the total quantity of foods consumed, thereby impacting chronic diseases such as obesity and diabetes. Finally, the influence of plant-based foods on the gut microbiome is important, as the nature of the microorganisms in the colon is known to have a major impact on human health and wellbeing. The nutritional profile and health effects of plant-based foods should therefore be taken into account when creating plant-based analogs of meat, fish, egg, or dairy products. Indeed, the transition to more plant-based diets provides the food industry with an excellent opportunity to address many of the adverse health effects currently associated with the modern Western diet. In this chapter, we focus on some of the factors that need to be considered when designing the next generation of plant-based products.

5.2 Macronutrients

In this section, we provide an overview of the nutritional attributes of the key macronutrients found in plant-based foods: proteins, lipids, and carbohydrates. Each of these general classes of macronutrients has different molecular features, which influence their gastrointestinal fate and their impact on human health. In addition, within each macronutrient class there are different kinds of molecules that have different nutritional effects. For instance, lipids may be saturated, monounsaturated, or polyunsaturated. Consequently, it is important to know the total concentration of the different classes of macronutrients present (proteins, lipids, and carbohydrates), as well as the specific kinds of macronutrients within each class. Moreover, it is important to understand how different nutrients interact with each other and alter each other's gastrointestinal fate and nutritional effects. In general, an ingested plant-based food should be mechanically, chemically, and enzymatically disintegrated within the human gut, leading to the formation of small digestion products that are suitable for absorption by the epithelium cells that line the gastrointestinal tract (Fig. 5.1).

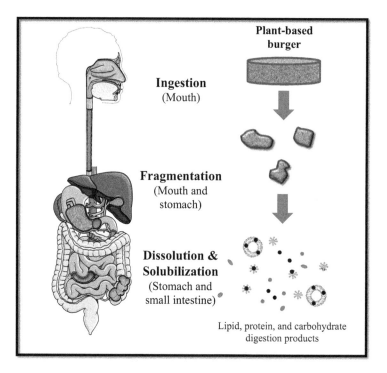

Fig. 5.1 The overall bioavailability of hydrophobic bioactives depends on numerous factors, including their bioaccessibility, absorption, distribution, metabolism and excretion. (Image of human gastrointestinal tract from Servier Medical Art (smart.servier.com) used under CC BY 3.0 (https://creativecommons.org/licenses/by/3.0/))

5.2 Macronutrients

5.2.1 Proteins

5.2.1.1 Introduction

Meat, fish, eggs, and dairy products are a major source of proteins in the human diet, particularly in developed countries. For instance, an analysis of data from the National Health and Nutrition Examination Survey (NHANES) found that the average protein intake by adults in the United States from 2007 to 2010 mainly came from animal (46%) and dairy (16%) sources (Pasiakos et al., 2015). Less than a third of the protein (30%) came from plant sources. Switching from an animal-based diet to a plant-based one could, therefore, have a major impact on the types and amounts of proteins consumed, which could have important nutritional and health consequences.

5.2.1.2 Amino Acid Profiles

Each type of protein has its own unique amino acid profile depending on its origin and function in nature (Loveday, 2019, 2020). The amino acid compositions of plant proteins are therefore different from those of animal proteins, which may influence their nutritional effects (Mathai et al., 2017). Some amino acids are categorized as indispensable amino acids (IAAs) because they cannot be synthesized at sufficiently high rates within the human body and so must be obtained from the diet. These nutrients are also commonly referred to as essential amino acids. The IAAs are isoleucine, leucine, lysine, methionine, phenylalanine, threonine, tryptophan, histidine, and valine. Animal proteins, such as those from meat, fish, egg, and milk, commonly have the full complement of IAAS, but some kinds of plant proteins have limited amounts of specific IAAs (Table 5.1). For instance, cereals like wheat, rice, corn, barley, and oats have relatively low levels of lysine, whereas legumes like soybeans, kidney beans, chickpeas, peas, and lentils have relatively low levels of sulfur-containing amino acids such as methionine and cysteine (Gorissen et al., 2018). If a person was to exclusively obtain all of their proteins from one of these plant-based sources, then they might suffer from nutritional deficiencies. In practice, people tend to combine proteins from different sources (*e.g.*, legumes and cereals) and so they can obtain all the essential amino acids required for good health and wellbeing (Herreman et al., 2020). Moreover, people in developed countries often eat much more protein than they actually need (around 0.8 g of protein per kg of bodyweight per day) to meet their basic nutritional requirements, and so even sources of proteins with low levels of some essential amino acids may meet the recommended daily allowance (RDA). It should be noted that dispensable amino acids (DAAs) can be synthesized within the human body, but their ingestion is still important to ensure good health (Wolfe et al., 2016).

As mentioned earlier, in most developed countries, the majority of the population consumes sufficient protein to meet their basic nutritional needs, but there may be

Table 5.1 Comparison of the amino acid compositions (g per 100 g of material) of animal and plant proteins derived from various sources

	Soy	Wheat	Pea	Oat	Lupin	Hemp	Potato	Brown rice	Corn	Microalgae	Milk	Whey	Casein	Egg	Muscle
Essential amino acids															
Threonine	2.3	1.8	2.5	1.5	1.6	1.3	4.1	2.3	1.8	2.1	3.5	5.4	2.6	2	2.9
Methionine	0.3	0.7	0.3	0.1	0.2	1	1.3	2	1.1	0	2.1	1.8	1.6	1.4	1.7
Phenylalanine	3.2	3.7	3.7	2.7	1.8	1.8	4.2	3.7	3.4	2.1	3.5	2.5	3.1	2.3	3.8
Histidine	1.5	1.4	1.6	0.9	1.2	1.1	1.4	1.5	1.1	0.7	1.9	1.4	1.7	0.9	2.8
Lysine	3.4	1.1	4.7	1.3	2.1	1.4	4.8	1.9	1	3.6	5.9	7.1	4.6	2.7	6.6
Valine	2.2	2.3	2.7	2	1.4	1.3	3.7	2.8	2.1	2.1	3.6	3.5	3	2	4.3
Isoleucine	1.9	2	2.3	1.3	1.5	1	3.1	2	1.7	1.2	2.9	3.8	2.3	1.6	3.4
Leucine	5	5	5.7	3.8	3.2	2.6	6.7	5.8	8.8	4	7	8.6	5.8	3.6	6.3
ΣEAA	19.9	18	23.6	13.7	13.1	11.6	29.3	22.1	21	15.7	30.3	34.1	24.8	16.5	31.8
Non-essential amino acids															
Serine	3.4	3.5	3.6	2.2	2.5	2.3	3.4	3.4	2.9	2.1	4	4	3.4	3.3	2.3
Glycine	2.7	2.4	2.8	1.7	2.1	2.1	3.2	3.4	1.6	2.6	1.5	1.5	1.2	1.4	3.1
Glutamic acid	12.4	26.9	12.9	11	12.4	7.4	7.1	12.7	13.1	5.7	16.7	15.5	13.9	5.1	13.1
Proline	3.3	8.8	3.1	2.5	2	1.8	3.3	3.4	5.2	2.3	7.3	4.8	6.5	1.8	0
Cysteine	0.2	0.7	0.2	0.4	0.2	0.2	0.3	0.6	0.3	0.1	0.2	0.8	0.1	0.4	0
Alanine	2.8	1.8	3.2	2.2	1.7	1.9	3.3	4.3	4.8	4	2.6	4.2	2	2.6	4.1
Tyrosine	2.2	2.4	2.6	1.5	1.9	1.3	3.8	3.5	2.7	1.2	3.8	2.4	3.4	1.8	2
Arginine	4.8	2.4	5.9	3.1	5.5	5.3	3.3	5.4	1.7	3.4	2.6	1.7	2.1	2.6	4.4
ΣNEAA	31.9	48.9	34.4	24.7	28.2	22.4	27.8	36.8	32.3	21.4	38.6	34.9	32.5	19	29

Data obtained from previous studies (Gorissen et al., 2018). The data are given as grams per 100 grams of raw material. The aspartic acid, asparagine, glutamine and tryptophan concentrations were not measured. The data do not add up to 100% because some amino acids were not measured and because the total protein content of the samples analyzed varied.

some subpopulations that do not, such as the elderly (Joye, 2019). In contrast, protein deficiency can be a major health concern in many developing countries due to the lack of sufficiently high quality protein sources in the average diet. Deficiencies in essential amino acids are unlikely to be a problem in current consumers of next-generation plant-based foods (such as meat, fish, egg, and dairy analogs) because these people are mainly found in developed countries where the overall protein consumption is sufficiently high but the diet should still be carefully planned and monitored. However, as next-generation plant-based foods reach a larger percentage of the global population, it is imperative that they are formulated to provide well-balanced, high-quality proteins.

5.2.1.3 Digestibility

The bioavailability of some indispensable amino acids may be limited by the relatively poor digestibility of some plant proteins. In general, the digestion of proteins typically occurs within the human stomach and small intestine due to the presence of gastric and pancreatic proteases, like pepsin, trypsin, chymotrypsin, elastase, and carboxypeptidases (Joye, 2019). These enzymes hydrolyze the peptide bonds between the amino acids in the protein chains, which releases free amino acids or peptides that are small enough to be absorbed (only amino acids, di- and tripeptides are absorbed into the blood). The proteases in the human gut may be either exopeptidases or endopeptidases depending on whether they break peptide bonds at the exterior or interior of the polypeptide chain, respectively. The ability of proteases to hydrolyze food proteins and convert them into peptides and amino acids in the upper gastrointestinal tract may be retarded due to several reasons:

- *Protein structure*: The molecular structure of proteins, such as their amino acid sequence, conformation, and crosslinking, influence the ability of digestive enzymes to access the peptide bonds and hydrolyze them (Joye, 2019). For instance, proteins containing high levels of proline-rich sequences (such as gluten) tend to be relatively resistant to enzyme digestion because these sequences limit the access of the proteases to the peptide bonds. The conformation of proteins can also influence the access of proteases to the polypeptide chains. For instance, native β-lactoglobulin is highly resistant to hydrolysis by pepsin in the stomach but the heat-denatured form is rapidly digested. In general, proteins containing high amounts of β-sheet secondary structure tend to be more difficult to digest (Carbonaro et al., 2012). Extensive intramolecular or intermolecular covalent crosslinking of proteins, such as *via* disulfide bonds, may also limit their digestibility under gastrointestinal conditions.
- *Aggregation state*: The proteins in foods may exist as individual molecules or small clusters (*e.g.,* dimers) like the whey proteins in milk, or they may be present in the form of large aggregates held together by physical or covalent interactions like the whey proteins in heat-set gels. Typically, individual proteins are digested more rapidly than highly aggregated ones because it is easier for the

proteases to reach their surfaces (Deng et al., 2020; Guo et al., 2014). Moreover, the rate and extent of dissociation of the aggregates under gastrointestinal conditions impact protein digestibility (Guo et al., 2017a).

- *Food matrix effects*: The proteins in many plant-based foods are embedded within cellular tissues that make it difficult for the proteases to access them, *e.g.*, cell membranes or organelles (Becker & Yu, 2013; Bhattarai et al., 2017). These tissues are often not fully broken down in the upper gastrointestinal tract, which means that the proteins are not completely hydrolyzed, thereby reducing the bioavailability of the amino acids.
- *Dietary fibers*: The high levels of dietary fibers in some plant-based foods may also inhibit protein digestion (McClements, 2021; Williams et al., 2019). Dietary fibers may do this through a range of mechanisms, including increasing the viscosity of the gastrointestinal fluids and thereby reducing mixing and mass transport processes, forming protective coatings around proteins and thereby inhibiting the access of the proteases to the protein surfaces, or binding to proteases and thereby reducing their activity.
- *Antinutritional factors*: Some plant-based foods contain significant levels of antinutritional factors (ANFs), including trypsin inhibitors, tannins and phytates, that can inhibit the digestion and absorption of proteins, as well as other nutrients (Sarwar Gilani et al., 2012). These ANFs may reduce the activity of digestive enzymes (trypsin inhibitors), promote the precipitation of proteins and peptides (tannins), or bind to essential minerals (phytates). ANFs may also be formed within foods during processing, such as Maillard reaction products that can lead to a reduction in lysine absorption.

The nutritional benefits of plant proteins can often be improved by carrying out suitable processing operations, such as mechanical disruption, enzymatic treatment, thermal processing, or acid/alkaline hydrolysis, which break down the cellular structures or deactivate ANFs. Alternatively, it may be possible to remove the ANFs from the food source prior to consumption, *e.g.*, by soaking or washing the protein-rich raw materials.

The digestibility of proteins can be measured using standardized *in vitro* digestion models, such as the INFOGEST model that simulates the human gut, which enables one to assess the extent of hydrolysis and the type of peptides formed (Santos-Hernandez et al., 2020). These methods are important tools for establishing the nutritional benefits and potential allergenicity of plant proteins. If proteins are not digested and absorbed within the small intestine, then they will reach the colon where they may be metabolized by colonic bacteria (Joye, 2019). Here, decarboxylation and deamination reactions can occur that convert the peptides and amino acids into short chain fatty acids and amines. The presence of proteins, peptides, and amino acids in the colon may then impact human health by changing the gut microflora composition or by interacting with the molecules that the gut microflora generates (Ma et al., 2017; Peled & Livney, 2021). The types and amounts of peptides and amino acids reaching the colon would be expected to be different for plant and animal proteins, which would therefore be expected to lead to different effects on the gut microflora and human health. Further research is required in this area to establish the potential beneficial or adverse effects of consuming plant proteins rather than animal ones.

5.2.1.4 Protein Quality

The overall nutritional quality of proteins depends on their amino acid composition and digestibility, which is governed by their molecular structure and gastrointestinal fate. A number of standardized methods have been designed to determine the amount of amino acids that actually get absorbed by the body, which may be appreciably different from the amount that was consumed because some of the proteins are not completely digested and therefore not all of the amino acids are absorbed. As discussed in the previous section, proteins may not be fully digested because they have structures that are resistant to enzymatic hydrolysis or because the food contains antinutrients that interfere with the normal digestion process.

An expert panel at the Food and Agriculture Organization (FAO) of the United Nations (UN) recommended characterizing the nutritional quality of food proteins based on their Digestible Indispensable Amino Acid Score (DIAAS) (FAO, 2013). This method is based on measuring the amount of individual amino acids from an ingested protein that are absorbed within the small intestine and comparing them to a reference. The DIAAS score of a specific indispensable amino acid is expressed as:

$$\mathrm{DIAAS}(\%) = 100 \times M_S / M_R \qquad (5.1)$$

Here, M_S is the lowest amount in mg of digestible dietary indispensable amino acid in 1 g of the dietary protein that is absorbed by the end of the small intestine (ileum), whereas M_R is the mass in mg of the same dietary indispensable amino acid in 1g of the reference protein. The reference protein is taken to be an idealized protein that would provide adequate amounts of all the indispensable amino acids assuming that a person ate the average amount of protein recommended in the human diet. This average amount is usually taken to be the estimated average requirement (EAR) of protein (0.66 g/kg/d), which is based on the minimum amount needed to ensure 50% of the population have sufficient nitrogen in their diet (Wolfe et al., 2016). However, it has been suggested that this value is too low and should be replaced with a higher value, such as the recommended daily allowance (RDA) of 0.8 g/kg bodyweight/day, since this value should ensure that 98% of the population met their nutritional protein needs (Wolfe et al., 2016). Values for M_R for each kind of indispensable amino acid can be found in reference tables for different age groups *e.g.*, infants, children, and adults (Table 5.2). As an example, a DIAAS value of 50%

Table 5.2 Suggested indispensible amino acid requirements for different age groups (mg/g protein)

Indispensible amino acid	Infant (< 2 years)	Children (2–5 years)	Children (10–12 year)	Adults (> 12 years)
Histidine	26	19	19	16
Isoleucine	46	28	28	13
Leucine	93	66	44	19
Lysine	66	58	44	16
Methionine	42	25	22	17
Phenylalanine	72	63	22	19
Threonine	43	34	28	9
Tryptophan	17	11	9	5
Valine	55	35	25	13

Data taken from the Food and Agriculture Organization of the United Nations/World Health Organization. Protein Quality Evaluation: Report of the Joint FAO/WHO Expert Consultation. Rome, Italy: Food and Agriculture Organization of the United Nations; 1991. FAO Food and Nutrition Paper 51.

for a particular amino acid (*e.g.*, lysine) in a specific food protein source (*e.g.*, barley) would mean that a person would need to eat double the amount of protein compared to the reference protein/pattern to consume enough of the particular limiting amino acid per day (assuming they only consumed barley and eat the recommended amount of total protein). A list of the limiting indispensable amino acids in a number of protein sources is shown in Table 5.3. Methods for measuring the amounts of bioavailable amino acids present in the small intestine (M_S), as well as for determining the M_R values for different amino acids, have been critically reviewed elsewhere (Rieder et al., 2021; Wolfe et al., 2016).

The overall DIAAS of a particular protein is determined by the lowest DIAAS of the different indispensable amino acids it contains. For instance, if the lowest DIAAS is for lysine (*e.g.*, 36%), then it is considered to be the limiting indispensable amino acid from this dietary source and the overall DIAAS of this protein would be 36%. It is important to note, however, that there may be more than one amino acid that falls below the desired DIAAS value (100%) in a food source but if the person eats enough protein to have sufficient intake of the limiting amino acid then automatically also the intake of the other amino acids is sufficient. In reality, however, people tend to eat proteins from a variety of different sources, such as cereals and legumes (*e.g.*, rice and beans), as well as eating considerably more total protein per day than the EAR, which means they usually get sufficient levels of indispensable amino acids in their diet.

Finally, it should be noted that there are other measures of protein quality that are still widely used, such as the protein digestibility-corrected amino acid score (PDCAAS). The PDCAAS of a food protein is also based on its amino acid profile and digestibility within the human gut. However, this method has largely been replaced by the DIAAS in the assessment of the nutritional profile of proteins. There are several reasons for this change. For example, the PDCAAS only allows a

5.2 Macronutrients

Table 5.3 Comparison of the digestible indispensable amino acid scores (DIAAS) and essential amino acids lacking for different plant and animal sources

Protein source	DIAAS	Limiting amino acid
Cereals		
Corn	38	Lysine
Rice	52	Lysine
Wheat	39	Lysine
Oat	44	Lysine
Barley	50	Lysine
Legumes		
Soy	92	Methionine + Cysteine
Fava bean	67	Methionine + Cysteine
Lupin	68	Methionine + Cysteine
Pea	66	Methionine + Cysteine
Chickpeas	69	Methionine + Cysteine
Lentils	75	Methionine + Cysteine
Kidney beans	61	Methionine + Cysteine
Root vegetables		
Potato	85	Histidine
Animal proteins		
Gelatin	2	Tryptophan
Whey	85	Histidine
Casein	117	None
Milk	108	None
Egg	101	None
Pork	117	None
Chicken	108	None
Beef	112	None

Note, there are often other amino acids that are also below the 100% value but only the lowest (limiting) one is listed. Data from various sources (Ertl et al., 2016; Han et al., 2020; Herreman et al., 2020; Hertzler et al., 2020).

maximum score of 100% for any protein, meaning it meets basic nutritional requirements. However, some proteins have amino acid profiles that exceed these basic requirements, which is accounted for in the DIAAS system where a protein can score higher than 100%. Moreover, the DIAAS system is based on measurements of the amino acids remaining in the ileum (end of the small intestine) after ingestion of a protein-containing food, whereas the PDCAAS is based on measurements made in the colon. The level of amino acids remaining in the ileum is more representative of the non-adsorbed amino acids than the level remaining in the colon because the bacteria in the colon break some of the remaining amino acids down, which can result in an overestimation of absorbed amino acids if a sample after the colon is taken.

5.2.1.5 Bioactivity

As well as being important for general nutrition, proteins may also have a range of other health benefits. Ingested food proteins generate different kinds of peptides in the human gut when they are hydrolyzed by gastric and pancreatic proteases (Bhandari et al., 2020; Chakrabarti et al., 2018; Karami & Akbari-adergani, 2019). Some of these peptides have been shown to exhibit beneficial biological activities, such as antioxidant, antimicrobial, or blood pressure lowering effects. The potency of these peptides is governed by the number, type, and sequence of amino acids in their chains, which depends on the type of protein they come from and how they are hydrolyzed within the gut (Daliri et al., 2017). Consequently, there may be different biological activities depending on the nature of the proteins consumed, which may have important health implications when switching from an animal- to a plant-based diet. Nevertheless, there have been few systematic studies on the relative efficacy of bioactive peptides derived from plant or animal sources on human health and wellbeing. Consequently, this is an important area where further research is needed.

5.2.1.6 Allergenicity

It is important to consider the potential allergenicity of any proteins used to formulate plant-based foods, especially when novel sources of proteins are being used that may not be common in the human diet (Fasolin et al., 2019; Pali-Scholl et al., 2019). A growing number of people are becoming susceptible to specific substances in foods (particularly proteins) leading from mild to potentially life-threatening responses such as anaphylactic shocks (De Martinis et al., 2020; Valenta et al., 2015). However, a large fraction of the cases reported as food allergies are often caused by other factors, such as non-immune-mediated food intolerances (Solymosi et al., 2020). A true allergic reaction is typically due to an interaction between a specific protein or peptide fragment and the host's immunoglobin E (IgE). IgE is an antibody present in mammals that plays an important role in the immune response to infections but may also cause undesirable allergic reactions in some people. Around 10.8% of the people surveyed in a study carried out in the US reported that they suffered from some form of food allergy (Gupta et al., 2019).

The Asthma and Allergy Foundation of America reports that the most common food allergies are for cow's milk, soy, eggs, wheat, peanuts, tree nuts, fish, and shellfish (www.aafa.org). Other proteins can also cause allergies in some people but are less commonly consumed. In some cases (*e.g.,* cow's milk, egg, fish, and shellfish) it may therefore be beneficial to replace animal-based foods with plant-based ones. In other cases (*e.g.,* soy, wheat, peanuts, and tree nuts), it is important for people with allergies to be careful which plant-based foods they eat. However, food matrix and processing effects influence the allergenicity of proteins (Lafarga & Hayes, 2017; Vanga et al., 2017). Indeed, studies have shown that it may be possible to reduce the allergenicity of some plant proteins using fermentation and other

5.2 Macronutrients

processing technologies (Pi et al., 2021) but there is still not a complete molecular understanding of the critical features of food proteins that lead to allergic effects (Valenta et al., 2018). Consequently, it is difficult to predict whether a new protein source will be allergenic or not from the knowledge of its molecular structure. It is therefore important to have reliable empirical methods to characterize the potential allergenicity of new protein sources before they are introduced to a wide market (Krutz et al., 2020).

5.2.2 Lipids

5.2.2.1 Introduction

Lipids play an important role in determining the physicochemical, sensory, and nutritional attributes of foods. The major class of lipids in both animals and plants is triacylglycerols (also called triglycerides), which consist of a glycerol backbone with three fatty acids attached (Akoh, 2017; Leray, 2014). The fatty acids vary in their position on the glycerol backbone, the number of carbon atoms they contain, and the number, location, and isomeric form of the double bonds they contain

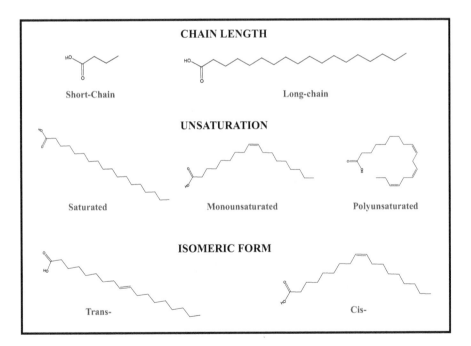

Fig. 5.2 The fatty acids in triacylglycerol molecules vary in their chain length, degree of unsaturation, and isometric form, which influences their functionality and health effects. (Chemical structures kindly drawn by Yuting Wang)

(Fig. 5.2). The position and nature of the fatty acids attached to the glycerol backbone have a pronounced impact on the nutritional attributes of triglycerides. The fatty acid profiles of animal and plant lipids vary considerably, which impacts their functionality in foods and their influence on human health. As well as triacylglycerols, other classes of lipids are also present in animal- and plant-based foods (usually) at relatively low levels, such as phospholipids, waxes, and sterols, which also play an important role in determining their nutritional attributes.

The fatty acid profiles of several representative animal and plant sources of lipids are summarized in Table 2.4. This data shows that animal lipids tend to contain more saturated fatty acids than plant ones, although some sources of plant lipids do have high saturated fatty acid levels (such as coconut oil). The fatty acid profiles also show that some animal lipids (notably fish) have relatively high levels of long chain omega-3 polyunsaturated fatty acids. Animal lipids also tend to contain more cholesterol than plant lipids. These differences in the composition of animal and plant lipids may have important health implications. In the remainder of this section, we provide a brief overview of the different kinds of fatty acids found in animal- and plant-derived food lipids and highlight some of the potential differences between them in terms of their nutritional aspects.

5.2.2.2 Saturated Fatty Acids

The nutritional advice given by most major health organizations is to limit the amount of saturated fatty acids in the human diet. Indeed, the World Health Organization (WHO) recommends that saturated fatty acids should make up less than 10% of total energy consumption based on a perceived link between saturated fatty acid consumption, blood cholesterol levels (LDL), and coronary heart disease (www.who.int/news-room/fact-sheets/detail/healthy-diet). In particular, the WHO recommends replacing saturated fats with polyunsaturated ones or whole grain carbohydrates to improve human health. These recommendations are based on the assumption that consuming excessive amounts of saturated fats leads to heart disease and other chronic diseases (NAS, 2005). Evidence for the link between saturated fats and heart disease mainly comes from epidemiological (observational) studies and randomized controlled trials (RCTs). Epidemiological studies have shown that populations that consume less saturated fats (such as in the Mediterranean) have lower levels of coronary heart disease than populations that consume more saturated fats (such as those in Northern Europe and the United States) (Menotti & Puddu, 2015). Moreover, meta-analyses of RCTs suggest that saturated fat consumption increases blood cholesterol levels (LDL) and heart disease (Hooper et al., 2020). However, any potential benefits of reducing saturated fat in the diet depend on what it is replaced with, *e.g.*, proteins, refined carbohydrates, dietary fibers, monounsaturated fats, or polyunsaturated fats (Briggs et al., 2017).

Some researchers have reported that the impact of saturated fatty acids on human health depends on their chain length, and so they should not all be considered to be nutritionally equivalent (Bloise et al., 2021). In particular, short and medium chain saturated fatty acids may have health benefits, whereas long chain ones do not. This may be important when utilizing saturated fats derived from plants in formulating plant-based foods, such as coconut oil, which is rich in medium chain saturated fatty acids. In contrast, the saturated fats in meat products tend to be long chain, whereas those in dairy products tend to be a mixture of short and long chain. Consequently, there is a need for more research on the impact of specific kinds of saturated fatty acids on human health and wellbeing.

It should be noted that the potential negative impacts of saturated fat on human health have been questioned by some nutrition scientists (Astrup et al., 2021; Harcombe et al., 2016). A large epidemiology study, known as the Prospective Urban Rural Epidemiology (PURE) study, reported that total fat and fat type was not linked to cardiovascular disease, and that saturated fat intake was inversely related to the risk of stroke (Dehghan et al., 2017). A meta-analysis of RCTs and observational studies reported that there was no evidence of reducing cardiovascular disease and total mortality by reducing the total amount of saturated fatty acids consumed (Astrup et al., 2020). It was suggested that one of the reasons for this observation was the impact of saturated fats on the type of low-density lipoprotein (LDL) cholesterol produced in the bloodstream after eating a fatty meal. Consuming high levels of saturated fatty acids increases the level of relatively large LDL particles in the blood, rather than the relatively small ones that have been linked to an increased risk of heart disease (Astrup et al., 2021). However, a recent meta analysis that also included the PURE study came to the conclusion that diets high in saturated fat were associated with higher mortality from all-causes, CVD, and cancer (Kim et al., 2021).

Overall, it seems that there is still much controversy about the role of saturated fats on human health. In general, the impact of saturated fats may also depend on the nature of the food matrix that surrounds them. Clearly, more research is needed to understand the impact of saturated fatty acid type (chain length) and food matrix effects on human health. As mentioned earlier, this will be important when transitioning to a more plant-based diet because plant lipids often contain different types and concentrations of saturated fats than animal lipids.

5.2.2.3 Unsaturated Fatty Acids

Unsaturated fatty acids contain one (monounsaturated) or more (polyunsaturated) double bonds on the hydrocarbon chain (Akoh, 2017; Leray, 2014). The location of these double bonds on the fatty acid chains may also vary. The double bonds in natural polyunsaturated lipids are present in a non-conjugated pentadiene system (Fig. 5.3). As a result, it is only necessary to specify the location of the first double bond relative to the methyl end of the fatty acid chain to know the position of all of the double bonds. Typically, the first double bond in unsaturated fatty acids is located three (omega-3), six (omega-6) or nine (omega-9) carbon atoms from the methylene end. The omega-3 and omega-6 amino acids are considered to be essential nutrients because they cannot be synthesized within the human body, whereas

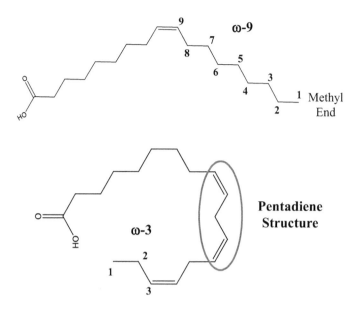

Fig. 5.3 Polyunsaturated fatty acids have a pentadiene system. The position of the first double bond is specified by the number of carbon atoms from the methyl end

the omega-9 ones are not because they can be synthesized. In nature, the double bonds in fatty acids usually have a cis-configuration, which leads to a highly bent chain (Fig. 5.2). This is important because it affects the melting point of the fatty acids, as well as the fluidity of the biological membranes that they are incorporated into within the human body. However, some animal fats naturally contain significant levels of unsaturated fatty acids with a trans-configuration, while some food processing operations promote cis- to trans-isomerization, such as partial hydrogenation (Bloise et al., 2021). The number, position, and isomeric form of the double bonds in fatty acids have a major impact on their nutritional effects, which has to be considered when assessing the nutritional impacts of replacing animal lipids with plant ones.

Monounsaturated fatty acids: Monounsaturated fatty acids (MUFAs) have a single double bond in their hydrocarbon chain (Fig. 5.3). Plant-derived lipids tend to have higher levels of MUFAs than animal-derived ones, which may have important health implications. It has been reported that consumption of diets rich in MUFAs, such as the Mediterranean diet, may have numerous health benefits, such as promoting healthy blood lipid profiles, mediating blood pressure levels, improving insulin sensitivity, and regulating glucose levels (Gillingham et al., 2011; Hammad et al., 2016). Consequently, consuming plant-based diets may have health benefits by increasing the amount of MUFAs in the diet relative to SFAs. Nevertheless, there are still debates about the relative amounts of MUFAs and PUFAs that should be included in the human diet to promote good health (Hammad et al., 2016).

Polyunsaturated fatty acids: PUFAs have numerous double bonds in their hydrocarbon chains (Fig. 5.3). They differ from each other according to the number and position of these double bonds. Most PUFAs can be classified as either omega-3 or omega-6 depending on the position of the first double bond relative to the methyl end of the fatty acid chain. Studies have shown that different kinds of PUFAs have different effects on human health. Omega-3 fatty acids are claimed to exhibit anti-inflammatory effects, which may lead to health benefits by reducing inflammatory diseases, cardiovascular diseases, brain diseases, and cancer (Saini & Keum, 2018; Shahidi & Ambigaipalan, 2018). In contrast, it is claimed that omega-6 fatty acids exhibit pro-inflammatory effects, which may have adverse effects on human health. The ratio of omega-6 to omega-3 PUFAs in the human diet is therefore believed to have important health consequences (Candela et al., 2011; Simopoulos, 2016; Zarate et al., 2017). It has been reported that the omega-6/omega-3 ratio in foods has changed from around 1:1 during most of human evolution to around 20:1 in the present day, which has been linked to negative health outcomes due to differences in the effects of these two types of PUFAs on various physiological processes inside the human body. In particular, a high dietary omega-6/omega-3 ratio has been linked to an increased prevalence of inflammation, heart disease, obesity, and various kinds of cancer (Zarate et al., 2017). Consequently, there may be health benefits associated with increasing the amount of omega-3 PUFAs in the human diet (Saini & Keum, 2018; Shahidi & Ambigaipalan, 2018).

Land animals (such as cows, pigs, and sheep) contain relatively low levels of omega-3 PUFAs, but fatty fish (such as salmon and tuna) contain relatively high levels, especially eicosapentaenoic acid (EPA) and docosahexaenoic acid (DHA). In contrast, the lipids isolated from most plant sources contain relatively high levels of MUFAs and omega-6 PUFAs. However, there are some plant sources that do contain relatively high levels of omega-3 PUFAs, such as the alpha-linolenic acid (ALA) found in flaxseed, walnut, soybean, and canola oils (Rajaram, 2014). It may therefore be beneficial to include these lipid sources into plant-based foods to improve their nutritional profile. However, EPA and DHA have been reported to be more beneficial to human health than ALA, and only a small fraction of ingested ALA is converted into EPA and DHA inside the human body (Baker et al., 2016). Consequently, it may be better to utilize alternative sources of omega-3 PUFAs in plant-based foods, such as the microalgae oils that are rich in DHA or the oils derived from genetically modified agricultural crops that are engineered to contain DHA and EPA (Tocher et al., 2019). A potential advantage of using these sources of omega-3 PUFAs is that they do not contain high levels of heavy metals, like mercury, which are sometimes found in wild fish.

It should be noted that some nutrition and medical scientists are questioning the health benefits of polyunsaturated lipids (Lawrence, 2021). A large meta-analysis of RCTs found that increasing consumption of omega-3 fatty acids had little or no effect on mortality or cardiovascular disease (Abdelhamid et al., 2018). Moreover, polyunsaturated lipids are highly prone to oxidation, which has been linked to oxidative stress, inflammation, atherosclerosis, and cancer (Lawrence, 2021). If this proves to be the case, then there may be adverse health effects associated with replacing saturated or monounsaturated fats with polyunsaturated ones (especially if they are highly oxidized).

Trans fatty acids: There is strong evidence that consumption of trans fatty acids is harmful to human health, leading to an increase in cholesterol levels and heart disease (Anand et al., 2015; Oteng & Kersten, 2020). As a result, nutrition experts recommend avoiding consuming trans fatty acids, particularly those produced by industrial manufacturing processes, such as partial hydrogenation (if not carefully controlled). Indeed, the WHO recommends that less than 1% of the energy in the human diet should come from trans fatty acids. However, some reviews suggest that the origin of the trans-fatty acids (natural *versus* industrial) has an impact on their effects on human health (Dawczynski & Lorkowski, 2016; Oteng & Kersten, 2020). In particular, the ones produced by partial hydrogenation during food processing appear to have strong adverse effects, whereas the ones naturally produced by some ruminants may actually have beneficial effects. Overall, current nutritional knowledge suggests that the use of industrially-produced trans fatty acids in formulating plant-based foods should be avoided. This often introduces challenges when formulating plant-based foods because most plant-derived lipids tend to be liquid at room temperature (such as corn, canola or sunflower oils). In many products, it is desirable to have a partially crystalline fat phase so as to provide desirable textural attributes to foods, such as butter, cheese, whipped cream, or ice cream analogs. Traditionally, this was done by partial hydrogenation of plant-based oils, such as sunflower, soybean, or palm oil. However, this led to the formation of high levels of trans fatty acids especially at the early stages when this technology was used, whereas nowadays the process has been optimized and results in much lower amounts of trans fatty acids. However, new strategies are required to create desirable textural attributes in plant-based foods without introducing high levels of trans fats or introducing other negative health effects.

5.2.2.4 Cholesterol

It is often assumed that cholesterol has adverse effects on human health. However, cholesterol is essential for maintaining human health as it plays a number of critical roles in many cellular and systemic functions (Luo et al., 2020; Yu et al., 2019). In particular, it impacts the rigidity and permeability of cell membranes, regulates the function of some membrane proteins, is involved in various signaling processes, and serves as a precursor for vitamin D, bile salts, and steroid hormones. Thus, it is not cholesterol itself that causes health problems such as heart disease, neurodegeneration, and cancer, but the dysregulation of normal cholesterol homeostasis, such as uptake, synthesis, metabolism, storage, and excretion (Luo et al., 2020). This homeostasis is influenced by the amount of cholesterol in the diet. The cholesterol in foods is usually incorporated into mixed micelles after lipid digestion in the small intestine, then adsorbed by specific proteins in the intestinal enterocytes (Ko et al., 2020). The cholesterol is then packaged into chylomicrons and transported to the liver, where it is stored and then repackaged into lipoproteins that carry it through the bloodstream so it can be taken up by various cells (Luo et al., 2020). In addition, cholesterol can be synthesized and metabolized within the human body. Appropriate cholesterol levels are attained in the cell through the coordination of various molecular processes including uptake, biosynthesis, metabolism, transport, and excretion.

Numerous studies have shown that there is a direct correlation between the concentration of the low density lipoprotein (LDL) - which transports lipids including cholesterol - in the human bloodstream and the incidence of cardiovascular disease (Mach et al., 2020; Yu et al., 2019). It is therefore beneficial to adopt a diet that will reduce blood cholesterol levels (*i.e.*, the concentration of the lipoproteins). It should be noted that the cholesterol in the foods we consume is only one contributor to the overall cholesterol levels in our bodies. The type and amount of lipids, carbohydrates, and proteins we consume, as well as various other dietary components, also impact blood cholesterol levels, as they alter cholesterol homeostasis. It has been recommended to reduce the dietary intake of cholesterol to below 300 mg per day, especially for individuals with high blood cholesterol levels (Mach et al., 2020). There are therefore benefits from consuming a plant-based diet, since plants typically contain no cholesterol. However, it is also important to consider the nature of the other foods that comprise the overall diet.

A meta-analysis of nutritional studies has shown that a number of foods can significantly reduce LDL cholesterol levels: (*i*) foods high in unsaturated fatty acids and low in saturated and trans fatty acids; (*ii*) foods fortified with phytosterols and/or phytostanols; and (*iii*) foods high in soluble dietary fibers (Schoeneck & Iggman, 2021). Many plant-based foods are rich in phytosterols, phytostanols, and dietary fibers and should therefore have benefits on maintaining healthy cholesterol levels. The meta-analysis also indicated that the consumption of specific kinds of plant-based foods may appreciably reduce LDL cholesterol levels in the blood, including almonds, avocados, flaxseeds, hazelnuts, pulses, soy protein, tomatoes, turmeric, walnuts, and wholegrain foods. Consequently, consuming a plant-based diet that contains these foods may have potential health benefits, especially in terms of heart disease (Schoeneck, et al., 2021).

A number of plant-based foods, notably vegetable oils and to a lesser extent fruits, vegetables, nuts, grains, and legumes, contain appreciable quantities of phytosterols and phytostanols, which have similar chemical structures and biological functions as the cholesterol found in animals (Moreau et al., 2002). Consumption of relatively high levels of these phytochemicals has been shown to reduce blood LDL cholesterol levels (Ghaedi et al., 2020), which may have potential benefits on human health. For instance, ingestion of around 2 grams of phytosterols and/or phytostanols daily has been reported to lower LDL cholesterol levels by around 8 to 10% (Gylling & Simonen, 2015). For this reason, some companies have created functional foods fortified with these cholesterol-lowering phytochemicals with the aim of improving human health and wellbeing. Phytosterols and phytostanols are believed to compete with cholesterol for incorporation into mixed micelles within the small intestine, thereby reducing the amount of cholesterol that is absorbed by the body (Gylling, et al., 2015). A recent meta-analysis also suggests that phytosterols and phytostanols may have anticancer effects (Cioccoloni et al., 2021). Thus, there may be significant benefits from eating plant-based foods that are naturally rich in these substances or processed foods that are fortified with them.

5.2.3 Carbohydrates

5.2.3.1 Introduction

Most animal products (meat, fish, and eggs) do not contain appreciable quantities of carbohydrates, with the exception of milk and liver, which contains around 4% of the disaccharide lactose and glycogen, respectively (McClements & Grossmann, 2021b). In contrast, most natural plants contain relatively high levels of carbohydrates, which can typically be divided into sugars, oligosaccharides, starches, and dietary fibers (Fig. 5.4) (Mattila et al., 2018). Moreover, many processed plant-based foods, such as meat, seafood, egg, and dairy analogs, are often assembled from a mixture of functional ingredients, including carbohydrates (McClements & Grossmann, 2021a). However, the type and concentration of carbohydrates used can be manipulated by the food manufacturer, which means that it is possible to control their nutritional profiles. Carbohydrates are often added to plant-based foods as functional ingredients to provide desirable color, texture, or flavor characteristics, but their impact on the nutritional attributes of these foods should also be considered. In this section, we highlight some of the important nutritional characteristics of carbohydrates that should be considered when formulating plant-based foods.

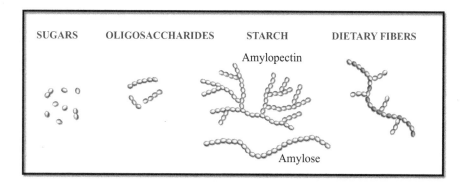

Fig. 5.4 The nutritional properties of carbohydrates depend on their chemical structure, which is influenced by the type, number, sequence, and bonding of the monosaccharides

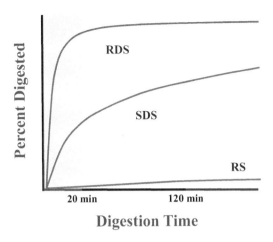

Fig. 5.5 The nutritional effects of starch depend on its digestibility. Starch can be categorized as rapidly digestible starch (RDS), slowly digestible starch (SDS), and resistant starch (RS)

5.2.3.2 Starches

In nature, starches are usually found in the form of small dense particles known as granules, which consist of concentric rings of crystalline and amorphous regions (Cornejo-Ramirez et al., 2018). At the molecular level, starch is made up of two homo-polysaccharides that are comprised of long chains of glucose molecules linked together by glycosidic bonds (Fig. 5.4). Amylose is a linear polymer consisting primarily of α-1,4 linked glucose units, whereas amylopectin is a highly branched polymer consisting of linear regions of α-1,4 linked glucose units with α-1,6 links in the branch points. From a nutritional perspective, starch can be divided into rapidly digestible starch (RDS), slowly digestible starch (SDS), and resistant starch (RS) depending on how quickly it is hydrolyzed by digestive enzymes in the upper regions of the human gut (Fig. 5.5), *i.e.,* mouth, stomach, and small intestine (Bello-Perez et al., 2020; Dhital et al., 2017). The glycosidic bonds in RDS molecules are easily accessible to amylases in the human gut, which leads to the rapid hydrolysis of the starch chains. As a result, there may be a spike of glucose in the bloodstream after ingestion of high quantities of RDS, which has been linked to an increase in the risk of diabetes and obesity. The glycosidic bonds in SDS are less available for hydrolysis by amylases, which leads to a slower release of glucose molecules in the gastrointestinal tract, thereby avoiding spikes in blood glucose levels. RS starch has a structure that is highly resistant to hydrolysis by digestive enzymes and therefore tends to resist digestion in the upper gastrointestinal tract. However, once it reaches the colon it can be fermented by the microbes residing there, leading to the generation of short chain fatty acids (SCFAs), which have beneficial effects on the gut microbiome and human health (Bello-Perez et al., 2020; Dhital et al., 2017). Our current understanding of the gastrointestinal fate and nutritional effects of starch suggests that it is beneficial to formulate plant-based foods from RS and SDS forms of starch rather than RDS forms. However, one must also take into account the impact of the different forms of starch on their functional attributes in foods and on the quality attributes of the end product.

5.2.3.3 Sugars and Oligosaccharides

Sugars typically consist of one (*e.g.,* glucose or fructose) or two (*e.g.,* sucrose, maltose, or lactose) monosaccharide units. Oligosaccharides are usually defined as having between 3 and 20 monosaccharide units. The monomers in disaccharides and oligosaccharides are held together by glycosidic bonds. Sugars are typically white crystalline substances that have a good water solubility. In terms of functionally, they are often incorporated as ingredients in foods to provide desirable flavors and colors, as well as various other roles. Most sugars are naturally sweet, but they may also contribute to other tastes and aromas due to their participation in chemical reactions during cooking, such as the Maillard reaction and caramelization (Elmore & Mottram, 2009). Sugars are also important for providing desirable brown colors during cooking, which occur due to the same chemical reactions. For instance, reducing sugars chemically react with proteins or peptides at high temperatures and intermediate moisture conditions to form melanonids, which contributes to the desirable brown color of cooked meat products (Shaheen et al., 2021). The sugar content of most animal products (such as meat, fish, and egg) is relatively low, although bovine milk does contain around 4% lactose. However, plant-based foods may contain added sugars to improve their appearance, flavor profile, or other functional attributes. For instance, sugars are often added to plant-based milks to improve their sweetness. Consequently, it is important to appreciate the potential impact of sugars on the nutritional attributes of plant-based foods. Added sugars may have a number of detrimental effects on the nutritional profiles of these foods. First, foods containing high levels of sugars have been linked to tooth decay (dental caries) (Moynihan & Kelly, 2014). Second, consumption of these types of foods has also been linked to an increased risk of a variety of chronic diseases, including diabetes, obesity, liver disease, and heart disease (O'Neil et al., 2020; Rippe & Angelopoulos, 2016). Third, some of the reaction products of the Maillard reaction between sugars and proteins (e.g., acrylamide) may have detrimental effects on human health. For instance, advanced glycation end products (AGE) have been reported to be linked to diabetes, neurological disorders, atherosclerosis, hypertension and some forms of cancer (Kuzan, 2021). Consequently, it is important to ensure that plant-based foods are formulated so that they do not contain high levels of sugars that might promote tooth decay or chronic diseases.

A number of oligosaccharides found in foods have been reported to have beneficial health effects, especially fructo-oligosaccharides (FOS) and galacto-oligosaccharides (GOS), which exhibit prebiotic effects (Bosscher et al., 2009; Davani-Davari et al., 2019). These oligosaccharides can promote a healthy gut microbiome by stimulating the growth of beneficial colonic bacteria. In general, they have been reported to have beneficial effects on the gastrointestinal, central nervous, immune, and cardiovascular systems (Davani-Davari et al., 2019). Human breast milk contains oligosaccharides that have been reported to have various health benefits to growing infants, such as protecting against pathogens, enhancing immune responses, modulating gut microbiomes, and enhancing mineral absorption (Al Mijan et al., 2011; Vandenplas et al., 2018). Consequently, it may be important to identify and utilize plant-derived oligosaccharides that can produce similar health benefits when developing plant-based foods for infants, such as infant formula. It should be stressed that the current generation of plant-based milks are unsuitable for infants because they have very different macronutrient and micronutrient contents than human or cow's milks. However, plant-based infant formula have been developed that contain the nutrients required for infants. In general, the incorporation of plant-derived prebiotics into meat, seafood, egg or dairy analogs may provide health benefits to the general population.

5.2.3.4 Dietary Fibers

Many sources of edible plants naturally contain high levels of dietary fibers, although these are often removed when isolating the functional ingredients (such as proteins and starches) used to formulate processed plant-based foods, like meat, fish, egg, or dairy analogs. However, plant-based foods can be fortified with dietary fibers isolated from plants, which may have health benefits but the structural organization and nutritional attributes of added dietary fibers are often very different from those of natural dietary fibers in whole foods (Augustin et al., 2020). Consequently, it may be beneficial from a nutritional perspective to leave as much as possible of the original structure of plant materials intact when formulating plant-based foods.

Dietary fibers are typically polysaccharides (but they may also include other associated substances like polyphenols, waxes, or proteins) that are not digested and absorbed in the mouth, stomach, and small intestine, and therefore pass through the upper GIT largely intact (Augustin et al., 2020). However, once they reach the colon, they may be fermented by the various kinds of bacteria that reside there. These fermentable dietary fibers can act as prebiotics that stimulate the growth of beneficial bacteria within the human gut, which can lead to several health benefits to humans (Roberfroid et al., 2010; Wang, 2009). Different kinds of dietary fibers can be classified as soluble/insoluble and fermentable/non-fermentable depending on their physicochemical and physiological properties. The precise nature of the dietary fibers present in foods has a major impact on their nutritional effects.

In general, dietary fibers can exhibit their health benefits through a number of mechanisms, which have been reviewed in detail elsewhere (Augustin et al., 2020; McClements, 2021):

- *Rheological modifications*: Ingested dietary fibers may increase the viscosity of gastrointestinal fluids, which alters mixing and diffusion processes, thereby retarding macronutrient digestion and absorption. Consequently, dietary fibers may inhibit spikes in glucose or lipid levels in the bloodstream that can lead to

dysregulation of metabolic systems. The presence of dietary fibers in the non-adsorbed materials traveling through the human gut can also reduce its solidity, thereby increasing its rate of passage, which can prevent constipation and improve gut function.
- *Binding interactions*: Dietary fibers may bind to other components in the gastrointestinal tract, thereby altering digestion, transportation and absorption processes. For instance, they may bind to digestive enzymes, which reduces their ability to hydrolyze fats, proteins, or starches. They may also bind to bile salts, calcium ions, or free fatty acids, thereby interfering with lipid digestion, mixed micelle formation, and mixed micelle movement to the epithelium cells, which can reduce the absorption of lipids and oil-soluble vitamins.
- *Aggregation state*: Dietary fibers may alter the aggregation state of macronutrients in the GIT, such as fat droplets, protein particles, or starch granules, which alters the ability of digestive enzymes to reach their surfaces and hydrolyze them. For instance, dietary fibers can promote the aggregation of fat droplets through bridging or depletion mechanisms, which reduces their digestion and absorption.
- *Coating and embedding*: Dietary fibers may form an indigestible coating or matrix around macronutrients that inhibits the ability of digestive enzymes to reach their surfaces, again slowing down their digestion and absorption.
- *Gastrointestinal barrier properties*: The presence of dietary fibers in the human gut can alter the barrier properties of the mucus and epithelium layers, which can alter the rate and extent of absorption of nutrients.
- *Fermentation and the gut microbiome*: Some dietary fibers are not digested within the upper GIT but are fermented in the colon, where they can be utilized by bacteria to produce short chain fatty acids (SCFAs), which can improve colonic health. Moreover, the consumption of dietary fibers can change the composition and function of the gut microbiome in a manner that improves human health.

The potentially beneficial effects of consuming dietary fibers on human health have been compared in a meta-analysis of various nutritional studies (Reynolds et al., 2019). This meta-analysis was based on data equivalent to 135 million person-years obtained from 185 prospective studies and 58 clinical trials on 4635 adults. The authors reported that consumption of high levels of dietary fiber led to a 15 to 30% decrease in all-cause mortality, heart disease, stroke, diabetes, and colorectal cancer. Those people who consumed the highest levels of dietary fiber also had lower body weights, blood pressures, and cholesterol levels than those who consumed the lowest levels. The largest decrease in risk from chronic disease was achieved by consuming about 25 to 29 g of dietary fiber per day. This level is considerably higher than the average amount currently consumed by the majority of the population in developed countries.

Animal-based foods like meat, fish, egg, and dairy products contain very low levels of dietary fibers. Consequently, switching to a plant-based diet containing high levels of dietary fibers could have important health benefits. Consumers can increase the total quantity of dietary fibers they consume by eating more fruits, vegetables, whole grains, legumes, nuts or foods prepared from these components. But many people do not consume these kinds of products on a regular basis because they are too expensive, too time consuming and laborious to prepare, or they do not like their taste. Consequently, it may be advantageous to improve the nutritional profile of processed plant-based foods (*i.e.,* meat, seafood, egg, or dairy analogs) by fortifying them with appropriate types and amounts of dietary fibers. It should be noted, however, that dietary fibers may behave differently in the GIT when consumed as part of whole foods than when isolated and incorporated into processed foods as functional additives (Grundy, Edwards, et al., 2016a; Grundy, Lapsley, & Ellis, 2016b; Guo et al., 2017b). This effect is mainly attributed to the fact that whole foods contain intact plant cell walls that can inhibit the digestion of entrapped macronutrients (such as starches and fats), as well as to the fact that they contain phytochemicals that might have other health benefits (such as polyphenols and carotenoids).

It should also be noted that overconsumption of certain kinds of dietary fibers can have undesirable effects on human health, such as gastrointestinal discomfort, bloating, flatulence, and loose stools, particularly in people suffering from bowel disorders (Nyyssola et al., 2020). The severity of these effects depends on the quantity and kind of dietary fibers ingested, as well as the nature of the surrounding food matrix. It is therefore important to consider the potential risks of incorporating high levels of dietary fibers into plant-based foods, as well as their potential benefits.

Finally, it should be noted that a broad range of substances can be considered to be dietary fibers, which have different molecular, physicochemical, and physiological characteristics. At present, there is still a relatively poor understanding of the relationship between the molecular features and nutritional effects of dietary fibers. As knowledge in this area increases, it may be possible to develop dietary fiber ingredients that can be used in plant-based foods to provide specific health benefits.

5.3 Micronutrients

Micronutrients (vitamins and minerals) are minor components found in foods that are essential for human wellbeing and health but cannot be synthesized in adequate levels by the human body (Gropper et al., 2021). As a result, they must be obtained from the diet. There are numerous types of vitamins and minerals in foods that all have different molecular characteristics, physicochemical properties, and physiological effects. In this section, we focus on those vitamins and minerals that may be lacking in a predominantly plant-based diet, *i.e.,* vegan or vegetarian. The main micronutrients of concern are vitamin B_{12}, vitamin D, omega-3 fatty acids, calcium, iron, and zinc (Bakaloudi et al., 2021; Craig, 2010). The nutritional impact of omega-3 fatty acids was discussed in the section on lipids and so will not be considered further here. A summary of the micronutrients that may be lacking from a plant-based diet is given in Table 5.4.

Table 5.4 Summary of the major vitamins and minerals that might be lacking in a plant-based diet and their role in human health

Micronutrient	Function	Source	RDA
Vitamin B_{12}	Conversion of food into energy, nervous system function, red blood cell formation	Dairy products, eggs, meat, seafood, fortified cereals, fortified plant-based foods	2.4 µg
Vitamin D	Blood pressure regulation, bone growth, calcium balance, hormone production, immune function, nervous system function	Eggs, fish, fish oil and cod liver oil, pork, fortified dairy products, fortified margarine, fortified orange juice, fortified plant-based beverages, fortified breakfast cereals, mushrooms	15 µg
Iron	Energy production, growth and development, immune function, red blood cell formation, reproduction, wound healing	Meat, seafood, eggs, beans, fruits, green vegetables, nuts, peas, seeds, tofu, whole grain	8–18 mg
Calcium	Blood clotting, bone and teeth formation, constriction and relaxation of blood vessels, hormone secretion, muscle contraction, nervous system function	Dairy products, canned seafood with bones, fortified orange juice, fortified plant-based beverages, fortified breakfast cereals green vegetables, tofu	1000 mg
Zinc	Growth and development, immune function, nervous system function, protein formation, reproduction, taste and smell, wound healing	Meat, seafood, dairy products, beans, peas, nuts, whole grains, fortified cereals	8–11 mg

Adapted from USDA Vitamins and Minerals Chart. RDA is the recommended daily amount for adults, which varies with sex, age, and pregnancy status (NIH).

Many manufacturers of plant-based foods are already adding these micronutrients to their products to improve their nutritional profiles, which should help to prevent any potential nutritional deficiencies that may occur from adopting an exclusively plant-based diet. It should be noted that the practice of fortifying foods with specific nutrients has been practiced for decades (such as vitamin D in cow's milk and folate in flour) and has been highly successful in reducing health problems associated with nutrient deficiencies. Moreover, depending on the animal husbandry system (especially non-ruminant), the animal feed is often enriched with specific micronutrients to enrich the animal products derived from them. Thus, supplementing plant-based foods directly would be a more efficient way to incorporate micronutrients in the human diet.

5.3 Micronutrients

5.3.1 Vitamins

5.3.1.1 Vitamin B_{12}

Vitamin B_{12} is a group of water-soluble molecules that are essential for converting food into energy, for keeping the nerve and blood cells functioning properly, and for creating genetic material within our cells (Rizzo et al., 2016). A diet deficient in this micronutrient may therefore cause health problems, especially anemia, neurological problems, and fatigue. The National Institutes of Health (USA) recommends a daily intake of 2.4 micrograms of vitamin B_{12} for adults and slightly more for pregnant and lactating women. In the human diet, vitamin B_{12} is mainly obtained from animal-derived foods, such as meat, fish, eggs, and milk, which is because vitamin B_{12} is produced from bacteria that reside in the gut of animals (and on the surface of some plant foods but amounts are lower and washing removes the vitamin). Thus, vegetarians who consume eggs and dairy products should have sufficient levels of this vitamin in their diet, whereas vegans who have a completely plant-based diet may be deficient in vitamin B_{12} (Pawlak et al., 2014). This problem can be overcome by taking supplements or fortified foods, such as some breakfast cereals and processed plant-based foods (Butola et al., 2020). In this case, the vitamin B_{12} is typically obtained from microbial sources, rather than from animals, so that it is suitable for people who do not want to eat any animal-derived products. In addition, nutritional yeast is another good source of vitamin B_{12} that is widely used as a seasoning in vegan recipes. Moreover, there has been interest in developing plant-based food ingredients that are naturally rich in vitamin B_{12}, such as seaweed, mushrooms, and some fermented foods (Rizzo et al., 2016). These ingredients could then be utilized to formulate processed plant-based foods so as to avoid any deficiencies associated with this vitamin.

5.3.1.2 Vitamin D

Vitamin D refers to a group of oil-soluble molecules mainly found in animal foods that are also essential for human health and wellbeing (Gropper et al., 2021). In particular, vitamin D promotes calcium absorption, healthy bone formation, regulating blood pressure, and the proper functioning of the muscular, hormone, nervous, and immune systems. There are two main types of this micronutrient in the human diet: vitamin D_2 and D_3. Vitamin D_3 is usually only found in significant quantities in animal-derived foods, such as meat, milk, and eggs, but it can also be synthesized in the human skin in the presence of sunlight. In contrast, vitamin D_2 is predominantly found in plant-derived foods, such as fungi, but these sources only make up a small proportion of the human diet. As a consequence, people following a strictly plant-based diet (vegans) may be deficient in vitamin D, which may have adverse health consequences, including osteoporosis, weak bones, growth retardation, and muscle weakness. For this reason, vegetarians and vegans may need to take supplements or

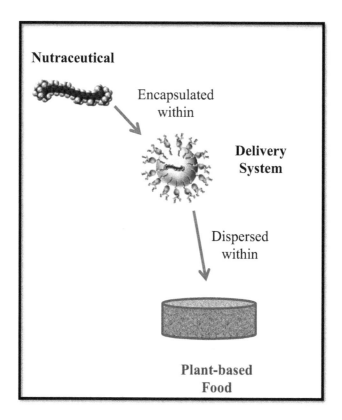

Fig. 5.6 Oil-soluble vitamins and nutraceuticals can be encapsulated within colloidal delivery systems, which can then be incorporated into plant-based foods to improve their nutritional profile

include foods fortified with vitamin D in their diets. The vitamin D_2 used in supplements is always from non-animal sources, but the vitamin D_3 may come from either animal or non-animal sources. For this reason, consumers should be careful when choosing supplements to strictly conform to a vegan or vegetarian diet.

Vitamin D is a highly hydrophobic molecule that has a relatively low bioavailability when present in many food matrices (Maurya et al., 2020). Consequently, there has been considerable interest in utilizing colloidal delivery systems, particularly emulsion-based ones, to increase the bioavailability of this oil-soluble micronutrient (Ozturk, 2017). In these systems, the vitamin D is trapped inside small triacylglycerol-rich droplets that can be incorporated into plant-based foods (Fig. 5.6). These fat droplets are typically rapidly digested within the gastrointestinal tract. The vitamin D is then released from the fat droplets and solubilized within the mixed micelles formed by the hydrolyzed triacylglycerol molecules and transported to the epithelium cells, thereby enhancing its bioaccessibility (Fig. 5.7). These kinds of delivery systems are particularly suitable for increasing the bioavailability of vitamin D in fortified plant-based foods, such as meat, seafood, egg or dairy analogs.

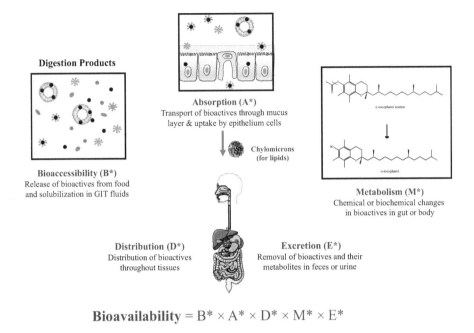

Fig. 5.7 The overall bioavailability of hydrophobic bioactives depends on numerous factors, including their bioaccessibility, absorption, distribution, metabolism and excretion. (Image of human gastrointestinal tract from Servier Medical Art (smart.servier.com) and used under CC BY 3.0 (https://creativecommons.org/licenses/by/3.0/))

5.3.2 Minerals

5.3.2.1 Iron

Iron is a trace mineral that plays multiple roles in the human body, and so a lack of it in the human diet over prolonged periods causes severe health problems, especially anemia (Gropper et al., 2021). In particular, iron plays an important role in the proper functioning of hemoglobin and myoglobin, which store and transport oxygen, as well as in reproduction, wound healing, red blood cell formation, growth and development, energy production, and immune function. The National Institutes of Health (NIH, USA) recommends that adult men and women (19–50 years) should consume 8 and 18 mg per day, respectively, with pregnant women requiring considerably more (27 mg). Iron can be obtained from a variety of natural and processed foods, including meat, seafood, eggs, fortified breakfast cereals, fortified bread, beans, lentils, spinach, kidney beans, peas, nuts, and dried fruits. The iron in these sources may be present in either a heme or non-heme form. Meat and seafood contain heme iron, whereas plant foods and fortified foods contain non-heme iron. The bioavailability of heme iron is relatively high because there are specific receptors on the intestinal enterocytes that facilitate the uptake of this form of iron (Rousseau et al., 2020). In contrast, the bioavailability of non-heme iron is relatively low because of its low water-solubility and its tendency to strongly bind to mineral antinutrients in many plant-based foods, such as phytic acid, polyphenols, and dietary fibers (Rousseau et al., 2020). For this reason, the NIH recommends that those on a strictly plant-based diet should have around twice as much iron in their diets as those who eat meat.

Researchers are developing methods of increasing the bioavailability of iron in plant-based foods. The iron-carrying leghemoglobin used to create some plant-based meat analogs should behave similarly in our bodies as the hemoglobin found in red meats, ensuring that it is in a highly bioavailable form. In addition, there is interest in utilizing specific mineral forms, colloidal delivery systems, or food matrix design to increase the bioavailability of iron in supplements and fortified foods (Pastore et al., 2020; Trivedi & Barve, 2021; Zhang et al., 2021; Zuidam, 2012). Application of these technologies in plant-based foods may help to prevent iron deficiency in vegans and vegetarians.

5.3.2.2 Zinc

Zinc is another trace mineral that may be deficient in the vegan or vegetarian diet since it is normally obtained from animal sources, such as meat and seafood (Grungreiff et al., 2020). However, it can also be found in some natural plant-derived foods, such as beans, peas, nuts, and whole grains, although its bioavailability is often lower than from animal sources. Zinc plays a critical role in a wide range of physiological processes critical to human health and wellbeing, including growth and development, immune function, nervous system function, protein synthesis, wound healing, reproduction, taste and smell. Prolonged zinc deficiency can therefore have adverse health effects, including increased susceptibility to infection, stunted growth, skin problems, and smell, taste, and eye disorders. The National Institute of Health recommends that adults should consume around 8 to 11 mg of zinc per day, with women requiring lower levels than men (unless they are pregnant or breastfeeding). The relatively low bioavailability of zinc in many plant-based foods is due to the fact they contain significant levels of mineral antinutrients, such as phytic acid, polyphenols, and dietary fibers, that can strongly bind to it and reduce its absorption (Rousseau et al., 2020). Vegans and vegetarians on diets that do not contain sufficient levels of this micronutrient should therefore take zinc supplements or consume processed plant-based foods that are fortified with zinc (Bakaloudi et al., 2021). In particular, it may be important for women who are strict vegans to take zinc supplements or consume fortified foods during their pregnancy (Sebastiani et al., 2019). The bioavailability of zinc can be improved by using processing technologies to remove or deactivate mineral antinutrients. In addition, specific mineral forms, delivery systems, or food matrices can be utilized to boost zinc bioavailability in fortified foods and supplements (Pastore et al., 2020; Trivedi, et al., 2021;

Zhang et al., 2021; Zuidam, 2012). The application of these technologies in plant-based foods is particularly important for preventing zinc deficiency in vegans and vegetarians.

5.3.2.3 Calcium

Calcium is a mineral that is required at much higher levels in the human diet than other minerals because of its importance in promoting bone health, as well as various other functions (Grungreiff et al., 2020). For instance, it also plays an important role in muscle, nerve, blood vessel, hormone, and enzyme functions. The NIH recommends that adult men and women (19–50 years) should consume around 1000 mg of calcium per day, but higher levels are recommended for the elderly and pregnant women. Calcium deficiency over a long period can lead to health problems, such as reduced bone mass and an increased risk of osteoporosis and bone fractures. A serious calcium deficiency can lead to numbness, tingling in the fingers, convulsions, and abnormal heart rhythms. Dairy products, such as milk, yogurt, and cheese are the main sources of calcium in the diets of consumers in many developed countries (Gao et al., 2006; Romanchik-Cerpovicz & McKemie, 2007). As a result, vegans are susceptible to calcium deficiencies if they do not consume supplements or other sources of calcium-rich foods. Calcium is also naturally present in various plant-based sources, such as certain green vegetables (kale, broccoli, collard greens) and grains. Even though calcium is not present in high concentrations in grains, they are still a good source because they are eaten in such high quantities. Calcium is also present in many fortified foods, such as breakfast cereals, fruit juices, and plant-based milks. It should be noted that the bioavailability of calcium is often limited in natural plant-based foods due to the presence of antinutrients that bind to it within the human gut and inhibit its absorption, such as oxalate and phytic acid (White & Broadley, 2005). Calcium ingredients come in a variety of forms that can be used for fortifying foods, such as calcium carbonate, citrate, gluconate, lactate, and phosphate, each with their own physicochemical and bioavailability characteristics (Fairweather-Tait & Teucher, 2002). Consequently, it is important to select the most appropriate form for a particular application. Moreover, there are also a number of other factors that need to be considered when fortifying foods with calcium: its presence may induce aggregation, precipitation, or gelation of other ingredients (especially anionic ones); it often gives an unpleasant mouthfeel; and its bioavailability is often relatively low and variable depending on the nature of the food matrix (Romanchik-Cerpovicz, et al., 2007). Consequently, it is important to select the most appropriate form of calcium and food matrix when fortifying plant-based foods with this important micronutrient.

5.4 Nutraceuticals

Plant-based foods contain numerous kinds of nutraceuticals, which are substances found in foods that are not essential for human health (like vitamins and minerals) but may still provide important health benefits (Santini et al., 2018; Santini et al., 2017). Common nutraceuticals include substances like carotenoids (*e.g.,* from carrots, peppers, and kale), curcumin (*e.g.,* from turmeric), anthocyanins (*e.g.,* from red cabbage and berries), and other kinds of polyphenol substances (*e.g.,* from tea, coffee, and many plants). The regular consumption of sufficiently high quantities of nutraceutical-rich foods may reduce the risk of chronic diseases, such as heart disease, cancer, diabetes, hypertension, stroke, brain diseases, or eye diseases (Gul et al., 2016). Alternatively, they may promote human performance, such as endurance, energy levels, mood, and attention. A number of mechanisms of action have been proposed for the health benefits of these bioactive food components, including antioxidant, antimicrobial, and anti-inflammatory activities. Nevertheless, many of the claims made for nutraceuticals still need to be verified using well designed randomized controlled studies and meta-analyses (McClements, 2019). Nutraceuticals may be naturally present in whole foods or they may be isolated and then incorporated into processed foods as functional ingredients, which may lead to a change in their biological activity (Fardet, 2015a, 2015b). Consequently, it is important to carefully design the food matrix to obtain the expected health benefits of nutraceuticals when used as functional ingredients in processed foods. For isolated nutraceuticals, it is important that they can be dispersed within the food matrix, are chemically stable within the food and gastrointestinal tract, and have a high bioavailability after ingestion (McClements, 2018b). This can often be achieved by encapsulating them within well-designed edible delivery systems, such as oil-in-water emulsions (McClements, 2015; Ting et al., 2014; Velikov & Pelan, 2008). The fat droplets in these emulsions have a hydrophobic interior and a hydrophilic exterior. As a result, hydrophobic nutraceuticals can be trapped inside them and then dispersed into aqueous-based food matrices, such as those in plant-based meat, seafood, egg, and dairy products (Fig. 5.6). Moreover, the small dimensions and large surface areas of the oil droplets lead to their rapid digestion within the gastrointestinal tract (McClements et al., 2015). As a result, the nutraceuticals are rapidly released from the oil droplets and solubilized within mixed micelles formed from the lipid digestion products. These mixed micelles then transport the nutraceuticals to the enterocytes where they are absorbed. The bioavailability of nutraceuticals in whole foods, such as fruits and vegetables, can be increased by consuming them with excipient emulsions or by controlling food matrix effects (Aboalnaja et al., 2016; McClements et al., 2015). In this case, the nutraceutical-rich whole food is consumed with fat droplets or food matrices that are rapidly digested in the small intestine and create mixed micelles that can again solubilize and transport the nutraceuticals to the enterocytes where they can be absorbed. The composition and structure of emulsion-based delivery and excipient systems should be optimized to increase the bioavailability of the nutraceuticals (McClements, 2018a). Typically, this involves designing

5.4 Nutraceuticals

Table 5.5 Examples of nutraceuticals found in plant-based foods and the health claims associated with them

Nutraceutical	Natural source	Claimed health benefits	Strength of evidence
Omega-3 fatty acids (DHA, EPA, ALA)	Algae oil, flaxseed oil	Reduced heart disease, inflammation, immune disorders, and mental disorders	Moderate to strong
Carotenoids – β-carotene, lycopene, lutein, zeaxanthin	Carrots, kale, mangoes, peppers, spinach, tomatoes, watermelon, yams	Pro-vitamin A activity, anticancer activity, improved eye health	Moderate to strong
Curcumin	Turmeric (*Curcuma longa*)	Reduced cancer, diabetes, depression, obesity, pain, and stroke,	Moderate
Resveratrol	Grape seeds, wine, berries, peanuts, cocoa	Reduced cancer, heart disease, diabetes, and brain disease	Moderate
Polyphenols	Coffee, tea, cocoa, fruit, berries, beans	Reduced cancer, inflammation, obesity, diabetes, heart disease, brain disease	Moderate to strong
Phytosterols/ phytostanols	Fruits, legumes, nuts, seeds, vegetables, whole grains	Reduction in cholesterol levels	Strong

Summary of the strength of evidence behind their efficacy: Strong = supported by clinical trials, epidemiology studies, and mechanistic studies. Moderate = some evidence, but not conclusive; Weak = little evidence of effect. Adapted from McClements et al. (2015).

a system that increases the bioavailability, chemical stability, and/or absorption of the nutraceutical (Fig. 5.7).

An analysis of numerous foods (>3,000) showed that plant-based ones contained a wide variety of constituents that exhibit good antioxidant activity (Carlsen et al., 2010). Moreover, the plant-based foods were found to have much higher antioxidant contents than the animal-based ones. Consequently, they might be expected to exhibit stronger health benefits by inhibiting oxidative reactions within the human body, which would otherwise damage key biochemical components, such as lipids, proteins, or DNA. In general, plant-based foods contain a wider diversity and higher concentration of nutraceuticals than animal-based ones. For instance, plants contain carotenoids, curcuminoids, tocopherols, tocotrienols, polyphenols, phytosterols, phytostanols, isoflavones, organosulphur compounds, prebiotics, and bioactive peptides (Abuajah et al., 2015). As a result, they would be expected to exhibit stronger health benefits than foods derived from animals. However, some animal-based foods do contain significant levels of certain kinds of nutraceutical ingredients, such as bioactive peptides, conjugated linolenic acid (CLA), polyunsaturated fatty acids, and prebiotics, which may provide some health benefits (Abuajah et al., 2015). As mentioned earlier, nutraceuticals isolated from plants may behave differently than they do when they are present in the whole plant, so that many of their health-promoting properties may be lost (Fardet, 2015a, 2015b). In this case, it may be important to carefully design the food matrix to ensure that the nutraceuticals exhibit the desired health-promoting properties.

A number of nutraceuticals found in plant-based foods and the health benefits claimed for them are listed in Table 5.5 (McClements, 2019). Each nutraceutical has its own unique molecular and physicochemical characteristics that need to be taken into account when formulating a suitable food matrix to ensure its stability and bioavailability (McClements, 2018b).

5.5 Gastrointestinal Fate: Digestibility, Bioavailability, and Fermentability

An improved understanding of the different nutritional and health effects of animal- and plant-based diets require an understanding of how these foods behave inside the human gut (McQuilken, 2021a, 2021b, 2021c). Foods travel through a complex set of environments after they are ingested, including the mouth, esophagus, stomach, small intestine, and colon (Fig. 5.1). The human gastrointestinal tract is partly designed to protect us from any harmful substances we ingest, such as chemical toxins or pathogenic microorganisms, as well as to efficiently digest and absorb any nutrients we consume to provide the energy and building blocks required to grow and survive. It carries out these functions by utilizing a combination of natural acids, surfactants, enzymes, and mechanical forces to break down the foods into smaller fragments and eventually into molecules that can pass through the enterocytes lining our guts and be absorbed and utilized by our bodies (Fig. 5.1). A brief summary of the different regions of the human gut is given here (McQuilken, 2021c):

Mouth: Immediately after ingestion, foods enter the oral cavity. Fluid foods, such as cow's or plant-based milks, may be rapidly swallowed, whereas solid foods, such as animal or plant-based meat, seafood, eggs, or cheese, may be masticated for some time prior to swallowing. In the mouth, foods are mixed with saliva, which is a viscous fluid containing mucin, minerals, and enzymes (*e.g.,* amylases and lingual lipases) with an approximately neutral pH (Carpenter, 2013; McQuilken, 2021c). Typically, solid foods are chewed and mixed with saliva until they are in a form suitable for easy swallowing, which is referred to as the bolus. The reduction in particle size that occurs in the mouth makes them easier to digest further down the gastrointestinal tract because it increases the surface area of macronutrients exposed to the digestive enzymes. After swallowing, the bolus travels through the esophagus and enters the stomach.

Stomach: The stomach consists of a muscular cavity containing highly acidic gastric fluids (McQuilken, 2021c). The gastric contents are usually around pH 1 to 3 in the fasting state but may increase appreciably immediately after consuming foods, before falling back over time. The stomach mixes and disrupts the foods by applying mechanical forces through muscular contractions (Bornhorst & Singh, 2014; Hunt et al., 2015). In addition, the acids and enzymes (gastric lipases and proteases) in the gastric juices facilitate the chemical disintegration and digestion of the macronutrients in foods. After spending about 2 hours in the stomach, the partially digested food (known as the chyme) passes through the pyloric sphincter and enters the small intestine. However, the residence time of foods in the stomach (which is related to gastric emptying) can vary considerably depending on their composition, structure, and rheology, which can affect pharmacokinetics and metabolic responses to nutrients, thereby impacting human nutrition and health (Somaratne et al., 2020).

Small intestine: The small intestine consists of a long tube that consists of the duodenum, jejunum, and ileum, which links the stomach to the colon (McQuilken, 2021c). Partially digested foods move through this tube driven by peristaltic forces as a result of coordinated muscular contractions (McQuilken, 2021b). The pancreas and bile duct secrete fluids into the small intestine containing digestive enzymes (amylases, proteases, lipases, and phospholipases), bile salts, phospholipids, and mineral salts that promote further disintegration and digestion of the foods. The starch, proteins and lipids are converted into small molecules, such as glucose, amino acids, small peptides, free fatty acids, and monoacylglycerols, which are in a form suitable for absorption by the intestinal enterocytes (Fig. 5.1). In the case of lipids, the digestion products are mixed with bile salts and phospholipids to form mixed micelles, which are then transported to the enterocytes for absorption (Fig. 5.7). The majority of ingested nutrients are absorbed in the small intestine through either passive or active mechanisms, which can be partly attributed to the relatively high surface area of the small intestine. Indeed, it has been calculated that the small intestine can be considered to be a tube of about 3 m long and 25 mm wide, with an effective surface area of about 30 m^2 (Helander & Fändriks, 2014). This relatively high surface area is attributed to the presence of villi and microvilli on the surfaces of the small intestine walls, which give it a variegated surface texture. The pH of the intestinal fluids goes from slightly acidic in the duodenum to close to neutral in the jejunum and ileum (Fallingborg, 1999). Typically, foods spend around 2 hours in the small intestine but again this time varies depending on the nature of the food consumed.

Colon: Any materials that are not digested and absorbed eventually reach the colon, where they may be fermented and metabolized by enzymes secreted by the microorganisms residing there (Korpela, 2018; Shortt et al., 2018). Foods typically spend over 24 hours in the colon where they are subjected to anaerobic and slightly acidic pH conditions. The nature of the undigested material impacts the type of biochemical processes occurring in the colon, which impacts human health and wellbeing. In particular, some indigestible oligosaccharides and dietary fibers may be converted into short chain fatty acids (SCFAs), which can be used as an energy source or that may act as signaling molecules with the human body (Rios-Covian et al., 2016). Other kinds of undigested food components can also be considered to be prebiotics that stimulate the growth of microbial populations associated with good health, such as certain polyphenols and peptides in foods (Sanders et al., 2019).

There are differences in the way that animal- and plant-based foods behave as they travel through the human gastrointestinal tract, which have important nutritional and health implications. As mentioned earlier, humans do not naturally produce enzymes that can digest dietary fibers in the upper gastrointestinal tract. Animal-based foods contain little or no dietary fibers and so tend to be almost fully digested in the upper gastrointestinal tract. Conversely, the cell walls of plants contain structural polysaccharides that give them their mechanical strength and control their barrier properties, such as cellulose, hemicellulose, and pectin (Dhingra et al., 2012; Holland et al., 2020). These dietary fibers therefore pass through the upper gastrointestinal tract and into the colon. As mentioned earlier, in the section on dietary fibers, there may be numerous health benefits associated with foods containing high levels of these indigestible food components, including a decreased calorie intake, reduced rate of macronutrient digestion and absorption, reduced constipation, reduced cholesterol levels, improved blood glucose control, and decreased prevalence of colon cancer.

Meats, eggs, and milk, on the other hand, tend to be fully digested within the human gut, so that all of their nutrients are released and absorbed. In the ancient past, this would have been an important evolutionary advantage for the hominid species. But now, in most developed countries, there are so many calorie and nutrient dense foods in the diet that any additional ones from meat are not essential.

As mentioned earlier, there are some potential negative aspects associated with getting our nutrients solely from plant-based foods (but again, these were more of a problem in the ancient past than they are now, at least in developed countries). Some plants contain antinutrient factors that can interfere with the normal digestion of fats, proteins, or carbohydrates, thereby reducing the number of calories and nutrients that can be derived from them (Rousseau et al., 2020; Shahidi, 1997). A list of some of the most important antinutrients found in plant-based foods is given below:

- *Phytate*: Phytate, also referred to as phytic acid, is naturally present in many seeds, grains, and legumes. It is a negatively charged substance that can strongly bind to positively-charged mineral ions in our diets, such as iron, zinc, magnesium, and calcium. As a result, the absorption of these essential minerals by our bodies is reduced.
- *Tannins*: Tannins are also commonly found in many plant-based foods. These polyphenols can strongly bind to the digestive enzymes within the human gastrointestinal tract, thereby inhibiting the digestion of fats, proteins, and starches.
- *Lectins*: Lectins are a type of protein found in numerous plant-based foods, but particularly in seeds, legumes and grains, which may be harmful when consumed in high amounts. This is because they can interfere with nutrient absorption and promote inflammation by making our guts leaky.
- *Oxalate*: Many vegetables (such as spinach) contain relatively high levels of a substance known as oxalate, which can bind strongly to calcium, thereby decreasing its ability to be absorbed by the body.

These antinutrients are not a major problem for most people living in developed countries because of the abundant and diverse range of micronutrients found in the diet. However, they may be a problem for people living in developing countries with nutrient-deficient diets, or for those who live exclusively on certain plant-based diets. It should be noted that the adverse nutritional effects of antinutrients are often reduced by normal food preparation and cooking procedures (e.g., trypsin inhibitors, which are discussed in more detail in Chap. 3), such as soaking, washing, or heating.

One of the main differences between the nutrients in plants and animals is their bioavailability, which is the fraction of the ingested nutrient that is actually absorbed by our bodies. A food can contain high levels of a nutrient, but if the nutrient never gets absorbed, then it cannot exhibit its health benefits. The bioavailability of nutrients in animal- and plant-based sources may vary considerably depending on their nature and food matrix effects. As mentioned earlier, the iron in meat and fish is contained within a ring-like organic compound known as porphyrin, which is usually attached to the proteins responsible for storing and transporting oxygen in the blood (hemoglobin) and muscles (myoglobin). This form of iron is highly bioavailable, meaning that a large fraction of it is absorbed after ingestion. In contrast, in plant-based foods, like fruits, vegetables, cereals, nuts, seeds, and beans, the iron is not held by this kind of natural structure, which makes it much less bioavailable. However, the bioavailability of non-heme iron can be increased by eating it with foods containing high levels of vitamin C (ascorbic acid) such as oranges or lemons, by using appropriate delivery systems, or by controlling food matrix effects, as discussed earlier. As just discussed, the bioavailability of many minerals and proteins is reduced due to the presence of antinutrient factors in plant-based foods. Lastly, the bioavailability of many lipophilic bioactive agents, such as oil-soluble vitamins and nutraceuticals, is often limited because of their relatively low solubility in aqueous gastrointestinal fluids. This problem can be overcome by consuming them with appropriate types and amounts of digestible lipids that can form mixed micelles in the small intestine that can solubilize and transport them (Fig. 5.7) (McClements, 2018a).

5.6 Impact of Diet on the Gut Microbiome

There is growing evidence that the nature of the gut microbiome has a major impact on the health and wellbeing of human beings (Dahl et al., 2020; Warmbrunn et al., 2020; Zhou et al., 2020). The types and numbers of different kinds of bacteria and other microbes residing in the human colon has been linked to many health indicators and diseases, such as inflammation, immune response, obesity, diabetes, heart disease, stroke, and cancer (Hills et al., 2019; Singh et al., 2017; Zmora et al., 2019). Consequently, there has been a large research effort on identifying beneficial gut microbiomes and then designing strategies to foster the development of these microbiomes in individuals. A "good" microbiome is usually taken to be one that exhibits a high functional diversity. One of the simplest means of altering the gut microbiome is through diet. Studies have shown that the composition and function of the gut microbiome is influenced by the relative proportion of different nutrients in our diets, such as fats, proteins, carbohydrates, vitamins, minerals, dietary fibers, and phytochemicals (Dahl et al., 2020; Gentile & Weir, 2018). Currently, the precise relationship between diet, a healthy microbiome, and human health is not fully understood but there have already been some interesting findings:

- *Carbohydrates*: The consumption of high amounts of sugars and rapidly digestible starches appears to have an adverse effect on the gut microbiome, whereas the consumption of high amounts of dietary fibers appears to have a beneficial effect. The fermentation of dietary fibers within the colon produces short chain fatty acids (SCFAs) that can be utilized as an energy source and that can send signals to the human body that regulate metabolism and decrease inflammation. Plant-based foods contain much higher levels of dietary fibers than animal-based ones and should therefore have more beneficial effects on the gut microbiome and human health.
- *Fats*: The consumption of high amounts of fats, especially saturated fatty acids, has an adverse effect on the gut microbiome. Conversely, the consumption of high amounts of omega-3 fats may have a beneficial effect. As many plant-based foods tend to contain less saturated fats than animal-based ones, they may have more beneficial effects on the microbiome. Moreover, it may be advantageous to fortify plant-based foods with omega-3 fatty acids (such as algal or flaxseed oil) to further improve their positive impact on the gut microbiome.
- *Proteins*: The influence of proteins on the gut microbiome depends on the type and amount consumed. The proteins from meats contain relatively high amounts of L-carnitine, an amino acid that can be converted into trimethylamine N-oxide (TMAO) by the microorganisms residing in the human colon (Zeisel & Warrier, 2017). TMAO has been linked to an increased risk of heart disease and therefore consuming high levels of meat may have adverse effects on human health (Velasquez et al., 2016). In contrast, the proteins derived from plants are not a significant source of this amino acid and so they should have less adverse effects on cardiovascular health.

Studies indicate that vegans and vegetarians on a fiber-rich plant-based diet typically have healthier microbiomes (more stable and diverse) than meat eaters (Kumar et al., 2016; Tomova et al., 2019). Beneficial changes in the nature of the gut microbiome have been reported to occur only a few days after people switch from an omnivore to a vegetarian diet (Singh et al., 2016). The improvement in the gut microbiome when people adopt a plant-based diet has mainly been attributed to the higher level of dietary fibers and phytochemicals they contain (Tomova et al., 2019). For example, a randomized controlled study examined the impact of switching 5 meals a week that contained animal-based food items (such as burgers, sausages, sausage patties, mince, and meatballs) with ones containing plant-based alternatives on the gut microbiomes of human participants (Toribio-Mateas et al., 2021). The authors reported a slight increase in biomarkers associated with an improvement in gut health (using 16S rRNA analysis), such as an increase in microbial diversity and an increase in butyrate-producing pathways. In another study, that compared the impact of a plant-rich Mediterranean diet with a typical meat-rich Western diet, it was found that the plant-based diet also increased butyrate-producing pathways, stool volume, and flatulence (which is an indicator of different metabolic pathways) (Barber et al., 2021). Overall, studies on the impact of diet on the gut microbiome suggest that eating a healthy plant-based diet has numerous health benefits.

5.7 Nutritional Studies Comparing Plant- and Animal-Based Diets

Ideally, the impact of specific plant- and animal-based diets on human health and wellness should be established using well-designed nutritional studies. These may include epidemiology studies that relate the reported diets of people to their long-term health or mortality status, randomized controlled trials (RCT) where different groups of people are fed different diets and changes in their health biomarkers or health status are measured, or mechanistic studies using *in vitro* or *in vivo* methods that attempt to establish the mechanisms of action of specific nutrients, foods, or diets (McClements, 2019).

Recently, there have been a number of meta-analyses of epidemiology and clinical studies that have compared the health outcomes of individuals who consume different levels of plant- and animal-based foods in their diets. Some of these studies have also examined the impact of consuming either a healthful or unhealthful plant-based diet. A healthful plant-based diet is considered to be rich in fruits, vegetables, whole grains, nuts, legumes, vegetable oils, tea, and coffee, whereas an unhealthful one is rich in refined grains, potatoes, sweets, snacks, desserts, fruit juices, and sweetened beverages (Hu et al., 2019). In a survey of over 200,000 men and women who were not known to have chronic diseases at the beginning of the study, it was reported that consuming more healthful plant-based foods led to a substantially lower risk of developing Type 2 diabetes (Satija et al., 2016). This effect was mainly attributed to the fact that the foods in this diet contained high levels of dietary fibers, antioxidants, minerals, vitamins, and unsaturated fatty acids, as well as low levels of saturated fatty acids. On the other hand, consuming higher levels of unhealthful plant-based foods actually increased the risk of diabetes, which was mainly attributed to its high levels of rapidly digestible carbohydrates and lack of the health-promoting components just mentioned. The authors concluded that reducing the amount of animal-based foods in a person's diet and replacing them with healthy plant-based ones led to an appreciable decrease in their risk of diabetes.

Similar analyses of food consumption and health status of large populations of adults have also shown that the consumption of a healthful plant-based diet (rather than an animal-based one) can also have other benefits, such as a reduction in cardiovascular diseases (Baden et al., 2019; Guasch-Ferre et al., 2019; Song et al., 2016), a reduction in stroke (Baden et al., 2021), improved physical and mental health (Baden et al., 2020), and reductions in all-cause mortality (Baden et al.,

2019). In contrast, consumption of an unhealthful plant-based diet had the opposite effects.

In a recent study, researchers at the Stanford School of Medicine compared the nutritional effects of replacing animal-based meat products with plant-based analogs (Crimarco et al., 2020). They carried out a 16-week randomized crossover trial that compared the consumption of plant-based chicken, beef, and pork products to organic, animal-based versions of the same products. The participants consumed two or more servings of the products every day for eight weeks. There was a significant improvement in cholesterol levels and body weight for the participants who consumed the plant-based meat analogs. The amounts of protein and sodium consumed were fairly similar for the two diets but the plant-based one contained more dietary fibers and less saturated fat. This study suggests that simply replacing animal-based foods with plant-based analogs in one's diet can have appreciable health benefits but more such studies are needed to draw a final conclusion for plant-based products that have undergone more processing.

5.8 Evolution, Genetics and Meat Consumption

Some people believe that humans are genetically hard-wired to be meat eaters because of evolution – our ancient ancestors ate meat, so we also need to eat meat to stay healthy. An analysis of the evolutionary fit between our bodies and our diets has been carried out from an anthropological perspective (Mann, 2007). The author of this analysis quoted Prof. Boyd Eaten of Emory University (Atlanta, Georgia, USA):

> "We are the heirs of inherited characteristics accrued over millions of years, the vast majority of our biochemistry and physiology are tuned to life conditions that existed prior to the advent of agriculture. Genetically our bodies are virtually the same as they were at the end of the Paleolithic period. The appearance of agriculture over 10 thousand years ago and the Industrial Revolution some 200 years ago introduced new dietary pressures for which no adaptation has been possible in such a short time span. Thus, an inevitable discordance exists between our dietary intake and that which our genes are suited to".

Anthropologists have gathered evidence from a diversity of sources to try to understand what our ancient ancestors ate (Mann, 2007). They have looked at changes in our anatomy, particularly our brains, jaws, teeth, and gastrointestinal tract over time, as well as comparing our anatomies to our close relatives (such as gorillas, chimps, and bonobos). They have also examined the dietary pattern of existing hunter-gather societies from around the world. Moreover, they have examined the energy requirements associated with developing a large brain relative to body size. These findings suggest that our ancient ancestors switched from a diet that was not rich in energy and nutrients, and contained high levels of fibrous plant materials, to one that contained more energy and nutrient-rich animal foods. As a result, the hominids of the paleolithic periods became meat-eaters. The transition from a predominantly plant-based diet to a more omnivore one in hominids is believed to have occurred around

2 to 2.5 million years ago (Zaraska, 2016). When scientists study the fossil remains of ancient hominids they find that their teeth and jaws changed from being specially designed to consume coarse fibrous plant materials to become more generalized to eat a mixture of fruits, nuts, and animal flesh (Mann, 2007). Similarly, evidence from fossil isotope ratios ($^{13}C/^{12}C$) is consistent with hominids changing from eating large amounts of fibrous plant materials to eating grazing animals that lived off grass. Originally, they probably started as scavengers who ate meat killed by other animals but eventually they became hunters, killing animals themselves, often much bigger than themselves (Zaraska, 2016). Strong evidence for evolution towards eating meat also comes by comparing the anatomy of the human gastrointestinal tract with those of animals that only eat plants (herbivores like cattle, horses, or gorillas) or that only eat meat (carnivores like big cats). Typically, herbivores have large stomachs and colons, which helps them to break down fibrous plant materials and convert them into energy. On the other hand, carnivores have highly acidic stomachs and a long small intestine. The anatomy of the human gut fits somewhere between that of strict herbivores and carnivores, suggesting we are omnivores with gastrointestinal tracts designed to eat a wide range of different foods. The size of our brains relative to our bodies is another indication that the bodies of our ancient ancestors evolved to eat meat. Like other primates, our brains are relatively big compared to our bodies. The precise origin of this effect is currently unknown but a number of hypotheses have been advanced. The brain requires a lot of energy to run, and also requires certain unique food components to build it, such as omega-3 fatty acids, which traditionally come from animal sources. One hypothesis is that to have a big brain, the body had to reduce the size of another major body organ – the gut. When we started eating foods that had a higher nutritional and energy content, and which were easier to digest, we did not need such a big gut. As our guts shrank, our brains grew. Presumably, this was an incremental iterative process driven by small evolutionary advantages that accrued over millennia. It has also been proposed that tool processing and cooking of foods, especially meat, enabled us to grow bigger brains because then our foods were easier to break down inside our bodies, releasing more calories. The animal flesh consumed by early hominids contained high levels of proteins and fats, especially the omega-3 fats needed to build brains, and so it seems likely that as our ancestors started moving from forests to grassy savannas, they started to eat more meat. Alternatively, they may have started to eat more meat because climate change caused forests to become savannas. As a result, there were fewer sources of nutritious plant-based foods available (such as leaves, flowers, fruit, nuts, and seeds), which meant that hominids needed to get more of their calories from other sources, like meat. Analyses of the diets of some existing hunter-gatherer societies indicate that they got the majority (> 60%) of their calories from animal products, which is not surprising from a cost-benefit analysis of the foraging process. However, this depends on the geographical location and cultural practices of the specific hunter-gatherer society and there are also other traditional societies that predominantly eat plants (e.g., the !Kung people in Africa that only consume around 30% of animal-based foods). Overall, the amount of time and energy required to get a certain amount of calories and nutrients from an animal

source is typically much less than that required to get it from a plant source. This is because the plant-based foods available to our ancient ancestors were low energy-dense foods that were often difficult to get, such as wild fruit, vegetables, roots and nuts. There are also other clues from our metabolism that suggest that the human physiology is adapted to eat meat. For instance, there are certain food components (*e.g.*, vitamins B_{12} and D) that are essential for human health, that are normally only obtained from animal or fish sources. It is hypothesized that hominids did not need to synthesize these components themselves because they were regularly getting them from their diet. As a result, their metabolisms changed over time so that we are now no longer capable of synthesizing them. Finally, there is evidence from parasites, which tend to co-evolve with their hosts (Mann, 2007). Certain kinds of parasites (*Taeniidae*) are known to have co-evolved with carnivores, which are spread by eating meat. Some of the parasites found in this family use humans exclusively as their hosts, which suggests that there must have been a substantial period in history when early humans or their ancestors were meat eaters. At the end of his review of paleolithic diets, it was concluded: "there is no historical or valid scientific argument to preclude lean meat from the human diet, and a substantial number of reasons to suggest it should be a central part of a well-balanced diet" (Mann, 2007).

Our ancient ancestors did live in a very different environment than we did, where eating meat (especially cooked meat) gave them an evolutionary advantage that helped them to evolve into the species we are today. But our environments are very different now – we live in houses, travel in cars, busses, trains, and airplanes, watch TV, and surf the internet. Moreover, the types and quantities of foods available to us today are very different – they are much more abundant, readily available, energy dense, and more easily digestible than those available to our ancestors. So, it is not just meat that our genetic profile is not adapted to, it is almost everything. In addition, the diet that our ancestors adopted might not be designed for a long and healthy life because passing on genes to the next generation can already be accomplished at a much younger age and there was no need to reach the lifespan of people nowadays living in western societies for successful reproduction. Thus, an optimum diet for a long and healthy life might be somewhat different from that of our ancestors.

An important aspect of evolution is that it has made humans incredibly adaptable creatures, which is one of the reasons that we can live in so many different ecological niches around the globe. Our ancestors may have evolved to eat meat, but we can certainly live without it (as demonstrated by the millions of people who have been born, lived, and died vegetarian or vegan throughout history). We can get the nutrients that our ancestors got from animal sources, such as iron, zinc, omega-3 fatty acids, and vitamin B_{12} from other sources now. For instance, we can get omega-3 fatty acids from algal or flaxseed oil and can get vitamin B_{12} from microbial fermentation. Consequently, there does not appear to be a strong argument that humans have to eat animal products to remain healthy.

5.9 The Agricultural Revolution and Meat Consumption

Our species experienced a profound change in its diet around 10,000-12,000 years ago due to the Agricultural Revolution (Mann, 2007). Many humans made a transition from hunting and gathering in relatively small tribes to growing and cultivating wild cereal crops. As a result, they tended to live in a particular location so that they could cultivate and harvest their crops, which led to the development of the first large towns and cities. One might think that this transition to a more plant-based diet would have made our ancestors healthier, but in fact, there is strong evidence that they became less healthy. People living in early agriculture societies appeared to be shorter, have worse bones and teeth, more nutritional deficiencies, and more infectious diseases than the hunters and gatherers who proceeded them. This fact was attributed to a switch from a diverse diet consisting of wild animals (deer, antelope, and gazelle), nuts, fruits, root plants, and wild pulses, to a much more nutritionally narrow diet consisting largely of cereal crops, such as wheat, oats, barley, rice, or corn - depending on the geographical location. Thus, people went from eating a diet high in protein, omega-3 fats, and dietary fiber that contained a wide range of nutrients, to a diet consisting largely of digestible carbohydrates that only contained a narrow range of nutrients. As a result, our early ancestors who lived in these agricultural societies were much more prone to nutritional deficiencies and diseases.

There has been another dramatic change in our diets over the past 200 years or so, which started with the industrial revolution and has continued to the highly-processed mass-produced foods that many people eat today. Modern food processing methods, such as milling, refining, and thermal processing, often remove or destroy valuable dietary fibers, vitamins, and minerals, leading to foods that are highly palatable, but that are also calorie-dense and are rapidly digested, as well as being rich in fat, sugar, and salt. As a result, these highly processed foods are believed to promote diseases such as obesity, diabetes, heart disease, stroke, and cancer. Consequently, it will be important to design the next generation of plant-based foods, which are typically highly-processed foods, to ensure that they are nutritious and healthy.

5.10 Improving Healthiness of Plant-Based Foods

In this section, we highlight methods of improving the nutritional value and healthiness of plant-based foods. There are several approaches that can be used to achieve this goal. In this section, we highlight some of the most commonly used: fortification and reformulation.

5.10.1 Fortification

As discussed earlier in this chapter, plant-based foods have a different nutritional profile compared to animal-based ones. In particular, there are differences in the types and amounts of micronutrients, macronutrients, dietary fibers, and nutraceuticals present. In some cases, plant-based foods lack some of the essential micronutrients required in the human diet, such as vitamin B_{12}, vitamin D, iron, zinc and calcium. However, as mentioned before, the industrial livestock production also often results in a deficiency of micronutrients in animals because the natural feed may not provide enough of these nutrients and therefore the animal feed is also often fortified. As a result it may be more efficient to directly fortify plant-based foods but in general, it may be necessary to fortify plant-based foods with bioavailable forms of some micronutrients to prevent undernutrition. Moreover, some food ingredients have been linked to improved human health and wellbeing, such as dietary fibers and nutraceuticals. The nutritional value of plant-based foods may therefore be enhanced by including these bioactive ingredients. There are a number of factors that should be taken into account when fortifying plant-based foods with micronutrients, nutraceuticals and dietary fibers (collectively referred to as bioactive ingredients), which are listed here:

- *Solubility*: Bioactive ingredients vary in their polarities and solubility characteristics. They may have an appreciable solubility in oil, water, both or neither depending on their molecular properties. For instance, vitamin D is mainly soluble in oil, whereas vitamin B_{12} is mainly soluble in water. The solubility of many minerals depends on the salt form and solution conditions used, *e.g.,* calcium carbonate is insoluble under neutral conditions but soluble under acidic conditions, whereas calcium chloride is soluble in water. It is therefore important to select the most appropriate form of this essential mineral ion for the intended application. In general, it is important to ensure that the bioactive agent is dissolved or dispersed within an appropriate medium.
- *Partitioning:* The oil-water partition coefficient (K_{OW} or $\log P$) of nutraceuticals determines their relative concentrations in different phases, which is important in food matrices that contain both oil and water domains. Substances with a positive partition coefficient ($\log P > 0$) are mainly hydrophobic and tend to be predominantly located in the oil phase (such as vitamin D or β-carotene). In contrast, substances with a negative partition coefficient ($\log P < 0$) are mainly hydrophilic and tend to be predominantly located in the water phase (such as vitamin B_{12}). In general, substances tend to be distributed between the oil and water phases, with the relative amounts in each phase depending on their partition coefficients and the oil-water ratio in the system. Typically, the more positive the partition coefficient and the greater the oil-water ratio, the more of the substance will be located in the oil phase.

- *Chemical stability:* Many micronutrients and nutraceuticals have a tendency to chemically degrade over time or during processing, which depends on the solution and environmental conditions (such as pH, temperature, solvent polarity, oxygen, light, and prooxidant levels). For instance, many carotenoids degrade rapidly when exposed to acidic conditions, heat, and light. Consequently, it is important to select micronutrient or nutraceutical forms that are more stable and/or to design food matrices that limit their degradation during storage and processing. This typically requires understanding the nature of the chemical reactions involved and identifying the key factors that promote or inhibit chemical degradation.
- *Food matrix compatibility:* It is important that any micronutrient, nutraceutical, or dietary fiber is compatible with the food matrix it is introduced into. In particular, it should not have adverse effects on the appearance, texture, flavor, or shelf-life of the plant-based food it is used to fortify. For example, some dietary fibers form highly viscous solutions when dissolved in water, which may adversely affect the flowability of low viscosity products like plant-based milks. In this case, it may be important to use dietary fibers with lower molecular weights so that they do not thicken the aqueous phase too much. Some nutraceuticals (such as bioactive peptides) may have a bitter or astringent mouthfeel, which negatively impacts the flavor profile of foods. In this case, it may be necessary to encapsulate the nutraceuticals so that they are released in the stomach or small intestine, rather than the mouth.
- *Bioavailability:* The concentration of a bioactive agent that actually gets absorbed into the body is more important than the concentration in the original food. Some bioactive agents normally have a low bioavailability because of their low water solubility, rapid metabolism, or poor absorption in the human gut. In these cases, it may be important to use well-designed delivery systems or food matrices to increase the bioavailability of the bioactive agents.
- *Other factors:* In addition, it is important to consider the regulatory status, cost, sustainability, and label friendliness of any bioactive food component used to fortify plant-based foods.

5.10.2 *Reformulation: Reduced Fat, Salt, Sugar, and Digestibility*

Many of the current generation of plant-based foods accurately simulate the look, feel, and taste of the animal-based foods they are designed to replace. In some cases, however, their nutritional profiles may not be consistent with a healthy diet. For instance, they may contain too many calories, or too much fat, salt, and sugar.

Moreover, they may be highly processed foods that are rapidly digested within the gastrointestinal tract, thereby leading to spikes in blood sugar and lipid levels that can dysregulate the hormonal system. Consequently, there is a need to reformulate many plant-based foods to improve their nutrient profile and digestibility profiles.

Reduced Fat: Fats play numerous important roles in determining the desirable physicochemical and textural properties of foods, contributing to their appearance, texture, mouthfeel, and flavor profile (Chung et al., 2016). Consequently, removing the fats can have adverse effects on product quality. It is therefore important to develop effective strategies to replace the desirable attributes that fats normally provide. The creamy appearance provided by fat droplets in plant-based foods may be mimicked using other kinds of food-grade particles that scatter light, such as proteins or indigestible carbohydrates. The textural attributes provided by fat droplets may be simulated using hydrocolloids or food-grade particles that increase the viscosity of the aqueous phase, such as xanthan gum, guar gum, locust bean gum, microcrystalline cellulose, or protein particles. The ability of the fat phase to hold oil-soluble flavors, vitamins, or other substances may be obtained by using non-digestible forms of lipids (such as mineral oils or sugar-fatty acid esters) or using lipids with lower calories (such as SALATRIM) rather than digestible lipids.

Reduced Sugar: Sugars may come in the form of simple carbohydrates like monosaccharides (glucose and fructose) and disaccharides (such as sucrose and lactose). Alternatively, they may come from digestible starches, which are broken down by amylases in the human gut and release glucose. Many nutrition scientists believe that frequently consuming foods containing high levels of rapidly adsorbable sugars may lead to health problems, such as overeating, obesity, and diabetes (Lang et al., 2021). As a result, food scientists are trying to identify effective methods of reducing the sugar content of foods without altering their desirable physicochemical or sensory attributes (Hutchings et al., 2019). This requires identifying ingredients or structures that can mimic the attributes normally supplied by sugars. Simple sugars are usually used to provide sweetness, or more complex flavors and colors during cooking (through the Maillard reaction). Starches are often used as thickening, gelling, or water holding agents to provide desirable textural and mouthfeel properties. The sweetness provided by simple sugars can often be replaced by using natural or synthetic low calorie sweeteners.

In general, there are two major classes of sugar replacers that are used in foods that are categorized according to the concentration required to produce a certain sweetness: low- and high-intensity sweeteners (Table 5.6). Low-intensity sweeteners typically have to be used at a concentration that is fairly similar to sugars, but they have fewer calories and/or adverse effects on health because they are typically absorbed at lower rates and metabolized in a different way compared to glucose or sucrose. This class of sweeteners includes natural sugars (like allulose found in fruits) or polyols that are formed by chemically modifying sugars (like sorbitol, mannitol, and xylitol). High-intensity artificial sweeteners have a much stronger perceived sweetness than sugars so that they can be utilized at much lower concentrations. This class of sweeteners may be chemically synthesized, such as aspartame

5.10 Improving Healthiness of Plant-Based Foods

Table 5.6 Relative sweetness of different kinds of sugar substitutes. The relative sweetness is compared to that of sucrose, which is taken to have a sweetness of 100. Here HFCS is high fructose corn syrup

Ingredient	Relative sweetness (kCalories/gram)	Relative sweetness	Relative sweetness (kCalories/gram)
Sugars		**Low intensity**	
Sucrose	100 (3.9)	Sorbitol	60 (2.6)
Fructose	150 (3.6)	Maltitol	70 (2.1)
Glucose	70 (3.8)	Mannitol	60 (1.6)
Lactose	20 (3.9)	Lactitol	40 (2.0)
HFCS	100 (2.7)	Xylitol	100 (2.4)
Honey	100 (3.0)	Erythritol	70 (0.2)
		Allulose	70 (0.3)
High intensity (natural)		**High intensity (artificial)**	
Glycyrrhizin	75	Acesulfame K	200
Monk fruit	200	Advantame	20,000
Stevia	300	Aspartame	200
		Neotame	10,000
		Saccharin	400
		Sucralose	600

The numbers in brackets are the kilocalories per gram.
Adapted from NutrientsReview.com and other sources.

Table 5.7 Artificial sweeteners used to mimic the sweetness of real sugars often have different flavor profiles, which influences their acceptance by consumers. This table shows sensory scores for sweet, bitter, acid, and metallic tastes for different natural and artificial sweeteners.

Sweetener	Sweet	Bitter	Acid	Metallic
Sucrose	82	7	7	11
Saccharin	62	25	12	28
Acesulfame K	69	31	7	33
Aspartame	69	4	6	15
Neotane	76	22	14	29

Data adapted from McClements (2019).

(Equal®), saccharin (Sweet'n Low®), and sucralose (Splenda®) or it may be derived from natural plant-based sources (such as stevia or glycyrrhizin).

There are a number of factors that need to be considered when utilizing these sugar replacers in plant-based foods. First, as well as providing a desirable sweetness, these sugar replacers may also have undesirable flavor notes, including astringency, bitterness, and metallic tastes (Table 5.7). Second, the sweetness intensity *versus* time profile of sugar replacers is often different from that of natural sugars. For instance, some sugar replacers provide a rapid initial sweetness that quickly fades, whereas others provide a more gradual sweetness that is perceived over a

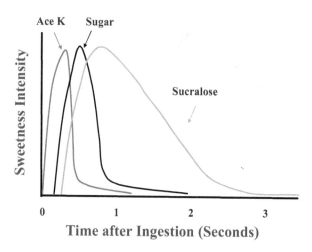

Fig. 5.8 The change in sweetness intensity with time is different for different sweeteners, which influences their perceived perception

long period (Fig. 5.8). As a result, sugar replacers cannot exactly match the flavor profiles of natural sugars, which can lead to poor acceptance by consumers. Third, sugar replacers may not be involved in the same chemical reactions in foods during cooking (such as the Maillard or caramelization reactions), and so they do not produce the same reaction products that generate the desirable color and flavor profile. Fifth, replacing sugars with sugar replacers may lead to changes in the bulkiness, textural attributes, or water activity of a food, which may alter its physicochemical, stability, and sensory attributes. Finally, some consumers are concerned about the potential adverse effects of synthetic sweeteners on human health and therefore avoid products with these ingredients in them.

Starches are often used to provide desirable textures and mouthfeels in foods. These attributes can sometimes be replaced by using other ingredients, such as hydrocolloids. These can be used in a molecular form or converted into microgels that are designed to mimic starch granules. The utilization of hydrocolloids as thickening, gelling, and water holding ingredients are discussed in detail in Chap. 2. Many hydrocolloids are dietary fibers. Consequently, they can have health benefits when used to replace starches in foods because they reduce the calorie content, do not lead to spikes in blood glucose levels, promote a healthy gut microbiome, increase laxation, and have various other desirable bioactive effects (discussed earlier).

Sweeteners and hydrocolloids behave differently from sugars and starches in foods. Consequently, the creation of plant-based foods with reduced sugar contents requires careful reformulation to ensure they have desirable quality attributes, such as appearance, textures, mouthfeels, and flavors.

Reduced Salt: The presence of salt in foods is important because it contributes to their desirable flavor profile ("saltiness"), as well as performing various other roles (Jaenke et al., 2017; Kuo & Lee, 2014). For instance, the presence of salt influences the electrostatic interactions within a food matrix, which can change the solubility

5.10 Improving Healthiness of Plant-Based Foods

and functionality of other ingredients. The salt may be added by the food manufacturer as a flavoring agent, or it may be present within other ingredients used to formulate the product. In particular, plant proteins often contain high salt levels because they are isolated using salting-out methods. Consuming high levels of salt over long periods has been linked to an increased risk of hypertension and stroke in individuals with high blood pressure. Consequently, it would be advantageous to reduce the levels of salt in plant-based foods (especially meats) to improve their nutritional profile.

A number of strategies have been examined for their potential for reducing the salt content of foods without altering their desirable physicochemical and sensory attributes, such as stealth salt reduction, saltiness potentiation, multisensorial effects, salt crystal design, and sodium replacement (Kuo, et al., 2014). The salt content of commercial food products can be gradually reduced over time so that consumers do not notice a dramatic change in their saltiness. Specific ingredients can be added to foods that can increase the perceived saltiness of foods, such as monosodium glutamate (MSG). In this case, a lower amount of salt can be used to create the same saltiness. Researchers are looking for other kinds of natural plant-based salt enhancers that can also be used for this purpose. Some researchers have manipulated the size and morphology of salt crystals in dried foods to increase their dissolution rate and therefore their perceived flavor intensity. For instance, researchers at Tate & Lyle (London, UK) produced small hollow salt crystals that rapidly dissolve in saliva, which allowed them to use 25 to 50% less salt to create the same perceived saltiness. Another strategy is to trap salts inside small particles that are released in bursts inside the mouth. This is because rapid fluctuations of the salt concentration within the mouth are perceived as being more intense than having a constant salt concentration over time. As a result, less total salt can be used to create the same perceived saltiness. It would also be useful to develop new approaches of isolating plant proteins that do not lead to high salt levels in the final powdered ingredient.

Reduced Digestibility: As mentioned earlier, the rapid digestion and absorption of macronutrients such as starch and fat can lead to spikes in blood sugar and lipid levels that may lead to a dysregulation of the metabolism. Consequently, it may be important to reformulate plant-based foods so as to reduce the rate of macronutrient digestion. In the case of starch, this can be achieved by using slowly digestible or resistant starch instead of rapidly digestible starch, provided this substitution does not adversely affect product quality. Alternatively, it may be possible to inhibit macronutrient digestion by incorporating dietary fibers that increase the viscosity of the gastrointestinal fluids, thereby slowing down mixing and mass transport processes. In addition, some plant-based ingredients (such as dietary fibers or polyphenols) can bind to key gastrointestinal components involved in lipid digestion, such as enzymes, bile salts, or calcium, thereby retarding digestive processes. Finally, it may be advantageous to leave plant-based ingredients in a less-processed form, *e.g.*, leaving the cell walls intact can inhibit the ability of digestive enzymes to reach the fats, starches, or proteins inside the plant tissues.

5.10.3 Agricultural and Processing Approaches

The nutritional quality of plant-based foods may also be improved by developing new agricultural and processing practices. Selective breeding or modern genetic engineering approaches could be used to create agricultural crops containing high levels of desirable nutrients, such as good quality proteins, polyunsaturated lipids, vitamins, minerals, or nutraceuticals. Moreover, the concentration of these desirable nutrients within the crops may also be enhanced by optimizing agronomic conditions such as soil health and fertilizer use. Food processing operations could also be redesigned to increase the levels of desirable nutrients remaining in the final product by reducing the removal and/or chemical degradation of the nutrients during the manufacturing process.

5.11 Microbiological and Chemical Toxins

There are also some additional health concerns that should be considered when switching from an animal to a plant-based diet. The microbiological and chemical contaminants in plants are different from those in animals, and so it is important to develop appropriate mitigation strategies and testing protocols to account for this. The food ingredients derived from plants may have pesticides or fertilizers associated with them, which should be minimized to avoid adverse effects on human health. This can be achieved by judicious use of pesticides and fertilizers during the growing of crops, as well as their proper handling and processing. Algae and microalgae can absorb harmful heavy metals from the water in their environment, which could cause problems if they become part of the food chain (Cavallo et al., 2021). Consequently, it will be important to grow and harvest these products away from polluted waters and to have good analytical methods to establish the degree of contamination. There are also microbiological issues that should be addressed. For instance, some plant-derived ingredients may be contaminated with bacteria, such as *Bacillus cereus, Salmonella* or *Escherichia coli,* with these latter two microorganisms occurring due to birds or rodents that come into contact with the agricultural crops during storage. They may also be contaminated with various kinds of mycotoxins in the field or during transport, such as certain *Aspergillus, Fusarium,* or *Penicillium* species, which can cause acute or chronic toxicity in humans (Alshannaq & Yu, 2017). Consequently, food ingredients derived from agricultural crops should be carefully handled, cleaned, and processed before being used to formulate plant-based foods to avoid these adverse health effects. In summary, there are different food safety challenges with plant-based foods than with animal-based ones, which should be carefully considered when developing these products.

5.12 Conclusions

There are clearly nutritional and health implications associated with switching from an animal- to a plant-based diet. As stated by Prof. Hu from the School of Public Health at Harvard University a "plant-based diet is not synonymous with a healthy diet" (Hemler, et al., 2019). It is important to choose healthful plant-based foods, such as fruits, vegetables, legumes, nuts, and whole grains, rather than unhealthful ones like refined grains, potatoes, snacks, sweets, and sugary beverages. It may also be important when planning a well controlled plant-based diet to avoid any potential nutritional deficiencies, such as vitamin B_{12} (Hemler, et al., 2019). Studies suggest that even relatively small reductions in the amount of animal foods consumed in the diet can have major improvements on human health. Indeed, the EAT-Lancet commission estimated that there would be 11 million fewer deaths per year globally if people adopted a more plant-based diet, as well as having major environmental benefits. Nonetheless, many of the current generation of highly processed foods may be detrimental to human health because they contain high levels of calories, saturated fats, sugars, and salt, as well as being rapidly digested and absorbed. As food manufacturers develop new plant-based alternatives to traditional meat, fish, egg, and dairy products it will be important that they consider nutritional attributes, such as macronutrient composition, digestion rate, and bioavailability, and the impact of the overall food matrix on influencing these attributes. Then, it will be possible to create food products that are good for both human health and the planet.

References

Abdelhamid, A. S., Brown, T. J., Brainard, J. S., Biswas, P., Thorpe, G. C., Moore, H. J., Deane, K. H. O., AlAbdulghafoor, F. K., Summerbell, C. D., Worthington, H. V., Song, F. J., & Hooper, L. (2018). Omega-3 fatty acids for the primary and secondary prevention of cardiovascular disease. *Cochrane Database of Systematic Reviews, 7*.

Aboalnaja, K. O., Yaghmoor, S., Kumosani, T. A., & McClements, D. J. (2016). Utilization of nanoemulsions to enhance bioactivity of pharmaceuticals, supplements, and nutraceuticals: Nanoemulsion delivery systems and nanoemulsion excipient systems. *Expert Opinion on Drug Delivery, 13*(9), 1327–1336.

Abuajah, C. I., Ogbonna, A. C., & Osuji, C. M. (2015). Functional components and medicinal properties of food: a review. *Journal of Food Science and Technology-Mysore, 52*(5), 2522–2529.

Akoh, C. C. (2017). *Food lipids: Chemistry, nutrition, and biotechnology*. CRC Press.

Al Mijan, M., Lee, Y. K., & Kwak, H. S. (2011). Classification, structure, and bioactive functions of oligosaccharides in milk. *Korean Journal for Food Science of Animal Resources, 31*(5), 631–640.

Alshannaq, A., & Yu, J. H. (2017). Occurrence, toxicity, and analysis of major mycotoxins in food. *International Journal of Environmental Research and Public Health, 14*(6).

Anand, S. S., Hawkes, C., de Souza, R. J., Mente, A., Dehghan, M., Nugent, R., Zulyniak, M. A., Weis, T., Bernstein, A. M., Krauss, R. M., Kromhout, D., Jenkins, D. J. A., Malik, V., Martinez-Gonzalez, M. A., Mozaffarian, D., Yusuf, S., Willett, W. C., & Popkin, B. M. (2015). Food consumption and its impact on cardiovascular disease: Importance of solutions focused on the globalized food system a report from the workshop convened by the world heart federation. *Journal of the American College of Cardiology, 66*(14), 1590–1614.

Astrup, A., Magkos, F., Bier, D. M., Brenna, J. T., Otto, M. C. D., Hill, J. O., King, J. C., Mente, A., Ordovas, J. M., Volek, J. S., Yusuf, S., & Krauss, R. M. (2020). Saturated fats and health: A reassessment and proposal for food-based recommendations JACC state-of-the-art review. *Journal of the American College of Cardiology, 76*(7), 844–857.

Astrup, A., Teicholz, N., Magkos, F., Bier, D. M., Brenna, J. T., King, J. C., Mente, A., Ordovas, J. M., Volek, J. S., Yusuf, S., & Krauss, R. M. (2021). Dietary saturated fats and health: Are the U.S. guidelines evidence-based? *Nutrients, 13*(10), 3305.

Augustin, L. S. A., Aas, A. M., Astrup, A., Atkinson, F. S., Baer-Sinnott, S., Barclay, A. W., Brand-Miller, J. C., Brighenti, F., Bullo, M., Buyken, A. E., Ceriello, A., Ellis, P. R., Ha, M. A., Henry, J. C., Kendall, C. W. C., La Vecchia, C., Liu, S. M., Livesey, G., Poli, A., ... Jenkins, D. J. A. (2020). Dietary fibre consensus from the International Carbohydrate Quality Consortium (ICQC). *Nutrients, 12*(9).

Baden, M. Y., Liu, G., Satija, A., Li, Y. P., Sun, Q., Fung, T. T., Rimm, E. B., Willett, W. C., Hu, F. B., & Bhupathiraju, S. N. (2019). Changes in plant-based diet quality and total and cause-specific mortality. *Circulation, 140*(12), 979–991.

Baden, M. Y., Kino, S., Liu, X. R., Li, Y. P., Kim, Y., Kubzansky, L. D., Pan, A., Okereke, O. I., Willett, W. C., Hu, F. B., & Kawachi, I. (2020). Changes in plant-based diet quality and health-related quality of life in women. *British Journal of Nutrition, 124*(9), 960–970.

Baden, M. Y., ..., Z. L., Wang, F. L., Li, Y. P., Manson, J. E., Rimm, E. B., Willett, W. C., Hu, F. B., & Re..., K. M. (2021). Quality of plant-based diet and risk of total, ischemic, and hemorrhagic stroke. *Neurology, 96*(15), E1940–E1953.

Bakaloudi, D. R., H..., Rippin, H. L., Oikonomidou, A. C., Dardavesis, T. I., Williams, J., Wickramasinghe, ..., J., & Chourdakis, M. (2021). Intake and adequacy of the vegan diet. A systematic review of the evidence. *Clinical Nutrition, 40*(5), 3503–3521.

Baker, E. J., Miles, E. A., Burdge, G. C., Yaqoob, P., & Calder, P. C. (2016). Metabolism and functional effects of plant-derived omega-3 fatty acids in humans. *Progress in Lipid Research, 64*, 30–56.

Barber, C., Mego, M., Sabater, C., Vallejo, F., Bendezu, R. A., Masihy, M., Guarner, F., Espín, J. C., Margolles, A., & Azpiroz, F. (2021). Differential effects of Western and mediterranean-type diets on gut microbiota: A metagenomics and metabolomics approach. *Nutrients, 13*(8), 2638.

Becker, P. M., & Yu, P. Q. (2013). What makes protein indigestible from tissue-related, cellular, and molecular aspects? *Molecular Nutrition & Food Research, 57*(10), 1695–1707.

Bello-Perez, L. A., Flores-Silva, P. C., Agama-Acevedo, E., & Tovar, J. (2020). Starch digestibility: past, present, and future. *Journal of the Science of Food and Agriculture, 100*(14), 5009–5016.

Bhandari, D., Rafiq, S., Gat, Y., Gat, P., Waghmare, R., & Kumar, V. (2020). A review on bioactive peptides: Physiological functions, bioavailability and safety. *International Journal of Peptide Research and Therapeutics, 26*(1), 139–150.

Bhattarai, R. R., Dhital, S., Wu, P., Chen, X. D., & Gidley, M. J. (2017). Digestion of isolated legume cells in a stomach-duodenum model: three mechanisms limit starch and protein hydrolysis. *Food & Function, 8*(7), 2573–2582.

Bloise, A., Simoes-Alves, A. C., Santos, A. D., Morio, B., & Costa-Silva, J. H. (2021). Cardiometabolic impacts of saturated fatty acids: Are they all comparable? *International Journal of Food Sciences and Nutrition*.

Bornhorst, G. M., & Singh, R. P. (2014). Gastric digestion in vivo and in vitro: How the structural aspects of food influence the digestion process. In M. P. Doyle & T. R. Klaenhammer (Eds.), *Annual review of food science and technology, vol 5* (Vol. 5, pp. 111–132).

Bosscher, D., Breynaert, A., Pieters, L., & Hermans, N. (2009). Food-based strategies to modulate the composition of the intestinal microbiota and their associated health effects. *Journal of Physiology and Pharmacology, 60*, 5–11.

Briggs, M. A., Petersen, K. S., & Kris-Etherton, P. M. (2017). Saturated fatty acids and cardiovascular disease: Replacements for saturated fat to reduce cardiovascular risk. *Healthcare, 5*(2).

References

Butola, L. K., Kute, P. K., Anjankar, A., Dhok, A., Gusain, N., & Vagga, A. (2020). Vitamin B12 - Do you know everything? *Journal of Evolution of Medical and Dental Sciences-Jemds, 9*(42), 3139–3146.

Candela, C. G., Lopez, L. M. B., & Kohen, V. L. (2011). Importance of a balanced omega 6/omega 3 ratio for the maintenance of health. Nutritional recommendations. *Nutricion Hospitalaria, 26*(2), 323–329.

Carbonaro, M., Maselli, P., & Nucara, A. (2012). Relationship between digestibility and secondary structure of raw and thermally treated legume proteins: A Fourier transform infrared (FT-IR) spectroscopic study. *Amino Acids, 43*(2), 911–921.

Carlsen, M. H., Halvorsen, B. L., Holte, K., Bohn, S. K., Dragland, S., Sampson, L., Willey, C., Senoo, H., Umezono, Y., Sanada, C., Barikmo, I., Berhe, N., Willett, W. C., Phillips, K. M., Jacobs, D. R., & Blomhoff, R. (2010). The total antioxidant content of more than 3100 foods, beverages, spices, herbs and supplements used worldwide. *Nutrition Journal, 9*.

Carpenter, G. H. (2013). The secretion, components, and properties of saliva. In M. P. Doyle & T. R. Klaenhammer (Eds.), *Annual review of food science and technology, vol 4* (Vol. 4, pp. 267–276).

Cavallo, G., Lorini, C., Garamella, G., & Bonaccorsi, G. (2021). Seaweeds as a "Palatable" challenge between innovation and sustainability: A systematic review of food safety. *Sustainability, 13*(14).

Chakrabarti, S., Guha, S., & Majumder, K. (2018). Food-derived bioactive peptides in human health: Challenges and opportunities. *Nutrients, 10*(11).

Chung, C., Smith, G., Degner, B., & McClements, D. J. (2016). Reduced fat food emulsions: Physicochemical, sensory, and biological aspects. *Critical Reviews in Food Science and Nutrition, 56*(4), 650–685.

Cioccoloni, G., Soteriou, C., Websdale, A., Wallis, L., Zulyniak, M. A., & Thorne, J. L. (2021). Phytosterols and phytostanols and the hallmarks of cancer in model organisms: A systematic review and meta-analysis. *Critical Reviews in Food Science and Nutrition*.

Cornejo-Ramirez, Y. I., Martinez-Cruz, O., Del Toro-Sanchez, C. L., Wong-Corral, F. J., Borboa-Flores, J., & Cinco-Moroyoqui, F. J. (2018). The structural characteristics of starches and their functional properties. *Cyta-Journal of Food, 16*(1), 1003–1017.

Craig, W. J. (2010). Nutrition concerns and health effects of vegetarian diets. *Nutrition in Clinical Practice, 25*(6), 613–620.

Crimarco, A., Springfield, S., Petlura, C., Streaty, T., Cunanan, K., Lee, J., Fielding-Singh, P., Carter, M. M., Topf, M. A., Wastyk, H. C., Sonnenburg, E. D., Sonnenburg, J. L., & Gardner, C. D. (2020). A randomized crossover trial on the effect of plant-based compared with animal-based meat on trimethylamine-N-oxide and cardiovascular disease risk factors in generally healthy adults: Study With Appetizing Plantfood- Meat Eating Alternative Trial (SWAP-MEAT). *American Journal of Clinical Nutrition, 112*(5), 1188–1199.

Dahl, W. J., Mendoza, D. R., & Lambert, J. M. (2020). Diet, nutrients and the microbiome. In J. Sun (Ed.), *Microbiome in health and disease* (Vol. 171, pp. 237–263).

Daliri, E. B. M., Oh, D. H., & Lee, B. H. (2017). Bioactive peptides. *Foods, 6*(5).

Davani-Davari, D., Negahdaripour, M., Karimzadeh, I., Seifan, M., Mohkam, M., Masoumi, S. J., Berenjian, A., & Ghasemi, Y. (2019). Prebiotics: Definition, types, sources, mechanisms, and clinical applications. *Foods, 8*(3).

Dawczynski, C., & Lorkowski, S. (2016). Trans-fatty acids and cardiovascular risk: Does origin matter? *Expert Review of Cardiovascular Therapy, 14*(9), 1001–1005.

De Martinis, M., Sirufo, M. M., Suppa, M., & Ginaldi, L. (2020). New perspectives in food allergy. *International Journal of Molecular Sciences, 21*(4).

Dehghan, M., Mente, A., Zhang, X. H., Swaminathan, S., Li, W., Mohan, V., Iqbal, R., Kumar, R., Wentzel-Viljoen, E., Rosengren, A., Amma, L. I., Avezum, A., Chifamba, J., Diaz, R., Khatib, R., Lear, S., Lopez-Jaramillo, P., Liu, X. Y., Gupta, R., ... Investigators, P. S. (2017). Associations of fats and carbohydrate intake with cardiovascular disease and mortality in 18 countries from five continents (PURE): a prospective cohort study. *Lancet, 390*(10107), 2050–2062.

Deng, R. X., Mars, M., Van der Sman, R. G. M., Smeets, P. A. M., & Janssen, A. E. M. (2020). The importance of swelling for in vitro gastric digestion of whey protein gels. *Food Chemistry, 330*.

Dhingra, D., Michael, M., Rajput, H., & Patil, R. T. (2012). Dietary fibre in foods: A review. *Journal of Food Science and Technology-Mysore, 49*(3), 255–266.

Dhital, S., Warren, F. J., Butterworth, P. J., Ellis, P. R., & Gidley, M. J. (2017). Mechanisms of starch digestion by alpha-amylase-Structural basis for kinetic properties. *Critical Reviews in Food Science and Nutrition, 57*(5), 875–892.

Elmore, J. S., & Mottram, D. S. (2009). 5 - Flavour development in meat. In J. P. Kerry & D. Ledward (Eds.), *Improving the sensory and nutritional quality of fresh meat* (pp. 111–146). Woodhead Publishing.

Ertl, P., Knaus, W., & Zollitsch, W. (2016). An approach to including protein quality when assessing the net contribution of livestock to human food supply. *Animal, 10*(11), 1883–1889. https://doi.org/10.1017/s1751731116000902

Fairweather-Tait, S. J., & Teucher, B. (2002). Iron and calcium bioavailability of fortified foods and dietary supplements. *Nutrition Reviews, 60*(11), 360–367.

Fallingborg, J. (1999). Intraluminal pH of the human gastrointestinal tract. *Danish Medical Bulletin, 46*(3), 183–196.

FAO. (2013). Dietary protein quality evaluation in human nutrition. In *FAO food and nutrition paper*. Food and Agricuture Organization of the United Nations.

Fardet, A. (2015a). Complex foods versus functional foods, nutraceuticals and dietary supplements: differential health impact (Part 1). *Agro Food Industry Hi-Tech, 26*(2), 20–24.

Fardet, A. (2015b). Complex foods versus functional foods, nutraceuticals and dietary supplements: Differential health impact (Part 2). *Agro Food Industry Hi-Tech, 26*(3), 20–22.

Fasolin, L. H., Pereira, R. N., Pinheiro, A. C., Martins, J. T., Andrade, C. C. P., Ramos, O. L., & Vicente, A. A. (2019). Emergent food proteins - Towards sustainability, health and innovation. *Food Research International, 125*.

Gao, X., Wilde, P. E., Lichtenstein, A. H., & Tucker, K. L. (2006). Meeting adequate intake for dietary calcium without dairy foods in adolescents aged 9 to 18 years (National Health and Nutrition Examination Survey 2001–2002). *Journal of the American Dietetic Association, 106*(11), 1759–1765.

Gentile, C. L., & Weir, T. L. (2018). The gut microbiota at the intersection of diet and human health. *Science, 362*(6416), 776.

Ghaedi, E., Kord-Varkaneh, H., Mohammadi, H., Askarpour, M., & Miraghajani, M. (2020). Phytosterol supplementation could improve atherogenic and anti-atherogenic apolipoproteins: A systematic review and dose-response meta-analysis of randomized controlled trials. *Journal of the American College of Nutrition, 39*(1), 82–92.

Gillingham, L. G., Harris-Janz, S., & Jones, P. J. H. (2011). Dietary monounsaturated fatty acids are protective against metabolic syndrome and cardiovascular disease risk factors. *Lipids, 46*(3), 209–228.

Gorissen, S. H. M., Crombag, J. J. R., Senden, J. M. G., Waterval, W. A. H., Bierau, J., Verdijk, L. B., & van Loon, L. J. C. (2018). Protein content and amino acid composition of commercially available plant-based protein isolates. *Amino Acids, 50*(12), 1685–1695. https://doi.org/10.1007/s00726-018-2640-5

Gropper, S. S., Smith, J. L., & Carr, R. P. (2021). *Advanced nutrition and human metabolism*. Cengage Learning.

Grundy, M. M. L., Edwards, C. H., Mackie, A. R., Gidley, M. J., Butterworth, P. J., & Ellis, P. R. (2016a). Re-evaluation of the mechanisms of dietary fibre and implications for macronutrient bioaccessibility, digestion and postprandial metabolism. *British Journal of Nutrition, 116*(5), 816–833.

Grundy, M. M. L., Lapsley, K., & Ellis, P. R. (2016b). A review of the impact of processing on nutrient bioaccessibility and digestion of almonds. *International Journal of Food Science and Technology, 51*(9), 1937–1946.

Grungreiff, K., Gottstein, T., & Reinhold, D. (2020). Zinc deficiency-An independent risk factor in the pathogenesis of haemorrhagic stroke? *Nutrients, 12*(11).

Guasch-Ferre, M., Satija, A., Blondin, S. A., Janiszewski, M., Emlen, E., O'Connor, L. E., Campbell, W. W., Hu, F. B., Willett, W. C., & Stampfer, M. J. (2019). Meta-analysis of randomized controlled trials of red meat consumption in comparison with various comparison diets on cardiovascular risk factors. *Circulation, 139*(15), 1828–1845.

Gul, K., Singh, A. K., & Jabeen, R. (2016). Nutraceuticals and functional foods: The foods for the future world. *Critical Reviews in Food Science and Nutrition, 56*(16), 2617–2627.

Guo, Q., Ye, A. Q., Lad, M., Dalgleish, D., & Singh, H. (2014). Effect of gel structure on the gastric digestion of whey protein emulsion gels. *Soft Matter, 10*(8), 1214–1223.

Guo, Q., Bellissimo, N., & Rousseau, D. (2017a). Role of gel structure in controlling in vitro intestinal lipid digestion in whey protein emulsion gels. *Food Hydrocolloids, 69*, 264–272.

Guo, Q., Ye, A. Q., Bellissimo, N., Singh, H., & Rousseau, D. (2017b). Modulating fat digestion through food structure design. *Progress in Lipid Research, 68*, 109–118.

Gupta, R. S., Warren, C. M., Smith, B. M., Jiang, J., Blumenstock, J. A., Davis, M. M., Schleimer, R. P., & Nadeau, K. C. (2019). Prevalence and severity of food allergies among us adults. *JAMA Network Open, 2*(1), e185630–e185630.

Gylling, H., & Simonen, P. (2015). Phytosterols, phytostanols, and lipoprotein metabolism. *Nutrients, 7*(9), 7965–7977.

Hammad, S., Pu, S. H., & Jones, P. J. (2016). Current evidence supporting the link between dietary fatty acids and cardiovascular disease. *Lipids, 51*(5), 507–517.

Han, F., Moughan, P. J., Li, J. T., & Pang, S. J. (2020). Digestible Indispensable Amino Acid Scores (DIAAS) of Six Cooked Chinese Pulses. *Nutrients, 12*(12). https://doi.org/10.3390/nu12123831

Harcombe, Z., Baker, J. S., DiNicolantonio, J. J., Grace, F., & Davies, B. (2016). Evidence from randomised controlled trials does not support current dietary fat guidelines: A systematic review and meta-analysis. *Open Heart, 3*(2).

Helander, H. F., & Fändriks, L. (2014). Surface area of the digestive tract - Revisited. *Scandinavian Journal of Gastroenterology, 49*(6), 681–689.

Hemler, E. C., & Hu, F. B. (2019). Plant-based diets for cardiovascular disease prevention: All plant foods are not created equal. *Current Atherosclerosis Reports, 21*(5).

Herreman, L., Nommensen, P., Pennings, B., & Laus, M. C. (2020). Comprehensive overview of the quality of plant- and animal-sourced proteins based on the digestible indispensable amino acid score. *Food Science & Nutrition, 8*(10), 5379–5391. https://doi.org/10.1002/fsn3.1809

Hertzler, S. R., Lieblein-Boff, J. C., Weiler, M., & Allgeier, C. (2020). Plant proteins: Assessing their nutritional quality and effects on health and physical function. *Nutrients, 12*(12). https://doi.org/10.3390/nu12123704

Hills, R. D., Pontefract, B. A., Mishcon, H. R., Black, C. A., Sutton, S. C., & Theberge, C. R. (2019). Gut microbiome: Profound implications for diet and disease. *Nutrients, 11*(7).

Holland, C., Ryden, P., Edwards, C. H., & Grundy, M. M. L. (2020). Plant cell walls: Impact on nutrient bioaccessibility and digestibility. *Foods, 9*(2).

Hooper, L., Martin, N., Jimoh, O. F., Kirk, C., Foster, E., & Abdelhamid, A. S. (2020). Reduction in saturated fat intake for cardiovascular disease. *Cochrane Database of Systematic Reviews, 8*.

Hu, F. B., Otis, B. O., & McCarthy, G. (2019). Can plant-based meat alternatives be part of a healthy and sustainable diet? *Jama-Journal of the American Medical Association, 322*(16), 1547–1548.

Hunt, R. H., Camilleri, M., Crowe, S. E., El-Omar, E. M., Fox, J. G., Kuipers, E. J., Malfertheiner, P., McColl, K. E. L., Pritchard, D. M., Rugge, M., Sonnenberg, A., Sugano, K., & Tack, J. (2015). The stomach in health and disease. *Gut, 64*(10), 1650–1668.

Hutchings, S. C., Low, J. Y. Q., & Keast, R. S. J. (2019). Sugar reduction without compromising sensory perception. An impossible dream? *Critical Reviews in Food Science and Nutrition, 59*(14), 2287–2307.

Jaenke, R., Barzi, F., McMahon, E., Webster, J., & Brimblecombe, J. (2017). Consumer acceptance of reformulated food products: A systematic review and meta-analysis of salt-reduced foods. *Critical Reviews in Food Science and Nutrition, 57*(16), 3357–3372.

Joye, I. (2019). Protein digestibility of cereal products. *Foods, 8*(6).

Karami, Z., & Akbari-adergani, B. (2019). Bioactive food derived peptides: a review on correlation between structure of bioactive peptides and their functional properties. *Journal of Food Science and Technology-Mysore, 56*(2), 535–547.

Kim, Y., Je, Y., & Giovannucci, E. L. (2021). Association between dietary fat intake and mortality from all-causes, cardiovascular disease, and cancer: A systematic review and meta-analysis of prospective cohort studies. *Clinical Nutrition, 40*(3), 1060–1070. https://doi.org/10.1016/j.clnu.2020.07.007

Ko, C. W., Qu, J., Black, D. D., & Tso, P. (2020). Regulation of intestinal lipid metabolism: Current concepts and relevance to disease. *Nature Reviews Gastroenterology & Hepatology, 17*(3), 169–183.

Korpela, K. (2018). Diet, microbiota, and metabolic health: Trade-off between saccharolytic and proteolytic fermentation. In M. P. Doyle & T. R. Klaenhammer (Eds.), *Annual review of food science and technology, vol 9* (Vol. 9, pp. 65–84).

Krutz, N. L., Kimber, I., Maurer-Stroh, S., & Gerberick, G. F. (2020). Determination of the relative allergenic potency of proteins: hurdles and opportunities. *Critical Reviews in Toxicology, 50*(6), 521–530.

Kumar, M., Babaei, P., Boyang, J., & Nielsen, J. (2016). Human gut microbiota and healthy aging: Recent developments and future prospective. *Nutrition and Health Aging, 4*, 3–16.

Kuo, W. Y., & Lee, Y. S. (2014). Effect of food matrix on saltiness perception-implications for sodium reduction. *Comprehensive Reviews in Food Science and Food Safety, 13*(5), 906–923.

Kuzan, A. (2021). Toxicity of advanced glycation end products (Review). *Biomedical Reports, 14*(5).

Lafarga, T., & Hayes, M. (2017). Bioactive protein hydrolysates in the functional food ingredient industry: Overcoming current challenges. *Food Reviews International, 33*(3), 217–246.

Lang, A., Kuss, O., Filla, T., & Schlesinger, S. (2021). Association between per capita sugar consumption and diabetes prevalence mediated by the body mass index: results of a global mediation analysis. *European Journal of Nutrition, 60*(4), 2121–2129.

Lawrence, G. D. (2021). Perspective: The saturated fat-unsaturated oil dilemma: Relations of dietary fatty acids and serum cholesterol, atherosclerosis, inflammation, cancer, and all-cause mortality. *Advances in Nutrition, 12*(3), 647–656.

Leray, C. (2014). *Lipids: Nutrition and health*. CRC Press.

Loveday, S. M. (2019). Food proteins: Technological, Nutritional, and sustainability attributes of traditional and emerging proteins. In M. P. Doyle & D. J. McClements (Eds.), *Annual review of food science and technology, vol 10* (Vol. 10, pp. 311–339).

Loveday, S. M. (2020). Plant protein ingredients with food functionality potential. *Nutrition Bulletin, 45*(3), 321–327.

Luo, J., Yang, H. Y., & Song, B. L. (2020). Mechanisms and regulation of cholesterol homeostasis. *Nature Reviews Molecular Cell Biology, 21*(4), 225–245.

Ma, N., Tian, Y., Wu, Y., & Ma, X. (2017). Contributions of the interaction between dietary protein and gut microbiota to intestinal health. *Current Protein & Peptide Science, 18*(8), 795–808.

Mach, F., Baigent, C., Catapano, A. L., Koskinas, K. C., Casula, M., Badimon, L., Chapman, M. J., De Backer, G. G., Delgado, V., Ference, B. A., Graham, I. M., Halliday, A., Landmesser, U., Mihaylova, B., Pedersen, T. R., Riccardi, G., Richter, D. J., Sabatine, M. S., Taskinen, M. R., ... European Atherosclerosis, S. (2020). 2019 ESC/EAS Guidelines for the management of dyslipidaemias: lipid modification to reduce cardiovascular risk The Task Force for the management of dyslipidaemias of the European Society of Cardiology (ESC) and European Atherosclerosis Society (EAS). *European Heart Journal, 41*(1), 111–188.

Mann, N. (2007). Meat in the human diet: An anthropological perspective. *Nutrition and Dietetics, 64*, S102–S107.

References

Mathai, J. K., Liu, Y. H., & Stein, H. H. (2017). Values for digestible indispensable amino acid scores (DIAAS) for some dairy and plant proteins may better describe protein quality than values calculated using the concept for protein digestibility-corrected amino acid scores (PDCAAS). *British Journal of Nutrition, 117*(4), 490–499.

Mattila, P., Makinen, S., Eurola, M., Jalava, T., Pihlava, J. M., Hellstrom, J., & Pihlanto, A. (2018). Nutritional value of commercial protein-rich plant products. *Plant Foods for Human Nutrition, 73*(2), 108–115.

Maurya, V. K., Bashir, K., & Aggarwal, M. (2020). Vitamin D microencapsulation and fortification: Trends and technologies. *Journal of Steroid Biochemistry and Molecular Biology, 196*.

McClements, D. J. (2015). Nanoscale nutrient delivery systems for food applications: Improving bioactive dispersibility, stability, and bioavailability. *Journal of Food Science, 80*(7), N1602–N1611.

McClements, D. J. (2018a). Enhanced delivery of lipophilic bioactives using emulsions: A review of major factors affecting vitamin, nutraceutical, and lipid bioaccessibility. *Food & Function, 9*(1), 22–41.

McClements, D. J. (2018b). Recent developments in encapsulation and release of functional food ingredients: Delivery by design. *Current Opinion in Food Science, 23*, 80–84.

McClements, D. J. (2019). *Future foods: How modern science is transforming the way we eat.* Springer Scientific.

McClements, D. J. (2021). Food hydrocolloids: Application as functional ingredients to control lipid digestion and bioavailability. *Food Hydrocolloids, 111*.

McClements, D. J., & Grossmann, L. (2021a). A brief review of the science behind the design of healthy and sustainable plant-based foods. *Npj Science of Food, 5*(1).

McClements, D. J., & Grossmann, L. (2021b). The science of plant-based foods: Constructing next-generation meat, fish, milk, and egg analogs. *Comprehensive Reviews in Food Science and Food Safety, 20*(4), 4049–4100.

McClements, D. J., Zou, L. Q., Zhang, R. J., Salvia-Trujillo, L., Kumosani, T., & Xiao, H. (2015). Enhancing nutraceutical performance using excipient foods: Designing food structures and compositions to increase bioavailability. *Comprehensive Reviews in Food Science and Food Safety, 14*(6), 824–847.

McQuilken, S. A. (2021a). Digestion and absorption. *Anaesthesia and Intensive Care Medicine, 22*(5), 336–338.

McQuilken, S. A. (2021b). Gut motility and its control. *Anaesthesia and Intensive Care Medicine, 22*(5), 339–342.

McQuilken, S. A. (2021c). The mouth, stomach and intestines. *Anaesthesia and Intensive Care Medicine, 22*(5), 330–335.

Menotti, A., & Puddu, P. E. (2015). How the seven countries study contributed to the definition and development of the Mediterranean diet concept: A 50-year journey. *Nutrition, Metabolism, and Cardiovascular Diseases, 25*(3), 245–252.

Moreau, R. A., Whitaker, B. D., & Hicks, K. B. (2002). Phytosterols, phytostanols, and their conjugates in foods: structural diversity, quantitative analysis, and health-promoting uses. *Progress in Lipid Research, 41*(6), 457–500.

Moynihan, P. J., & Kelly, S. A. M. (2014). Effect on caries of restricting sugars intake: Systematic review to inform WHO guidelines. *Journal of Dental Research, 93*(1), 8–18.

NAS. (2005). *Dietary reference intakes for energy, carbohydrate, fiber, fat, fatty acids, cholesterol, protein, and amino acids.* The National Academies Press.

Nyyssola, A., Ellila, S., Nordlund, E., & Poutanen, K. (2020). Reduction of FODMAP content by bioprocessing. *Trends in Food Science & Technology, 99*, 257–272.

O'Neil, C. E., Nicklas, T. A., Saab, R., & Fulgoni, V. L. (2020). Relationship of added sugars intakes with physiologic parameters in adults: an analysis of national health and nutrition examination survey 2001–2012. *Aims Public Health, 7*(3), 450–468.

Oteng, A. B., & Kersten, S. (2020). Mechanisms of Action of trans Fatty Acids. *Advances in Nutrition, 11*(3), 697–708.

Ozturk, B. (2017). Nanoemulsions for food fortification with lipophilic vitamins: Production challenges, stability, and bioavailability. *European Journal of Lipid Science and Technology, 119*(7).

Pali-Scholl, I., Verhoeckx, K., Mafra, I., Bavaro, S. L., Mills, E. N. C., & Monaci, L. (2019). Allergenic and novel food proteins: State of the art and challenges in the allergenicity assessment. *Trends in Food Science & Technology, 84*, 45–48.

Pasiakos, S. M., Agarwal, S., Lieberman, H. R., & Fulgoni, V. L., 3rd. (2015). Sources and amounts of animal, dairy, and plant protein intake of US adults in 2007–2010. *Nutrients, 7*(8), 7058–7069.

Pastore, P., Roverso, M., Tedesco, E., Micheletto, M., Mantovan, E., Zanella, M., & Benetti, F. (2020). Comparative evaluation of intestinal absorption and functional value of iron dietary supplements and drug with different delivery systems. *Molecules, 25*(24).

Pawlak, R., Lester, S. E., & Babatunde, T. (2014). The prevalence of cobalamin deficiency among vegetarians assessed by serum vitamin B12: A review of literature. *European Journal of Clinical Nutrition, 68*(5), 541–548.

Peled, S., & Livney, Y. D. (2021). The role of dietary proteins and carbohydrates in gut microbiome composition and activity: A review. *Food Hydrocolloids, 120*.

Pi, X. W., Yang, Y. L., Sun, Y. X., Cui, Q., Wan, Y., Fu, G. M., Chen, H. B., & Cheng, J. J. (2021). Recent advances in alleviating food allergenicity through fermentation. *Critical Reviews in Food Science and Nutrition*.

Rajaram, S. (2014). Health benefits of plant-derived alpha-linolenic acid. *American Journal of Clinical Nutrition, 100*(1), 443S–448S.

Reynolds, A., Mann, J., Cummings, J., Winter, N., Mete, E., & Te Morenga, L. (2019). Carbohydrate quality and human health: A series of systematic reviews and meta-analyses. *The Lancet, 393*(10170), 434–445.

Rieder, A., Afseth, N. K., Bocker, U., Knutsen, S. H., Kirkhus, B., Maehre, H. K., Ballance, S., & Wubshet, S. G. (2021). Improved estimation of in vitro protein digestibility of different foods using size exclusion chromatography. *Food Chemistry, 358*.

Rios-Covian, D., Ruas-Madiedo, P., Margolles, A., Gueimonde, M., de los Reyes-Gavilan, C. G., & Salazar, N. (2016). Intestinal short chain fatty acids and their link with diet and human health. *Frontiers in Microbiology, 7*.

Rippe, J. M., & Angelopoulos, T. J. (2016). Relationship between added sugars consumption and chronic disease risk factors: Current understanding. *Nutrients, 8*(11).

Rizzo, G., Lagana, A. S., Rapisarda, A. M. C., La Ferrera, G. M. G., Buscema, M., Rossetti, P., Nigro, A., Muscia, V., Valenti, G., Sapia, F., Sarpietro, G., Zigarelli, M., & Vitale, S. G. (2016). Vitamin B12 among vegetarians: Status, assessment and supplementation. *Nutrients, 8*(12).

Roberfroid, M., Gibson, G. R., Hoyles, L., McCartney, A. L., Rastall, R., Rowland, I., Wolvers, D., Watzl, B., Szajewska, H., Stahl, B., Guarner, F., Respondek, F., Whelan, K., Coxam, V., Davicco, M. J., Leotoing, L., Wittrant, Y., Delzenne, N. M., Cani, P. D., ... Meheust, A. (2010). Prebiotic effects: Metabolic and health benefits. *British Journal of Nutrition, 104*, S1–S63.

Romanchik-Cerpovicz, J. E., & McKemie, R. J. (2007). Research and professional briefs - Fortification of all-purpose wheat-flour tortillas with calcium lactate, calcium carbonate, or calcium citrate is acceptable. *Journal of the American Dietetic Association, 107*(3), 506–509.

Rousseau, S., Kyomugasho, C., Celus, M., Hendrickx, M. E. G., & Grauwet, T. (2020). Barriers impairing mineral bioaccessibility and bioavailability in plant-based foods and the perspectives for food processing. *Critical Reviews in Food Science and Nutrition, 60*(5), 826–843.

Saini, R. K., & Keum, Y. S. (2018). Omega-3 and omega-6 polyunsaturated fatty acids: Dietary sources, metabolism, and significance - A review. *Life Sciences, 203*, 255–267.

Sanders, M. E., Merenstein, D. J., Reid, G., Gibson, G. R., & Rastall, R. A. (2019). Probiotics and prebiotics in intestinal health and disease: From biology to the clinic. *Nature Reviews Gastroenterology & Hepatology, 16*(10), 605–616.

Santini, A., Tenore, G. C., & Novellino, E. (2017). Nutraceuticals: A paradigm of proactive medicine. *European Journal of Pharmaceutical Sciences, 96*, 53–61.

Santini, A., Cammarata, S. M., Capone, G., Ianaro, A., Tenore, G. C., Pani, L., & Novellino, E. (2018). Nutraceuticals: Opening the debate for a regulatory framework. *British Journal of Clinical Pharmacology, 84*(4), 659–672.

References

Santos-Hernandez, M., Alfieri, F., Gallo, V., Miralles, B., Masi, P., Romano, A., Ferranti, P., & Recio, I. (2020). Compared digestibility of plant protein isolates by using the INFOGEST digestion protocol. *Food Research International, 137*.

Sarwar Gilani, G., Wu Xiao, C., & Cockell, K. A. (2012). Impact of antinutritional factors in food proteins on the digestibility of protein and the bioavailability of amino acids and on protein quality. *British Journal of Nutrition, 108*(S2), S315–S332.

Satija, A., Bhupathiraju, S. N., Rimm, E. B., Spiegelman, D., Chiuve, S. E., Borgi, L., Willett, W. C., Manson, J. E., Sun, Q., & Hu, F. B. (2016). Plant-based dietary patterns and incidence of type 2 diabetes in US men and women: Results from three prospective cohort studies. *PLoS Medicine, 13*(6).

Schoeneck, M., & Iggman, D. (2021). The effects of foods on LDL cholesterol levels: A systematic review of the accumulated evidence from systematic reviews and meta-analyses of randomized controlled trials. *Nutrition, Metabolism, and Cardiovascular Diseases, 31*(5), 1325–1338.

Sebastiani, G., Herranz Barbero, A., Borras-Novell, C., Casanova, M. A., Aldecoa-Bilbao, V., Andreu-Fernandez, V., Tutusaus, M. P., Martinez, S. F., Roig, M. D. G., & Garcia-Algar, O. (2019). The effects of vegetarian and vegan diet during pregnancy on the health of mothers and offspring. *Nutrients, 11*(3).

Shaheen, S., Shorbagi, M., Lorenzo, J. M., & Farag, M. A. (2021). Dissecting dietary melanoidins: Formation mechanisms, gut interactions and functional properties. *Critical Reviews in Food Science and Nutrition*.

Shahidi, F. (1997). Beneficial health effects and drawbacks of antinutrients and phytochemicals in foods - An overview. In F. Shahidi (Ed.), *Antinutrients and phytochemicals in food* (Vol. 662, pp. 1–9).

Shahidi, F., & Ambigaipalan, P. (2018). Omega-3 polyunsaturated fatty acids and their health benefits. In M. P. Doyle & T. R. Klaenhammer (Eds.), *Annual review of food science and technology, vol 9* (Vol. 9, pp. 345–381).

Shortt, C., Hasselwander, O., Meynier, A., Nauta, A., Fernandez, E. N., Putz, P., Rowland, I., Swann, J., Turk, J., Vermeiren, J., & Antoine, J. M. (2018). Systematic review of the effects of the intestinal microbiota on selected nutrients and non-nutrients. *European Journal of Nutrition, 57*(1), 25–49.

Simopoulos, A. P. (2016). An increase in the omega-6/omega-3 fatty acid ratio increases the risk for obesity. *Nutrients, 8*(3).

Singh, V., Yeoh, B. S., & Vijay-Kumar, M. (2016). Gut microbiome as a novel cardiovascular therapeutic target. *Current Opinion in Pharmacology, 27*, 8–12.

Singh, R. K., Chang, H. W., Yan, D., Lee, K. M., Ucmak, D., Wong, K., Abrouk, M., Farahnik, B., Nakamura, M., Zhu, T. H., Bhutani, T., & Liao, W. (2017). Influence of diet on the gut microbiome and implications for human health. *Journal of Translational Medicine, 15*.

Solymosi, D., Diczig, B., Sardy, M., & Ponyai, G. (2020). Food allergy? Intolerance? - Examination of adverse reactions to foods in 406 adult patients. *Orvosi Hetilap, 161*(25), 1042–1049.

Somaratne, G., Ferrua, M. J., Ye, A. Q., Nau, F., Floury, J., Dupont, D., & Singh, J. (2020). Food material properties as determining factors in nutrient release during human gastric digestion: A review. *Critical Reviews in Food Science and Nutrition, 60*(22), 3753–3769.

Song, M. Y., Fung, T. T., Hu, F. B., Willett, W. C., Longo, V. D., Chan, A. T., & Giovannucci, E. L. (2016). Association of animal and plant protein intake with all-cause and cause-specific mortality. *JAMA Internal Medicine, 176*(10), 1453–1463.

Ting, Y. W., Jiang, Y., Ho, C. T., & Huang, Q. R. (2014). Common delivery systems for enhancing in vivo bioavailability and biological efficacy of nutraceuticals. *Journal of Functional Foods, 7*, 112–128.

Tocher, D. R., Betancor, M. B., Sprague, M., Olsen, R. E., & Napier, J. A. (2019). Omega-3 long-chain polyunsaturated fatty acids, EPA and DHA: Bridging the gap between supply and demand. *Nutrients, 11*(1).

Tomova, A., Bukovsky, I., Rembert, E., Yonas, W., Alwarith, J., Barnard, N. D., & Kahleova, H. (2019). The effects of vegetarian and vegan diets on gut microbiota. *Frontiers in Nutrition, 6*.

Toribio-Mateas, M. A., Bester, A., & Klimenko, N. (2021). Impact of plant-based meat alternatives on the gut microbiota of consumers: A real-world study. *Foods, 10*(9), 2040.

Trivedi, R., & Barve, K. (2021). Delivery systems for improving iron uptake in anemia. *International Journal of Pharmaceutics, 601*.

Valenta, R., Hochwallner, H., Linhart, B., & Pahr, S. (2015). Food allergies: The basics. *Gastroenterology, 148*(6), 1120–U1164.

Valenta, R., Karaulov, A., Niederberger, V., Gattinger, P., van Hage, M., Flicker, S., Linhart, B., Campana, R., Focke-Tejkl, M., Curin, M., Eckl-Dorna, J., Lupinek, C., Resch-Marat, Y., Vrtala, S., Mittermann, I., Garib, V., Khaitov, M., Valent, P., & Pickl, W. F. (2018). Molecular aspects of allergens and allergy. In F. Alt (Ed.), *Advances in immunology, vol 138* (Vol. 138, pp. 195–256).

Vandenplas, Y., Berger, B., Carnielli, V. P., Ksiazyk, J., Lagstrom, H., Luna, M. S., Migacheva, N., Mosselmans, J. M., Picaud, J. C., Possner, M., Singhal, A., & Wabitsch, M. (2018). Human Milk Oligosaccharides: 2-Fucosyllactose (2-FL) and Lacto-N-Neotetraose (LNnT) in Infant Formula. *Nutrients, 10*(9).

Vanga, S. K., Singh, A., & Raghavan, V. (2017). Review of conventional and novel food processing methods on food allergens. *Critical Reviews in Food Science and Nutrition, 57*(10), 2077–2094.

Velasquez, M. T., Ramezani, A., Manal, A., & Raj, D. S. (2016). Trimethylamine N-Oxide: The good, the bad and the unknown. *Toxins, 8*(11).

Velikov, K. P., & Pelan, E. (2008). Colloidal delivery systems for micronutrients and nutraceuticals. *Soft Matter, 4*(10), 1964–1980.

Wang, Y. B. (2009). Prebiotics: Present and future in food science and technology. *Food Research International, 42*(1), 8–12.

Warmbrunn, M. V., Herrema, H., Aron-Wisnewsky, J., Soeters, M. R., Van Raalte, D. H., & Nieuwdorp, M. (2020). Gut microbiota: A promising target against cardiometabolic diseases. *Expert Review of Endocrinology and Metabolism, 15*(1), 13–27.

White, P. J., & Broadley, M. R. (2005). Biofortifying crops with essential mineral elements. *Trends in Plant Science, 10*(12), 586–593.

Williams, B. A., Mikkelsen, D., Flanagan, B. M., & Gidley, M. J. (2019). "Dietary fibre": moving beyond the "soluble/insoluble" classification for monogastric nutrition, with an emphasis on humans and pigs. *Journal of Animal Science and Biotechnology, 10*.

Wolfe, R. R., Rutherfurd, S. M., Kim, I. Y., & Moughan, P. J. (2016). Protein quality as determined by the Digestible Indispensable Amino Acid Score: Evaluation of factors underlying the calculation. *Nutrition Reviews, 74*(9), 584–599.

Yu, X. H., Zhang, D. W., Zheng, X. L., & Tang, C. K. (2019). Cholesterol transport system: An integrated cholesterol transport model involved in atherosclerosis. *Progress in Lipid Research, 73*, 65–91.

Zaraska, M. (2016). *Meathooked: The history and science of our 2.5-million-year obsession with meat*. Basic Books.

Zarate, R., el Jaber-Vazdekis, N., Tejera, N., Perez, J. A., & Rodriguez, C. (2017). Significance of long chain polyunsaturated fatty acids in human health. *Clinical and Translational Medicine, 6*.

Zeisel, S. H., & Warrier, M. (2017). Trimethylamine N-oxide, the microbiome, and heart and kidney disease. In P. J. Stover & R. Balling (Eds.), *Annual review of nutrition, vol 37* (Vol. 37, pp. 157–181).

Zhang, Y. Y., Stockmann, R., Ng, K., & Ajlouni, S. (2021). Opportunities for plant-derived enhancers for iron, zinc, and calcium bioavailability: A review. *Comprehensive Reviews in Food Science and Food Safety, 20*(1), 652–685.

Zhou, W. Y., Cheng, Y. Y., Zhu, P., Nasser, M. I., Zhang, X. Y., & Zhao, M. Y. (2020). Implication of gut microbiota in cardiovascular diseases. *Oxidative Medicine and Cellular Longevity, 2020*.

Zmora, N., Suez, J., & Elinav, E. (2019). You are what you eat: Diet, health and the gut microbiota. *Nature Reviews Gastroenterology & Hepatology, 16*(1), 35–56.

Zuidam, N. J. (2012). An industry perspective on the advantages and disadvantages of iron micronutrient delivery systems. In N. Garti & D. J. McClements (Eds.), *Encapsulation technologies and delivery systems for food ingredients and nutraceuticals* (pp. 505–540).

Chapter 6
Meat and Fish Alternatives

6.1 Introduction

Plant-based meat alternatives have been consumed since ancient times, especially in Asian regions such as India and China, to provide protein-rich foods. Prominent examples are tofu (made from soy), tempeh (made from soy), and seitan (made from wheat). The production of tofu for human consumption can be traced back to the Chinese Han Dynasty 2000 years ago (He et al., 2020). However, these plant-based foods were not designed to accurately simulate the sensory and nutritional attributes of real meat products, such as their appearance, texture, flavor, and nutrient profile. Over the past few decades, many food companies have produced vegetarian and vegan foods designed to mimic the properties of real meat products, but they were often of relatively poor quality and had poor consumer acceptance. More recently, however, food companies have created plant-based meat and seafood analogs that much more accurately simulate the desirable quality attributes (Fig. 6.1, Table 6.1). The availability of these products has been one of the main reasons for the growing adoption of plant-based foods by many consumers. Companies such as 'Impossible Foods' and 'Beyond Meat' have introduced a diverse range of commercially successful plant-based meat products, including burgers, sausages, ground meat, and nuggets. In addition, many new companies are being established in this area, and many traditional food companies are adding plant-based alternatives to their existing meat product portfolio, which is increasing the diversity of foods available to vegans, vegetarians, and flexitarians.

The creation of high-quality meat and seafood analogs has been enabled by several technical innovations. Texturized vegetable protein (TVP) was released on the market in the 1960s and was one of the first ingredients used to create plant-based products designed to mimic the textural attributes of meat (Riaz, 2011). Since then, the science and technology of plant-based meat production has advanced considerably, especially the availability of innovative functional ingredients and processing

Plant-based Nuggets **Plant-based Ground beef** **Plant-based Salmon** **Plant-based Sausages** **Plant-based Burger**

Fig. 6.1 Examples of commercial plant-based meat and seafood products. The photographs of the plant-based nuggets, ground beef, sausages, and burgers were kindly supplied by Beyond Meat. The photograph of the salmon was kindly provided by Revo Foods

technologies, as well as the development and application of structural design principles. These advances have contributed to the creation of the high-quality meat and seafood analogs mentioned earlier (Fig. 6.1). However, the success of these products is also a result of the growing consumer demand for plant-based foods due to their concerns about environmental, health, and animal welfare issues associated with the modern food supply (Chap. 1). The popularity of these products is also reflected by their economic growth. The sales of plant-based meat analogs were reported to be worth around $1.4 billion in 2020 in the USA alone, which was a 45% increase from the previous year (GFI, 2020).

Consumer studies have shown that many western consumers who identify as meat-eaters are more willing to switch to plant-based meat products if they resemble real meat products (Elzerman et al., 2011; Hoek et al., 2011). This highlights the importance of creating meat analogs that accurately mimic the taste, texture, appearance, and cooking experience of real meat. Consumers can then simply substitute existing meat-based products with plant-based alternatives without having to greatly change their eating patterns, *e.g.,* by replacing beef burgers with plant-based ones. However, there are still several social factors that are acting as hurdles to the more widespread adoption of plant-based meats (Michel et al., 2021).

The current generation of plant-based meat analogs typically uses pea, soy, or wheat gluten as their main protein source, but other proteins are also being explored and adopted for this purpose. Numerous other functional ingredients are added to the formulations to enhance their flavor (*e.g.,* flavors, herbs, spices, and salts), appearance (*e.g.,* coloring and lightning agents), texture (*e.g.,* thickening, gelling, or binding agents), and nutritional profile (*e.g.,* vitamins and minerals). As a result, the ingredient lists of plant-based analogs are typically much longer than those of real

6.1 Introduction

Table 6.1 Selected meat analog products on the market in 2021

Product	Ingredients	Company	Reference
Beyond burger	Water, pea protein, expeller-pressed canola oil, refined coconut oil, rice protein, natural flavors, dried yeast, cocoa butter, methylcellulose, and less than 1% of potato starch, salt, potassium chloride, beet juice color, apple extract, pomegranate concentrate, sunflower lecithin, vinegar, lemon juice concentrate, vitamins and minerals (zinc sulfate, niacinamide [vitamin B3], pyridoxine hydrochloride [vitamin B6], cyanocobalamin [vitamin B12], calcium pantothenate).	Beyond Foods, USA	https://www.beyondmeat.com/products/the-beyond-burger
Impossible burger	water, soy protein concentrate, coconut oil, sunflower oil, natural flavors, 2% or less of: potato protein, methylcellulose, yeast extract, cultured dextrose, food starch modified, soy leghemoglobin, salt, soy protein isolate, mixed tocopherols (vitamin e), zinc gluconate, thiamine hydrochloride (vitamin b1), sodium ascorbate (vitamin c), niacin, pyridoxine hydrochloride (vitamin b6), riboflavin (vitamin b2), vitamin b12.	Impossible Foods, USA	https://faq.impossiblefoods.com/hc/en-us/articles/360018937494-Whatarethe-ingredients
MorningStar farms grillers original burger	water, wheat gluten, soy flour, vegetable oil (corn, canola and/or sunflower oil), egg whites, calcium caseinate, corn starch, contains 2% or less of: onion powder, soy sauce powder (soybeans, salt, wheat), methylcellulose, cooked onion and carrot juice concentrate, salt, natural flavor, soy protein isolate, garlic powder, spices, sugar, gum acacia, whey, yeast extract, xanthan gum, potato starch, tomato paste (tomatos), onion juice concentrate.	MorningStar Farms, Kellogg's, USA	http://smartlabel.kelloggs.com/Product/Index/00028989100801
Gardein meatless meat balls	Water, Textured Soy Protein Concentrate, Canola Oil, Vital Wheat Gluten, Soy Protein Isolate, Enriched Wheat Flour, Wheat Flour, Niacin, Reduced Iron, Thiamine Mononitrate, Riboflavin, Folic Acid, 2% Or Less Of: Methylcellulose, Yeast Extract, Onion Powder, Salt, Barley Malt Extract, Spices, Garlic Powder, Sugar, Fennel, Natural Flavors, Crushed Red Pepper, Yeast	Garden Protein International, Pinnacle Foods, Canada	https://www.gardein.com/products/meatless-meatballs/

(continued)

Table 6.1 (continued)

Product	Ingredients	Company	Reference
Tofurky ham roast with glaze	water, vital wheat gluten, organic tofu (water, organic soybeans, magnesium chloride, calcium chloride), expeller pressed canola oil, contains less than 2% of sea salt, spices, granulated garlic, cane sugar, natural flavors, natural smoke flavor, color (lycopene, purple carrot juice), oat fiber, carrageenan, dextrose, konjac, potassium chloride, xanthan gum.	Tofurky, Turtle Island Foods, USA	https://tofurky.com/what-we-make/roasts/ham-roast/
Quorn Meatless Nuggets	Wheat Flour, Mycoprotein (34%), Water, Egg White, Wheat Starch, Canola Oil, Contains 2% Or Less Of Milk Proteins, Sugar, Textured Wheat Protein (Wheat Protein, Wheat Flour), Sage, Potato Dextrin, Onion Powder, Yeast Extract, Guar Gum, Wheat Gluten, Pectin, Salt, Calcium Chloride, Modified Corn Starch, Calcium Acetate, Pea Fiber, Yeast, Dextrose, Pepper, Turbinado Sugar.	Quorn Foods, Inc.	https://www.quorn.us/products/quorn-meatless-chicken-nuggets
Garden Gorumet Sensational Filet Pieces Asian Seasoning	water, 25,3% protein SOY, vegetable oils (sunflower, rapeseed), vinegar acid, sugar, yeast extract, tamari sauce (SOY beans, water, salt), cornstarch, salt, flavouring, spices (garlic, chili, paprika, cinnamon, ginger), tomato powder, flaxseed flour, colour (paprika extract).	Tivall Deutschland GmbH, Germany, part of Nestlé	https://www.gardengourmet.com/product/sensational-filet-pieces-asian-seasoning
planted. pulled – Nature	Water, vegetable proteins 31% (pea, sunflower, oat), pea fiber, canola oil, spice preparation, vitamin B12.	Planted Foods GmbH – Germany	https://shop.eatplanted.de/collections/newshop-planted-pulled-de/products/newshop-planted-pulled-natur-400g

Adapted and extended from (Bohrer, 2019)

meat products. Consequently, research is needed to understand how different functional ingredients behave in plant-based foods, which will allow formulators to reduce the total number of ingredients required and to select ingredients that are more label-friendly. To better understand what kind of properties plant-based meat analogs should possess, this chapter begins with a short overview of the most important structural, physicochemical, functional, and sensory attributes of real meat.

6.2 Properties of Meat and Fish

Land-based animals, such as cows, pigs, sheep, and chickens have been used as a source of food throughout human history and pre-history (Standage, 2009). Most of the organs and tissues found in animals have been used for this purpose, including limbs, breast, back, abdomen, kidneys, liver, heart, brain, tongue, skin, and bones. However, the most often consumed meat products in many countries are comprised of muscle fibers, connective tissue, and adipose tissue, and so these constituents will be the focus of this section. Muscle tissue typically consists of around 75 % water, 19 % protein, 2.5 % lipids, 1.2 % carbohydrates, and 2.3 % of other compounds (*e.g.,* non-protein nitrogen and minerals) but these values vary depending on animal and muscle type (López-Bote, 2017). Fish has also been consumed by humans throughout history and contains mainly water (70–80 %), proteins (15–20 %), and lipids (2–5 %) (Kazir & Livney, 2021).

6.2.1 Muscle Structure and Composition

Animal tissue is organized into complex hierarchical structures (Fig. 6.2), which contribute to the unique physicochemical and sensory properties of meat and fish products (Prayson et al., 2008). The following section mainly focuses on the structure and composition of meat from land animals but fish has comparable properties with some important differences (Ochiai & Ozawa, 2020). The main difference is that fish contains layers of muscle structures (myotomes) that are separated by sheaths of connective tissue (myocommata) instead of muscle bundles. The connective tissue is also softer because land animals mainly have fibrillar collagen types I and III, whereas fish mainly have collagen types I and V (Listrat et al., 2016). This structural difference in the muscle tissues is primarily because fish do not need to continuously act against gravity. Moreover, fish store fat subcutaneously in the perimysium and myosepta, whereas mammals store fat subcutaneous, intermuscular, and intramuscular (Listrat et al., 2016). Nevertheless, both muscle types consist of contractile elements, which will be reviewed in this section.

In living animals, the main function of muscle fibers is to enable the contraction and expansion of muscles so that the creature can move around (Lawrie & Ledward, 2006). Muscles are connected *via* tendons that transmit the force from the muscle to

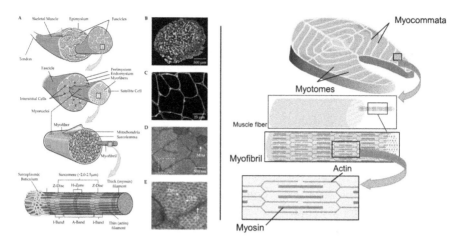

Fig. 6.2 The muscles in meat (left) and fish (right) have a complex hierarchical structure that is difficult to mimic using plant proteins. In particular, muscles contain several muscle bundles (fascicles, myotomes) that consist of muscle fibers, which are structured by myofibrils. All these structures are connected and stabilized by layers of connective tissue. Additionally, fat is deposited between the fascicles (not shown).
Key: B = stained myofibers; C = sarcolemma (white) and nuclei (green); D = sarcoplasmic reticulum that surrounds individual myofibrils, as well as the mitochondria; E = thick and thin myofilaments. Adapted and modified from (Betts et al., 2017; Jorgenson et al., 2020; Kazir & Livney, 2021) under CC-BY 4.0 https://creativecommons.org/licenses/by/4.0/

the skeletal structure. Most of these tendons are typically trimmed before eating, leaving mainly the muscle tissue for consumption. The muscle itself consists of bundles of fascicles that are themselves assembled from bundles of muscle fibers (Fig. 6.2). Each skeletal muscle fiber is a single cylindrical muscle cell that possesses several nuclei that produce all the different types of proteins needed for the muscle structure. The muscle fibers themselves have a complex architecture of packed protein bundles called myofibrils. They are assembled from sarcomeres consisting of several different kinds of muscle proteins, including myosin, actin, titin, nebulin, and troponin. Myosin (5.5 %) and actin (2.5 %) make up the major share of the total myofibrillar protein (11.5 %) in the system (López-Bote, 2017). These proteins make a major contribution to the gelation, emulsification, adhesion, and fluid-binding (oil, water) properties of meat products. Moreover, the sarcomeres are the basic contractile units, which enable muscle contraction with ATP upon activation.

Fat cells (adipocytes) are another major constituent of animal muscle. Fat cells can grow to more than 100 μm in diameter. In fish, an increase in fat content results in an increase in the thickness of myosepta (Listrat et al., 2016). In general, animals store fats in the form of triglycerides. Several adipocytes are grouped together by connective tissue to form adipose tissue, which serves as an energy reserve and as insulation for the animal. It also plays an important role in maintaining nutrient and energy homeostasis by releasing various hormones (Stern et al., 2016). Meat mainly contains three types of white fat: subcutaneous, intermuscular, and intramuscular fat (Purslow, 2017). Subcutaneous fat is the thick adipose tissue layer between the muscle and the skin. Intermuscular fat is the tissue that is stored between different muscles, while intramuscular fat (marbling) is situated within the muscle tissue. These tissues typically have a high proportion of 16:0 (palmitic acid), 18:0 (stearic acid), and 18:1 (oleic acid) fatty acids, while fat that is found in the muscle of pigs also contains considerable amounts of 18:2 (linoleic acid) fatty acids (López-Bote, 2017). The high levels of saturated fatty acids and connective tissue in animal fat give it its unique viscoelastic properties, which is an important part of the textural and sensory attributes of meat and its products. This structure is hard to mimic with plant-based ingredients because the combination of structuring proteins from connective tissue and saturated lipids is not found in plants.

The last major component found in muscle is connective tissue. Connective tissue surrounds the different muscle structures to support and protect them. Connective tissue also separates the different hierarchical structures in muscle, leading to its characteristic anisotropic structure. Without connective tissue, the muscle fibers would not be separated as individual fibers and the fascicles would not appear as individual muscle bundles. The fibrous, anisotropic structure of muscles is one of the main factors contributing to the sensory and textural attributes of real meat. The different layers of connective tissue in muscles are called epimysium, perimysium, and endomysium (Fig. 6.2). The epimysium surrounds the whole muscle, the perimysium protects the muscle fiber bundles, and the endomysium surrounds each muscle fiber. The major proteins found in connective tissue are collagen and elastin. Collagen comprises around 1.0 % of the total muscle mass. At the molecular level, collagen forms supercoiled triple-helical polypeptide structures containing a relatively high proportion of glycine, proline, and hydroxyproline (Purslow, 2017).

The connective tissue found in muscles plays a critical role in meat tenderness and depends on a variety of properties, such as muscle type, age, and type of animal. Connective tissue considerably decreases the tenderness of meat before the gelatinization of collagen because the triple helices contribute to the stiffness. When heated, collagen denatures at around 64 to 68 °C, which supports the shrinkage of the fibrils and results in an increase in fluid loss. This initially decreases the tenderness of the meat, especially in the muscles obtained from older animals. However, once the temperature is further increased between 80 and 90°C, the collagen gelatinizes, which increases the tenderness (Weston et al., 2002). The changes in collagen structure during heating are one of the reasons that meat changes its texture during cooking. Consequently, it is important to try to mimic this behavior using plant-based ingredients.

The muscle fibers, connective tissue, and fat cells all play an important role in determining the overall structure and properties of meat. From a rheological perspective, meat can be classified as a viscoelastic solid, which exhibits both viscous and elastic behavior when deformed. Moreover, meat has anisotropic structural properties, which means that it reacts differently when stress is applied in parallel or perpendicular directions. The change in the dynamic shear moduli of fish (salmon) when the strain is increased is shown in Fig. 6.3. These measurements are consistent

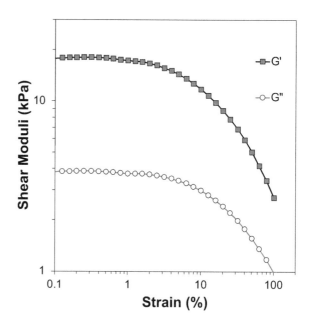

Fig. 6.3 Change in the dynamic elastic modulus (G') and viscous modulus (G") of salmon when the shear strain is increased. The elastic modulus is greater than the viscous modulus because fish is predominantly solid-like. The shear moduli decrease at high strains due to disruption of the 3D network structure in the muscle tissues. The authors thank Zhiyun (Kevin) Zhang (UMass) for providing this data.

with tissues having a 3D network structure of muscle fibers, connective tissue, and fat tissue. As a result, the elastic modulus is greater than the viscous modulus (G'>G"). Once a certain shear strain is exceeded the shear modulus falls appreciably, which can be attributed to disruption of this 3D network structure.

Thus, an important aim when producing plant-based meat alternatives is to mimic the complex anisotropic structure of meat (Fig. 6.2), as this structure plays an important role in determining the textural and sensory attributes of meat (Krintiras et al., 2014).

6.2.2 Appearance

Whole meats are optically opaque semi-solid materials, whose color depends on the type and level of natural pigments present (such as myoglobin). Meat appears opaque because it contains heterogeneities with dimensions similar to the wavelength of light, such as protein fibers or fat cells, which diffusely scatter the incoming light. In fish, the superficial lateral red muscle has a high content of myoglobin and has a rich color (typically brown), whereas the white muscle is almost translucent (Listrat et al., 2016). The orange to red color of salmonids is not caused by muscle pigments but by the carotenoid intake of the fish. Myoglobin and oxymyoglobin (myoglobin that holds an O_2 molecule and contains Fe^{2+}) typically absorb light intensely at wavelengths around 476 nm (blue) and 572 nm (yellow) but not at higher wavelengths (red). As a result, the red light waves are reflected from the surface of raw meat, thereby giving it its characteristic reddish appearance (Table 6.5) (Bjelanovic et al., 2013; Wright & Davis, 2015).

The appearance of meat products changes in a particular fashion during food preparation. For example, a beef steak changes from shiny red/pink when it is raw to matt brown when it is cooked, whereas a chicken breast changes from shiny pink when it is raw to beige when it is cooked. These thermally induced color changes typically occur over a specific time and temperature range due to chemical reactions that occur within the product during heating (Fig. 6.4). The alteration in the appearance of meat products during cooking is also governed by changes in the physical state and structural organization of the different components, which alters the degree of light scattering (Toldra, 2017). Before cooking, raw whole muscle tissues have a shiny and opaque appearance with a pinkish to red color (depending on the animal). Shininess occurs because the light waves are specularly reflected from the smooth wet surfaces of the meat (Chap. 4). Cooking leads to a matt appearance because

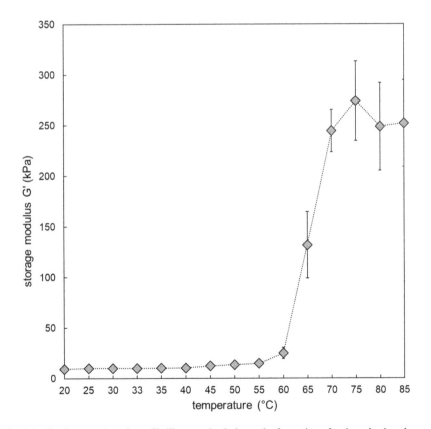

Fig. 6.4 The denaturation of myofibrillar proteins induces the formation of a viscoelastic gel upon heating of a meat batter during the production of emulsified sausages, as indicated by the sharp increase in G' above 60 °C. The meat batter shown here consists of 75 % pork meat and 25 % water. Adopted from (Katz et al., 2021)

some of the surface water evaporates, leading to a dried surface with a rougher texture. Additionally, protein denaturation leads to the formation of irregular structures that also increase the surface roughness. As a result, light waves are scattered diffusely from the surface of the meat, leading to a matt appearance.

6.2.3 Textural Attributes

As discussed earlier, the semi-solid texture (Fig. 6.3) of meat is due to the presence of a structurally complex 3D network of muscle fibers, connective tissue, and adipose tissue, which is mainly held together by physical interactions, such as van der Waals forces, hydrogen bonding, electrostatic attraction, and hydrophobic bonds (Acton & Dick, 1984; Xiong, 1994). The structural organization of this gel network changes when the product is heated, due to the thermal denaturation and aggregation of the proteins, as well as because of water evaporation. As a result, the textural properties of meat change in a characteristic way during cooking. In particular, softening and hardening events typically occur at temperatures where thermal transitions of the meat proteins occur.

6.2.4 Cooking Loss and Heat-induced Changes

Cooking loss is a measure of the reduction in the mass of a meat product during cooking due to the expulsion and evaporation of fluids, which typically contain water, proteins, minerals, and lipids. The fluid holding properties of meat depend on biopolymer-solvent interactions, the elastic modulus of the biopolymer gel network, and the osmotic pressure associated with the uneven distribution of mineral ions inside and outside the biopolymer gel network (Cornet et al., 2021). Depending on their side chains, amino acids can bind 1 water molecule (nonpolar), 2–3 water molecules (polar), or up to 7 water molecules (ionic) (Zayas, 1997). Under environmental conditions that favor a high charge on the protein molecules (*i.e.,* pH ≠ pI and a low ionic strength), the uptake of water into the matrix is more favorable. This is because the charged protein residues can interact with water, which can be approximated by the following equation according to Schnepf (Schnepf, 1992):

$$A = f_e + 0.4 f_p + 0.2 f_n \tag{6.1}$$

Here, A is the amount of bound water (g water/g protein), f_e is the fraction of charged side chains, f_p is the fraction of polar side chains, and f_n is the fraction of non-polar side chains. This equation assumes that all the amino acid side chains in a protein can interact with water, but it seems likely that any side chains located within the interior of a protein may not be able to interact directly with water. As expected, this equation shows that charged amino acids have the greatest impact on water binding, and consequently on the water holding capacity.

Several physicochemical phenomena have been linked to the ability of meat products to retain fluids (Cornet et al., 2021). First, the protein molecules within the 3D biopolymer network have diverse surface groups that can directly interact with the molecules in the fluids inside the meat. These fluids mainly consist of water, but they may also contain dissolved or dispersed proteins, salts, and lipid droplets. The interactions between the water molecules in the fluids and the protein molecules in the biopolymer network are particularly important in determining the fluid holding properties of meat products. Second, the imbalance of mineral ions inside and outside a meat product generates an osmotic stress, which favors a certain amount of water being retained by the biopolymer network (to reduce the concentration gradient). Third, the elastic modulus of the biopolymer network is important because it puts a limit on the expansion or collapse of the meat product in the presence of water.

An alternative approach to interpreting the fluid holding properties of meat products is to consider the capillary pressure. If the protein networks in a meat product were dehydrated, there would be a large positive free energy change associated with the released air-protein interfaces. If this dehydrated meat product were then placed in water, water molecules would move into the protein network to reduce the contact area between the proteins and air. The driving force for this process is the fact that the air-protein interfacial tension is higher (more thermodynamically unfavorable) than the water-protein interfacial tension. As a result, water would be absorbed into the pores in the protein network. The capillary pressure p_c can be expressed as:

$$pc = \frac{2\gamma\theta}{r} \tag{6.2}$$

Here, γ is the interfacial tension between the two phases, θ is the wetting angle of the liquid on the surface of the material (*i.e.*, protein), and r is the radius of the pores. Thus, for an optimum liquid retention in the pores, the pores should be small and have good wettability. In the case of water, this would mean that hydrogen bonds can be formed between the protein surfaces and the water molecules. Moreover, the pores should possess mechanical resistance to minimize their disruption during cooking (*e.g.*, disappearance or growth), as this would promote fluid expulsion.

Water is typically absorbed by proteins in several layers. Some of the water is directly bound to the protein (constitutional water), but further layers are more loosely bound (multilayer water) and the further away the water is located from the protein surface, the less interaction takes place. In meat, around 1 % of water is bound to the proteins, whereas 75 % are intramyofibrillar water, 10 % are extramyofibrillar water, and the remaining 15 % are extracellular water (Warner, 2017). The loss of these fluids is a major reason for the visible shrinking of the meat and a change in its tenderness and juiciness. The texture of meat also changes when it is cooked as a result of changes in the structural organization and interactions of the various kinds of proteins present (Hughes et al., 2014; Yu et al., 2017).

Two processes mainly govern the structural changes of meat during cooking: protein denaturation and fat melting. Different kinds of meat proteins have different denaturation temperatures, which affects the cooking behavior of meat. The myosin proteins typically denature between 40 to 60 °C, with the myosin S1 sub-fragment denaturing between 42 to 48 °C and the myosin tail denaturing at around 55 °C (Wu et al., 2009). Actin proteins have considerably higher denaturation temperatures of about 70 to 80 °C (Warner, 2017). Sarcoplasmic proteins commonly denature somewhat between the myosin and actin proteins, which is typically in the range of about 50 to 70 °C (Yu et al., 2017). The thermal transitions of the meat proteins lead to alterations in the structural organization of the muscle fibers, causing them to shrink in width (50–65 °C) and then in length (70–75 °C), which leads to some of the entrained fluids being forced out, as well as to toughening of the meat (Purslow et al., 2016). The denaturation of myofibrillar proteins causes meat batters to change from predominantly viscoelastic liquids into viscoelastic solids upon heating, *e.g.,* during emulsified sausage production (Fig. 6.4).

Conversely, softening of the meat may occur during heating due to the unraveling of the triple helix structure of the collagen molecules in the connective tissue. As mentioned earlier, collagen typically denatures between 64 to 68°C but lower or higher denaturation temperatures can be observed depending on the species, age, and health status of the animal (Park et al., 2012). Collagen denaturation leads to a shrinkage of the fibrils, which causes some of the fluids to be squeezed out. This initially decreases the tenderness of the meat, especially in muscles from older animals. If the temperature is further increased to around 80 to 90°C, the collagen gelatinizes, which then increases the tenderness (Weston et al., 2002). In fish, collagen has a low thermal stability and therefore does not have a major influence on the tenderness. Instead, the tenderness of fish is mainly influenced by the actin and myosin filaments (Listrat et al., 2016).

Another phase transition that impacts the texture of meats during cooking is the melting of fats. At refrigerator temperatures, a significant fraction of the lipid phase in meat is usually crystallized, which contributes to its mechanical strength. Upon heating, the solid fat content (SFC) decreases as more and more fat crystals melt, which results in a disruption and softening of the semi-solid lipid phase. Major peak transitions are typically observed at around 30 °C for pork fat, and the SFC has been reported to decrease from around 30–35 % at 0 °C to 0% at 40–45 °C (pork lard) (Manaf et al., 2014). Beef tallow has a somewhat higher SFC under refrigerator conditions and melts at a higher temperature than pork lard (Pang et al., 2019). However, the SFC *versus* temperature profiles differ from animal to animal and also depend on the location of the fat in the animal (Wood et al., 2008). The melting of the fat phase during heating typically leads to a softening of the fatty parts of meat. In contrast, the lipid phase in fish is usually more fluid because of the high level of polyunsaturated fatty acids it contains.

The thermal denaturation of myoglobin during cooking is another important phase transition that impacts meat quality. The red/pink color of uncooked meat is caused by selective light absorption by myoglobin as discussed earlier. Myoglobin is a globular protein that contains a heme group. The heme group has an iron atom at its center that may be present in one of two oxidation states: Fe^{2+} or Fe^{3+} (*i.e.*, the iron has 'released' two or three electrons). The physiological role of myoglobin is to accept an oxygen molecule from hemoglobin in the blood and then transport it to the muscle cells to facilitate energy generation. In meat, the color of myoglobin (red, purple, or brown) depends on the oxidation state of the iron in the heme group, as well as whether the heme binds oxygen or not. At ambient temperature, the heme group contains Fe^{2+} that has four unpaired electrons, which enables the binding of oxygen, leading to a bright red color (Fig. 6.5).

The absorbance of light in the visible range decreases when meat is heated from around 55 to 83 °C (Trout, 1989). This color change has been attributed to the denaturation of the myoglobin at elevated temperatures, which makes it more susceptible to oxidation and results in the release of oxygen and the conversion of Fe^{2+} to Fe^{3+}. Indeed, it has been reported that more than 90 % of myoglobin is denatured when a beef muscle is heated to 80 °C (Trout, 1989). As a result, the color of the meat changes from red to brown during cooking.

On a final note, the dark brown crust of well-cooked meat is mainly caused by the Maillard reaction, which involves a complex series of reactions between proteins and reducing sugars at high temperatures and intermediate moisture contents ($a_w \sim 0.4$–0.8). These conditions mainly occur on the surfaces of the meat during cooking, which is why the exterior is dark brown (less moisture, higher temperature) whereas the interior is light brown (more moisture, lower temperature).

6.2.5 Flavor Profile and Oral Processing

The unique flavor of meat products is a combination of odor, taste, and mouthfeel attributes. Raw meat has a rather bland flavor, which indicates that the characteristic meaty flavor of cooked meat develops because of chemical changes that occur

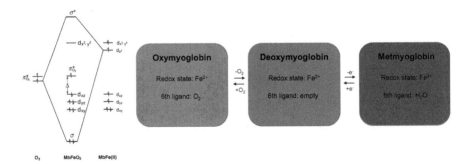

Fig. 6.5 Molecular orbital diagram for FeO_2 bonding in myoglobin ($MbFeO_2$) with deoxymyoglobin in a high-spin ferrous state and dioxygen in a triplet ground state as well as the schematic color appearance of myoglobin at different oxygenation and redox states. The iron ion interacts with 6 ligands in myoglobin: 4 nitrogens in the porphyrin ring, the imidazole side chain of His93, and the sixth is the binding site for oxygen. Adopted and extended with the permission of Elsevier from (Shikama, 2006).

during the cooking process. Thousands of different compounds are responsible for the taste and smell of different kinds of meats, but most of them originate from only a few ingredients (Flores, 2017):

- *Thiamin* (vitamin B1) → Heat → *Thiols, S-compounds, Furans, Thiophene, Thiazoles*
- *Lipids* → Lipolysis → *Unsaturated Fatty Acids* → Oxidation → *Hydrocarbons, Aldehydes, Alcohols, Ketones, etc.*
- *Proteins* → Proteolysis → *Free amino acids* → Maillard reaction → *Aldehydes, Acetaldehyde, Pyrazines, Furfurals, Thiophenes, Pyrroles, etc.*
- *Nucleotides (e.g., ATP, DNA)* → Heat → *5'-ribonucleotides*

Thiamin degrades into several sulfur-containing compounds, whereas nucleotides transform into 5'-ribonucleotides that are important for the typical umami taste when consuming meat. Moreover, the remaining ribose molecules interact with amino acids during the Maillard reaction. Proteins and lipids undergo hydrolysis to release amino acids and fatty acids, respectively. Unsaturated free fatty acids react with oxygen, producing a variety of volatile oxidation products that contribute to the flavor profile of cooked meat. Phospholipids have been shown to play a key role in the development of volatile compounds in meat (Huang et al., 2010). Overall, the flavor profile of a meat product depends on the composition of the raw meat, as well as the cooking technique used (*e.g.,* frying, baking, grilling, steaming, and boiling).

During chewing, the complex internal structure of meat is progressively disassembled, which contributes to its desirable mouthfeel (Lillford, 2016). The anisotropic, fibrous structure of meat is broken down into smaller fragments that are mixed with saliva and air. The fragmentation and softening of the meat in the mouth during mastication means that lower mechanical forces are needed at longer chewing times (Mioche et al., 2003). Interestingly, a tender juicy steak has a different breakdown path compared to tough dry meat. Lillford (Lillford, 2011) reported that the fibrous mouthfeel of meat is not caused by the primary contractile fibers, but is related to the muscle bundles themselves. As muscle bundles are held together by connective tissue, this indicates that collagen plays an essential role in the textural sensation of meat during mastication. During chewing, the connective tissue retains some of the muscle bundle's structure, which means that the meat is not completely fractured. After a certain time, the meat is sufficiently broken down into small fragments and then the saliva in combination with these small fragments (the 'bolus') is swallowed.

6.3 Ingredients for Formulating Plant-based Meat Analogs

The previous section highlighted the complex structure and composition of meat, which is responsible for its textural characteristics and sensorial attributes. Food scientists are using a variety of ingredients to create plant-based products that accurately simulate the unique physicochemical and sensory characteristics of meat, while still maintaining a roughly similar nutritional profile. Some of the most common functional ingredients used to achieve this goal are proteins, polysaccharides, lipids, phospholipids, water, and additives (Chap. 2), which may be utilized as gelling agents, texturing agents, emulsifiers, binders, preservatives, flavors, or colorants (Table 6.1). A specific combination of ingredients is typically used depending on the textural attributes required in the meat product category that is being simulated (Kyriakopoulou et al., 2021):

- **Sausage-type products**: Texturized and non-texturized plant proteins, binders, fats, colorants, and spices
- **Patty- and nugget-type products**: Texturized plant proteins, binders, fats, colorants, and flavorings
- **Whole muscle-type products**: Non-texturized plant proteins (will be texturized during process), polysaccharides, fats, colorants, and flavors

For each product category, the ingredients need to fulfill certain specific functionalities, for example, structuring, gelling, binding, emulsifying, and water-holding. Plant proteins are one of the key ingredients used in all these products because of their versatile functional attributes and because real meat products typically have a relatively high protein content.

6.3.1 Plant Proteins

Plants produce a variety of proteins with different molecular, physicochemical, and functional characteristics that play different roles during the development, growth, and functioning of the plant. These proteins are distributed throughout plant tissues, including the roots, stems, leaves, and seeds. The protein concentration is particularly high in the seeds of many plants (such as soybeans, peas, corn kernels, rice, and nuts). In principle, every protein-rich plant or plant seed that contains proteins with the desired functionality can be used as an ingredient in meat alternatives. In practice, however, the large-scale production of meat analogs requires protein ingredients that are safe, affordable, nutritional, and available at a large scale, which greatly limits the plant proteins that can be utilized for this purpose. For this reason, the most common proteins currently used for formulating meat analogs are proteins derived from soy, pea, wheat, potato, and oilseeds (such as those from sunflower and canola) (Table 6.1). Nevertheless, active research and development programs are being carried out in academic, industrial, and government laboratories to identify alternative sources. Soybeans are currently the most frequently used protein source in meat alternatives. The main reasons for this are its low price (one metric ton cost $615 in June 2021 (EC, 2021)), relatively high protein quality (DIAAS of 91 (Herreman et al., 2020)), versatile functionality, abundance, high yield, and stable production output. However, pea proteins are finding increasing utilization in plant-based meat formulations because of their lower allergenicity (Yuliarti et al., 2021). More detailed information about protein extraction and properties is given in Chap. 2.

The proteins used to produce meat analogs need to fulfill a specific set of functional properties to obtain the required product attributes. An overview of the reported functional properties of soy, pea, and gluten proteins is given in Table 6.2. The main functional attributes that proteins should possess in meat alternatives are: extrudability, gelling, texturizing, emulsifying, binding, fluid holding, cohesion, and adhesion. The relative importance of these different attributes depends on the product category being simulated (*e.g.*, burger, sausage, nugget, ground meat, or whole muscle).

Soy proteins mainly consist of globulins, with 7S (β-conglycinin) and 11S (glycinin) being the major protein types present (Grossmann & Weiss, 2021). In general, soy proteins have good structuring, gelling, emulsification, and fluid holding properties (Nishinari et al., 2014). However, as they are processed during meat alternative production, their conformation and aggregation state are altered, which changes these properties. Soy proteins can be readily structured using extrusion and shear cell processing methods, which is advantageous for producing plant-based meat analogs with semi-solid fibrous textures. Soy proteins are often combined with other proteins or carbohydrates during extrusion or shear cell processing to obtain the required structures, which is discussed in more detail in Chap. 3 (MacDonald et al., 2009; Pietsch et al., 2019).

Pea proteins are also being used to formulate meat analogs because of their abundance, low cost, functional versatility, and low allergenicity. The most important pea proteins are 7S vicilin and 11S legumin, which are both globular proteins. Pea proteins can form gels but they may be weaker than soy protein gels (Batista et al., 2005). Pea proteins have been shown to exhibit good emulsifying, foaming, and fluid holding properties (Raikos et al., 2014; Sridharan et al., 2020; Zayas, 1997). They have also been shown to form anisotropic fibrous structures during extrusion and shear cell processing (Osen et al., 2014; Schreuders et al., 2019). The good functional performance of pea proteins is highlighted by the number of commercial meat analogs that are formulated with this type of protein (Table 6.1).

Wheat proteins such as gluten (gliadin and glutenin) are also commonly used in meat analogs because of their unique texturing and other functional properties. In general, gluten is a predominantly hydrophobic protein that has a low water solubility, which is important for its functional performance in meat analogs. Gluten is rich in glutamine (~35 %) and proline (~10 %), which facilitates the formation of intra- and intermolecular interactions *via* hydrogen and hydrophobic bonding between the protein molecules (Iwaki et al., 2020). Gliadins are relatively small monomeric proteins (28,000–55,000 Da) with strong intramolecular disulfide bonds, whereas glutenins are relatively large multimeric proteins that are connected by disulfide bonds that form large supramolecular structures with high molecular weights (> 10 million Da) (Wieser, 2007). Pure gliadins are rather viscous when mixed with water, whereas pure glutenins develop a tough stretchable dough with low elasticity and high moduli. However, when these two kinds of protein are combined, the resulting gluten develops a characteristic viscoelastic texture, which has been attributed to the ability of the small gliadins to plasticize the large glutenins. The high hydrophobicity and large molecular weight of gluten lead to its low water solubility, which also

6.3 Ingredients for Formulating Plant-based Meat Analogs

Table 6.2 Overview of composition and functionality of common protein-rich ingredients in meat alternatives

Protein ingredient	Composition (%w/w)	Functionality	Application in meat analogues
Soy isolate (alkaline/acid precipitation treatment)	~90% protein	Good solubility, gelling and emulsification	Structuring process: Extrusion, shear cell, spinning, freeze structuring Role: Protein source, texture, binder, base for fat substitutes, emulsifier Products: Burger patties, minced meat, sausages
Soy isolate (additional heat treatment/toasted isolate)	~90% protein, denatured due to heat treatment	Decreased solubility, increased water holding capacity, good gelling	Structuring process: Extrusion, shear cell Role: Protein source, texture, binder, base for fat substitutes Products: Burger patties, minced meat, sausages
Soy concentrate	~70% protein	Good texturization properties	Process: Extrusion, Shear cell Role: Protein source, texture, binder Products: Burger patties, minced meat, sausages, muscle-type products
Soy milk (spray-dried powder)	>45% protein, ~30% fat	High solubility, good emulsification properties	Process: Freeze structuring Role: Emulsifier, texture Products: Tofu and yuba production
Soy flour/meal (defatted)	~43–56% protein, ~0.5–9% fat, ~3–7% crude fibre, >30% total carbohydrate	Water binding capacity and fat retention, native protein	Process: Extrusion Role: Texture, Binder Products: Burger patties, minced meat, sausages, muscle-type products
Wheat Gluten isolate	75–80% protein, 15–17% carbohydrates, 5–8% fat	Binding, Dough forming/Cross-linking capacity via S-S bridges, low solubility	Structuring process: Extrusion, shear cell Role: Adhesion, texture Products: Burger patties, muscle-type products

(continued)

Table 6.2 (continued)

Protein ingredient	Composition (%w/w)	Functionality	Application in meat analogues
Pea isolate	~85% protein	Water and fat binding, emulsification, and firm texture after thermal processing	Process: Extrusion, shear cell, spinning Role: Emulsifier, texture, Binder Products: Burger patties, minced meat, sausages, muscle-type products

Adopted from (Kyriakopoulou et al., 2021) under Creative Commons Attribution (CC BY 4.0) license (https://creativecommons.org/licenses/by/4.0/)

contributes to its unique textural attributes in meat analogs. Moreover, in the presence of water, there are strong hydrophobic, hydrogen, and disulfide bonds between the protein molecules that also contribute to these textural attributes. Indeed, although gluten is insoluble in water it can absorb around 225–350 % of its own weight of water (Kaushik et al., 2015; Zayas, 1997). Gluten proteins have been shown to produce anisotropic structures at high barrel temperatures during extrusion, which is useful for simulating the structure and texture of meat (Krintiras et al., 2014; Pietsch et al., 2017). In addition, the adhesive properties of gluten facilitate the binding of different ingredients, which help, for example, during the production of meat analog burgers.

6.3.2 Lipids

The incorporation of lipids (fats and oils) into meat analogs is important to achieve the desired physicochemical, sensory, and nutritional properties (Kyriakopoulou et al., 2021). Depending on the product, the fat content may range from less than 5 % to more than 20 % (Bohrer, 2019). It is sometimes difficult to accurately mimic the functional properties of animal fat tissue, especially their textural behavior. The fats from land animals often contain relatively high levels of saturated and monounsaturated fatty acids, which means they tend to be partially crystallized at refrigerator and ambient temperatures (Table 6.3). Consequently, the adipose tissue has some solid-like characteristics at these temperatures due to the formation of a 3D network of fat crystals. When this type of tissue is heated, the fat crystals melt and the lipid phase becomes liquid, leading to softening of the adipose tissue. Consequently, the melting and crystallization behavior of the fat phase contributes to the characteristic textural attributes of real meat products. It should be noted that certain kinds of meat (some pork products) and most fatty fish contain high levels of polyunsaturated fatty acids, which means they tend to have lower melting points and are more fluid like. It is therefore important to select a plant-based lipid phase

6.3 Ingredients for Formulating Plant-based Meat Analogs

Table 6.3 Median and median 95% confidence interval fatty acid (FA) composition (as % of total FAs) of commercial plant-based ($n=3$) and meat-based ($n=4$) burger patties

	Median	Meat-based 95%	Plant-based Median	95%
SFAs	48.8	45.6–53.4	52.2	40.5–61.9
MUFAs	45.7	38.2–50.6	32.3	16.1–41.3
PUFAs	4.9	3.9–10.5	20.1	15.4–23.0
n-3	0.64	0.4–0.9	3.6	0.3–4.0
n-6	3.9	3.2–8.4	15.8	11.7–22.3
n-6/n-3 ratio	7.3	5.3–9.5	3.5	3.2–85.0
CLA	0.55	0.45–0.79	0.044	0.04–0.05
cis-FAs	2.8	2.4–2.5	0.93	0.32–1.85
trans-FAs	0.13	0.06–0.18	0.079	0.004–0.099
SCFAs	0.18	0.14–0.40	7.2	5.3–8.9
MCFAs	35.9	35.1–37.6	41.9	32.2–49.7
LCFAs	64.5	63.5–65.1	50.9	41.4–62.5

Modified from (De Marchi et al., 2021) under CC BY 4.0 (http://creativecommons.org/licenses/by/4.0/)
SFAs saturated FAs, *MUFAs* monounsaturated FAs, *PUFAs* polyunsaturated FAs, *CLA* conjugated linoleic acid, *n*-3 Omega-3 FAs, *n*-6 Omega-6 FAs, *SCFAs* short-chain FAs, *MCFAs* medium-chain FAs, *LCFAs* long-chain FAs, *cis-FAs* cis stereoisomers of FAs, *trans-FAs* trans stereoisomers of FAs excluding CLA

that can accurately simulate the melting and crystallization behavior of the lipid phase in the meat or fish product being mimicked.

As described earlier, the lipids in muscle tissue may be located within different locations in and around the muscle: subcutaneous fat is the thick adipose tissue layer between the muscle and the skin; intermuscular fat is the tissue that is stored between different muscles; intramuscular fat (marbling) is located within the muscle tissue. Some parts of animals contain relatively high lipid levels that are stabilized and structured by a protein (connective tissue) matrix (Fig. 6.6). For instance, pork back fat contains around 80 % fat, with the remainder being water (17 %) and protein (< 3%) (Olsen et al., 2005). Plant-based meat analogs should therefore be designed to contain lipids that mimic the concentration, distribution, and behavior of animal fatty tissue (Table 6.4).

Most plant-derived lipids contain relatively high levels of unsaturated fatty acids and are therefore liquid at room temperature. As a result, they cannot mimic the ability of animal fats to form semi-solid textures that melt upon heating. For this reason, plant-derived lipids that do have relatively high melting points, such as coconut oil, cocoa butter, shea butter, and palm oil, are often used for this purpose (Herz et al., 2021; Wang et al., 2018). These solid fats are often blended with liquid oils (such as sunflower or canola oil) to obtain the desired solid fat content *versus* temperature profile (Chap. 2). When plant-derived solid fats (such as coconut or palm oil) are used to formulate plant-based meat analogs, the final products may have a higher saturated fatty acid content than the real meat products (Table 6.3).

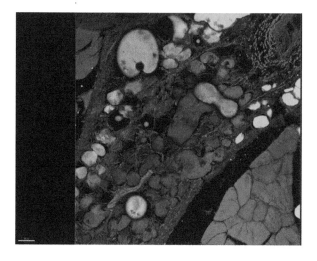

Fig. 6.6 Animal adipose tissue is a combination of fat cells (green) that are surrounded by connective tissue (red). Plant-based fat analogs aim to mimic this structure to obtain similar physicochemical and textural attributes. Microscopy image was kindly provided by Dominic Oppen (University of Hohenheim, Germany)

Table 6.4 Selected texture profile analysis values (50 % compression) of transglutaminase crosslinked emulsion gels (lipid phase 70 %) prepared with canola oil and partial replacement with hydrogenated canola oil to increase the solid fat content (SFC)

SFC (%)	Hardness (N)	Cohesiveness (–)	Springiness (mm)
0	2.3 ± 0.5[a]	0.5 ± 0.3[a]	7.1 ± 0.7[a]
5	4.1 ± 0.4[b]	0.5 ± 0.2[b]	7.4 ± 0.8[a]
10	5.3 ± 0.7[b]	0.5 ± 0.1[b]	7.1 ± 0.6[a]
30	13.9 ± 2.2[d]	0.1 ± 0.1[d]	2.8 ± 0.5[d]
Fatty tissue pork	215 ± 71	0.3 ± 0.2	4.1 ± 1.0

Modified from (Dreher et al., 2020)
Means ± standard error was determined by triplicate analysis. Values followed by different superscripts are significantly different within the same column ($P < 0.05$)

Moreover, the textural attributes obtained by plant-derived lipids are usually quite different from those produced by animal-derived ones. For instance, mixtures of high-melting plant-derived fats with liquid oil (e.g., 75 % solid fat, 25 % liquid oil) behave like a perfect elastoplastic material (Gonzalez-Gutierrez & Scanlon, 2018), which means they can deform elastically until they irreversibly break at a certain strain. Typically, a higher content of long-chain saturated fatty acids will result in higher hardness values and lower yield strains (material breakage), which can be different compared to animal fatty tissue. This is shown by the decreasing springiness (how much the material bounces back) and cohesiveness (change in the amount of work for a second compression) with increasing solid fat content for a plant-based fat mimetic based on a semisolid emulsion gel (Table 6.4). Thus, when mimicking the fat tissue of animals, it is important to achieve similar hardness values but also consider the elastic properties of the material.

Several approaches have been developed to mimic the properties of animal fat tissue in vegan and flexitarian meats, including their spatial distribution, microstructure, and textural attributes:

6.3 Ingredients for Formulating Plant-based Meat Analogs

Table 6.5 pH and $L^*a^*b^*$-values from 3 different commercial plant-based burger patties and four meat-based burger patties. Data is reported as Median values and median 95% confidence interval

	Meat-based burger		Plant-based burger	
	Median	95%	Median	95%
pH	5.48	5.28–5.70	5.81	5.58–7.29
L^*	44.9	42.4–48.6	48.0	39.9–48.9
a^*	19.8	17.0–20.9	16.8	15.6–17.5
b^*	14.5	13.6–15.9	11.2	9.6–11.8

Modified from (De Marchi et al., 2021) under CC BY 4.0 (http://creativecommons.org/licenses/by/4.0/)

- **Blending**: By mixing fats and oils from sources that differ in their solid fat content, the hardness, yield stress/strain, and melting behavior of the fat can be adjusted (Motamedzadegan et al., 2020; Piska et al., 2006). The blended fats can then be incorporated into the meat alternative matrices by emulsification (liquid oils, *e.g.*, sausage-type) or by mixing (solid fats, *e.g.*, burger patties).
- **Emulsification**: Oils can be directly emulsified during the production process, or they can be pre-emulsified and then added during a mixing stage. Emulsification is commonly used to produce sausage-type products, during which the proteins act as emulsifiers to facilitate the formation and stabilization of small oil droplets. The oil droplets are subsequently entrapped in the gelled matrix.
- **Marinating**: Whole-muscle-like products are marinated with oil, during which the oil penetrates the protein matrix by diffusion and capillary effects.
- **Injecting**: Oils can be directly injected into the protein matrix during extrusion at one of the earlier barrel sections or just before the die. However, high oil contents prevent the formation of anisotropic structures during extrusion (Kendler et al., 2021). Oils may also be injected after the formation of the protein matrix to create a marbling effect.
- **Gelling**: Oils can be gelled and incorporated into meat analog matrices, such as emulsified or raw-fermented product types (Fig. 6.7). The gelling of oil can be obtained by producing oleogels that are structured by ethylcellulose, or by forming emulsion gels from proteins that can be crosslinked by transglutaminase (Davidovich-Pinhas et al., 2015; Dreher et al., 2020).

Depending on the final product, different approaches can be used to incorporate fats and oils into the biopolymer matrix of meat analogs. For example, semi-solid lipid phases can be incorporated into a plant-based burger in the form of small particles consisting of partially solidified fats or structured lipids. These particles may be formed before they are added to the product, or they may be formed during the blending process. Similarly, liquid oils can be emulsified before or during blending with the biopolymer matrix.

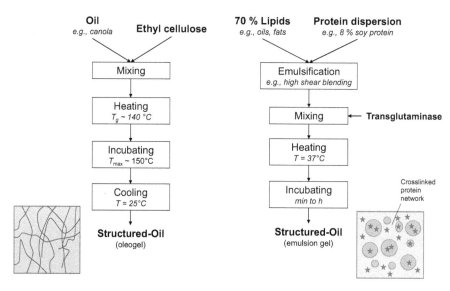

Fig. 6.7 Workflow for producing structured lipids that have semi-solid properties from liquid oils. Oleogels *(left)* can be produced by using ethyl cellulose as an oleogelator, which has a glass transition temperature T_g of around 140 °C. Ethyl cellulose entraps the oil upon cooling in a 3-dimensional polysaccharide network. Emulsion gels *(right)* can be obtained by employing crosslinked high internal phase emulsions in which the oil droplets and the continuous phase are interconnected by protein-protein crosslinks, which facilitate the formation of a network with or without solid fat crystals (shown in red)
Modified from (Dreher et al., 2020; Gravelle et al., 2012)

6.3.3 Binders

Biopolymers are often used as binders in meat analogs, such as plant-based patties, sausages, and nuggets. The main task of binders is to act as an adhesive agent between different ingredients or as a cohesive agent between similar ingredients and to increase the fluid holding properties (Kyriakopoulou et al., 2021). They, therefore, hold the different ingredients together within the meat-like solid matrix.

The efficacy of binders can be illustrated by examining the processes used to create meat analogs. Many of these products are produced by blending texturized proteins, lipids, water, and other functional ingredients. During the subsequent production steps, the texturized proteins are chopped into the desired particle size range. For example, during the preparation of plant-based burgers, the extruded proteins are chopped into smaller fragments to resemble the particles found in conventional minced meat. Subsequently, the particles need to be 'glued' together so that they become an integral part of the food matrix. Because the proteins have been processed using high temperature and shearing conditions, they often lose most of their cohesive properties because the bonds responsible for attractive protein-protein interactions (non-covalent and covalent bonds) have been depleted during processing. For instance, they may have undergone thermal denaturation and aggregation during processing, so the number of exposed non-polar and sulfhydryl groups on their surfaces are greatly reduced, which decreases their ability to form hydrophobic and disulfide bonds with their neighbors. Moreover, they have a reduced molecular mobility, which reduces their ability to encounter other proteins and prevents efficient binding. However, it has to be noted that it is not well understood how such interactions and bonds change during thermomechanical processing and current research tries to shed more light on this area.

Thus, a binding agent must be added that can interact with the different ingredients and hold them together, which is commonly some kind of hydrocolloid (Chap. 2). This could be a native protein that is able to unfold and aggregate upon heating. As a result, it can form a gel between the protein particles that binds them together, as well as contributing to the texture and fluid holding properties. In addition, several kinds of polysaccharides can be used as binding agents in plant-based meat analogs. Some of these polysaccharides exhibit adhesive and/or cohesive properties at room temperature but also during heating (*e.g.,* methylcellulose), which is advantageous for some applications. In general, binding agents tend to be high molecular weight, predominantly hydrophilic molecules, which have functional groups that enable them to interact with various kinds of ingredients in meat analogs.

The pores in meat analogs are often rather large (Fig. 6.8), which reduces their ability to retain fluids because the capillary pressure is relatively low (see $pc = \dfrac{2\gamma\theta}{r}$ (6.2). To overcome this problem, high molecular weight binders are added that penetrate into these large pores, where they form 3D networks with smaller pore sizes that can successfully retain the fluids. In addition, the presence of these binders increases the viscosity of the material inside the large pores, thereby reducing water mobility and decreasing fluid expulsion. Depending on the nature of the binder used, it may undergo a sol-gel-transition in response to changes in temperature

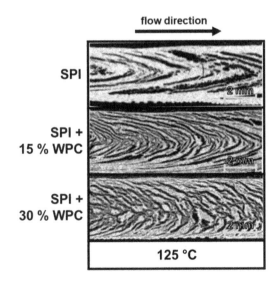

Fig. 6.8 X-ray tomography image of a freeze-dried soy protein extrudate that shows the porous structure of extruded protein that reduces its ability to retain water. SPI = soy protein isolate, WPC = whey protein concentrate **Adapted from (Wittek et al., 2021) under CC BY 4.0 (http://creativecommons.org/licenses/by/4.0/)**

(heat-set or cold-set) or salt addition (ion-set). Proteins, pectin, methylcellulose, starch, and carrageenan have all been used as binding agents. The characteristics of several binding agents commonly used in plant-based meat analogs are discussed further here:

- **Potato protein**: The main protein fraction in potato protein is patatin but it may also contain some protease inhibitors, with the relative amounts depending on the extraction method used. Potato protein unfolds and gels at relatively low temperatures (T_d = 60–70 °C) compared to many other plant proteins (Schmidt et al., 2019). Consequently, it more closely simulates the thermal denaturation behavior of the myofibrillar proteins in real meat.
- **Wheat gluten**: The two main protein fractions in wheat gluten are gliadin and glutenin. These proteins can bind to various other ingredients in the food matrix and create a viscoelastic network because they can form hydrogen bonds (glutamine), hydrophobic bonds (proline), and disulfide bonds (cysteine) with their neighbors.
- **Methylcellulose**: Methylcellulose is formed by chemically attaching methyl groups to cellulose, which is usually obtained from wood pulp or cotton. The addition of methyl groups makes the cellulose molecules more hydrophobic, but it also allows water molecules to penetrate between the cellulose chains and disrupt the hydrogen bonds that normally hold them together (which makes natural cellulose insoluble in water). Methylcellulose forms a reversible gel upon heating above 52 °C, which is due to the increase in the strength of the hydrophobic attractive forces between the methyl groups (Murray, 2009). It has both non-polar (methyl) and polar (hydroxyl) groups on its surfaces, which means that it can interact with both hydrophobic and hydrophilic substances within its environment, thereby contributing to its binding properties.
- **Pectin**: The term pectin refers to a group of polymers usually found between the cell walls of plants that have some similar molecular features, such as a high content of galacturonic acid residues and some other characteristic chemical groups. After extraction, the functional attributes of pectin ingredients can be modified by using chemical or enzymatic methods to alter the degree of methylation. High methoxy pectin can form gels under acidic conditions and high sugar contents. In contrast, low methoxy pectin can form gels in the presence of calcium ions due to the ability of the cationic divalent ions to link together two anionic carboxyl groups on different molecules. The attachment of amidated galacturonic residues to pectin molecules can be used to tune their calcium sensitivity. Some kinds of pectin also exhibit surface activity because they contain non-polar side groups (such as methyl groups and ferulic acid) (Bindereif et al., 2021).
- **Fiber**: Non-digestible plant materials contain various components that can act as binders or enhance the fluid holding properties of meat analogs. For example, citrus fiber, which mainly contains pectin and cellulose, increases the water retention in meat products, so could also be used for this purpose in meat analogs (Powell et al., 2019). Some of the polysaccharides in these crude fiber ingredients may also be able to form gels and bind different ingredients together, such as pectin.

- **Modified starch**: Starch is comprised of two homo-polysaccharides that consist of glucose units linked together by glycosidic bonds: amylose (linear, α-1–4 bond) and amylopectin (branched, α-1–4 and α-1–6 bonds). In nature, starch is usually present as small granules, which can absorb water and form gels upon heating (gelatinization) and upon cooling (retrogradation). Some starches, such as those from tapioca and potato, have pasting and gelatinization temperatures similar to the denaturation temperatures of meat proteins, *i.e.*, around 60 to 70 °C (Taggart & Mitchell, 2009). Native starches can be modified using various physical, chemical, and enzymatic methods to alter their functional properties (Klemaszewski et al., 2016; Taggart & Mitchell, 2009). Acid hydrolysis can be used to decrease the paste viscosity during gelatinization and increase the gel strength upon retrogradation. Octenyl succinate derivatization can be used to attach non-polar side groups to the polar starch molecules, which reduces oiling-off effects. Crosslinking increases the heat and shear stability of starch granules during processing and cooking.
- **Carrageenan**: Carrageenan-based ingredients contain linear sulfated polysaccharides consisting of alternating 3-linked β-D-galactopyranose and 4-linked α-galactopyranose or 3,6-anhydro-α-galactopyranose units. A number of different carrageenan ingredients are available with different molecular and functional attributes: κ-, ι-, and λ-carrageenan contain one, two, and three sulfate groups per disaccharide unit, respectively. The κ- and ι-carrageenan ingredients can form thermo-reversible gels when they are heated and then cooled below 40 to 70 °C in the presence of potassium or calcium ions, respectively. In contrast, λ-carrageenan does not gel, but it can still be used as a thickener and binding agent.
- **Xanthan gum**: Xanthan gum is obtained from the bacteria *Xanthomonas campestris* using a microbial fermentation process. Molecularly, it consists of a β-D-glucose backbone with anionic trisaccharide branches attached. It does not undergo a coil-helix transition but exhibits high thickening properties over a broad pH-range. When incorporated into food matrices, it may also be able to bind water by forming a 3D network with small pores.
- **Transglutaminase**: Transglutaminase is a food-grade crosslinking enzyme normally obtained using microbial fermentation methods. It crosslinks the γ-carboxamide group on glutamine to the ε-amine group on lysine, which leads to the formation of isopeptide bonds between proteins and the release of ammonia. The protein molecules or particles within a food matrix can therefore be covalently crosslinked by adding sufficient transglutaminase and then incubating under optimum pH (5–8) and temperature (25–50 °C) conditions. The enzyme can then be deactivated by heating the food matrix above 75 °C. This enzyme has been widely used as a binder and gelling agent in plant-based meat analogs.

In summary, the binding agents have a diverse range of functional groups on their surfaces, including polar, non-polar, charged, and chemically reactive groups, which means they can be involved in a diverse range of molecular interactions with other molecules. For instance, polar groups can be involved in hydrogen bonding, non-polar groups in hydrophobic attraction, charged groups in electrostatic attraction, and sulfhydryl groups in disulfide bonds. The relatively high molecular weight of some binding agents facilitates their ability to bind ingredients that are spatially separated from each other (a few to tens of nanometers). Overall, these molecular features are largely responsible for the adhesive and cohesive as well as liquid holding properties of these binding agents.

6.3.4 Coloring Agents

Real meat and meat products may exhibit a variety of different colors depending on the product category, including red, pink, beige, and brownish tones. Raw meat and some meat products (*e.g.*, raw ham and salami) have a reddish-pink color. These meat products retain their color during heating because of the use of nitrites as a curing salt. In contrast, raw meat undergoes a color change from red to brown during heating because of the degradation of myoglobin at elevated temperatures. Other meat products, like bologna-type sausages, have a light beige-pinkish color. The structuring ingredients used to produce meat analogs, such as proteins, polysaccharides, and lipids, do not tend to exhibit these meat-like colors. Consequently, coloring agents are required to mimic the colors of meat and meat products (Chap. 2). Most producers of plant-based meat products prefer to use coloring agents that are natural in origin. For this reason, we give a brief overview of these kinds of colors in this section.

Heat-stable Red Color Carotenoids (*e.g.*, lycopene and canthaxanthin) and iron oxides can be used to produce a heat-stable red color in meat analogs (Kyriakopoulou et al., 2021). These compounds are insoluble in water, which means they must either be solubilized in an oil phase (carotenoids) or dispersed as small particles in the formulation (iron oxides).

- *Carotenoids*: Purified carotenoids extracted from plants (such as lycopene from tomatoes and β-carotene from carrots), as well as mixed carotenoids extracted from plants (such as paprika oleoresin or annatto extract), have been shown to be reasonably stable against heat treatment even up to 90–100 °C (Bolognesi & Garcia, 2018; Gheonea et al., 2020). This characteristic is important in products that should not turn from red to brown during cooking or processing, *e.g.*, salami analogs. However, the heat stability of carotenoids depends on the formulation and the environmental conditions (*e.g.*, pH and presence of transition metals). Moreover, degradation is usually accelerated in the presence of light and oxygen, which needs to be considered during the packaging and storage of the product (Boon et al., 2009). Carotenoid degradation can also be inhibited by adding antioxidants or chelating agents to the formulation.
- *Iron oxide red:* Iron oxide red (E 172, Fe_2O_3) is produced from iron(II) sulfate using the Penniman–Zoph method (EFSA, 2016). These pigments are relatively

stable over a wide range of environmental conditions, which facilitates their application in meat analogs. However, they are insoluble in water, which means they must be used in the form of small particles that can be evenly dispersed throughout the food matrix.

Commonly, different pigments are combined to obtain a desired color profile. For example, a mixture of iron oxide red and β-carotene results in a color that resembles nitrosomyoglobin, which is the predominant color in cured meat products, like salami.

Heat-labile Red Color In some plant-based meat analogs it is desirable to use a coloring agent that changes from red to brown during heating, to mimic the cooking behavior of many real types of meat. Commonly, red beet extract or juice is used for this purpose, which contains the natural pigment betalain or 'betanin'. This pigment is susceptible to degradation during cooking and can therefore be used as a heat-labile red coloring agent in meat analogs (Cejudo-Bastante et al., 2016; Rolan et al., 2008). It is relatively stable at most pH values found in meat analogs (pH 3-7), which means that it is not very susceptible to fading during storage prior to cooking.

Leghemoglobin is another coloring agent that has been utilized in commercial meat analogs, such as those produced by 'Impossible Foods'. This compound is a symbiotic hemoglobin found in the root nodules of leguminous plants, such as soy. The 3D structure and function of this protein are very similar to hemoglobin found in animals, although the amino acid sequence does differ significantly (Fraser et al., 2018). Leghemoglobin contains a porphyrin ring that carries iron, which is also known as heme (Kumar et al., 2015). In nature, leghemoglobin binds oxygen and can control its concentration in the root nodule, which allows nitrogen-fixing bacteria to thrive. In practice, it is not economically feasible to extract enough leghemoglobin from soy root nodules for commercial purposes. For this reason, it is usually produced using genetically modified yeast cells using fermentation processes (Brown et al., 2019; Fraser et al., 2018). As an example, the yeast *Pichia pastoris* has been reported to produce 65 % of its total protein as leghemoglobin (Fraser et al., 2018). Typically, leghemoglobin is used in concentrations of around 0.8 % in meat analogs. It thermally denatures around 64 °C, which is fairly similar to the temperatures (70–74 °C) at which animal myoglobin denatures (FDA, 2017). Moreover, it undergoes a similar color change from red to brown as animal myoglobin during cooking (Fig. 6.9). In addition, it is generally recognized as safe (GRAS) and so accepted for utilization as a food ingredient within the United States.

Beige-red Color Many types of sausages are primarily beige with a slight red/pink appearance, e.g., Bologna or Frankfurter sausages. The beige color in these products is obtained by emulsifying the product in a bowl chopper, which results in emulsified fat globules that have particle sizes ranging from a few to several thousand micrometers (Youssef & Barbut, 2010). These droplets scatter light waves, which leads to a creamy beige appearance. The slight red to pink color of these products is the result of nitroso-myoglobin that selectively reflects light waves in the red region of the electromagnetic spectrum. Nitrite curing salts are added to these

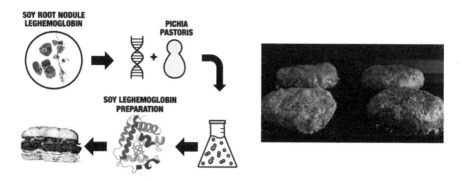

Fig. 6.9 Leghemoglobin is produced by leguminous plants to optimize the oxygen concentration in the roots. The genes that encode the 145 amino acids are incorporated into the yeast *Pichia pastoris*, which then produces leghemoglobin. The heme-containing proteins enable plant-based meats to turn from a reddish to a brownish color during cooking (right picture)
Modified from (Jin et al., 2018) under CC BY 4.0 (https://creativecommons.org/licenses/by/4.0/). Plant-based burger patty image was kindly provided by Impossible Foods (Redwood City, USA)

products, which are converted to nitric oxide due to their interactions with the myoglobin in meat, thereby leading to a stable red/pink color.

In meat analogs, a beige color can also be obtained by incorporating emulsified oils into the plant protein matrix. Ideally, the oils used should be free of pigments, but a slightly yellow-colored oil can contribute to the desired beige tone. This yellowish color can be obtained using coloring agents like curcumin, turmeric, and caramel colors (Kyriakopoulou et al., 2021). To obtain a slightly reddish tone, low levels of carotenoids or iron oxide red particles can be added to the formulation, as described earlier.

6.3.5 Flavoring Agents

The flavor of cooked meat is composed of a complex set of compounds that originate from thiamin, lipids, proteins, sugars, and nucleotides. Most of these compounds are generated during cooking by the different kinds of chemical reactions as described earlier (Sect. 6.2.5). Ideally, it would be advantageous for plant-based meat analogs to contain similar volatile and non-volatile compounds. This is challenging because many of the plant-derived ingredients used to formulate meat analogs do not naturally contain meat-like flavors. For this reason, flavor companies have been creating vegan and vegetarian flavoring agents that simulate the characteristic taste and odor of products such as beef, pork, chicken, turkey, or fish. Based on knowledge of the types of flavor molecules present in specific meat products, the flavor companies may be able to identify equivalent flavors in plant sources.

Alternatively, they may be able to create them from plant-derived ingredients using carefully controlled chemical reactions or fermentation processes (Chap. 2). Typically, the manufacturers of plant-based foods work closely with flavor companies to identify a suitable vegan or vegetarian flavoring agent that can be used in the products to give the required meaty flavors.

Another major challenge when formulating meat analogs is that many plant proteins carry off-flavors, such as aldehydes, alcohols, ketones, acids, pyrazines, sulfur compounds, saponins, phenolic compounds, and sometimes alkaloids (Roland et al., 2017). Some of these compounds are generated due to the oxidation of lipids catalyzed by endogenous enzymes, such as the lipoxygenase found in legumes (Duque-Estrada et al., 2020). The peptides generated by hydrolysis of plant proteins during their processing and extraction can also contribute to their off-flavor. For example, 14 short bitter peptides were identified in pea protein isolates, which contributed to the perceived off-flavor of pea protein products (Cosson et al., 2022). Consequently, it is important to avoid, remove, or mask these off-flavors. For this reason, many food ingredient manufacturers are improving the flavor profiles of their plant-based protein ingredients by breeding new plant varieties, removing off-flavor precursors, deactivating lipoxygenases, and utilizing fermentation strategies (Kyriakopoulou et al., 2021).

Moreover, it has been reported that certain volatile compounds can interact with plant proteins, which decreases their volatility and therefore their perceived flavor intensity (Wang & Arntfield, 2017). Most flavoring compounds attach reversibly to a protein *via* hydrophobic interactions but also hydrogen bonds, electrostatic interactions, van der Waals forces, and covalent bonds may be responsible for flavor retention by some proteins. For example, ketones and aldehydes are known to interact with proteins by hydrophobic interactions, whereas vanillin can also form covalent bonds (Wang & Arntfield, 2017). This leads to different amounts of volatile compounds when meat analogs are formulated with different proteins. For example, a study showed that higher retention rates of volatile substances were obtained when meat analogs were formulated with higher wheat gluten and lower moisture contents (Guo et al., 2020). Moreover, analyses of the volatile compounds in low- and high-moisture extruded pea proteins showed significant differences. For example, hexanal, a characteristic green odorant, was reduced up to six-fold compared to the initial pea protein powder when pea proteins were extruded at high moisture conditions using a cooling dye. Moreover, high-moisture extrusion retains more volatile compounds compared to low-moisture extrusion (Ebert et al., 2022).

The removal of off-flavors can enhance the sensorial acceptance of plant-based ingredients in meat analogs, but it does not generate a meaty flavor. Consequently, flavor companies are developing flavoring agents from plant-derived ingredients that can provide meaty flavors, such as beef-, pork-, chicken-, turkey-, or fish-like flavors. These are often produced by using controlled chemical reactions (such as the Maillard, caramelization, or enzymatic reactions) or by fermentation processes using different microorganisms (such as yeasts, molds, or bacteria) using plant proteins and other substances as substrates (Chap. 2). To achieve a meaty aroma and taste, these flavoring agents or flavor precursors are added to plant-based meat analogs or are produced in the product during cooking (Kyriakopoulou et al., 2021). Different strategies are employed to develop these flavors. Several of these strategies are highlighted here:

- **Flavor precursors**: Flavor precursors are blended with heme-iron-containing proteins (*e.g.*, leghemoglobin), which catalyzes the formation of meat-like flavors during cooking (Fraser et al., 2017). Common flavor precursors used for this purpose include amino acids (*e.g.*, cysteine, glutamic acid, or lysine), unsaturated lipids (*e.g.*, oleic acid or linoleic acid), thiamin, lactic acid, sugars (*e.g.*, glucose or ribose), and nucleotides (inosine monophosphate or guanosine monophosphate). As an example, a patent described the use of cysteine (10 mM), glutamic acid (10 mM), glucose (20 mM), thiamine (1 mM), and 1 % leghemoglobin that were added to a plant-based burger patty that was cooked at 150 °C for 5 min (Fraser et al., 2017). As a result, the patty developed a 'beefy' and 'bloody' flavor that was caused by compounds like 5-thiazoleethanol, 4-methyl-furan, 3,3'-dithio bis 2-methyl-thiazole, and 4-methylthiazole. In contrast, 'beefy' and 'savory' flavors were much less developed without the addition of leghemoglobin. Moreover, the addition of lysine increased the roasted browned notes in similar formulations as described in this patent (Fraser et al., 2017).
- **Maillard reaction:** One of the most important reactions responsible for the generation of flavors during cooking is the Maillard reaction. For this reaction to occur in foods, a reducing sugar and amino acid need to be present at an optimum water activity ($a_w \approx 0.7$). Previous studies have shown that heating a cysteine/ribose mixture at 145 °C for 20 min results in the production of various chemical compounds that provide meat-like flavors, such as 2-furfurylthiol, 2-methyl-3-furanthiol, 2-thenyl mercaptan, and ethyl mercaptan (Hofmann & Schieberle, 1995).
- **Hydrolysis:** Hydrolyzing plant proteins is one of the oldest techniques for generating meat-like flavors in plant-based materials (*e.g.*, soy sauce). Typically, microbial fermentation, enzymes, or acids are used to initiate protein hydrolysis, which results in free amino acids and the generation of various volatile compounds *via* Strecker degradation reactions or decomposition of sulfur-containing amino acids (Aaslyng et al., 1998). Moreover, glutamic acid may be produced, which provides an umami taste that is characteristic in many meat products (Jo & Lee, 2008).

Finally, it should be noted that ingredient interactions and processing conditions also influence the nature of the flavors produced during cooking, which should be considered when formulating plant-based meat analogs with different compositions and processing requirements.

6.4 Processing Methods

After an appropriate combination of ingredients is selected, they must be converted into a meat analog using a combination of food processing operations. The type and sequence of these processing operations depend on the nature of the product being manufactured. Ultimately, the manufacturing process should be designed to reliably produce a high-quality product that is affordable, safe, and delicious. In this section, we discuss some of the most common processing operations used to create plant-based meat analogs.

To classify such analogs, it is helpful to understand how real meat preparations and meat products have been defined in official regulations (Regulation (EC) No 852, 2004):

- *'Meat preparations' means fresh meat, including meat that has been reduced to fragments, which has had foodstuffs, seasonings or additives added to it or which has undergone processes insufficient to modify the internal muscle fibre structure of the meat and thus to eliminate the characteristics of fresh meat.'*
- *'Meat products' means processed products resulting from the processing of meat or from the further processing of such processed products, so that the cut surface shows that the product no longer has the characteristics of fresh meat.'*

Based on these definitions, whole muscle, ground meat, burger, nugget, and strip analogs can be considered as 'plant-based meat preparations', whereas sausage analogs can be considered as 'plant-based meat products'. The processing operations used to create such meat-like structures and textures can be categorized as either bottom-up or top-down methods depending on their underlying principles (Dekkers et al., 2018). Bottom-up methods produce meat-like structures by creating anisotropic structural elements that are then combined into macroscopic materials, which include wet spinning and electrospinning methods. Conversely, top-down methods produce meat-like structures by applying forces to macroscopic materials to induce structure formation, which includes extrusion, shear-cell, and freeze-structuring technologies.

In the remainder of this section, we focus on the extrusion and shear cell technologies because they are currently the most economically viable techniques for the large-scale production of plant-based meat analogs. Moreover, we describe the processes used to obtain processed products from these texturized proteins.

6.4.1 Protein Texturization

Extrusion is commonly used to produce anisotropic raw materials from globular plant proteins using a combination of thermal and mechanical processing. A more in-depth discussion on extrusion can be found in Chap. 3. Typically, a co-rotating twin-screw extruder is used that either results in a low- or high-moisture texturized protein being produced (Grossmann & Weiss, 2021). Low-moisture (< 50 %) texturized protein is produced using a short-shaping die that involves a pressure drop and release of moisture at the die. In contrast, high-moisture (50–70 %) texturized protein is produced using a cooling die that prevents expansion of the product after the barrel section, cools down the protein suspension, and leads to the formation of a layered and anisotropic meat-like structure (Fig. 6.10).

Commonly, low-moisture texturized protein is used as a meat extender (it can be used in plant-based meats as well), whereas high-moisture texturized protein is used to formulate plant-based meats, such as burger, sausage, or nugget analogs (Fellows, 2017). In most commercial high-moisture extruders, the product has a rectangular shape after exiting the cooling die (Pietsch et al., 2017). The textured protein is then chopped or minced to obtain particles or chunks in the desired size range (Fig. 6.11).

One disadvantage of extrusion is that the thickness of the textured protein sample that can be produced is limited by the cooling efficiency of the die. Consequently, it is challenging to produce plant-based analogs similar to whole muscle tissue, like beef steaks, pork chops, or chicken breasts. This problem can be addressed by using the shear cell technology. In this batch process, a protein suspension is sheared in a cone-in-cone or Couette-based geometry with simultaneous heating (Krintiras et al., 2016). The proteins become aligned along the shear gradient and form anisotropic structures. Because the height of the gap can be adjusted, thicker plant-based protein pieces can be produced, which more realistically mimic that of a whole-cut from real meat or a fish product (Kyriakopoulou et al., 2021). For example,

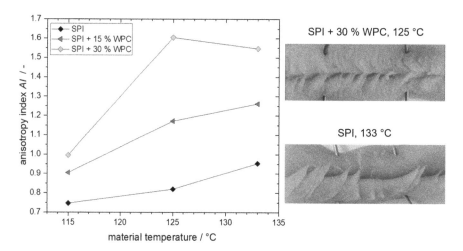

Fig. 6.10 Plant proteins are extruded to form anisotropic structures similar to meat. The formation of anisotropic structures can be analyzed by tensile tests in perpendicular and parallel directions to obtain the anisotropy index. Shown here is the formation of anisotropic structures at different soy protein isolate (SPI) to whey protein concentrate (WPC) ratios produced by high-moisture extrusion with an L/D of 40:1 at a water content of 57%
Modified from (Wittek et al., 2021) under CC BY 4.0 (https://creativecommons.org/licenses/by/4.0/)

6.4 Processing Methods

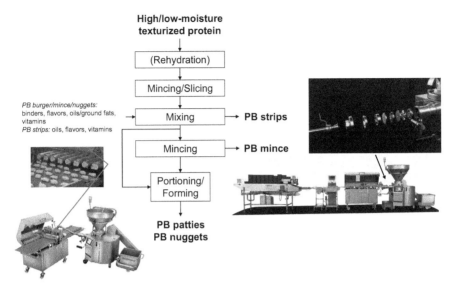

Fig. 6.11 Plant-based (PB) burger, nugget, and strip analogs can be produced from high-moisture extruded plant protein that is sliced or minced to obtain the desired shape and particle size (low-moisture extruded protein requires a rehydration step). The small pieces formed are mixed with any other functional ingredients during this process. For plant-based burgers and nuggets, the minced plant-based texturized protein is formed to obtain the typical burger/nugget shape. Fine-grinding can be introduced to skip the second grinding step for PB patties and nuggets. Pictures show PB nugget and mince production lines after the mixing step using a vacuum filler system VF 848/838 S. Images were kindly provided by Handtmann GmbH (Biberach an der Riß, Germany).

length-to-height dimensions of 596 × 332 mm have been reported for a Couette shear cell system, which is considerably higher than the dimensions that can be achieved using high-moisture extrusion (Krintiras et al., 2016; Palanisamy et al., 2019). However, the overall post-processing scheme remains the same (Fig. 6.11).

6.4.2 Plant-based Meat Preparations

The starting point in the production of plant-based ground meat, burgers, nuggets, or strips is low- or high-moisture extruded protein. The main unit operations involved in producing these products are chopping, mixing, forming, and portioning (Fig. 6.11). The equipment needed to produce these types of products is similar to real meat preparations: grinders, mixers, fillers, cutters, conveyor belts, packaging machines, and others (see Chap. 3). The most commonly produced products are plant-based strips, ground meat, burger patties, and nuggets:

- Plant-based strips are the most straightforward to produce. These products are usually created from texturized protein (commonly obtained using high-moisture extrusion) that is sliced to achieve the desired dimensions. Fibers (such as citrus fiber) are often added to these products during extrusion to enhance their water holding capacity during heating (Table 6.1). In addition to the proteins and fibers, the slices may also contain other functional ingredients, such as lipids, vitamins, flavors, and colors.

- Plant-based ground meat is usually produced by mincing texturized protein into a dough-like batter and then incorporating the other functional ingredients (e.g., fats). The batter is transferred into a vacuum filler with an inline grinding system with hole plates to obtain the final particle size and overall shape (see Chap. 3). The mince is portioned by cutting, transferred into trays, and sealed for cold storage.
- For plant-based burgers and nuggets, the minced textured protein is often mixed with soluble polysaccharides (*e.g.,* methylcellulose or starch) and proteins (*e.g.,* potato protein) to bind the protein particles together and simulate the cooking behavior of real meat. Again, various other kinds of functional ingredients can be incorporated during the mixing process. Depending on the particle size obtained during the initial mincing step, the batter might be minced again after mixing or is directly transferred into a vacuum filler that transports it to the portioning and forming machine (Fig. 6.11). Subsequently, the product is portioned and formed by pushing it through a nozzle of the desired shape (nuggets) and then sliced using a mechanical cutting system to obtain the required dimensions.

6.4.3 Plant-based Meat Products

Plant-based sausages are becoming increasingly popular alternatives to real meat sausages. Currently, most of the products on the market resemble either raw fermented sausages (*e.g.,* salami) or emulsified sausages (*e.g.,* Bologna). In this section, we review the main production principles of these two types of sausage and how their plant-based counterparts can be produced.

Fermented Sausages Salami is a popular raw fermented sausage that is produced using several operations: (*i*) chopping raw lean meat and animal fat to produce small fat and muscle pieces; (*ii*) adding curing salts, spices, and starter cultures (*e.g.,* lactic acid bacteria and catalase-positive cocci); and, (*iii*) filling the mass into water vapor permeable cases. The sausages are then ripened and dried until they reach a pH below 5.2 and a water activity below 0.91. The resulting product is shelf-stable in its raw state when stored at room temperature.

Plant-based salami analogs are produced using different principles. First, the product is thermally processed to ensure microbial safety because plant-derived ingredients can contain appreciable amounts of undesirable microorganisms (Filho et al., 2005). This heating process may also facilitate the binding of the different ingredients used to formulate the product. Second, salami analogs are not fermented with starter cultures, but the pH is lowered by adding glucono-delta-lactone (GDL).

6.4 Processing Methods

GDL is primarily used because the microorganisms used for the fermentation of meat have not yet been adopted for the fermentation of plant-derived ingredients.

An example of a method to produce a plant-based salami analog is shown in Fig. 6.12. Initially, a textured plant protein is combined with a solidified fat phase (Fig. 6.7) and then chopped in a bowl chopper to achieve the desired particle size. The textured plant protein can be pre-heated to decrease the microbial load prior to chopping. The chopped lipids and proteins are then blended with GDL, gluten, water, salt, spices, and pigments. The GDL promotes a decrease in pH, which enhances the shelf life and provides a desirable acidic taste. The gluten contributes to the texture and binds the different components together. The salt and spices provide a desirable flavor profile. The pigments are designed to simulate the reddish-pink color of salami. After mixing, the mass is filled into water vapor-permeable cellulose casings and heat-treated to ensure product safety and to increase the binding effectiveness of the gluten. The heat-treated product can then be smoked and dried, if desired. A product produced using these processing steps was reported to have a final pH of around 5.85 and a water activity of around 0.93 (Dreher et al., 2021). These values are both higher than those found in real salami. The higher pH-value may result in a different taste perception by consumers. The heat treatment used to produce the plant-based salami analog should ensure that it is safe to consume, even though the pH and water activity are relatively high. However, the heating step may promote oiling off, which could lead to undesirable changes in the appearance and texture of the product (Dreher et al., 2021).

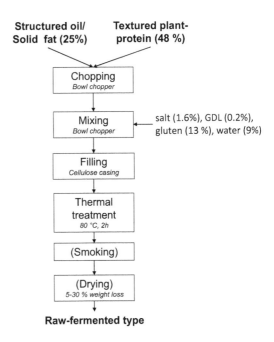

Fig. 6.12 Processing steps for manufacturing a plant-based salami analog. The product might be smoked or dried to enhance its shelf life and add more flavor
Modified from (Dreher et al., 2021)

Emulsified Sausages Bologna-type sausages are also a popular meat product, which is categorized as an emulsified sausage. These types of sausage are produced by grinding meat and fat, followed by chopping both parts in a bowl chopper to form a uniform emulsified batter. In this case, the fat particles are reduced in their size and finely dispersed in the meat matrix, in contrast to the salami-type sausages discussed earlier (there are also some sausage types with larger particles present, *e.g.*, coarse meatloaves). After the addition of nitrite curing salts, phosphates, ice, and spices, the meat batter is further chopped, filled into casings, and heat-treated. The core temperature of the sausages reaches around 75 °C during this process, which induces the formation of a thermally irreversible heat-set gel because of protein denaturation and aggregation. In addition, the heat treatment is needed to ensure the microbial safety of these products because they are not fermented or dried. In short, this process mainly involves preparing a homogenous protein matrix containing small fat particles using mechanical processing and then treating the batter with a heating step.

Plant-based analogs of emulsified sausages can be prepared using similar processing steps. The plant proteins used need to fulfill several functions: emulsify the added oils, retain fluids, and bind the different ingredients together. An example of a process that can be used to produce this kind of sausage is summarized in Fig. 6.13.

This process mainly involves mixing and hydrating the various ingredients, as well as transforming the oils into finely dispersed droplets. Binders are added to increase the water-binding and obtain a heat-stable product with textural attributes like animal-based emulsified sausages. Methylcellulose (~1.5 %), starches (~2.5 %), gluten, or other hydrocolloids are incorporated to provide desirable textural attributes, as well as for water- and oil-binding (Cavallini et al., 2006).

Ideally, a plant protein with a denaturation temperature close to that of myofibrillar meat proteins should be used in emulsified-type sausage analogs, as this allows for similar processing conditions to be used. However, most common plant proteins have higher denaturation and gelling temperatures than meat proteins. For example, the proteins in soy denature between about 80 and 93 °C, and those in pea denature between about 75 and 79 °C, which is considerably higher than myosin in real meat, which denatures between about 40 and 60 °C (McClements & Grossmann, 2021). Thus, higher cooking temperatures are usually needed to form a viscoelastic gel in plant-based emulsified sausages.

6.5 Key Properties

Plant-based meat analogs are formulated to mimic the properties of the real meat products they are designed to replace, such as sausages, burgers, or nuggets. Ideally, the meat analogs should look, feel, taste, and behave like real meat products so that consumers can simply substitute one for the other. Consequently, there is a need to understand how to assemble plant-derived ingredients, which are very different in

6.5 Key Properties

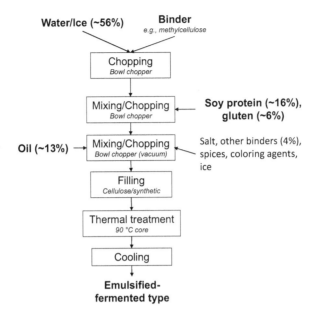

Fig. 6.13 Process for creating a plant-based emulsified sausage analog. Commonly, plant protein isolates or concentrates (pea, soy) are used as the main ingredient. Gluten might be added to increase the binding properties. Textured plant protein from extrusion processes can be used to increase the elastic properties of the product. In some instances, it might be useful to hydrate the protein before the chopping operation. Plant oils such as sunflower or canola oil are added and emulsified in the bowl chopper followed by the addition of other ingredients such as salt, binders, and coloring agents. The thermal treatment denatures the protein and solidifies the product into a 3D viscoelastic network. Adopted from patented process (Cavallini et al., 2006)

their molecular characteristics than animal-derived ones, into products with meat-like characteristics. In this section, we, therefore, discuss the most important physicochemical, sensory, and nutritional attributes of plant-based meat analogs and compare these to real meat products.

6.5.1 Color

The appearance of meat analogs is the first sensory impression that consumers use to assess their quality and desirability. The color of meat analogs should therefore be designed to accurately mimic the appearances of specific meat-based products. For example, plant-based Bologna-type sausages should appear beige/pink, burgers should appear brownish after cooking, and salmon should appear a pinky/orange color. In some cases, a product should change from one color to another during cooking. For instance, a ground beef analog is expected to turn from reddish/pink to brown upon heating. These temperature-induced changes may be caused by various physical or chemical processes, such as protein denaturation, Maillard reaction, oxidation, and condensation reactions.

Chicken-like analogs have a light beige color that can sometimes be mimicked without the need for added pigments. Many plant protein ingredients naturally have a light beige or brown color, which can simulate the color of real chicken. During extrusion, however, the brown color of proteins can increase because of chemical reactions promoted by the relatively high temperatures employed, such as the Maillard or caramelization reactions (Samard & Ryu, 2019). These color changes should therefore be considered when selecting appropriate ingredients and extrusion conditions. For red meat analogs, such as plant-based beef products, it is usually necessary to use coloring agents that are heat sensitive and/or change color during cooking (see earlier). The appearance of these products is also influenced by the presence of any reducing sugars (such as glucose or fructose) that can participate in the Maillard reaction, thereby generating a brownish color. The various kinds of coloring agents that can be used to formulate plant-based meat analogs are discussed in more detail in Sect. 6.3.4 and Chap. 2.

A comparison of the color (L^*, a^*, b^*) of commercially available plant-based burgers and real burgers is shown in Table 6.5. In general, the colors of these two kinds of products are quite similar, with intermediate lightness (L^* around 45 to 48) and moderate redness (a^* around + 17 to + 20), and yellowness (b^* around + 11 to + 15) values. This shows that it is possible to accurately mimic the color appearance in such burger patties but it has to be mentioned that it is still difficult to mimic the color change during cooking.

6.5.2 Texture

Ideally, the textural characteristics of meat analogs should also be designed to mimic those of real meat products. The texture of plant-based meat analogs is mainly influenced by the nature of the ingredients and processes used to create them. The textural attributes of plant-based burgers and real beef burgers measured using a texture analyzer are compared in Table 6.6. Although there are some

Table 6.6 Comparison of textural properties obtained by texture profile analysis from different commercial plant-based (PB) and animal-based burger patties

	PB-patty 1	PB-patty 2	PB-patty 3	Beef patty
Hardness (g)	1300 ± 250	1500 ± 180	270 ± 21	2400 ± 450
Adhesiveness (g.sec)	−0.34 ± 0.27	−1.2 ± 0.65	−1.6 ± 0.7	−0.47 ± 0.19
Resilience (%)	20 ± 1.6	16 ± 1.0	5.8 ± 0.68	24 ± 1.3
Cohesion	0.52 ± 0.03	0.46 ± 0.02	0.21 ± 0.02	0.61 ± 0.02
Springiness (%)	79 ± 5.0	64 ± 3.2	32 ± 3.8	88 ± 2.8
Chewiness	530 ± 120	450 ± 62	18 ± 4.0	1300 ± 250

Modified from (Zhou et al., 2022)

similarities, the burger analogs tended to be softer, less elastic, and less chewy than the real meat ones, which would be expected to alter their perceived mouthfeel during oral processing. These results suggest that more research is required to create textures in plant-based burgers that more closely match those in real burgers. Other studies have also shown that the cutting strengths of meat analogs in both the transverse and longitudinal directions are lower than those of real meat (Samard & Ryu, 2019). This study also highlights the challenge of accurately simulating the anisotropic structure and texture of real meat, which can mainly be attributed to the complex hierarchical structure of the muscle fibers, connective tissue, and adipose tissue in animal flesh, as well as changes that occur after the slaughter of the animal (Bhat et al., 2018).

The textural and sensory attributes of meat analogs can be controlled by adding binders, which increase the hardness and juiciness by binding different ingredients together and by enhancing the water holding capacity (Kyriakopoulou et al., 2021). Moreover, the inclusion of proteins with strong adhesive properties, such as gluten, can facilitate the formation of more meat-like structures and textures (Fiorentini et al., 2020). Other researchers have explored the impact of including κ-carrageen, konjac mannan, or xanthan gum in plant-based sausages made from soy protein isolate. The addition of relatively low concentrations of κ-carrageen (0.3-0.6 %) or konjac mannan (0.6 %) improved the sensory acceptance values but the addition of xanthan gum did not (Majzoobi et al., 2017). These authors also showed that the addition of κ-carrageenan or konjac mannan increased the water holding capacity of the sausage analogs. In another study, the impact of the moisture content and temperature on the sensory attributes of meat analogs were investigated (Lin et al., 2002). The water content (60 %, 65 %, or 70 %) was found to have a major impact on the sensory attributes of the product, whereas the extrusion temperature used had a lower effect. For example, products with lower moisture contents were tougher, chewier, more cohesive, more layered, and more fibrous.

6.5.3 Fluid Holding

Ideally, plant-based meat analogs should be able to retain fluids during their production, storage, and preparation, as these fluids contribute to the texture, cookability, and juiciness of the product (Wi et al., 2020). Typically, the fluids involved primarily consist of water, but they may also contain dissolved proteins, carbohydrates, and salts, as well as lipid droplets. The fluid holding properties of meat analogs are mainly governed by biopolymer-solvent interactions, the elastic strength of the biopolymer gel network, and the osmotic pressure associated with an uneven distribution of ions inside and outside the biopolymer gel network (Chap. 4) (Cornet et al., 2021). Typically, the fluid holding properties of a meat analog increase as the size of the pores in the biopolymer gel network decreases and the number of pores increases. In addition, the wetting properties are important because they determine the affinity of the biopolymer matrix for the fluids. Binders, such as methylcellulose, citrus fibers or other hydrocolloids, are often incorporated into meat analogs to enhance their fluid holding capacity (Kyriakopoulou et al., 2021). The presence of these binders can reduce the pore size, increase the pore number, and improve the wetting properties of the meat analog matrix.

The fluid holding properties of meat analogs also influence their reduction in mass ('cooking loss') during thermal processing. The cooking losses of some plant-based and real burgers are compared in Table 6.7. Interestingly, the cooking loss was higher in the real burgers than the plant-based ones. This result suggests that the plant-based burgers may have contained a stronger biopolymer network that was more resistant to heating and had a larger number of small pores. The presence of binders, such as starch, methylcellulose, and potato proteins, may also have contributed to this effect, as discussed earlier. The fibrils in real meat are known to shrink during heating, which leads to the expulsion of fluids, which may also contribute to their higher cooking losses (Section 6.2.4). The cooked plant-based burgers exhibited less resistance to cutting than the real ones, which indicated that there were also important textural differences, which may have been related to differences in their fluid holding properties.

In another study, the water-holding properties (measured by centrifugation) of a plant-based meat prepared from soy proteins were investigated. In general, the water-holding capacity of texturized (low-moisture extruded) soy protein has been shown to increase after it was cooked (Wi et al., 2020). This effect was mainly attributed to a strengthening of the 3D protein network after heating due to increased unfolding and aggregation of the proteins, which allows more water to be trapped inside the meat analog structure. Interestingly, the addition of oil increased the water holding capacity of the uncooked texturized protein, thereby decreasing the cooking loss. This effect was attributed to the ability of the oil to penetrate into the pores in the texturized protein and inhibit the mass transport of fluids. Conversely, soaking the uncooked texturized protein in a soy protein dispersion decreased its water holding capacity especially for long soaking times, thereby increasing the

Table 6.7 Cooking loss (%), volume loss (%), and Warner-Bratzler shear force (N) of cooked commercial plant-based (n=3) and meat-based burgers (n=4). Results are shown as median and median 95% confidence interval

	Meat-based burger		Plant-based burger	
	Median	95%	Median	95%
Cooking loss[a] (%)	25.7	24.2–32.5	16.0	8.6–20.4
Volume loss[a] (%)	26.0	17.9–34.4	20.2	13.5–25.0
Shear force[a] (N)	12.9	10.7–16.9	6.3	5.9–6.7
Cooking loss[b] (%)	23.1	21.2–29.2	21.0	18.8–26.1
Volume loss[b] (%)	28.0	25.3–48.1	21.8	14.2–35.1
Shear force[b] (N)	13.7	10.6–18.3	10.1	8.8–15.0

Modified from (De Marchi et al., 2021) under CC BY 4.0 (http://creativecommons.org/licenses/by/4.0/)
[a]Water bath method (cooking in a bag and collecting the released fluids)
[b]Cooking plate method (frying in a pan and measuring the weight reduction of the patty)

cooking loss. Presumably, the native soy proteins were able to bind less strongly to the texturized protein than the water molecules. However, once the textured protein was cooked, the water holding capacity was less influenced by the type and concentration of fluids used for soaking before cooking.

The factors impacting the swelling behavior of plant-based meat analogs formulated from soy and gluten proteins have also been investigated, which is helpful to understand how marinating can be optimized (Cornet et al., 2021). The swelling of the meat analogs in the presence of a marinade was found to be influenced by pH, ionic strength, and the degree of crosslinking. Under environmental conditions that favored a high protein charge (*i.e.*, pH $\neq$ pI and low ionic strength), the uptake of water into the meat analog was favored, which lead to increased swelling. This effect is probably related to the increased electrostatic repulsion between the protein molecules in the meat analog matrix under these conditions and to the increased water-protein interactions present, which means that the size of the individual pores increases and so they can take up more water.

6.5.4 Flavor

As discussed earlier, the characteristic flavor profiles of meat products are a result of various kinds of volatile and non-volatile molecules, which are present in the original product or are generated during cooking by a complex series of chemical reactions (Sect. 6.2.5). Ideally, plant-based meat analogs should have flavor profiles that accurately mimic those of the meat products they are designed to replace. This is often challenging because the types of aromatic molecules naturally occurring in animal-derived products are different from those typically found in the ingredients used to formulate plant-based foods, such as soy, pea, or wheat proteins. Consequently, it is important to identify meat-like flavor compounds that can be found in other sources of plant materials or that can be produced from them. Typically, manufacturers of plant-based foods or flavoring agents quantify the type and concentration of flavor molecules that contribute to the meat-like taste and aroma of meat products. They then use this information to identify plant-based flavors that can generate a similar flavor profile. These flavors are often produced by flavor companies using controlled chemical reactions or fermentation processes (Sect. 6.3.5), or by identifying suitable natural sources that contain them (such as yeast, mushroom, or seaweed).

A major challenge to the use of many plant-derived ingredients used to formulate meat and seafood analogs (especially proteins) is that they contain off-flavors. These off-flavors may arise due to chemical degradation of the proteins or other constituents in the ingredients, *e.g.*, due to lipid oxidation (lipoxygenase, reactive oxygen) or hydrolysis reactions (Cosson et al., 2022; Fiorentini et al., 2020). Alternatively, they may be a result of minor components associated with the ingredients that have undesirable tastes or aromas, such as saponins, phenolics, and alkaloids (Roland et al., 2017). To reduce these off-flavors, ingredient manufactures are breeding new plant varieties and developing processing technologies to avoid, remove, or neutralize them. For example, the concentration of volatile off-flavors in chick-pea seeds has been reduced by combining thermal and infrared treatments to decrease the lipoxygenase activity (Shariati-Ievari et al., 2016). More information about flavoring agents suitable for use in meat and seafood analogs is provided in Sect. 6.3.5 and Chap. 2.

6.5.5 Nutritional Value

Ideally, meat and seafood analogs should have similar or better nutritional profiles than real meat and seafood products. This is often challenging because meat and seafood come from living animals that contain a diverse range of nutrients essential to human health and wellbeing, including good quality proteins, vitamins, and minerals (Herreman et al., 2020). Moreover, these nutrients are often present in a bioavailable form. For example, consuming 100 g of red meat a day makes an important contribution to an average person's daily nutrient requirements: vitamin B_{12} (67 %), selenium (37 %), zinc (26 %), riboflavin (25 %), niacin (25 %), vitamin B_6 (25 %), pantothenic acid (25 %), and potassium (20 %) (de Pereira & dos Vicente, 2013). However, meat and meat products can also be a significant source of saturated fatty acids (Bohrer, 2019). Many nutrition scientists and organizations recommend reducing the consumption of saturated fatty acids because they may be linked to an increased risk of coronary heart diseases (Hooper et al., 2015), although some scientists are now questioning this link (Astrup et al., 2020; Lawrence, 2021), which is discussed in more detail in Chap. 5. Many of the current plant-based meat analogs also contain considerable amounts of saturated fatty acids (Table 6.3). If required, these products could be reformulated to reduce the concentration of these fatty acids, as discussed in Sect. 6.3.2.

Nutritional studies suggest that the replacement of meat products with plant-based alternatives could have both beneficial and detrimental effects on human nutrition. For instance, increasing the amount of plant-based meat analogs consumed would increase the intake of dietary fiber, magnesium, folate, and polyunsaturated fatty acids, and total iron, but could lead to deficiencies in zinc, vitamin B_{12}, as well as to a lower overall protein intake (Bohrer, 2019; Vatanparast et al., 2020). However, the nutritional effects of replacing animal with plant-based meat products still need to be carefully evaluated in clinical studies. As mentioned earlier, the bioavailability of nutrients plays a critical role in determining their nutritional impact, and plants contain ingredients (such as phytates) that can decrease the availability of important nutrients (such as iron and zinc). Consequently, the amount of these minerals consumed may not reflect the actual amount that reaches the bloodstream and can be used by the human body.

The nutritional profile of a commercial plant-based burger is compared to that of a real beef burger in Table 6.8. The two kinds of products have quite similar protein and fat contents, but the plant-based burgers contain more carbohydrates from

6.5 Key Properties

Table 6.8 Nutritional composition reported as median and median 95% confidence interval of commercial plant-based (*n*=3) and meat-based (*n*=4) burger patties

	Meat-based		Plant-based	
	Median	95%	Median	95%
Moisture (%)	65.9	61.5–69.6	60.9	52.8–64.0
Ash (%)	1.79	1.77–1.82	2.52	1.87–3.47
Protein (%)	18.0	15.9–18.8	18.0	13.3–18.4
–*Collagen** (%)	2.49	1.37–2.83	–	–
Fat (%)	12.5	8.0–20.3	11.1	8.8–19.1
–*Cholesterol* (mg/100 g of the raw product)	50.6	48.8–54.3	3.98	3.88–4.55
Carbohydrates (%)	2.09	2.00–2.76	8.4	7.6–10.0
–*Starch* (%)	0.93	0.86–1.14	0.31	0.10–1.21
–*Fructose* (%)	0.022	0.019–0.055	0.013	0.010–0.056
–*Total dietary fiber* (%)	0.74	0.48–0.98	4.27	2.90–5.02
Gross energy (MJ/kg dry matter)	28.4	26.1–30.3	24.9	24.0–28.6
Gross energy (MJ/kg of the raw product)	9.7	7.9–12.2	9.4	8.9–13.5

Modified from (De Marchi et al., 2021) under CC BY 4.0 (http://creativecommons.org/licenses/by/4.0/).
*Collagen percentage was calculated as (Hydroxyproline × 8)/10^3

added sugars and more fibers than their animal-based counterparts. Presumably, the sugars are added to create desirable colors and flavors through the Maillard reaction, while the fibers are added to promote the formation of fibrous structures, to give desirable textural attributes, and to ensure fluid retention.

Comparison of the amino acid compositions of the two products shows that the plant-based burgers are deficient in methionine, which is an essential amino acid (Table 6.9). This is not surprising because pulse proteins, such as those derived from peas and soybeans, are low in methionine (Grossmann & Weiss, 2021). Having said this, the protein quality of these proteins is still reasonably high, *e.g.*, the DIAAS values of soy and pea protein are 91 and 70, respectively (Herreman et al., 2020). Amino acids that may be lacking in the pulse proteins used to formulate plant-based meat products (such as cysteine and methionine) can be obtained from cereal proteins in the diet, such as from wheat and oat. However, the absence of specific amino acids in plant-based analog foods needs to be considered when designing plant-based food analogs to ensure an adequate amino acid supply in the population since a lower DIAAS means that more protein needs to be consumed to achieve an adequate amino acid uptake. More information can be found in relevant publications of the FAO, e.g., in (Food and Agriculture Organization of the United Nations, 2013).

Comparison of the mineral compositions of the two products shows that the plant-based burgers have lower zinc levels than the real beef burgers (Table 6.9). Conversely, the plant-based burgers were higher in iron. However, the bioavailability of the non-heme iron in plant-based foods is often lower than that of the heme iron in meat products. Other researchers have reported that plant-based nuggets were lower in K, Zn, Cu, and Fe than meat products (Kumar et al., 2017). However, these findings have to be treated with caution because each manufacturer has

Table 6.9 Amino acid (AA) composition (mg/100 g of the raw product) and mineral composition (mg/kg of the raw product) of commercial plant-based (n=3) and meat-based (n=4) burger patties

AA	Amino acid profile					Mineral profile			
	Meat-based		Plant-based			Meat-based		Plant-based	
	Median	95%	Median	95%		Median	95%	Median	95%
Alanine	1096	816–1215	687	606–819	Na	4267	4215–4445	4285	3653–7153
Arginine	1086	913–1489	1061	930–1478	K	2718	2524–2960	3457	3280–5449
Aspartic acid	1581	1232–1886	1925	1549–2502	P	1270	1191–1416	2099	1558–3648
Cysteine	154	142–188	252	175–308	S	1325	1297–1538	1443	1113–1692
Glycine	1305	949–1910	689	640–762	Mg	159	152–174	615	181–1404
Glutamic acid	3027	2276–3516	4352	4018–5805	Ca	85.6	79.7–141.5	715	177–854
Hydroxyproline	311	171–354	–	–					
Isoleucine*	555	464–750	508	472–619	Zn	30.7	27.6–36.3	21.4	8.4–25.1
Histidine*	582	499–774	592	417–869	Si	20.9	19.3–23.6	20.4	16.2–98.8
Leucine*	1164	963–1497	1215	872–1500	Fe	13.1	11.6–14.1	26.5	23.9–33.1
Lysine*	1391	1046–1836	928	707–1604	Cu	1.52	0.80–4.28	3.52	2.62–9.12
Methionine*	300	244–392	13.8	9.6–51.1	Sr	0.50	0.46–0.53	2.76	0.61–3.03
Phenylalanine*	662	613–892	899	630–1067	Mn	0.18	0.17–0.27	10.5	2.96–11.3
Proline	856	696–1047	775	534–1160	Li**	0.10	–	0.28	0.26–0.38
Serine	673	536–885	903	593–1039	Cr	0.08	0.07–0.11	0.18	0.15–0.31
Tyrosine	482	437–643	591	366–675	Ba	0.07	0.06–0.26	0.53	0.14–2.36
Threonine*	686	579–927	572	468–648	Ni	0.06	0.06–0.10	0.22	0.19–0.33
Tryptophan*	103	93–133	108	87–128	Ti	0.05	0.04–0.07	0.17	0.11–0.81
Valine*	612	576–826	559	517–672					

Modified from (De Marchi et al., 2021) under CC BY 4.0 (http://creativecommons.org/licenses/by/4.0/)

* Essential amino acids; **Li was only detected in 1 sample of meat-based burgers, and in 4 samples of the same plant-based burgers brand

6.5 Key Properties

different formulations and thus this statement does not hold true for all products on the market. Moreover, in future studies, the bioavailability of iron and other essential minerals in plant- and animal-based meat products should be compared.

To overcome these challenges, researchers are investigating innovative approaches to increase the concentration and bioavailability of important nutrients in plant-based foods, including crop breeding, genetic engineering, and processing approaches to increase nutrient levels or decrease the antinutrient content. In addition, meat and seafood analogs can be fortified with micronutrients that might be lacking in a plant-based diet, such as vitamins (especially vitamins B_{12} and D) and minerals (especially iron and zinc) (Table 6.1).

6.5.6 Environmental Sustainability

A major concern associated with consuming meat-rich diets is their negative environmental impact, especially in terms of greenhouse gas emissions (Scarborough et al., 2014; Willett et al., 2019). The ingredients typically used to formulate plant-based meats have been shown to generate lower greenhouse gas emissions than meat derived from beef, lamb, pork, and poultry (Table 6.10).

However, the manufacturing processes used to convert plant-derived ingredients into meat analogs are often energy-intensive (*e.g.*, extrusion and thermal processing). In addition, the ingredients must be transported and stored prior to use, which also involves the utilization of energy and other resources. For this reason, scientists have carried out more comprehensive life cycle assessments of the environmental impacts associated with creating meat and plant-based products by considering all stages of production, including primary production, processing, distribution, and

Table 6.10 Summary of greenhouse gas emissions (GHG) per kg product of selected meat products (boneless) and ingredients used to formulate plant-based meat analogs

Source	GHG emissions (kg CO_2-eq/kg)
Chicken	4.12
Turkey	6.04
Pork (world average)	5.85
Beef (world average)	28.7
Lamb (world average)	27.9
Pea	0.60
Soybean	0.58
Potatoes	0.20
Sunflower oil	0.76
Rapeseed oil	0.26
Soy protein isolate	2.4[a]

Taken from (Braun et al., 2016; Clune et al., 2017; Schmidt, 2015)
[a]The values for soy protein isolate are reported per kg of protein

Table 6.11 Selected environmental impacts of real meat and plant-based burger patties per kg of fresh, ready-to pack burger patties

	Beef	Chicken	Pork	Soy	Pumpkin
Global warming potential (kg CO_2 eq)	26.6	6.05	5.83	**0.53**	**0.75**
Ozone depletion (mg CFC-11 eq)	0.088	0.074	0.072	**0.033**	**0.041**
Agricultural land occupation (m^2a^*)	5.91	4.28	5.24	**0.79**	**0.49**
Particulate matter formation (kg PM_{10} eq)	0.1	0.01	0.01	**0.001**	**0.002**
Terrestrial acidification (kg SO_2 eq)	0.7	0.07	0.09	**0.003**	**0.007**
Urban land occupation (m^2a^*)	**0.002**	**0.002**	**0.002**	0.01	0.01
Freshwater eutrophication (kg P eq)	0.006	0.0006	0.0006	**0.0002**	0.001

Taken from (Saerens et al., 2021)
* m^2a = square meters-year; land-use per specific time

storage (Saerens et al., 2021). The results of a recent study comparing the environmental impacts of producing different kinds of meat and plant-based burgers are summarized in Table 6.11.

This study indicates that plant-based meat analogs have lower environmental impacts in most categories when compared to animal-based products, even after processing, distribution, and storage are considered. Particularly, greenhouse gas emissions, agricultural land occupation, and ozone depletion are considerably lower for the plant-based foods (Table 6.11). The results of studies commissioned by food companies, such as Beyond Meat, also suggest that replacing beef burgers with their plant-based analogs can have large environmental benefits (Heller & Keoleian, 2018). Overall, the life cycle analyses that have been carried out so far are encouraging and are in line with larger studies that advocate for the reduction of meat product consumption for environmental sustainability and health reasons (Willett et al., 2019).

6.6 Future Directions

The science and technology behind plant-based meat and seafood analogs is rapidly advancing. Plant-derived ingredients with improved functional attributes and reliability are being introduced. Innovative processing technologies are being developed and established ones are being optimized. Structural design principles are being employed to fabricate plant-based foods with improved or novel properties. The application of these findings will facilitate the creation of a more diverse range of high-quality plant-based foods that accurately simulate real meat or seafood products, as well as being delicious, affordable, convenient, healthy, and sustainable. Nevertheless, there are still significant challenges that need to be solved to ensure a broader acceptance and deeper market penetration of plant-based meat products. Most of the commercially successful products that are currently available are designed to simulate products made from comminuted meat, such as burgers, sausages, nuggets, and ground beef (Fig. 6.1). In the future, it will be important to create a broader range of meat and seafood analogs, such as beef steaks, chicken breasts, pork chops, and fish fillets.

Another major challenge is to overcome social barriers to the adoption of plant-based meat analogs. A recent study reported that meat is typically linked with positive associations, whereas meat analogs are linked with negative ones (Michel et al., 2021). Interestingly, this study also indicated that peer pressure often makes people more resistant to consuming plant-based meats, especially in social settings. The authors reported that eating meat analogs alone and not together with others produced the highest acceptance ratings – meaning the participants felt more comfortable eating these kinds of foods in the absence of other people. In contrast, eating meat analogs on a Sunday with the family, when invited for dinner at a restaurant, for a business meal, or at a barbecue party resulted in lower acceptance values. This was attributed to the fear of being negatively judged by others about their eating behavior. This study also showed that meat-eaters tend to prefer plant-based products that are like real meat. Thus, mimicking real meat products more accurately will be important for the development of the next generation of plant-based foods. Some consumers consider plant-based meat analogs to be highly processed foods, which makes them less likely to adopt them (Michel et al., 2021). Consequently, it will also be important to change the consumers' perceptions about these kinds of products. For instance, by challenging the idea that just because something is highly processed it is bad for you.

Finally, further research is required to establish the nutritional and health implications of incorporating more plant-based meat and seafood products into the human diet. This knowledge can then be used to create meat and seafood analogs that have beneficial effects on human health, which will require them to be designed to contain an appropriate balance of nutrients, as well as to control their digestibility and bioavailability.

References

Aaslyng, M. D., Elmore, J. S., & Mottram, D. S. (1998). Comparison of the aroma characteristics of acid-hydrolyzed and enzyme-hydrolyzed vegetable proteins produced from soy. *Journal of Agricultural and Food Chemistry, 46*(12), 5225–5231. https://doi.org/10.1021/jf9806816

Acton, J. C., & Dick, R. L. (1984). Protein-protein interaction in processed meats. *American Meat Science Association*, 8.

Astrup, A., Magkos, F., Bier, D. M., Brenna, J. T., de Oliveira Otto, M. C., Hill, J. O., … Krauss, R. M. (2020). Saturated fats and health: A reassessment and proposal for food-based recommendations: JACC state-of-the-art review. *Journal of the American College of Cardiology, 76*(7), 844–857. https://doi.org/10.1016/j.jacc.2020.05.077

Batista, A. P., Portugal, C. A. M., Sousa, I., Crespo, J. G., & Raymundo, A. (2005). Accessing gelling ability of vegetable proteins using rheological and fluorescence techniques. *International Journal of Biological Macromolecules, 36*(3), 135–143. https://doi.org/10.1016/j.ijbiomac.2005.04.003

Betts, J. G., Desaix, P., Johnson, E., Johnson, J. E., Korol, O., Kruse, D., … Young, K. A. (2017). *Anatomy & physiology* (p. 1420). Rice University.

Bhat, Z. F., Morton, J. D., Mason, S. L., & Bekhit, A. E.-D. A. (2018). Role of calpain system in meat tenderness: A review. *Food Science and Human Wellness, 7*(3), 196–204. https://doi.org/10.1016/j.fshw.2018.08.002

Bindereif, B., Eichhöfer, H., Bunzel, M., Karbstein, H. P., Wefers, D., & van der Schaaf, U. S. (2021). Arabinan side-chains strongly affect the emulsifying properties of acid-extracted sugar beet pectins. *Food Hydrocolloids, 121*, 106968. https://doi.org/10.1016/j.foodhyd.2021.106968

Bjelanovic, M., Sørheim, O., Slinde, E., Puolanne, E., Isaksson, T., & Egelandsdal, B. (2013). Determination of the myoglobin states in ground beef using non-invasive reflectance spectrometry and multivariate regression analysis. *Meat Science, 95*(3), 451–457. https://doi.org/10.1016/j.meatsci.2013.05.021

Bohrer, B. M. (2019). An investigation of the formulation and nutritional composition of modern meat analogue products. *Food Science and Human Wellness, 8*(4), 320–329. https://doi.org/10.1016/j.fshw.2019.11.006

Bolognesi, V. J., & Garcia, C. E. R. (2018). Chapter 12 - annatto carotenoids as additives replacers in meat products. In A. M. Holban & A. M. Grumezescu (Eds.), *Alternative and replacement foods* (pp. 355–384). Academic. https://doi.org/10.1016/B978-0-12-811446-9.00012-5

Boon, C. S., McClements, D. J., Weiss, J., & Decker, E. A. (2009). Role of iron and hydroperoxides in the degradation of lycopene in oil-in-water emulsions. *Journal of Agricultural and Food Chemistry, 57*(7), 2993–2998. https://doi.org/10.1021/jf803747j

Braun, M., Muñoz, I., Schmidt, J. H., & Thrane, M. (2016). Sustainability of Soy protein from life cycle assessment. *The FASEB Journal, 30*(S1), 894.5–894.5. https://doi.org/10.1096/fasebj.30.1_supplement.894.5

Brown, P. O., Eisen, M., Fraser, R., Holz-Schietinger, C., Karr, J., Solomatin, S., … Vrljic, M. (2019). Methods and compositions for consumables. Patent number: AU2018200726B2. Retrieved from: https://patents.google.com/patent/AU2018200726B2/en?q=leghemoglobin+vegan&oq=leghemoglobin+vegan. Accessed Sept 2021

Cavallini, V., Hargarten, P. G., & Joehnke, J. (2006). Vegetable protein meat analog. Patent number: US7070827B2. Retrieved from: https://patents.google.com/patent/US7070827/fi. Accessed Sept 2021

Cejudo-Bastante, M. J., Hurtado, N., Delgado, A., & Heredia, F. J. (2016). Impact of pH and temperature on the colour and betalain content of Colombian yellow pitaya peel (Selenicereus megalanthus). *Journal of Food Science and Technology, 53*(5), 2405–2413. https://doi.org/10.1007/s13197-016-2215-y

Clune, S., Crossin, E., & Verghese, K. (2017). Systematic review of greenhouse gas emissions for different fresh food categories. *Journal of Cleaner Production, 140*, 766–783. https://doi.org/10.1016/j.jclepro.2016.04.082

Cornet, S. H. V., Snel, S. J. E., Lesschen, J., van der Goot, A. J., & van der Sman, R. G. M. (2021). Enhancing the water holding capacity of model meat analogues through marinade composition. *Journal of Food Engineering, 290*, 110283. https://doi.org/10.1016/j.jfoodeng.2020.110283

Cosson, A., Oliveira Correia, L., Descamps, N., Saint-Eve, A., & Souchon, I. (2022). Identification and characterization of the main peptides in pea protein isolates using ultra high-performance liquid chromatography coupled with mass spectrometry and bioinformatics tools. *Food Chemistry, 367*, 130747. https://doi.org/10.1016/j.foodchem.2021.130747

Davidovich-Pinhas, M., Barbut, S., & Marangoni, A. G. (2015). The role of surfactants on ethylcellulose oleogel structure and mechanical properties. *Carbohydrate Polymers, 127*, 355–362. https://doi.org/10.1016/j.carbpol.2015.03.085

De Marchi, M., Costa, A., Pozza, M., Goi, A., & Manuelian, C. L. (2021). Detailed characterization of plant-based burgers. *Scientific Reports, 11*(1), 2049. https://doi.org/10.1038/s41598-021-81684-9

de Pereira, P. M. C. C., & dos Vicente, A. F. R. B. (2013). Meat nutritional composition and nutritive role in the human diet. *Meat Science, 93*(3), 586–592. https://doi.org/10.1016/j.meatsci.2012.09.018

References

Dekkers, B. L., Boom, R. M., & van der Goot, A. J. (2018). Structuring processes for meat analogues. *Trends in Food Science & Technology, 81*, 25–36. https://doi.org/10.1016/j.tifs.2018.08.011

Dreher, J., Blach, C., Terjung, N., Gibis, M., & Weiss, J. (2020). Formation and characterization of plant-based emulsified and crosslinked fat crystal networks to mimic animal fat tissue. *Journal of Food Science, 85*(2), 421–431. https://doi.org/10.1111/1750-3841.14993

Dreher, J., König, M., Herrmann, K., Terjung, N., Gibis, M., & Weiss, J. (2021). Varying the amount of solid fat in animal fat mimetics for plant-based salami analogues influences texture, appearance and sensory characteristics. *LWT, 143*, 111140. https://doi.org/10.1016/j.lwt.2021.111140

Duque-Estrada, P., Kyriakopoulou, K., de Groot, W., van der Goot, A. J., & Berton-Carabin, C. C. (2020). Oxidative stability of soy proteins: From ground soybeans to structured products. *Food Chemistry, 318*, 126499. https://doi.org/10.1016/j.foodchem.2020.126499

Ebert, S., Michel, W., Nedele, A.-K., Baune, M.-C., Terjung, N., Zhang, Y., ... Weiss, J. (2022). Influence of protein extraction and texturization on odor-active compounds of pea proteins. *Journal of the Science of Food and Agriculture., n/a*(n/a). Doi:https://doi.org/10.1002/jsfa.11437

EC. (2021). Commodity Price Dashboard European Commission June 2021. EC. Retrieved from: https://ec.europa.eu/info/sites/default/files/food-farming-fisheries/farming/documents/commodity-price-dashboard_2021-07_en.pdf Accessed Aug 2021

EFSA. (2016). Safety and efficacy of iron oxide black, red and yellow for all animal species. *EFSA Journal, 14*(6), e04482. https://doi.org/10.2903/j.efsa.2016.4482

Elzerman, J. E., Hoek, A. C., van Boekel, M. A. J. S., & Luning, P. A. (2011). Consumer acceptance and appropriateness of meat substitutes in a meal context. *Food Quality and Preference, 22*(3), 233–240. https://doi.org/10.1016/j.foodqual.2010.10.006

FDA. (2017). GRAS Notice 737: GRAS Notification for Soy Leghemoglobin protein preparation derived from pichia pastoris. Retrieved from: https://www.fda.gov/media/124351/download. Accessed Sept 2021

Fellows, P. J. (2017). Extrusion cooking. In P. J. Fellows (Ed.), *Food processing technology* (Vol. 17, 4th ed., pp. 753–780). Woodhead Publishing. https://doi.org/10.1016/B978-0-08-100522-4.00017-1

Filho, G. C. S., Penna, T. C. V., & Schaffner, D. W. (2005). Microbiologial quality of vegetable proteins during the preparation of a meat analog. *Italian Journal of Food Science, 17*(3), 269–283.

Fiorentini, M., Kinchla, A. J., & Nolden, A. A. (2020). Role of sensory evaluation in consumer acceptance of plant-based meat analogs and meat extenders: A scoping review. *Foods, 9*(9), 1334. https://doi.org/10.3390/foods9091334

Flores, M. (2017). Chapter 13 - the eating quality of meat: III—flavor. In F. Toldrá (Ed.), *Lawrie´s meat science* (8th ed., pp. 383–417). Woodhead Publishing. https://doi.org/10.1016/B978-0-08-100694-8.00013-3

Food and Agriculture Organization of the United Nations (Ed.). (2013). *Dietary protein quality evaluation in human nutrition: report of an FAO expert consultation*. Presented at the FAO Expert Consultation on Protein Quality Evaluation in Human Nutrition, Rome: Food and Agriculture Organization of the United Nations.

Fraser, R., Brown, P. O., Karr, J., Holz-Schietinger, C., & Cohn, E. (2017). Methods and compositions for affecting the flavor and aroma profile of consumables. Patent number: US9700067B2. Retrieved from: https://patents.google.com/patent/US9700067B2/en. Accessed Sept 2021

Fraser, R. Z., Shitut, M., Agrawal, P., Mendes, O., & Klapholz, S. (2018). Safety evaluation of soy leghemoglobin protein preparation derived from Pichia pastoris, intended for use as a flavor catalyst in plant-based meat. *International Journal of Toxicology, 37*(3), 241–262. https://doi.org/10.1177/1091581818766318

GFI. (2020). *2020 State of the Industry Report. The Good Food Institute*. Retrieved from: https://gfi.org/resource/plant-based-retail-report/. Accessed Aug 2021

Gheonea, I., Aprodu, I., Enachi, E., Horincar, G., Bolea, C. A., Bahrim, G. E., ... Stănciuc, N. (2020). Investigations on thermostability of carotenoids from tomato peels in oils using a kinetic approach. *Journal of Food Processing and Preservation, 44*(1), e14303. https://doi.org/10.1111/jfpp.14303

Gonzalez-Gutierrez, J., & Scanlon, M. G. (2018). Chapter 5 - rheology and mechanical properties of fats. In A. G. Marangoni (Ed.), *Structure-function analysis of edible fats* (2nd ed., pp. 119–168). AOCS Press. https://doi.org/10.1016/B978-0-12-814041-3.00005-8

Gravelle, A. J., Barbut, S., & Marangoni, A. G. (2012). Ethylcellulose oleogels: Manufacturing considerations and effects of oil oxidation. *Food Research International, 48*(2), 578–583. https://doi.org/10.1016/j.foodres.2012.05.020

Grossmann, L., & Weiss, J. (2021). Alternative protein sources as technofunctional food ingredients. *Annual Review of Food Science and Technology, 12*(1), 93–117. https://doi.org/10.1146/annurev-food-062520-093642

Guo, Z., Teng, F., Huang, Z., Lv, B., Lv, X., Babich, O., ... Jiang, L. (2020). Effects of material characteristics on the structural characteristics and flavor substances retention of meat analogs. *Food Hydrocolloids, 105*, 105752. https://doi.org/10.1016/j.foodhyd.2020.105752

He, J., Evans, N. M., Liu, H., & Shao, S. (2020). A review of research on plant-based meat alternatives: Driving forces, history, manufacturing, and consumer attitudes. *Comprehensive Reviews in Food Science and Food Safety, 19*(5), 2639–2656. https://doi.org/10.1111/1541-4337.12610

Heller, M., & Keoleian, G. (2018). Beyond meat's beyond burger life cycle assessment: A detailed comparison between a plant- based and an animal-based protein source. Retrieved from: https://css.umich.edu/sites/default/files/publication/CSS18-10.pdf. Accessed Oct 2021

Herreman, L., Nommensen, P., Pennings, B., & Laus, M. C. (2020). Comprehensive overview of the quality of plant- And animal-sourced proteins based on the digestible indispensable amino acid score. *Food Science & Nutrition, 8*(10), 5379–5391. https://doi.org/10.1002/fsn3.1809

Herz, E., Herz, L., Dreher, J., Gibis, M., Ray, J., Pibarot, P., ... Weiss, J. (2021). Influencing factors on the ability to assemble a complex meat analogue using a soy-protein-binder. *Innovative Food Science & Emerging Technologies, 73*, 102806. https://doi.org/10.1016/j.ifset.2021.102806

Hoek, A. C., Luning, P. A., Weijzen, P., Engels, W., Kok, F. J., & de Graaf, C. (2011). Replacement of meat by meat substitutes. A survey on person- and product-related factors in consumer acceptance. *Appetite, 56*(3), 662–673. https://doi.org/10.1016/j.appet.2011.02.001

Hofmann, T., & Schieberle, P. (1995). Evaluation of the key odorants in a thermally treated solution of ribose and cysteine by aroma extract dilution techniques. *Journal of Agricultural and Food Chemistry, 43*(8), 2187–2194. https://doi.org/10.1021/jf00056a042

Hooper, L., Martin, N., Abdelhamid, A., & Smith, G. D. (2015). Reduction in saturated fat intake for cardiovascular disease. *Cochrane Database of Systematic Reviews, 6*. https://doi.org/10.1002/14651858.CD011737

Huang, Y.-C., Li, H.-J., He, Z.-F., Wang, T., & Qin, G. (2010). Study on the flavor contribution of phospholipids and triglycerides to pork. *Food Science and Biotechnology, 19*(5), 1267–1276. https://doi.org/10.1007/s10068-010-0181-0

Hughes, J. M., Oiseth, S. K., Purslow, P. P., & Warner, R. D. (2014). A structural approach to understanding the interactions between colour, water-holding capacity and tenderness. *Meat Science, 98*(3), 520–532. https://doi.org/10.1016/j.meatsci.2014.05.022

Iwaki, S., Aono, S., Hayakawa, K., Fu, B. X., & Otobe, C. (2020). Changes in protein non-covalent bonds and aggregate size during dough formation. *Foods, 9*(11), 1643. https://doi.org/10.3390/foods9111643

Jin, Y., He, X., Andoh-Kumi, K., Fraser, R. Z., Lu, M., & Goodman, R. E. (2018). Evaluating potential risks of food allergy and toxicity of soy leghemoglobin expressed in Pichia pastoris. *Molecular Nutrition & Food Research, 62*(1), 1700297. https://doi.org/10.1002/mnfr.201700297

Jo, M.-N., & Lee, Y.-M. (2008). Analyzing the sensory characteristics and taste-sensor ions of MSG substitutes. *Journal of Food Science, 73*(5), S191–S198. https://doi.org/10.1111/j.1750-3841.2008.00769.x

Jorgenson, K. W., Phillips, S. M., & Hornberger, T. A. (2020). Identifying the structural adaptations that drive the mechanical load-induced growth of skeletal muscle: A scoping review. *Cells, 9*(7), 1658. https://doi.org/10.3390/cells9071658

Katz, F.-A., Grossmann, L., Gerhards, C., & Weiss, J. (2021). Inert hydrophilic particles enhance the thermal properties and structural resilience of meat protein gels during heating. *Food and Function, 12*(2), 862–872. https://doi.org/10.1039/d0fo02169e

Kaushik, R., Kumar, N., Sihag, M. K., & Ray, A. (2015). Isolation, characterization of wheat gluten and its regeneration properties. *Journal of Food Science and Technology, 52*(9), 5930–5937. https://doi.org/10.1007/s13197-014-1690-2

Kazir, M., & Livney, Y. D. (2021). Plant-based seafood analogs. *Molecules, 26*(6), 1559. https://doi.org/10.3390/molecules26061559

Kendler, C., Duchardt, A., Karbstein, H. P., & Emin, M. A. (2021). Effect of oil content and oil addition point on the extrusion processing of wheat gluten-based meat analogues. *Foods, 10*(4), 697. https://doi.org/10.3390/foods10040697

Klemaszewski, J. L., Fonteyn, D., Bouron, F., & Lemonnier, L. (2016). Cheese product with modified starches. Patent number: WO2016195814A1. Retrieved from: https://patents.google.com/patent/WO2016195814A1/en. Accessed March 2021

Krintiras, G. A., Göbel, J., Bouwman, W. G., Jan van der Goot, A., & Stefanidis, G. D. (2014). On characterization of anisotropic plant protein structures. *Food Function, 5*(12), 3233–3240. https://doi.org/10.1039/C4FO00537F

Krintiras, G. A., Gadea Diaz, J., van der Goot, A. J., Stankiewicz, A. I., & Stefanidis, G. D. (2016). On the use of the Couette Cell technology for large scale production of textured soy-based meat replacers. *Journal of Food Engineering, 169*, 205–213. https://doi.org/10.1016/j.jfoodeng.2015.08.021

Kumar, A., Alenkina, I. V., Zakharova, A. P., Oshtrakh, M. I., & Semionkin, V. A. (2015). Hyperfine interactions in soybean and lupin oxy-leghemoglobins studied using Mössbauer spectroscopy with a high velocity resolution. *Hyperfine Interactions, 230*(1), 131–139. https://doi.org/10.1007/s10751-015-1132-1

Kumar, P., Chatli, M. K., Mehta, N., Singh, P., Malav, O. P., & Verma, A. K. (2017). Meat analogues: Health promising sustainable meat substitutes. *Critical Reviews in Food Science and Nutrition, 57*(5), 923–932. https://doi.org/10.1080/10408398.2014.939739

Kyriakopoulou, K., Keppler, J. K., & van der Goot, A. J. (2021). Functionality of ingredients and additives in plant-based meat analogues. *Foods, 10*(3), 600. https://doi.org/10.3390/foods10030600

Lawrence, G. D. (2021). Perspective: The saturated fat–unsaturated oil dilemma: Relations of dietary fatty acids and serum cholesterol, atherosclerosis, inflammation, cancer, and all-cause mortality. *Advances in Nutrition, 0*, 1–10. https://doi.org/10.1093/advances/nmab013

Lawrie, R. A., & Ledward, D. (2006). *Lawrie's meat science* (7th ed.). Woodhead Publishing.

Lillford, P. J. (2011). The importance of food microstructure in fracture physics and texture perception. *Journal of Texture Studies, 42*(2), 130–136. https://doi.org/10.1111/j.1745-4603.2011.00293.x

Lillford, P. J. (2016). The impact of food structure on taste and digestibility. *Food & Function, 7*(10), 4131–4136. https://doi.org/10.1039/C5FO01375E

Lin, S., Huff, H. E., & Hsieh, F. (2002). Extrusion process parameters, sensory characteristics, and structural properties of a high moisture soy protein meat analog. *Journal of Food Science, 67*(3), 1066–1072. https://doi.org/10.1111/j.1365-2621.2002.tb09454.x

Listrat, A., Lebret, B., Louveau, I., Astruc, T., Bonnet, M., Lefaucheur, L., … Bugeon, J. (2016). How muscle structure and composition influence meat and flesh quality. *The Scientific World Journal, 2016*, 3182746. https://doi.org/10.1155/2016/3182746

López-Bote, C. (2017). Chapter 4 – chemical and biochemical constitution of muscle. In F. Toldrá (Ed.), *Lawrie´s meat science* (8th ed., pp. 99–158). Woodhead Publishing. https://doi.org/10.1016/B978-0-08-100694-8.00004-2

MacDonald, R. S., Pryzbyszewski, J., & Hsieh, F.-H. (2009). Soy protein isolate extruded with high moisture retains high nutritional quality. *Journal of Agricultural and Food Chemistry, 57*(9), 3550–3555. https://doi.org/10.1021/jf803435x

Majzoobi, M., Talebanfar, S., Eskandari, M. H., & Farahnaky, A. (2017). Improving the quality of meat-free sausages using κ-carrageenan, konjac mannan and xanthan gum. *International Journal of Food Science & Technology, 52*(5), 1269–1275. https://doi.org/10.1111/ijfs.13394

Manaf, Y. N. A., Marikkar, J. M. N., Musthafa, S., & Saari, M. M. (2014). Composition and thermal analysis of binary mixtures of mee fat and palm stearin. *Journal of Oleo Science, 63*(4), 325–332. https://doi.org/10.5650/jos.ess13193

McClements, D. J., & Grossmann, L. (2021). The science of plant-based foods: Constructing next-generation meat, fish, milk, and egg analogs. *Comprehensive Reviews in Food Science and Food Safety, 20*(4), 1–52. https://doi.org/10.1111/1541-4337.12771

Michel, F., Hartmann, C., & Siegrist, M. (2021). Consumers' associations, perceptions and acceptance of meat and plant-based meat alternatives. *Food Quality and Preference, 87*, 104063. https://doi.org/10.1016/j.foodqual.2020.104063

Mioche, L., Bourdiol, P., & Monier, S. (2003). Chewing behaviour and bolus formation during mastication of meat with different textures. *Archives of Oral Biology, 48*(3), 193–200. https://doi.org/10.1016/S0003-9969(03)00002-5

Motamedzadegan, A., Dehghan, B., Nemati, A., Tirgarian, B., & Safarpour, B. (2020). Functionality improvement of virgin coconut oil through physical blending and chemical interesterification. *SN Applied Sciences, 2*(9), 1513. https://doi.org/10.1007/s42452-020-03309-6

Murray, J. C. F. (2009). Cellulosics. In G. O. Phillips & P. A. Williams (Eds.), *Handbook of hydrocolloids* (Vol. 25, 2nd ed., pp. 710–723). Woodhead Publishing. https://doi.org/10.1533/9781845695873.710

Nishinari, K., Fang, Y., Guo, S., & Phillips, G. O. (2014). Soy proteins: A review on composition, aggregation and emulsification. *Food Hydrocolloids, 39*, 301–318. https://doi.org/10.1016/j.foodhyd.2014.01.013

Ochiai, Y., & Ozawa, H. (2020). Biochemical and physicochemical characteristics of the major muscle proteins from fish and shellfish. *Fisheries Science, 86*(5), 729–740. https://doi.org/10.1007/s12562-020-01444-y

Olsen, E., Vogt, G., Ekeberg, D., Sandbakk, M., Pettersen, J., & Nilsson, A. (2005). Analysis of the early stages of lipid oxidation in freeze-stored pork back fat and mechanically recovered poultry meat. *Journal of Agricultural and Food Chemistry, 53*(2), 338–348. https://doi.org/10.1021/jf0488559

Osen, R., Toelstede, S., Wild, F., Eisner, P., & Schweiggert-Weisz, U. (2014). High moisture extrusion cooking of pea protein isolates: Raw material characteristics, extruder responses, and texture properties. *Journal of Food Engineering, 127*, 67–74. https://doi.org/10.1016/j.jfoodeng.2013.11.023

Palanisamy, M., Franke, K., Berger, R. G., Heinz, V., & Töpfl, S. (2019). High moisture extrusion of lupin protein: Influence of extrusion parameters on extruder responses and product properties. *Journal of the Science of Food and Agriculture, 99*(5), 2175–2185. https://doi.org/10.1002/jsfa.9410

Pang, M., Ge, Y., Cao, L., Cheng, J., & Jiang, S. (2019). Physicochemical properties, crystallization behavior and oxidative stabilities of enzymatic interesterified fats of beef tallow, palm stearin and camellia oil blends. *Journal of Oleo Science, 68*(2), 131–139. https://doi.org/10.5650/jos.ess18201

Park, S.-H., Song, T., Bae, T. S., Khang, G., Choi, B. H., Park, S. R., & Min, B.-H. (2012). Comparative analysis of collagens extracted from different animal sources for application of cartilage tissue engineering. *International Journal of Precision Engineering and Manufacturing, 13*(11), 2059–2066. https://doi.org/10.1007/s12541-012-0271-4

Pietsch, V. L., Emin, M. A., & Schuchmann, H. P. (2017). Process conditions influencing wheat gluten polymerization during high moisture extrusion of meat analog products. *Journal of Food Engineering, 198*, 28–35. https://doi.org/10.1016/j.jfoodeng.2016.10.027

Pietsch, V. L., Bühler, J. M., Karbstein, H. P., & Emin, M. A. (2019). High moisture extrusion of soy protein concentrate: Influence of thermomechanical treatment on protein-protein interactions and rheological properties. *Journal of Food Engineering, 251*, 11–18. https://doi.org/10.1016/j.jfoodeng.2019.01.001

Piska, I., Zárubová, M., Loužecký, T., Karami, H., & Filip, V. (2006). Properties and crystallization of fat blends. *Journal of Food Engineering, 77*(3), 433–438. https://doi.org/10.1016/j.jfoodeng.2005.07.010

Powell, M. J., Sebranek, J. G., Prusa, K. J., & Tarté, R. (2019). Evaluation of citrus fiber as a natural replacer of sodium phosphate in alternatively-cured all-pork Bologna sausage. *Meat Science, 157*, 107883. https://doi.org/10.1016/j.meatsci.2019.107883

Prayson, B., McMahon, J. T., & Prayson, R. A. (2008). Fast food hamburgers: What are we really eating? *Annals of Diagnostic Pathology, 12*(6), 406–409. https://doi.org/10.1016/j.anndiagpath.2008.06.002

Purslow, P. P. (2017). Chapter 3 - the structure and growth of muscle. In F. Toldrá (Ed.), *Lawrie´s meat science (8)* (pp. 49–97). Woodhead Publishing. Doi:https://doi.org/10.1016/B978-0-08-100694-8.00003-0

Purslow, P. P., Oiseth, S., Hughes, J., & Warner, R. D. (2016). The structural basis of cooking loss in beef: Variations with temperature and ageing. *Food Research International, 89*, 739–748. https://doi.org/10.1016/j.foodres.2016.09.010

Raikos, V., Neacsu, M., Russell, W., & Duthie, G. (2014). Comparative study of the functional properties of lupin, green pea, fava bean, hemp, and buckwheat flours as affected by pH. *Food Science & Nutrition, 2*(6), 802–810. https://doi.org/10.1002/fsn3.143

Regulation (EC) No 852. (2004). Regulation (EC) No 852/2004 of the European parliament and of the council. European Union.

Riaz, M. N. (2011). Texturized vegetable proteins. In G. O. Phillips & P. A. Williams (Eds.), *Handbook of food proteins* (Vol. 15, pp. 395–418). Woodhead Publishing. https://doi.org/10.1533/9780857093639.395

Rolan, T., Mueller, I., Mertle, T. J., Swenson, K. J., Conley, C., Orcutt, M. W., & Mease, L. E. (2008). Ground meat and meat analog compositions having improved nutritional properties. Patent number: US20080268112A1. Retrieved from: https://patents.google.com/patent/US20080268112A1/en. Accessed Sept 2021

Roland, W. S. U., Pouvreau, L., Curran, J., van de Velde, F., & de Kok, P. M. T. (2017). Flavor aspects of pulse ingredients. *Cereal Chemistry, 94*(1), 58–65. https://doi.org/10.1094/CCHEM-06-16-0161-FI

Saerens, W., Smetana, S., Van Campenhout, L., Lammers, V., & Heinz, V. (2021). Life cycle assessment of burger patties produced with extruded meat substitutes. *Journal of Cleaner Production, 127177*. https://doi.org/10.1016/j.jclepro.2021.127177

Samard, S., & Ryu, G.-H. (2019). A comparison of physicochemical characteristics, texture, and structure of meat analogue and meats. *Journal of the Science of Food and Agriculture, 99*(6), 2708–2715. https://doi.org/10.1002/jsfa.9438

Scarborough, P., Appleby, P. N., Mizdrak, A., Briggs, A. D. M., Travis, R. C., Bradbury, K. E., & Key, T. J. (2014). Dietary greenhouse gas emissions of meat-eaters, fish-eaters, vegetarians and vegans in the UK. *Climatic Change, 125*(2), 179–192. https://doi.org/10.1007/s10584-014-1169-1

Schmidt, J. H. (2015). Life cycle assessment of five vegetable oils. *Journal of Cleaner Production, 87*, 130–138. https://doi.org/10.1016/j.jclepro.2014.10.011

Schmidt, J. M., Damgaard, H., Greve-Poulsen, M., Sunds, A. V., Larsen, L. B., & Hammershøj, M. (2019). Gel properties of potato protein and the isolated fractions of patatins and protease inhibitors – Impact of drying method, protein concentration, pH and ionic strength. *Food Hydrocolloids, 96*, 246–258. https://doi.org/10.1016/j.foodhyd.2019.05.022

Schnepf, M. I. (1992). Protein-Water Interactions. In B. J. F. Hudson (Ed.), *Biochemistry of food proteins* (pp. 1–33). Springer US. https://doi.org/10.1007/978-1-4684-9895-0_1

Schreuders, F. K. G., Dekkers, B. L., Bodnár, I., Erni, P., Boom, R. M., & van der Goot, A. J. (2019). Comparing structuring potential of pea and soy protein with gluten for meat analogue preparation. *Journal of Food Engineering, 261*, 32–39. https://doi.org/10.1016/j.jfoodeng.2019.04.022

Shariati-Ievari, S., Ryland, D., Edel, A., Nicholson, T., Suh, M., & Aliani, M. (2016). Sensory and physicochemical studies of thermally micronized Chickpea (Cicer arietinum) and Green Lentil (Lens culinaris) flours as binders in low-fat beef burgers. *Journal of Food Science, 81*(5), S1230–S1242. https://doi.org/10.1111/1750-3841.13273

Shikama, K. (2006). Nature of the FeO2 bonding in myoglobin and hemoglobin: A new molecular paradigm. *Progress in Biophysics and Molecular Biology, 91*(1), 83–162. https://doi.org/10.1016/j.pbiomolbio.2005.04.001

Sridharan, S., Meinders, M. B. J., Bitter, J. H., & Nikiforidis, C. V. (2020). Pea flour as stabilizer of oil-in-water emulsions: Protein purification unnecessary. *Food Hydrocolloids, 101*, 105533. https://doi.org/10.1016/j.foodhyd.2019.105533

Standage, T. (2009). *An edible history of humanity* (1st ed.). Walker Books.

Stern, J. H., Rutkowski, J. M., & Scherer, P. E. (2016). Adiponectin, leptin, and fatty acids in the maintenance of metabolic homeostasis through adipose tissue crosstalk. *Cell metabolism, 23*(5), 770–784. https://doi.org/10.1016/j.cmet.2016.04.011

Taggart, P., & Mitchell, J. R. (2009). 5 - Starch. In G. O. Phillips & P. A. Williams (Eds.), *Handbook of hydrocolloids* (2nd ed., pp. 108–141). Woodhead Publishing. https://doi.org/10.1533/9781845695873.108

Toldra, F. (Ed.). (2017). *Lawrie's meat science* (8th ed.). Woodhead Publishing.

Trout, G. R. (1989). Variation in myoglobin denaturation and color of cooked beef, pork, and turkey meat as influenced by pH, sodium chloride, sodium tripolyphosphate, and cooking temperature. *Journal of Food Science, 54*(3), 536–540. https://doi.org/10.1111/j.1365-2621.1989.tb04644.x

Vatanparast, H., Islam, N., Shafiee, M., & Ramdath, D. D. (2020). Increasing plant-based meat alternatives and decreasing red and processed meat in the diet differentially affect the diet quality and nutrient intakes of canadians. *Nutrients, 12*(7), 2034. https://doi.org/10.3390/nu12072034

Wang, K., & Arntfield, S. D. (2017). Effect of protein-flavour binding on flavour delivery and protein functional properties: A special emphasis on plant-based proteins. *Flavour and Fragrance Journal, 32*(2), 92–101. https://doi.org/10.1002/ffj.3365

Wang, Y., Wang, W., Jia, H., Gao, G., Wang, X., Zhang, X., & Wang, Y. (2018). Using cellulose nanofibers and its palm oil pickering emulsion as fat substitutes in emulsified sausage. *Journal of Food Science, 83*(6), 1740–1747. https://doi.org/10.1111/1750-3841.14164

Warner, R. D. (2017). Chapter 14 - The eating quality of meat—IV water-holding capacity and juiciness. In F. Toldrá (Ed.), *Lawrie's Meat Science* (8th ed., pp. 419–459). Woodhead Publishing. https://doi.org/10.1016/B978-0-08-100694-8.00014-5

Weston, A. R., Rogers, R. W., & Althen, T. G. (2002). Review: The role of collagen in meat tenderness. *The Professional Animal Scientist, 18*(2), 107–111. https://doi.org/10.15232/S1080-7446(15)31497-2

Wi, G., Bae, J., Kim, H., Cho, Y., & Choi, M.-J. (2020). Evaluation of the physicochemical and structural properties and the sensory characteristics of meat analogues prepared with various non-animal based liquid additives. *Foods, 9*(4), 461. https://doi.org/10.3390/foods9040461

Wieser, H. (2007). Chemistry of gluten proteins. *Food Microbiology, 24*(2), 115–119. https://doi.org/10.1016/j.fm.2006.07.004

Willett, W., Rockström, J., Loken, B., Springmann, M., Lang, T., Vermeulen, S., … Murray, C. J. L. (2019). Food in the Anthropocene: The EAT–Lancet Commission on healthy diets from sustainable food systems. *The Lancet, 393*(10170), 447–492. https://doi.org/10.1016/S0140-6736(18)31788-4

Wittek, P., Karbstein, H. P., & Emin, M. A. (2021). Blending proteins in high moisture extrusion to design meat analogues: Rheological properties, morphology development and product properties. *Foods, 10*(7), 1509. https://doi.org/10.3390/foods10071509

References

Wood, J. D., Enser, M., Fisher, A. V., Nute, G. R., Sheard, P. R., Richardson, R. I., ... Whittington, F. M. (2008). Fat deposition, fatty acid composition and meat quality: A review. *Meat Science, 78*(4), 343–358. https://doi.org/10.1016/j.meatsci.2007.07.019

Wright, T. J., & Davis, R. W. (2015). Myoglobin extraction from mammalian skeletal muscle and oxygen affinity determination under physiological conditions. *Protein Expression and Purification, 107*, 50–55. https://doi.org/10.1016/j.pep.2014.11.004

Wu, M., Xiong, Y. L., Chen, J., Tang, X., & Zhou, G. (2009). Rheological and microstructural properties of porcine myofibrillar protein–lipid emulsion composite gels. *Journal of Food Science, 74*(4), E207–E217. https://doi.org/10.1111/j.1750-3841.2009.01140.x

Xiong, Y. L. (1994). Myofibrillar protein from different muscle fiber types: Implications of biochemical and functional properties in meat processing. *Critical Reviews in Food Science and Nutrition, 34*(3), 293–320. https://doi.org/10.1080/10408399409527665

Youssef, M. K., & Barbut, S. (2010). Physicochemical effects of the lipid phase and protein level on meat emulsion stability, texture, and microstructure. *Journal of Food Science, 75*(2), S108–S114. https://doi.org/10.1111/j.1750-3841.2009.01475.x

Yu, T.-Y., Morton, J. D., Clerens, S., & Dyer, J. M. (2017). Cooking-induced protein modifications in meat. *Comprehensive Reviews in Food Science and Food Safety, 16*(1), 141–159. https://doi.org/10.1111/1541-4337.12243

Yuliarti, O., Kiat Kovis, T. J., & Yi, N. J. (2021). Structuring the meat analogue by using plant-based derived composites. *Journal of Food Engineering, 288*, 110138. https://doi.org/10.1016/j.jfoodeng.2020.110138

Zayas, J. F. (1997). Water holding capacity of proteins. In J. F. Zayas (Ed.), *Functionality of proteins in food* (pp. 76–133). Springer. https://doi.org/10.1007/978-3-642-59116-7_3

Zhou, H., Vu, G., Xiping, G., & McClements, D. (2022). Development of standardized tests for the characterization of beef burgers and their plant-based analogs. *Under review*.

Chapter 7
Eggs and Egg Products

7.1 Introduction

Eggs are a common part of many people's diets. They can be consumed as foods themselves, such as boiled, fried, poached, or scrambled eggs (Fig. 7.1), or they can be used as ingredients in other foods, such as mayonnaise, sauces, desserts, and baked goods (McGee, 2004). Like other animal foods, eggs have a complex composition and structure that determines many of their desirable physicochemical, functional, and sensory attributes (Stadelman et al., 2017). Consequently, it is important to understand the properties of real eggs when trying to design plant-based alternatives that can accurately simulate their desirable attributes. We therefore begin this chapter by providing an overview of the characteristics of real eggs before discussing the design and production of egg analogs. Chicken's eggs are the most common type of egg consumed in most developed countries, so they will be the primary focus of this chapter.

A report commissioned by the Good Food Institute (Washington D.C., USA) indicated that the size of the plant-based egg market in the United States was around $27 million in 2020, which was a 168% increase over the previous year (GFI, 2021). However, egg analogs only made up around 0.4% of the total egg product market (real and plant-based), highlighting the considerable room for growth in this sector. The large increase in sales can largely be attributed to the introduction of several high-quality plant-based egg analogs to the market, such as those by *Eat Just*, a Californian food company. Plant-based ingredients that behave like eggs may also be needed in other food categories that normally use hen's eggs as functional ingredients, such as plant-based baked goods ($152 million market in 2020 in the USA) and condiments and dressings ($81 million market in 2020 in the USA) (GFI, 2021). Clearly, there is a large potential market for plant-based eggs, which is stimulating research in this area.

Fig. 7.1 Eggs can be served in various ways. For example, as scrambled eggs (by Tom Ipr), boiled eggs (by Maria Eklind) and fried eggs (by Matthew Murdoch). (All images used under CC BY 2.0 (https://creativecommons.org/licenses/by/2.0/))

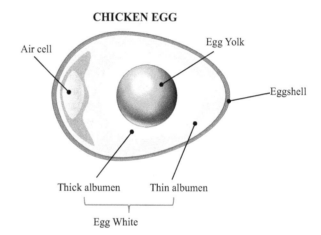

Fig. 7.2 Structure of a chicken egg, which mainly consists of egg yolk, egg white (thick and thin albumen), and egg shell

7.2 Properties of Hen's Eggs

7.2.1 Composition and Structure

Hen's eggs are actually large single cells that contain all the materials required for the initial development of the embryo (Stadelman et al., 2017). They provide the energy and building blocks required for this large single cell to split into numerous smaller cells, form an embryo, and then grow into a chick that can leave the egg and feed itself. An egg is therefore an extraordinary natural phenomenon – a self-contained unit capable of producing a living chick in around 3 weeks. To achieve this remarkable feat, the interior of eggs must contain all the nutrients and other substances needed to produce and nurture the embryo, such as proteins, lipids, carbohydrates, vitamins, and minerals dispersed within an aqueous medium. Proximate analysis of whole hen's eggs indicates that they contain about 75% water, 12% protein, 12% lipids, and smaller amounts of carbohydrates, vitamins, and minerals (Kovacs-Nolan et al., 2005). The internal structure of eggs is also designed to facilitate the formation and growth of the embryo. From inside to outside, a hen's egg consists of several concentric layers (Fig. 7.2). At the core, is the egg yolk, which has an opaque yellowish color, a semi-solid texture, and contains high levels of nutrients. It makes up around 28% of the whole egg. The yolk is surrounded by a clear shear-thinning viscous fluid, referred to as egg white, which makes up around 63% of the whole egg. The exterior of the egg consists of a hard calcium carbonate-rich shell that makes up about 10% of the whole egg. This shell plays an important role in protecting the growing embryo from the external environment, as well as modulating the flows of gases such as oxygen in and out of the egg. Eggs contain a diverse range of proteins - most of them are present in the egg yolk and egg white but a fraction of them is also present in the eggshell (Kovacs-Nolan et al., 2005). Egg proteins may consist entirely of polypeptide chains, or they may consist of polypeptide chains chemically bound or physically linked to lipids (lipoproteins) or carbohydrates (glycoproteins) (Table 7.1).

Egg Yolk Properties Egg yolk is a natural colloidal dispersion with a structure resembling that of an oil-in-water emulsion (Anton, 2013). Indeed, it contains various kinds of lipid-rich colloidal particles suspended in an aqueous medium containing soluble proteins, vitamins, and minerals. Centrifugation can be used to

Table 7.1 Molecular and physicochemical characteristics of some of the most important egg proteins

	% Protein in fraction	M_W (kDa)	pI	T_m (°C)
Egg yolk			6.0	
Livetins (α, β, γ)	12	33–203	4.3–7.6 (5.3)	83.3
Phosvitin	7	35	4.0	80.0
HDL	12	400 kDa subunits (4–20 nm) packed into larger particles	4.0	72–76
LDL	68	Colloidal particles (≈30 nm)	3.5	72–76
Egg white			4.5	
Ovalbumin	58	45	4.6	85
Conalbumin	13	80	6.6	63
Ovomucoid	11	28	3.9	70
Ovoglobulins	8	30–45	5.5–5.8	93
Lysozyme	3.5	14.6	10.7	78
Ovomucin	1.5	210	4.5–5.0	–

Data extracted from various sources (Le Denmat et al., 1999; McClements & Grossmann, 2021; Strixner et al., 2014; Tsutsui, 1988).
Key: M_W, molecular weight; pI, isoelectric point; T_m, thermal denaturation temperature; HDL, high density lipoprotein; and LDL low density lipoprotein

separate egg yolk into a granule fraction ($\approx$20%) and a plasma fraction ($\approx$80%) that have different compositions and structures (Strixner et al., 2014), as well as providing different functional attributes to egg products (Huang & Ahn, 2019). The proximate composition of egg yolk has been reported to be around 50% water, 23% lipids, 16% proteins, 9% phospholipids, 1.7% minerals, and 0.3% carbohydrates (Clark, 2012). These constituents are present as supramolecular structures or as individual molecules in solution. The major egg yolk proteins are low density lipoproteins (LDL, 68%), high density lipoproteins (HDL, 12%), livetins (12%) and phosvitin (7%) (Table 7.1). These proteins may be present in the granules or plasma (Anton, 2013): the HDL and phosvitin are mainly present in the granules, while the LDL and livetins are mainly present in the plasma. The granules consist of numerous HDL and phosvitin subunits that are held together by phosphocalcic bridges (Strixner et al., 2014). The diameters of the granules range from around a few hundred nanometers to a few micrometers (Anton, 2013). LDL and HDL are themselves colloidal particles ($d < 1,000$ nm), which have a hydrophobic core consisting of a mixture of triacylglycerol and cholesterol molecules, which is coated by an interfacial membrane consisting of phospholipids and proteins. The main unsaturated fatty acids in egg yolk have been reported to be oleic (47%), linoleic (16%), palmitoleic (5%) and linolenic (2%) acids, while the main saturated fatty acids are palmitic (23%), steric (4%) and myristic (1%) acids (NRC, 1976). It has been reported that around 9% of the fatty acids are in the form of egg lecithin. The structure and integrity of the granules in egg yolk may be changed when the pH or ionic composition is altered, because this changes the electrostatic interactions holding them together (Anton, 2013). The various forms of protein in egg yolk, in both molecular and colloidal form, play an important role in determining the emulsifying properties of eggs, such as in mayonnaise and dressings. Due to its biological role in supplying the growing embryo with nutrients, the yolk of the egg also contains a broad spectrum of oil- and water-soluble vitamins (*e.g.*, vitamins A, B_1, B_2, B_5, B_9, D, and choline), as well as minerals (*e.g.*, phosphorous, zinc, iron, and calcium).

Egg White Egg white is a transparent viscous solution that contains various kinds of proteins and glycoproteins dissolved in an aqueous medium containing water-soluble vitamins (riboflavin) and essential minerals (selenium) (Brady, 2013). The most abundant proteins in egg white are ovalbumin (58%), conalbumin (13%), ovomucoid (11%), ovoglobulins (8%), lysozyme (3.5%) and ovomucin (1.5%) (Table 7.1). Most of these proteins are globular proteins that exhibit a range of functional attributes in foods, particularly, foaming, emulsifying, thickening, gelling, binding, and antimicrobial properties. Compositional analysis indicates that egg white consists of about 90% water and 10% protein, with small amounts of carbohydrates, lipids, vitamins, and minerals.

7.2.2 Processing

The eggs utilized as foods or ingredients by consumers, restaurants, and the food industry come in numerous forms, including whole eggs, egg whites, egg yolks, and blends, which may be supplied as fresh products or as processed fluids or powders (McGee, 2004; Stadelman et al., 2017). Consequently, the type and degree of processing depend on the nature of the final product that will be created.

On a commercial scale, fresh whole eggs are typically produced by hens known as "breeders" that are raised in large numbers in carefully regulated facilities with automatic watering, feeding, and egg collection systems (Clauer, 2021). Typically, the eggs are collected on a daily basis and transported using belts and rollers to an egg processing facility that is connected to the hen houses. Specialized machines then wash, grade, sort, and package the eggs. The packaged eggs may be used as whole eggs, or they may undergo further processing. For instance, the eggs may be removed from their shells, homogenized, pasteurized, and placed into cartons prior to sale as whole liquid egg ingredients. Alternatively, the egg whites and yolks may be separated to create different kinds of food ingredients and products. Again, these products may be homogenized and pasteurized prior to sale. In some cases, the egg products are turned into powdered ingredients, as this increases their storage stability and facilitates their application in some foods. Moreover, pre-cooked egg products may be prepared in a food manufacturing operation that are then sold to consumers, restaurants, or institutions to increase convenience.

7.2.3 Physicochemical Properties

As mentioned in the previous section, eggs may be used as foods or ingredients in a variety of forms, such as fresh or processed whole eggs, egg whites, egg yolks, or their blends (McGee, 2004). Currently, it is not possible to construct commercially viable intact whole eggs from plant-based ingredients because of their composition and structural complexity, as well as economic and processing constraints. For this reason, researchers usually attempt to mimic the properties of blended whole eggs, which are prepared by mixing egg yolks and egg whites together. We therefore focus on the physicochemical and functional attributes of blended whole eggs in this section. At room temperature, blended whole eggs are viscous fluids that are optically opaque and have a creamy yellow color. Typically, their pH is slightly above neutral (pH 7.2–7.5) when fresh but may increase during storage due to certain kinds of chemical reactions (Panaite et al., 2019).

7.2.3.1 Appearance

The yellowish-orangish color of whole eggs is mainly because they contain significant levels of highly-pigmented carotenoids in their yolks, including lutein, zeaxanthin, canthaxanthin, apo-carotene-ester, citranxanthin, and cryptoxanthin (Grizio & Specht, 2021). Hens cannot produce carotenoids themselves, and so these pigments come from their diet, which is usually controlled by the egg producer by using feed that contains the required pigments. In large scale production, the feed usually consists of cereal-based pellets containing the macronutrients, vitamins, minerals, and pigments needed. In smaller farms, the hens may also be fed more natural diets containing various kinds of fruits, vegetables, and cereals. As a result, the color of egg yolks can range from light yellow to deep orange depending on the type and concentration of carotenoids in the chicken feed. These carotenoids are released from the feed after it is digested in the hen's gastrointestinal tract and are absorbed into the hen's body, before being incorporated into the colloidal particles (lipoproteins) in the egg yolk. Studies have shown that the carotenoids are mainly located inside the hydrophobic cores of the LDL particles in the plasma of the egg yolk (Anton, 2013). Moreover, fluid whole egg is optically opaque because the colloidal particles in the egg yolk (such as the granules, LDL, and HDL) scatter light strongly. Instrumental colorimetry measurements on egg yolk have reported that the L^* value is around 59, the a^* value is around -5 (slightly green) and the b^* value is around $+54$ (strongly yellow) (Panaite et al., 2019). For cooked whole eggs, it has been reported that the L^*, a^* and b^* values are around 77, -3, and $+21$ (Li et al., 2018b). The increase in lightness and decrease in color intensity can be ascribed to an increase in the degree of light scattering after heating the eggs due to the formation of protein aggregates.

7.2.3.2 Rheology

Fluid whole eggs are much more viscous than pure water because they contain various kinds of polymers and colloidal particles. Researchers have characterized the rheological properties of fluid whole eggs using shear viscosity measurements as a function of shear rate (Atilgan & Unluturk, 2008; Panaite et al., 2019). In general, the apparent shear viscosity of these products decreases with increasing shear rate, temperature, and storage time. Some researchers have reported that these kinds of fluid egg products have a small yield stress of around 0.2 Pa (Atilgan & Unluturk, 2008). The yield stress and shear thinning ($n = 0.65$ to 0.72) behavior of fluid whole egg products can be attributed to the formation of a delicate 3D network of colloidal particles and polymers that is disrupted at higher shear rates. For example, the change in shear viscosity with shear rate for blended whole eggs and for egg white was measured in our laboratory using an instrumental shear rheometer that is shown in Fig. 7.3. This data also highlights the shear thinning behavior of fluid eggs. The change in the rheological properties of eggs during and after cooking also plays a critical role in determining their desirable quality attributes, which is discussed in

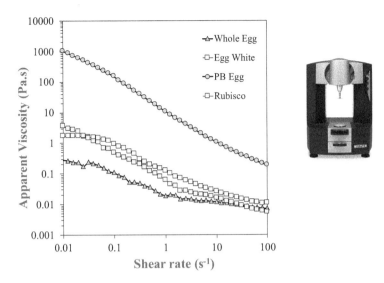

Fig. 7.3 Change in apparent shear viscosity with shear rate for whole hen's egg, hen's egg white, a commercial plant-based egg (PB Egg), and a model plant protein solution (10% rubisco). (Data kindly supplied by Hualu Zhou (UMASS). The image of the rheometer (Kinexus) used to make the measurements was kindly provided by Philip Rolfe of Netzsch (Selb, Germany) with permission)

the section on functional properties (Sect. 7.2.4.1). Moreover, the apparent shear viscosities of fluid whole eggs were reported to decrease from around 37 to 28 to 21 mPa·s as the temperature was raised from 4 to 25 to 60 °C and the rheology of fluid whole eggs has also been reported to depend on the feed given to the chickens (Panaite et al., 2019).

7.2.3.3 Stability

Fluid whole eggs may break down during storage due to physical, chemical, or biological instability mechanisms (Stadelman et al., 2017). These products are colloidal dispersions and so they are susceptible to physical breakdown due to the aggregation or gravitational separation of the various kinds of particles they contain, such as granules and lipoproteins. As with other colloidal dispersions, the resistance of the particles to aggregation depends on the strength of the attractive and repulsive interactions between them (McClements & Grossmann, 2021). Electrostatic interactions are strongly dependent on pH and ionic strength. In particular, the proteins and lipoproteins may aggregate if the pH moves too close to the isoelectric point or in the presence of high salt levels. Consequently, it may be important to control these factors when formulating fluid whole eggs. It may also be important to consider how the pH and ionic strength may change when these eggs are incorporated into other kinds of food products, as this will impact their aggregation behavior. The resistance of the fluid whole eggs to creaming or sedimentation depends on the size of the particles they contain, as well as the viscosity of the surrounding fluid. Some of the granules in egg yolk can be relatively large (a few micrometers big), which could lead to rapid creaming or sedimentation. Indeed, a centrifugation study found that egg yolk contained three main fractions of granules that were comprised of HDL, LDL, and phosvitin that had different diameters and densities: (*i*) 0.84 μm and 1200 kg m^{-3}; (*ii*) 1.8 μm and 1081 kg m^{-3}; and (*iii*) 4.9 μm and 1113 kg m^{-3} (Strixner & Kulozik, 2013). Consequently, it may be important to homogenize the products to reduce the particle size prior to packaging. However, the relatively high viscosity of the aqueous phase in fluid whole egg products may slow down this separation process.

There are also various kinds of chemical reactions that can occur within eggs during storage that can decrease their shelf life. In whole eggs, the pH of both the egg white and yolk tends to increase over time, which has been attributed to the loss of carbonic acid via CO_2 gas diffusing through the porous eggshells (Eke et al., 2013). As a result, the egg white may become visibly clearer and less viscous because the electrostatic repulsion between the protein molecules increases, which causes protein aggregates to dissociate. In addition, egg yolks may contain appreciable levels of omega-3 fatty acids, with the amount depending on the hen's feed (Javed et al., 2019). These polyunsaturated fats are highly susceptible to lipid oxidation, which can lead to off flavors (Galobart et al., 2001). Lipid oxidation of eggs containing high levels of polyunsaturated fats may also be accelerated during cooking (Cortinas et al., 2003). Storage studies have also shown that the triacylglycerols and phospholipids in whole eggs may be hydrolyzed during storage, leading to the formation of free fatty acids that can have undesirable sensory attributes (Wang et al., 2017). This degradation was attributed to the presence of lipase in the eggs, which hydrolyzed the ester bonds in the lipids. However, the lipase molecules would be deactivated if fresh eggs are thermally processed because this will denature them, thereby reducing their catalytic activity. Consequently, it is important to control processing conditions, storage conditions, packaging materials, and formulations to retard lipid oxidation, hydrolysis, and other chemical reactions. For instance, products may be stored at refrigerated temperatures in the absence of light to retard their chemical degradation.

Finally, eggs are nutrient-rich products that are highly susceptible to contamination with spoilage and pathogenic bacteria, such as *Salmonella* (Baron & Jan, 2011). Consequently, it is important to use appropriate handling and sanitation protocols to avoid contamination, as well as to use appropriate processing methods (such as pasteurization or irradiation) or additives (antimicrobials) to deactivate any microorganisms that contaminate the eggs (Silva & Gibbs, 2012; Whiley & Ross, 2015).

7.2.4 Functional Properties

Eggs are highly versatile functional ingredients that can be used for a variety of applications in foods, such as thickening, gelling, binding, emulsifying, and foaming (Vega & Mercade-Prieto, 2011). Many of these functional attributes are a result of the globular proteins present in both the egg white and egg yolk (Table 7.1).

Indeed, the presence of these proteins is largely responsible for the ability of eggs to form semi-solids when they are boiled, fried, scrambled or poached, as well as their ability to facilitate the formation and stabilization of oil droplets or air bubbles in dressings, mayonnaise, sauces, desserts, baked goods, and meringues (McGee, 2004). In this section, we highlight some of the most important functional attributes of the hen's eggs, including thickening, gelling, binding, emulsifying, and foaming properties. However, it should be noted, that eggs also play other important roles in some food products, such as providing water holding properties, controlling crystallization of sugar, fat, or ice crystals, exhibiting antimicrobial effects, or providing surface gloss (Grizio & Specht, 2021).

7.2.4.1 Thickening, Gelling, and Binding

One of the most important functional attributes of eggs in many food applications is their ability to thicken or gel solutions or to bind different components together. The molecular origin of these functional attributes is the ability of the globular proteins in egg white and egg yolk to undergo partial unfolding when heated above their thermal denaturation temperatures (T_m). The T_m values of egg proteins depend on protein type, as well as solution conditions (such as pH and ionic strength), but they usually lie somewhere between about 63 and 93 °C (Table 7.1). The thermal denaturation of egg proteins can conveniently be measured using differential scanning calorimetry (DSC) instruments that measure the change in heat-flow with temperature (Fig. 7.4). The partial unraveling of the polypeptide chains above the thermal denaturation temperature leads to the exposure of amino acids at the protein surfaces that have non-polar or sulfur-containing side-groups. These exposed groups can then promote the aggregation of the partially denatured proteins through hydrophobic attraction and disulfide bonds (Fig. 7.5). If the protein concentration is sufficiently high, which it naturally is in hen's eggs, then a network of aggregated protein molecules is formed that extends throughout the entire volume of the

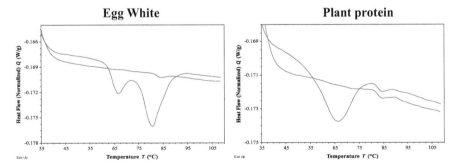

Fig. 7.4 Change in heat flow with temperature for egg white and a model plant protein (10% rubisco) measured using a differential scanning calorimeter (first run blue, second run red). (Data kindly supplied by Hualu Zhou and Giang Vu (UMass))

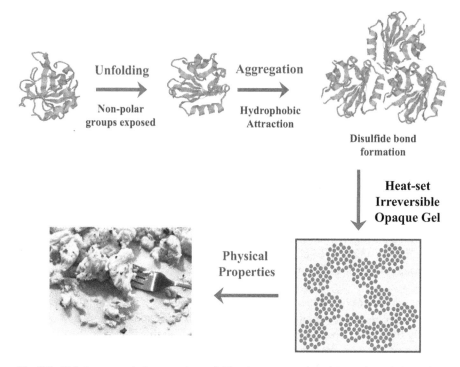

Fig. 7.5 Globular egg and plant proteins unfold and aggregate when they are heated above their thermal denaturation temperature leading to the formation of large aggregates that interact with each other and form a 3D network that provides some elastic properties. (Image of scrambled eggs kindly supplied by Eat Just Inc (San Francisco, CA))

system, thereby leading to the formation of a gel (Kiosseoglou & Paraskevopoulou, 2005). The temperature at which a gel is first formed is known as the gelation temperature (T_{gel}). If the protein concentration is not high enough to form a gel network, for example, when the egg is used in a diluted form as an ingredient in another product, then the aggregation of the proteins may lead to thickening of the aqueous phase. Typically, the globular proteins in eggs form a particulate-type gel that consists of protein clumps a few hundred nanometers to a few micrometers in diameter that are clumped together and hold the aqueous medium between them (Cordobes et al., 2004; Li et al., 2018b). Each protein clump contains multitudes of individual protein molecules aggregated with each other (Fig. 7.5). The gels formed by whole eggs also contain granules and lipoproteins from the egg yolk and white that are embedded within the protein-network. The presence of these different kinds of colloidal particles causes the egg to appear optically opaque because of light scattering effects. Gels with textures that vary from soft to hard can be produced by varying the heating times and temperatures (Vega & Mercade-Prieto, 2011), as well as by adjusting the pH or ionic strength (Croguennec et al., 2002; Raikos et al., 2007). Eggs can also be used as binders in food products (such as burgers or sausages) where they can hold the different constituents together through van der Waals,

7.2 Properties of Hen's Eggs

hydrophobic, electrostatic, hydrogen bonding and/or disulfide interactions. Again, this usually requires that they are heated above their thermal denaturation temperature so as to unfold them and expose the non-polar and sulfur-containing groups at their surfaces. The formation of a 3D network from the ingredients in eggs can also provide humectancy, *i.e.*, the ability of the product to retain water, which is an important contribution to their texture and mouthfeel (Grizio & Specht, 2021).

Hen's eggs behave in a particular fashion during cooking and thermal processing. As an example, the change in the complex shear modulus of egg white with temperature is shown in Fig. 7.6. Initially, the shear modulus is relatively low at ambient temperature because the globular proteins are in their native state and do not aggregate with each other. When the egg is heated, the shear modulus remains low until the proteins start to unfold and associate with each other, leading to the creation of a 3D gel network that extends throughout the entire system. Typically, the gelation temperature of eggs during heating is around 65 to 75 °C, which is just above the thermal denaturation temperature of the major egg proteins. When the egg is cooled, the gel strength increases, which can be attributed to the increase in the strength of the hydrogen bonds between the protein molecules with decreasing temperature. Eggs therefore form heat-set thermo-irreversible gels. In this study, the final gel strength of the cooled egg whites was around 20 kPa (Fig. 7.6). In addition, the textural attributes of cooked hen's eggs are also often characterized using texture profile analysis (TPA). Measurements made in our laboratory using a TPA instrument on cooked egg white are shown in Fig. 7.7. The samples were compressed/decompressed twice and the change in force with time was recorded during this process. The hardness and fracture properties of cooked eggs (as well as various other textural attributes) can be obtained from these force-time profiles (Chap. 4). The hardness and fracture strength of the cooked hen's egg whites determined from the data shown

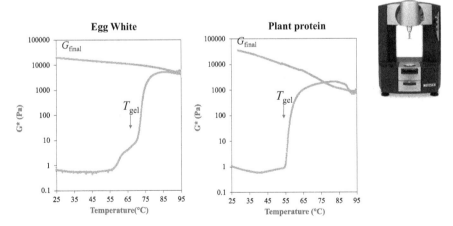

Fig. 7.6 Change in shear modulus with temperature for egg white and a model plant protein (10% rubisco). (Data kindly supplied by Hualu Zhou and Giang Vu (UMass). The image of the rheometer (Kinexus) used to make the measurements was kindly provided by Philip Rolfe from Netzsch (Selb, Germany) with permission)

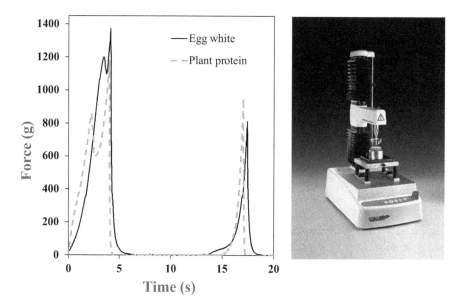

Fig. 7.7 Texture profile analysis of egg white and a model plant protein (10% rubisco). Cubed samples (1 cm^3) were compressed/decompressed at 2 mm/s to a final strain of 75% at ambient temperature using a stainless-steel cylindrical probe. (Data kindly supplied by Hualu Zhou and Giang Vu (UMAass). Photograph of texture analyzer kindly supplied by Marc Johnson, of Texture Technologies Corp. and used with permission)

in Fig. 7.7 were around 1350 N and 1040 N, respectively. The measured hardness was similar to the value of 1300 N reported by other researchers using a similar approach to characterize the texture of cooked hen's eggs (Kassis et al., 2010).

7.2.4.2 Emulsifying and Foaming

Eggs are used as ingredients in many foods because of their good emulsifying and foaming properties (McGee, 2004). For instance, they are used as emulsifiers to form mayonnaise, dressings, and sauces, whereas they are used as foaming agents in meringues and cakes. These properties are partly due to the ability of the globular proteins in egg white and egg yolk to adsorb to oil-water interfaces or air-water surfaces during homogenization or whipping, and then form protective coatings around the fat droplets or air bubbles (Fig. 7.8). However, the granules and lipoproteins from the egg yolk also play an important role in forming egg-based emulsions (Le Denmat et al., 2000). It was concluded that the LDL fraction from the plasma of the egg yolk was the most important emulsifier. The LDL particles are comprised of proteins and phospholipids, which are both amphiphilic molecules that may contribute to the overall surface activity. This is similar to the soluble globular proteins in egg white and egg yolk that are also surface-active molecules because they contain both non-polar and polar groups on their exteriors. However, proteins may

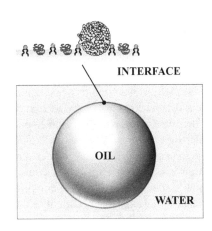

Fig. 7.8 Eggs are often used to stabilize emulsions and foams because of their ability to adsorb at oil-water interfaces. The interfaces may be comprised of egg proteins, phospholipids, or colloidal particles (like LDL, HDL, or granules). (Image of the lipoprotein is from Servier Medical Art (smart.servier.com) used under CC BY 3.0 (https://creativecommons.org/licenses/by/3.0/))

behave differently after becoming attached to an interface or surface, because the proteins may unfold (surface denaturation), which promotes their crosslinking with each other. The formation of such a 2D gel network around the fat droplets or air bubbles may be important in increasing their stability. The electrical characteristics of the interfacial layers formed by egg components are also important in determining emulsion and foam stability. The zeta-potential of the fat droplets in oil-in-water emulsions formed from eggs tend to go from positive at low pH to negative at high pH (Le Denmat et al., 2000), which can mainly be attributed to changes in the charge of the adsorbed protein molecules. The isoelectric point of egg white and egg yolk dispersions has been reported to be around pH 5 and 5.5, respectively (Li et al., 2018a). Under acidic conditions, the amino and carboxyl groups are protonated leading to a net positive charge (–COOH and –NH_3^+) but under neutral-to-alkaline conditions they are de-protonated leading to a net negative charge (–COO⁻ and –NH_2). Emulsions prepared using egg components as emulsifiers therefore tend to be relatively stable to aggregation at pH values well above and well below the isoelectric point where the electrostatic repulsion is strong but not around the isoelectric point where it is weak (Li et al., 2018a).

7.2.5 Flavor

In their raw state, hen's eggs only have a mild aroma whose character depends somewhat on the type of feed the hens are given, but the aroma tends to increase during storage due to various chemical changes (McGee, 2004; Plagemann et al., 2011). During cooking, egg proteins undergo further chemical reactions, which leads to a mild but distinct sulfur aroma (Jo et al., 2013; Warren et al., 1995). This sulfur aroma, which is attributed to hydrogen sulfide (H_2S) gas, is believed to mainly originate from the egg white and is attributed to unfolding of albumin molecules when they are heated above their thermal denaturation temperature, which exposes chemically reactive sulfur-containing amino acids that are normally buried in the protein interior. The buttery sweet aroma of scrambled eggs is believed to mainly originate from the egg yolk (Warren et al., 1995; Warren & Ball, 1991). The various kinds of volatile components in raw (Xiang et al., 2019) and cooked (Goldberg et al., 2012; Jo et al., 2013) eggs have been quantified using gas chromatography coupled with mass spectrometry (GC-MS). It has been reported that cooked eggs typically contain hundreds of different volatile components (MacLeod & Cave, 1975; McGee, 2004), which contribute to the overall flavor profile to different extents. In practice, the distinct flavor of eggs depends on the cooking method used, such as boiling, poaching, scrambling, or frying, since different heat-transfer media (water, oil) and time-temperature combinations are involved (Sheldon & Kimsey, 1985). Information about the aroma profile of hen's eggs before and after cooking is therefore important when trying to develop plant-based eggs that accurately simulate their flavor attributes.

7.3 Plant-Based Egg Analogs

7.3.1 Composition and Structure

As discussed in the previous section, hen's eggs naturally contain a wide variety of different constituents (*e.g.*, water, proteins, lipids, phospholipids, carbohydrates, minerals, vitamins, and carotenoids), which may be organized into colloidal structures (such as HDL, LDL, and granules) that all contribute to their desirable functional and sensory attributes. Typically, a number of different plant-based ingredients are therefore needed to accurately simulate their properties (Table 7.2). For instance, plant-based lipids may be used to create the hydrophobic core of fat droplets designed to mimic lipoproteins and granules. Lipids with different fatty acid compositions may be selected to create products with different nutritional profiles and chemical stabilities. For instance, omega-3 rich oils like algal oil or flaxseed oil could be used to create egg analogs fortified with these healthy lipids but special procedures may be required to inhibit their oxidation during storage, such as adding natural antioxidants or controlling storage and packaging conditions (Jacobsen, 2015; Jacobsen et al., 2013b). Moreover, amphiphilic plant-based proteins (such as soybean, pea, legume, or mung bean proteins) or phospholipids (such as sunflower or soybean lecithin) can be used as emulsifiers or foaming agents to stabilize emulsions or foams in applications where this is important. Alternatively, plant-based emulsifiers that are not normally found in eggs could be used for this purpose, like quillaja saponins or modified starches. Plant-based globular proteins also play an important role in determining the thickening, gelling, and binding characteristics of egg analogs during cooking, which is due to their ability to unfold and aggregate with their neighbors and other molecules. This kind of behavior is important in products such as plant-based scrambled eggs, desserts, or cakes.

Egg analogs intended to replace whole hen's eggs may also need one or more flavoring or coloring agents to mimic their characteristic sensory attributes. The flavoring agents should produce the sweet buttery flavor of egg yolk and the slight sulfur flavor of egg white when cooked but other flavors may also be required.

7.3 Plant-Based Egg Analogs

Table 7.2 Examples of some key ingredients used to formulate plant-based egg analogs and their functional attributes

Ingredient	Examples	Functions
Lipids	Algal, canola, corn, flaxseed, olive, vegetable, soybean, and sunflower oils	Form hydrophobic core of fat droplets; solubilize non-polar flavors, colors, and vitamins
Proteins	Soy, pea, mung bean, canola, chickpea, fava bean, lentil, lupin, and RuBisCO proteins	Emulsifiers, foaming agents, gelling agents, thickeners, binders
Phospholipids	Soy and sunflower lecithin	Emulsifiers, foaming agents
Polysaccharides	Gellan gum, starch, locust bean gum, xanthan gum, and gum Arabic	Thickeners, stabilizers
Pigments	Carotenoids, turmeric (curcumin), and amaranth	Provide yellow color
Flavor	Salts, spices, sugars, monosodium glutamate, sulfur-containing flavors	Taste and aroma profiles
Vitamins	Vitamins A, D, E, B_{12}	Nutrition
Minerals	Sodium, calcium, selenium…	Nutrition, crosslinking, taste
Crosslinkers	Transglutaminase	Crosslink proteins
Preservatives	Botanical extracts, EDTA, nisin	Antioxidants, chelating agents, antimicrobials
pH-modulators	Potassium citrate, citric acid, potassium carbonate, sodium bicarbonate, calcium carbonate, and tetrasodium pyrophosphate	pH adjustment, buffering agents

Definitions and descriptions of the functional attributes of plant-based ingredients are given in Chap. 2

Ideally, these flavors should be generated by the source of plant proteins and/or lipids used to formulate the product. However, in some cases, additional flavoring agents may be required to complete the flavor profile. The coloring agents should create the yellowish color of egg, which can be achieved by using natural pigments like carotenoids and curcumin. The micronutrient profile of egg analogs can be designed to match that of real eggs by adding suitable vitamins or minerals. Moreover, egg analogs can be fortified with similar health promoting molecules, such as nutraceuticals like carotenoids or curcumin, which can also act as natural colors and antioxidants. Acidulants, bases, or buffers may also need to be added to adjust and maintain the pH of the egg analogs. Finally, it may be necessary to add natural plant-based preservatives, such as antimicrobials, antioxidants, or chelating agents to inhibit microbial growth or undesirable chemical reactions.

Ideally, the manufacturers of plant-based eggs want to mimic the properties of real eggs using as few ingredients as possible and only those ingredients that consumers find acceptable ("label friendly"). Moreover, all the ingredients must meet the appropriate regulations for the country of sale, as well as being cost effective, available in an abundant and reliable supply, easy to use, have consistent functional attributes, and preferably be non-allergenic. Thus, there are many considerations manufacturers must account for when designing and developing these kinds of products. A more detailed discussion of different plant-based ingredients and their functionalities is given in Chap. 4.

7.3.2 Processing

Once an appropriate combination of functional ingredients has been selected, it is necessary to combine them and process them using a suitable manufacturing operation. In general, plant-based eggs can be produced in a variety of ways. In this section, we describe a simple approach that contains many of the key processing operations that may be used to manufacture egg analogs. This approach involves the production of a fluid whole egg analog that consists of protein-coated fat droplets dispersed within a viscous aqueous medium containing gelling globular proteins and thickening polysaccharides (Fig. 7.9). In this formulation, the fat droplets scatter light thereby leading to an opaque color, as well as contributing to the texture and acting as a solvent for any oil-soluble additives (such as colors, flavors, and vitamins). The globular proteins act as an emulsifier during emulsion formation, as well as a gelling agent during heating. Moreover, they may also provide foaming and binding properties in some food applications. The polysaccharides (hydrocolloids)

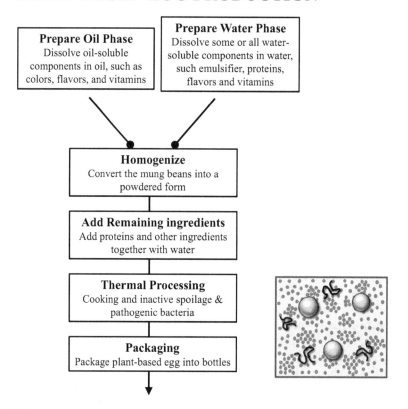

Fig. 7.9 A proposed method of producing plant-based eggs showing some of the key steps

7.3 Plant-Based Egg Analogs

thicken the aqueous phase, thereby inhibiting the creaming or sedimentation of any particulate matter (such as fat droplets, protein particles, or spices). A multistep process is required to form emulsion-based egg analogs using this approach:

- *Oil phase preparation*: The oil-soluble additives, such as hydrophobic flavors, pigments, or vitamins are dissolved in the oil phase prior to homogenization. The oil phase may need to be warmed to help solubilize any components that are crystalline at room temperature, such as carotenoids or curcumin.
- *Aqueous phase preparation*: An appropriate hydrophilic emulsifier, which is typically a plant-based protein and/or phospholipid, should be dissolved in water. Any other hydrophilic additives may also be added to the aqueous phase before homogenization, such as thickeners, gelling agents, colors, flavors, pH manipulators, and preservatives. However, there may also be advantages to adding some or all of these ingredients after homogenization, which have to be established depending on the precise nature of the ingredients and processes used. The pH of the aqueous phase should be controlled (typically just above neutral) to ensure the ingredients (especially the proteins) function as anticipated. In some cases, heating may be required to dissolve one or more of the ingredients (especially hydrocolloids), which can be carried out before or after they are mixed with the other ingredients. At this stage, it is critical to select an emulsifier with appropriate emulsion formation and stabilizing properties (McClements et al., 2017). The type and concentration of emulsifier used should be able to form small fat droplets during homogenization (to inhibit creaming), as well as generating strong repulsive forces between the droplets (to inhibit aggregation).
- *Homogenization*: The oil and water phases are usually blended together using a high-shear mixing device to form a coarse emulsion containing emulsifier-coated fat droplets (Fig. 7.10). The size of the fat droplets in the coarse emulsion can then be reduced further using a homogenizer, such as a high-pressure valve homogenizer or sonicator, which reduces their tendency to aggregate or cream during storage.
- *Formulation*: If required, any other water-soluble additives can be added to the aqueous phase of the final emulsion after homogenization. It is sometimes better to add these ingredients at this stage because they may interfere with droplet disruption inside the homogenizer, or they may block the homogenizer. In some cases, an enzymatic crosslinking agent can be introduced into the formulation, such as transglutaminase, to partially crosslink the proteins and increase the viscosity of the aqueous phase.
- *Thermal processing*: The product may then be thermally processed to promote unfolding and aggregation of the protein molecules and/or to deactivate enzymes or microorganisms, thereby extending the shelf life and ensuring it is safe. For fluid whole egg analogs, it is important to ensure that the proteins do not gel before the product is cooked. Consequently, the temperature-time profile of the thermal process may have to be carefully controlled.
- *Packaging*: Finally, the product is packaged into a suitable container, sealed, and then distributed.

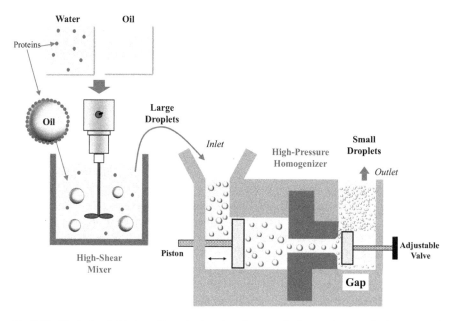

Fig. 7.10 The hydrophobic ingredients in plant-based eggs and other products can be incorporated in an emulsified form. Oil-in-water emulsions can be formed by blending oil, water, and emulsifier together and then further homogenizing the primary emulsion to reduce the droplet size

In some cases, it may not be necessary to carry out some of these steps because the oil-soluble ingredients (such as flavors, colors, and vitamins) can be purchased in a pre-emulsified form (often as a powder). These ingredients can simply be dispersed into the aqueous phase with the water-soluble ingredients and so there is no need for the oil phase preparation or homogenization steps.

7.3.3 Physicochemical Properties

7.3.3.1 Appearance

Fluid whole egg analogs should mimic the creamy yellow appearance of whole egg blends formed from hen's eggs (McClements & Grossmann, 2021). The creamy appearance of real eggs is due to colloidal particles that scatter light, such as LDL, HDL, and granules, which typically have dimensions ranging from a few hundred nanometers to a few micrometers. Consequently, it is important to include plant-based colloidal particles that can also scatter light in egg analogs. This can be achieved by creating oil-in-water emulsions containing emulsifier-coated fat droplets and/or including biopolymer-based particles, such as protein aggregates or starch granules. The size and concentration of these particles should be varied to match the lightness of the hen's egg product. The yellowish color of whole hen's

7.3 Plant-Based Egg Analogs

eggs can be mimicked by adding appropriate plant-based pigments, such as turmeric (curcumin), carotenoids (such as β-carotene, lutein, or lycopene), and/or amaranth. The type and concentration of these pigments should be optimized to match the $a*$ and $b*$ values of real egg. In some cases, it may be necessary to blend different pigments together to achieve the required color. It is also important to consider the storage stability of these colors, as well as the impact of food preparation conditions on their stability. For instance, color fading may occur during storage or when samples are held at elevated temperatures for extended periods. The tristimulus color coordinates of several plant-based eggs are compared with those of hen's eggs in Table 7.3. These measurements show that both types of eggs have a relatively high lightness (high $L*$) and yellowness (high positive $b*$) but low redness/greenness (low $a*$).

7.3.3.2 Rheology

Ideally, egg analogs should mimic the textural characteristics of both uncooked and cooked eggs, as well as how these characteristics change during the cooking process (McClements & Grossmann, 2021). Thus, they should be viscous shear thinning fluids at ambient temperatures but then change into gels with specific textural characteristics when heated above about 65 to 75 °C. The viscous properties of uncooked eggs can be simulated by including thickening agents in the formulation, such as hydrocolloidal gums (like gellan, locust bean, or xanthan gums) or protein aggregates (like crosslinked plant proteins). Ideally, the type and concentration of these thickening agents should be optimized to give an apparent shear viscosity *versus* shear rate profile that matches that of an uncooked whole hen's egg. In practice, measurements in our laboratory have shown that commercial plant-based eggs may have much higher shear viscosities than real eggs (Fig. 7.3 and Table 7.4). This is presumably because of the need to prevent creaming or sedimentation of the colloidal particles in the product during long-term storage, such as fat droplets, protein aggregates, or starch granules. Nevertheless, in general it is possible to closely match the rheology of liquid eggs by using plant-proteins. For instance, the shear viscosity profile of a plant-based protein solution (10% RuBisCo) measured in our laboratory was fairly similar to that of egg white (Fig. 7.3).

Table 7.3 Tristimulus color coordinates of whole hen's eggs and plant-based egg analogs (blended whole eggs)

Sample	$L*$	$a*$	$b*$	Reference
Hens whole egg (uncooked)	71	+6	+53	Measured in our laboratory
Hens whole egg (cooked)	74	−0.4	+23	Measured in our laboratory
Hens whole egg (cooked)	77	−3	+21	Li et al. (2018b)
Hens whole egg (cooked)	87	−4	+28	Kassis et al. (2010)
Plant-based egg (uncooked)	77	+0.6	+45	Measured in our laboratory
Plant-based egg (cooked)	74	+1.6	+42	Measured in our laboratory

The authors thank Dr. Hualu Zhou for kindly measuring the properties of the hen's and plant-based eggs in our laboratory (cooking was 90 °C for 30 min).

Table 7.4 Rheological properties of whole eggs and analogs

Sample	Yield stress (Pa)	K (Pa s^n)	n	Reference
Whole hen's egg	0.20	0.030	0.97	Panaite et al. (2019)
Whole hen's egg	0.009	0.013	0.97	Our laboratory
PB whole egg	9.7	0.11	0.95	Our laboratory

The data are the best-fit parameters fitted to the Herschel-Bulkley model, where K is the consistency index and n is the power index. The measurements in our laboratory were kindly done by Dr. Hualu Zhou (UMass).

The formation of a gel during heating can be achieved by adding a sufficient quantity of globular proteins that unfold and aggregate when heated above their thermal denaturation temperature. Ideally, these proteins should unfold around the same temperatures as egg proteins (like ovalbumin), which is around 63 to 85 °C (Table 7.1). The thermal denaturation temperatures of plant-proteins can be conveniently measured using different scanning calorimetry and compared to those of egg proteins (Fig. 7.4). A suitable plant protein can then be selected to mimic the unfolding behavior of egg proteins. The rheological properties of the gel formed, such as its gelling temperature, hardness, and fracture properties, should mimic those of hen's egg. The rheological properties depend on protein type and concentration, as well as pH and ionic strength, heating conditions (time-temperature), and ingredient interactions, which must all be considered when optimizing the formulation. For example, the change in shear modulus with temperature for a plant-based protein solution (10% RuBisCO) measured in our laboratory is compared to that of egg white in Fig. 7.6. These results show that the plant protein solution forms a gel when heated above about 60 to 70 °C and that the final gel strength is fairly similar to the gels formed by egg white proteins. Consequently, this plant protein may be useful for mimicking some of the desirable textural and cooking attributes of real eggs. Other plant proteins that unfold around the same temperature and form heat-set gels could also be used for this purpose, *e.g.*, mung bean or lupine proteins. Moreover, researchers have reported that the hardness of cooked whole hen's eggs measured by texture profile analysis was around 1300 N (Kassis et al., 2010). Consequently, it would be expected that plant-based egg analogs should exhibit somewhat similar textural properties after cooking. In our laboratory, we have found that hardness values measured by texture profile analysis for heat-set gels formed from model plant proteins (*e.g.*, 10 to 12.5% RuBisCo) are in this range (1100–1900 N).

7.3.3.3 Stability

When stored under refrigerator conditions, real eggs can be stored for around 60 days in their shells and around 4 days after being removed from their shells (Tetrick et al., 2019). Under similar conditions, cartons of processed fluid whole eggs have shelf lives of around 120 days and should be used within about 7 days after opening. Ideally, plant-based eggs should therefore have similar or better shelf lives than the real eggs they are designed to replace. Egg analogs can be created with shelf lives similar to cartons of whole fluid hen's eggs, but their composition and processing must be carefully controlled to prevent break down through physical, chemical, and microbial destabilization mechanisms.

Physical Stability Egg analogs may contain several different kinds of colloidal particles to provide their desirable optical, textural, and flavor characteristics, such as fat droplets, starch granules, protein aggregates, herbs, and spices. These particles may cream or sediment during storage leading to an undesirable visible layer at the top or bottom of the product, respectively. As discussed in Chap. 4, gravitational separation of colloidal particles can be inhibited by using a variety of strategies, with the most important one being reducing the particle size and increasing the aqueous phase viscosity. The particle size can be reduced by homogenization of the fat droplets or other particulate matter using mechanical devices such as colloid mills, high-pressure valve homogenizers, or sonicators (Fig. 7.10). Typically, the particle size should be less than a few hundred nanometers (<300 nm) to strongly retard creaming or sedimentation. In practice, it is often difficult or impractical to reach these small particle dimensions. For instance, any protein aggregates or cell fragments present may be difficult to disrupt using a conventional homogenizer because they block the valves or channels. Moreover, homogenizers are often expensive to purchase and operate. For these reasons, thickening agents, such as hydrocolloids, are often added to the aqueous phase to increase the viscosity and slow down particle movement. These hydrocolloids are typically hydrophilic polysaccharides that have extended molecular structures when dispersed in water, such as xanthan, guar, and locust bean gums. However, it is important that they provide a desirable flow behavior and mouthfeel. Some hydrocolloids lead to an undesirable slimy or lumpy texture that consumers may find unacceptable or undesirable.

The colloidal particles in egg analogs may also destabilize due to their tendency to aggregate with each other (Fig. 7.11). Aggregation tends to occur when the attractive forces acting between the particles outweigh the repulsive forces (Chap. 4). A number of phenomena can potentially lead to aggregation in egg analogs (McClements, 2015).

- *Van der Waals attraction*: This kind of colloidal interaction operates between all kinds of particles and tends to pull them together. Consequently, it is always important to ensure that there are some kinds of repulsive colloidal interactions that are strong enough to overcome the van der Waals attraction, such as steric or electrostatic repulsion.
- *Hydrophobic attraction*: Hydrophobic interactions are relatively strong and long-range and so will tend to pull particles together. The strength of hydrophobic interactions increases when the surface hydrophobicity of the particles increases. Consequently, any change in the system that increases the surface hydrophobicity of the particles may lead to destabilization by increasing the strength of the hydrophobic attractive interactions. For instance, heating globular

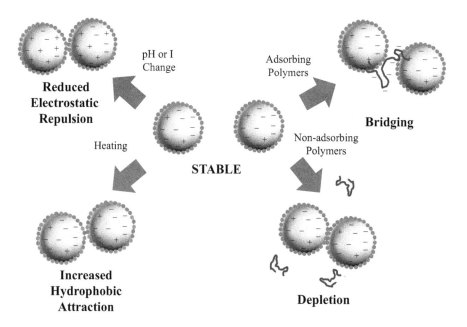

Fig. 7.11 The emulsifier-coated fat droplets in egg analogs may aggregate through several mechanisms, including bridging flocculation, depletion flocculation, reduced electrostatic repulsion, or increased hydrophobic attraction

proteins above their thermal denaturation temperature can promote the aggregation of free or adsorbed plant proteins in egg analogs.
- *Electrostatic repulsion*: Electrostatic repulsion is one of the most important colloidal interactions opposing the aggregation of the particles in egg analogs. The strength of the electrostatic repulsion usually increases as the surface charge increases and the ionic strength decreases. Consequently, any changes in the system that reduce the surface charge or increase the ionic strength may reduce the electrostatic repulsion and lead to particle aggregation. For example, if the pH changes towards the isoelectric point of the proteins, their surface charge will decrease, which can eventually lead to aggregation. Similarly, if high levels of salt, particularly multivalent counter-ions, are added to an egg analog then it may aggregate due to electrostatic screening or ion binding effects.
- *Steric repulsion*: The colloidal particles in egg analogs may be partially stabilized by the presence of hydrophilic polymers at their surfaces that generate a strong steric repulsion. For example, the adsorption of amphiphilic polymers, like gum arabic or modified starch, to the surfaces of fat droplets or other colloidal particles generates a strong and relatively long-range steric repulsion. If the thickness of the polymer layer is not enough, however, the steric repulsion may not be sufficient to stabilize the system. Globular plant proteins tend to form relatively thin interfacial layers around fat droplets because of their small molecular dimensions, leading to a relatively short-range steric repulsion. As a result,

they cannot prevent droplets from aggregating in the absence of other kinds of repulsive interactions, such as long-range electrostatic repulsion.
- *Bridging effects*: Egg analogs may contain polymers that are attracted to the surfaces of any particles present. As a result, they may form a bridge between a number of different particles that causes them to aggregate. The most common form of this bridging effect is due to electrostatic attraction between a polymer that has charged groups that are opposite in sign to those at the surfaces of the particles, *e.g.*, an anionic hydrocolloid may bind to cationic regions on the surfaces of proteins. This may happen even when the net charge on the proteins is neutral or negative, since there may still be some positively charged groups where the negatively charged polymers can attach.
- *Depletion effects*: Egg analogs may also contain non-adsorbed polymers (such as hydrocolloids) that are not attracted to the surfaces of the particles but can still promote their aggregation by generating an osmotic attraction between them *via* a depletion mechanism. In this case, the non-adsorbed polymers have a finite concentration in the aqueous bulk phase but a zero concentration in a narrow region surrounding each of the particles. The dimensions of this region are approximately equal to the hydrodynamic radius of the polymer molecules. This depletion zone exists because the center of the polymers cannot get closer to the polymer surface than their radius of hydration. As a result, there is a concentration gradient between the bulk aqueous phase and the depletion zone, which leads to an osmotic pressure. This osmotic pressure tends to push the colloidal particles together because when they aggregate with each other the total volume of the depletion zone in the system is reduced. The strength of the osmotic pressure increases with increasing number density of the polymers and tends to be stronger for polymers with more extended conformations. One must therefore be careful when choosing hydrocolloid-based functional ingredients (like thickening agents) when formulating plant-based egg analogs, otherwise they may promote aggregation.

Chemical Stability The physicochemical and sensory attributes of egg analogs may change during storage due to various types of chemical reactions. The nature of these reactions depends on the type of ingredients present, solution conditions such as pH and ionic composition, as well as external conditions, such as temperature, light, and oxygen levels. Some of the ingredients added to egg analogs are susceptible to oxidation, including unsaturated fatty acids, carotenoids, curcumin, and proteins (Boon et al., 2010; Hellwig, 2019; Waraho et al., 2011). Oxidation leads to rancidity for unsaturated fatty acids, color fading for natural pigments, and loss of nutritional value and functionality for proteins. Typically, oxidation rates increase with increasing temperature, oxygen levels, and light exposure, as well as with increasing pro-oxidant levels (such as transition metal ions, lipoxygenases, or photosensitizers). Consequently, it is important to control storage and transport conditions, use appropriate packaging materials, and control product composition, such as adding antioxidants and chelating agents, as well as avoiding, removing, or deactivating any prooxidants.

A number of ingredients that may be used in plant-based egg analogs are also susceptible to hydrolysis during storage, *e.g.*, polysaccharides, phospholipids, and lipids. For example, polysaccharides may be hydrolyzed by enzymes in plant materials, which could cause depolymerization, thereby reducing their thickening or gelling properties. Fatty acids may be released from phospholipids or lipids due to hydrolysis by esterases in plant materials. Consequently, it is often important to deactivate these enzymes, for example by thermal processing, prior to distributing the product. Hydrolysis reactions may also increase under some pH conditions, especially at elevated temperatures. In addition, proteins may react with reducing sugars at elevated temperatures due to the Maillard reaction leading to a brown color (Tamanna & Mahmood, 2015). The browning of an egg analog is important during certain cooking processes, such as frying, grilling, and baking, since it may lead to a desirable or undesirable brown color on the surface of the product. Consequently, it is important to understand the nature of the various chemical reactions occurring in egg analogs and to develop effective means of controlling them.

Microbial Stability Plant-based eggs are a rich source of nutrients and are therefore susceptible to microbial contamination. The shelf life and safety of egg analogs are influenced by the presence of any spoilage or pathogenic microorganisms that may contaminate them. The adverse effects of these microorganisms can be reduced by subjecting the egg analogs to appropriate thermal processing prior to distribution. For example, high temperature-short time (HTST) pasteurization could be used, which involves heating the product at 60 °C for around 7.6 min (assuming the same conditions as whole hen's egg) (USDA, 2020). These conditions should be sufficient to deactivate the microorganisms without causing gelation of the proteins in the product prior to utilization but this also depends on the nature of the microbial contamination present and needs to be evaluated individually for every product. Antimicrobial agents may also be added to the products to reduce the growth of microorganisms, such as nisin or some botanical extracts. In addition, good handling practices should be utilized throughout processing, packaging, storage, and distribution operations.

7.3.4 Functional Properties

7.3.4.1 Thickening, Gelling, and Binding

Hen's eggs are used as functional ingredients in many food products because of their thickening, gelling, or binding properties. For instance, eggs are used to thicken custards, to solidify flans and to bind the ingredients together in some meat, fish and bakery products (McGee, 2004). Consequently, it is important to ensure that egg analogs carry out similar functions when used in applications that require this kind of behavior. The ability of egg proteins to thicken, gel, and bind foods is due to protein unfolding when heated above their thermal denaturation temperature, and then aggregate with each other or with other substances in their environment

PLANT-BASED EGG

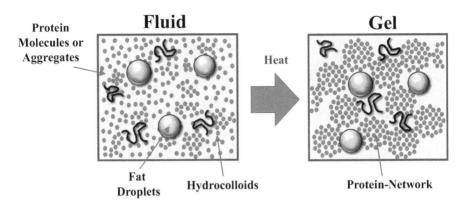

Fig. 7.12 Plant-based eggs can be formed by heating a mixture of proteins, fat droplets, and hydrocolloids dispersed in water above the thermal denaturation temperature of the proteins

(Fig. 7.12). Thickening occurs when the protein concentration is not high enough to form a 3D network throughout the system (*e.g.*, in a custard), whereas gelling occurs when it is high enough (*e.g.*, in a cooked egg or flan). Binding occurs when the egg proteins bind to other substances in their environment, such as meat or fish proteins.

Many plants contain globular proteins that are able to unfold and aggregate with their neighbors under suitable conditions, thereby giving them thickening, gelling, and binding properties. For instance, soybean, pea, chickpea, bean, and sunflower proteins have all been shown to unfold and aggregate when heated under appropriate conditions (Hettiarachchy et al., 2013). However, plant proteins may behave quite differently from egg proteins when heated. For instance, there may be differences in the temperature at which unfolding occurs, the amount of protein required to form a gel, the sensitivity of protein unfolding and aggregation to environmental conditions (such as pH and ionic strength), and the nature of the final gels formed (such as their appearances, texture, and water holding). Consequently, it is important to identify a suitable source of plant proteins and suitable processing conditions to simulate these functional attributes of egg.

The thermal denaturation temperatures of plant proteins (Table 7.5) are usually different from those of egg proteins (Table 7.1). Indeed, many plant proteins have T_m values much higher than those of egg proteins, which means they must be heated to a higher temperature and for a longer time before they unfold and aggregate. This difference in thermal behavior would affect the cookability and functionality of egg analogs. Many consumers are familiar with cooking real hen's eggs and expect eggs to behave in a certain way. As a result, they may find large differences in cookability of eggs and egg analogs unexpected and undesirable. In addition, egg white and egg yolk form gels with different characteristics because of their different compositions and structures. The gels formed from egg white tend

Table 7.5 Molecular characteristics of various kinds of plant proteins

	pI	T_m (°C)
Soy	4.5–5	80–93
Pea	4.5	75–79
Lentil	4.5	120
Chickpea	4.5	90
Lupin	4.5	79–101
Canola	4.5	84–102
Mung bean	4.7	81–83
RuBisCO	4.7	74

In nature, these proteins often exist as multimers consisting of one or more similar or different proteins. Adapted from McClements and Grossmann (2021) and Tang and Sun (2010). The RuBisCO protein data was measured in our laboratory.

Key: M_W = molar mass; pI = isoelectric point; and T_m = thermal denaturation temperature

to be glossy white and brittle, whereas those formed from egg yolk tend to be matt yellow and crumbly. The gels formed by blends of egg white and egg yolk have characteristics somewhere between those formed by the individual components. Consequently, plant-based alternatives may have to be designed to mimic the properties of the specific egg product they are intended to replace, such as egg white, egg yolk, or whole egg.

The nature of the gels formed by plant proteins when they are heated are often considerably different from those formed from egg proteins. In particular, the appearance, texture, and water holding properties may be very different. The gels formed by egg white have a glossy smooth white appearance that is difficult to mimic using plant proteins. However, the gels formed by egg yolk and whole egg have a matt yellowish appearance that is easier to simulate. It is important to simulate the hardness, fracture properties, and other textural attributes of the gels formed by eggs since these are expected by consumers. As mentioned earlier, the textural attributes of cooked whole hen's eggs have been determined by texture profile analysis: hardness (1300 N), springiness (2.04), cohesiveness (0.61), gumminess (980), chewiness (1600), and resilience (0.32) (Kassis et al., 2010). In general, it will be important to use instrumental methods, such as compression or shear tests, to quantify the rheological properties of cooked eggs and then optimize the formulations of egg analogs to create the same textural attributes. For example, measurements of the shear modulus *versus* temperature during heating and cooling can be useful to assess the potential behavior of a plant protein candidate as an egg analog during cooking (Fig. 7.6). Moreover, it is also important to carry out sensory analysis to ensure that the optimized egg analogs are perceived as having the same textures and mouthfeels as cooked hen's eggs (Rondoni et al., 2020).

7.3.4.2 Emulsifying and Foaming

Another key functional attribute of the ingredients found in hen's eggs in many foods is their ability to adsorb to oil-water or air-water interfaces where they create interfacial films that stabilize the fat droplets in emulsions or air bubbles in foams (Fig. 7.8). For instance, eggs are often used in mayonnaise, dressings, and sauces as emulsifiers, while they are used in meringues and cakes as foaming agents (McGee, 2004). Consequently, it is important for egg analogs to contain surface active substances that can promote the formation and stability of emulsions and foams. The most commonly used plant-based surface active substances for this purpose are proteins, phospholipids, polysaccharides, and saponins (McClements et al., 2017). These are typically water-soluble amphiphilic molecules that can adsorb to interfaces and form protective films that generate strong electrostatic and/or steric repulsion between the fat droplets or air bubbles. In addition, their ability to interact with each other at the interface and to form viscoelastic films with some mechanical rigidity may also contribute to their ability to stabilize fat droplets and air bubbles. It is also possible to use plant-based colloidal particles to stabilize emulsions and foams through a Pickering mechanism, *i.e.*, the oil droplets or air bubbles are coated by a layer of small particles rather than surface-active molecules (Amagliani & Schmitt, 2017). Particle-stabilized emulsions and foams are particularly resistant to coalescence because of the thick dense interfacial layer formed around the fat droplets or air bubbles.

As discussed in Chap. 2, there are a wide variety of surface-active plant-based substances that can be used as emulsifying and foaming agents, including proteins (*e.g.*, soy, pea, chickpea, lentil, faba bean, and mung bean proteins), polysaccharides (*e.g.*, gum arabic and modified starch), phospholipids (*e.g.*, sunflower or soy lecithin) and saponins (*e.g.*, quillaja saponin), which can be used in isolation or combination. Consequently, these plant-based substances may be suitable for applications where emulsification or foaming properties are important in egg analogs. However, it is important to select the most appropriate plant-based ingredient(s) for the requirements of the end product.

7.3.5 *Characterization of Egg Analogs*

In this section, we provide a brief overview of the analytical methods typically used to characterize egg analogs during the research and development stage. More details about the different methods available are given in Chap. 4. First, the optical properties of egg analogs can be characterized by taking digital images of their appearance under standardized lighting conditions. In addition, quantitative information about their appearance can be obtained by measuring their tristimulus color coordinates (*e.g.*, $L^*a^*b^*$) using an instrumental colorimeter. Second, the rheology of fluid egg analog products is usually assessed by measuring their apparent shear viscosity *versus* shear rate profile. The thermal gelation of egg analogs can be characterized by measuring their dynamic shear modulus *versus* temperature during heating and cooling at a controlled rate (Fig. 7.6). The texture of gelled egg analogs can be conveniently determined using textural profile analysis (TPA) where the sample is compressed/decompressed twice under controlled conditions and parameters such as hardness, cohesiveness, and springiness are obtained. Third, insights into the aggregation stability of egg analogs can be obtained by measuring their microstructure using light or electron microscopy or by measuring their particle size distribution using light scattering instruments. Moreover, measurements of the zeta-potential *versus* pH can provide useful insights into their stability as this provides information about the electrical charge on the particle's surfaces. Lastly, the stability of egg analogs to gravitational separation can be determined by taking digital photographs over time or by using laser scanning devices.

7.3.6 Commercial Egg Analogs

There are already several successful egg analogs on the market that are designed to simulate either uncooked or cooked whole egg products. For instance, *JUST Egg* is a product designed to mimic fluid whole eggs, which can be used to make scrambled eggs by cooking it in a frying pan (Fig. 7.13), whereas *JUST Egg Folded* is a product designed to mimic cooked egg slices that can be prepared using an oven, microwave, skillet, or toaster and then consumed as part of a breakfast sandwich (www.ju.st). The fluid eggs come in a plastic bottle, while the folded egg slices

Fig. 7.13 Plant-based scrambled egg produced by cooking a fluid egg analog (JUST Egg, shown). (Image kindly supplied by *Eat Just Inc.* (San Francisco, CA) and used with permission)

7.3 Plant-Based Egg Analogs

Fig. 7.14 Plant-based fluid egg (left) and pre-cooked egg slices (right) are commercially available, such as those sold by *Eat Just Inc.* (www.ju.st). (Image kindly supplied by *Eat Just Inc.* (San Francisco, CA) and used with permission)

come in the form of frozen yellowish blocks that are stored in a box (Fig. 7.14). In these products, mung bean protein is utilized as a gelling agent to provide the semi-solid textural attributes in the final products. Mung beans contain globular proteins that unfold and aggregate when heated above their thermal denaturation temperature, which is around 82 °C (Table 7.5), consequently they can be used to form gels when cooked. The fluid egg analogs also contain gellan gum, which will thicken the viscosity of the aqueous phase and prevent creaming and sedimentation from occurring during storage. Plant-based coloring agents, including carotenoids from carrots and curcumin from turmeric, are included in the egg analogs to provide an egg-like yellowish color. The creaminess or lightness of the egg analogs can be attributed to light scattering by colloidal particles, such as canola oil droplets and aggregated mung bean proteins. A number of other functional ingredients are also included in the egg analogs to enhance their properties, including natural flavors (such as onion, garlic, sugar, and salt), pH regulators (such as citrates, phosphates, and bicarbonates) and preservatives (such as nisin). The egg analogs also list transglutaminase on the label, which is presumably utilized to crosslink the mung bean proteins, which may provide an increased viscosity in the original product, as well as modify the strength of the gels formed after cooking (Gharibzahedi et al., 2018).

7.4 Comparative Nutrition, Sustainability, and Ethics of Eggs and Egg Analogs

As highlighted in Chap. 1, there are numerous reasons why people may switch from animal-based foods to plant-based alternatives. Three of the most common reasons that consumers give are (*i*) plant-based foods are healthier; (*ii*) plant-based foods are better for the environment; and (*iii*) plant-based foods are more ethical. In this section, we briefly compare the nutrition, sustainability, and ethical aspects of eggs *versus* egg analogs.

Nutrition The nutritional contents of hen's eggs and commercial whole egg analogs are compared in Table 7.6. On an equal weight basis, the egg analogs contain more calories, total fat, polyunsaturated fat, salt, and carbohydrates than the hen's eggs but contain less saturated fat and cholesterol. Both products contain similar

Table 7.6 Comparison of nutrient content of hen's eggs and plant-based hen's eggs

	Percentage content (w/w)		
	Plant-based egg (fluid)	Plant-based egg folded	Hen's egg
Calories per serving (kcal)	77 kcal/44 mL	100 kcal/57 g	72 kcal/50 g
Calorie density (kcal per 100 g)	175	175	144
Total fat	11.4	12.3	10
Saturated fat	0.0	0.9	3.2
Trans fat	0.0	0.0	0
Polyunsaturated fat	3.4	3.5	1.8
Monounsaturated fat	6.8	7.9	3.6
Cholesterol	0.0	0.0	0.411
Sodium	0.386	0.526	0.129
Total carbohydrate	2.3	5.3	1
Dietary fiber	0.0	0.0	0
Sugars	0.0	0.0	0.2
Protein	11.4	12.3	12

Values are given per 100 g of product. Information obtained from the USDA FoodData Central database (https://fdc.nal.usda.gov/). The plant-based egg products are both manufactured by *Eat Just* (San Francisco, CA). Note: there will be appreciable differences in these values depending on the origin of the animal and plant-based products compared.

Ingredients:

JUST Egg (Fluid): Water, Mung Bean Protein Isolate, Expeller-Pressed Canola Oil, Contains less than 2% of Dehydrated Onion, Gellan Gum, Natural Carrot Extractives (color), Natural Flavors, Natural Turmeric Extractives (color), Potassium Citrate, Salt, Soy Lecithin, Sugar, Tapioca Syrup, Tetrasodium Pyrophosphate, Transglutaminase, Nisin (preservative). (Contains soy)

JUST Egg Folded (Semi-solid): Water, Mung Bean Protein Isolate, Expeller-Pressed Canola Oil, Corn Starch, Contains less than 2% of Baking Powder (sodium acid pyrophosphate, sodium bicarbonate, corn starch, monocalcium phosphate), Dehydrated Garlic, Dehydrated Onion, Natural Carrot Extractives (color), Natural Turmeric Extractives (color), Salt, Transglutaminase

Hen's Egg: Grade A, Large whole egg

total protein contents. But the nutritional quality of hen's egg proteins is typically better than that of plant proteins because they contain all the essential amino acids needed for human health at high concentrations, whereas some kinds of plant proteins are lacking certain kinds of amino acids at high concentrations. This problem can be overcome by utilizing combinations of different plant proteins to provide the full complement essential amino acids in the final product. It may also be important to include fatty acids that are believed to provide health benefits to egg analogs to enhance their nutritional profiles, such as omega-3 fatty acids but it will be important to stabilize these against oxidation during storage and food preparation. Hen's eggs also contain a wide range of vitamins and minerals that are also beneficial to human health, which may be lacking in egg analogs, unless they are fortified. However, fortifying plant-based eggs increases ingredient and processing costs, as well as leading to a much longer ingredient label. It should be noted, however, that the feed of chickens is already often fortified with nutrients and nutraceuticals already and it would therefore be more efficient to directly fortify the plant-based product. Moreover, it is important that any vitamins and minerals added are present in a chemically stable and bioavailable form. It is also important to note that hen's eggs and their analogs may promote different metabolic and physiological responses after consumption, which could have health impacts that are not currently understood. Finally, the allergenicity of the proteins and other ingredients in egg analogs should be considered. A significant fraction of the population has allergies to hen's egg products, which limits their consumption. This problem could be overcome by formulating egg analogs using non-allergenic ingredients. However, many widely used plant-based ingredients are also allergenic, such as soy proteins. As nutritional science and food formulation evolve, it is likely that future generations of plant-based eggs will be designed to have better nutritional and health profiles.

Sustainability Another important driver for consumer adoption of plant-based eggs is their lower impact on the environment than hen's eggs. A life cycle analysis (LCA) carried out for *Eat Just* reported that there was 98% reduction in water use, 86% reduction in land use, and 93% reduction in carbon dioxide emissions for their egg analogs compared to chicken eggs (https://www.ju.st/learn). In a peer-reviewed study, a detailed analysis of the environmental impacts of many different kinds of plant- and animal-based foods was carried out (Poore & Nemecek, 2018). This analysis also showed that egg products had a much bigger impact on the environment than tofu, which is often used as a plant-based protein-rich alternative to egg (Table 7.7). For instance, there was a 61%, 82%, 53% and 80% reduction in land use, water use, greenhouse gas emissions, and eutrophication associated with switching from hen's eggs to tofu, respectively. However, there is still a need for comprehensive LCA studies that compare plant-based egg products with hen's egg along the value chain.

Ethics Many consumers are concerned about the welfare of the chickens used by the egg industry. Some farmers keep free range chickens that have a considerable degree of freedom, but most chickens are raised in large factory farms where they are housed in conditions that are unnatural and stressful to the birds. There were around 69 billion chickens slaughtered in 2018 around the world (https://ourworldindata.org/meat-

Table 7.7 Comparison of environmental impact of the production of chicken eggs and tofu reported per 100 g of protein

Parameter	Eggs	Tofu	% Reduction
Land use (m^2)	5.7	2.2	61%
Water use (L)	521	93	82%
GHE (kg CO$_2$–eq)	4.21	1.98	53%
Pollution (g PO$_4^{3-}$–eq)	19.6	3.9	80%

Here, GHE is greenhouse gas emissions, while pollution refers to eutrophication (in grams of phosphate equivalents).

production). Many of these birds are killed for their meat, but many of them are also killed because they have reached the end of their egg-laying lives or they are not suitable for egg laying (males). Switching from hen's eggs to egg analogs may therefore have considerable ethical 'benefits' by reducing the huge number of birds that are currently confined and slaughtered by the egg and chicken meat industries.

7.5 Egg Products

Eggs are used as ingredients in a wide range of food products because of their diverse range of functional attributes, such as emulsifying, foaming, thickening, binding, gelling, antioxidant, and antimicrobial properties. In this section, we focus on a few examples where plant-based ingredients are being utilized to replace eggs as functional ingredients in some products. It should be noted that, there are different approaches to using plant-based ingredients for this purpose. For instance, one could use an egg analog ingredient, such as one designed to mimic egg white, egg yolk, or whole egg, which can be used in all the same ways as the real hen's egg equivalent. Alternatively, one could use a simpler plant-based ingredient that only provides the functional attributes required for a particular product, such as a plant protein that can form and stabilize the fat droplets in a dressing or the air bubbles in a cake. In this section, we focus on a number of products where plant-based ingredients can be used to replace hen's egg-based ones. We note, there are several other food products where these ingredients could also be used that are commercially important but are not covered here, such as sauces, dips, soups, nutrition bars, pasta, and noodles (Grizio & Specht, 2021).

7.5.1 Emulsified Products: Mayonnaise and Salad Dressings

Eggs are commonly used in emulsified foods, such as mayonnaise and salad dressings (Fig. 7.15), to provide desirable textural, stability, and flavor aspects (Ma & Boye, 2013). Indeed, the standards of identity of these products stipulated by federal law in the US specifies that they should contain a certain amount of egg. An

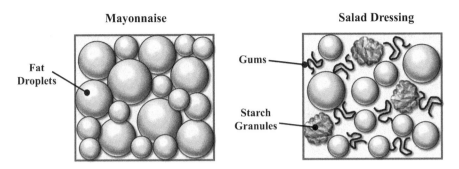

Fig. 7.15 Plant-based ingredients can be used to simulate the properties of traditional mayonnaise and salad dressing products

important role of the eggs in these emulsified products is their ability to form and stabilize the fat droplets. Both egg white and egg yolk contain surface active substances, such as proteins and phospholipids, that can adsorb to oil-water interfaces and form a protective coating around fat droplets that inhibits their aggregation (Fig. 7.8). Eggs also contain surface active colloidal particles, such as LDL, HDL, and granules, that can adsorb to interfaces and stabilize emulsions through a Pickering mechanism (Anton, 2013). In particular, the surface active molecules and colloidal particles in egg yolks play a critical role in the formation and stabilization of mayonnaise and salad dressing type products (Taslikh et al., 2021).

These emulsified products are usually produced by homogenizing an oil and water phase together in the presence of a hydrophilic emulsifier. A variety of homogenizers are used commercially for this purpose with the most common being high-shear mixers, colloid mills, and high-pressure valve homogenizers (McClements, 2015). The homogenizer used for a particular product depends on the fat droplet size required (small or large), as well as the physicochemical attributes of the end product (such as its viscosity). In many cases, a coarse emulsion premix is formed first using a high-shear mixer and then the droplet sizes are reduced further using a colloid mill or high-pressure valve homogenizer (Fig. 7.9).

7.5.1.1 Composition and Structure

Structurally, these food emulsions consist of emulsifier-coated fat droplets suspended in an aqueous medium, which may also contain various other ingredients, such as thickeners, flavors, preservatives, and pH modulators (McClements, 2015; McClements & Grossmann, 2021). The fat content in these products may vary from a few percent (low-fat dressings) to around 75% (mayonnaise), which influences the type and amount of emulsifier required. Typically, the concentration of emulsifier needed increases as the fat content increases and the droplet size decreases but it also depends on the emulsifier type. In particular, the amount of emulsifier required increases as its surface load increases, *i.e.*, the amount required to cover a unit surface area. It also has to be considered that the aqueous phase in mayonnaise and salad dressings is usually acidic (pH 3–4), which provides the characteristic sour taste and increases the microbial stability. The solubility and functionality of many plant-based emulsifiers (especially proteins) depend on the pH of the aqueous phase. Consequently, it is important that any plant-based emulsifiers used are compatible with the acidic conditions found in these products. Mayonnaise and salad dressings may also contain various other functional ingredients that could interact with the emulsifier used, such as thickeners, preservatives, chelating agents, pH modulators, and flavors. Consequently, it is important to ensure that any emulsifier-ingredient interactions do not have adverse effects on product quality and shelf life.

7.5.1.2 Physicochemical Properties

Knowledge of the physicochemical attributes of mayonnaise and salad dressings, such as their appearance, texture, and stability, is important when designing plant-based alternatives that accurately match their desirable functional and sensory attributes.

Appearance Traditional egg-based mayonnaise and salad dressings are optically opaque materials due to the strong light scattering caused by the relatively high concentration of fat droplets they contain (Sect. 7.2.3.1). In addition, other kinds of colloidal particles that scatter light may also contribute to their opacity, including lipoproteins, starch granules, and protein particles. As discussed in Chap. 4, the lightness of emulsified products tends to increase steeply from 0% to 5% fat but then remains relatively high and constant at higher fat contents, which is important when designing low-fat versions of these products. Egg-based emulsified products also have a distinct color due to selective absorption of light by chromophores. For instance, both mayonnaise and some salad dressing have a yellowish appearance due to the presence of carotenoids and other natural pigments arising from the egg yolk. The tristimulus color coordinates of selected egg-based emulsified foods measured using instrumental colorimetry are summarized in Table 7.8. These measurements show that the egg-based mayonnaise and salad dressings typically have a high lightness (>75) and yellowness (>+30). Researchers have shown that plant-based mayonnaise analogs can be formulated using plant-derived oils (75% soybean oil) and protein emulsifiers (3% chickpea, broad bean, or lupin isolate), as well as various other additives, including water, salt, xanthan gum, mustard, vinegar, garlic, onion, citric acid, potassium sorbate, and sodium benzoate (Alu'datt et al., 2017). The color coordinates ($L*a*b*$) of the mayonnaise analogs were fairly similar to those of traditional egg-based mayonnaise (Table 7.8). In another study, plant-based salad dressing consisting of 20% olive oil, 1.1% pea protein emulsifier, and various other ingredients (water, xanthan gum, starch, saffron powder, pepper, garlic, and salt) was shown to have fairly similar color coordinates as the egg-based ones (Table 7.8) (Kaltsa et al., 2018). These studies highlight the potential of creat-

7.5 Egg Products

Table 7.8 Tristimulus color coordinates of selected egg-based and egg analog emulsified foods

Emulsified food	L*	a*	b*	Reference
Egg-based mayonnaise	79.5	+7.7	+32.5	Huang et al. (2016)
Egg-based mayonnaise	73.4	+7.1	+35.5	Alu'datt et al. (2017)
Plant-based mayonnaise	74.4	+5.3	+26.5	Alu'datt et al. (2017)
Plant-based mayonnaise	85.4	+0.8	+19.6	Our laboratory
Egg-based salad dressing	77.8	+0.94	+32.6	Song and McClements (2021)
Plant-based salad dressing	79.8	−10.8	+45.3	Kaltsa et al. (2018)

The measurement of the plant-based mayonnaise was kindly made by Hualu Zhou (UMass).

ing plant-based mayonnaise and salad dressings that have appearances similar to those of egg-based ones. In general, the appearance of plant-based emulsified products can be matched to those of egg-based ones by controlling the size and concentration of fat droplets (and other colloidal particles) they contain to control their lightness, as well as by controlling the type and concentration of natural pigments (such as carotenoids, curcumin, saffron, or annatto) to control their color.

Texture Mayonnaise and salad dressings tend to be semi-solid materials that exhibit non-ideal plastic behavior (McClements, 2015). In mayonnaise, this is because the fat concentration is so high (>70%) that the fat droplets are packed tightly together and cannot easily move past each other when an external force is applied, thereby generating some mechanical rigidity. In salad dressings, a similar effect occurs, but in this case the disperse phase is made up of a mixture of fat droplets and hydrocolloids (such as starches or gums). These hydrocolloids contribute to the high disperse phase volume fraction of the system but can also increase the attraction between the fat droplets through a depletion mechanism, thereby forming a 3D network of aggregated droplets with some mechanical rigidity (Parker et al., 1995).

Measurements of the rheological properties of mayonnaise and salad dressings have shown that they exhibit non-ideal plastic behavior that can be described by the Herschel Bulkley model (Kaltsa et al., 2018):

$$\tau - \tau_0 = K\dot{\gamma}^n \tag{7.1}$$

Here, τ_0 is the yield stress, K is the consistency index, and n is the power index. The value of the power index is one for ideal fluids ($n = 1$), less than one for shear thinning fluids ($n < 1$), and greater than one for shear thickening fluids ($n > 1$). Below the yield stress, the emulsion behaves like an elastic solid but above the yield stress it flows. It has to be noted that the above equation is only appropriate once the yield stress has been exceeded ($\tau > \tau_0$). Measurements of the yield stress, consistency index, and power index for various egg- and plant-based mayonnaise and salad dressing products are shown in Table 7.9. These results indicate that traditional egg-based mayonnaise and salad dressings have an appreciable yield stress that must be exceeded before they flow, and that they exhibit strong shear thinning behavior

Table 7.9 Rheological properties of selected egg-based and egg analog emulsified foods.

Food	Yield stress (Pa)	K (Pa s^n)	n	Reference
Egg-based mayonnaise	22.3	36.7	0.40	Huang et al. (2016)
Egg-based mayonnaise	7.5	44.7	0.32	Yuceer et al. (2016)
Egg-based mayonnaise	85–198	8.5–25.8	0.38–0.49	Katsaros et al. (2020)
Plant-based mayonnaise	81.4	82.6	0.21	Our laboratory
Egg-based salad dressing	47	16.2	0.52	Hernandez et al. (2008)
Plant-based salad dressing	9.20	2.55	0.40	Kaltsa et al. (2018)

There are appreciable variations among products due to their different compositions and structures, as well as due to differences in the analytical methods used to determine their properties. The measurements in our laboratory were kindly carried out by Dr. Hualu Zhou (UMass).

($n \approx 0.4$). The precise values of the Herschel-Bulkley parameters of these emulsified egg-based products depend on their formulation and processing (such as oil content, droplet size, and thickener addition). Research has shown that plant-based emulsified products can be formulated that have similar rheological properties as egg-based ones. For instance, a plant-based salad dressing consisting of 20% olive oil, 1.1% pea protein and various other ingredients (water, xanthan gum, starch, saffron powder, pepper, garlic, and salt) had fairly similar rheological parameters as egg-based ones (Table 7.9) (Kaltsa et al., 2018). In addition, the shear stress versus shear rate profile of a commercial plant-based mayonnaise product is shown in Fig. 7.16, which clearly indicates that it exhibits non-ideal plastic behavior with a yield stress and shear thinning. In general, the rheological properties of plant-based emulsified foods can be matched to their egg-based counterparts by controlling the fat droplet concentration, size, and molecular/physical interactions, as well as by controlling the types and amounts of thickening agents added.

Physical Stability Emulsified food products are thermodynamically unstable materials that tend to break down through a variety of mechanisms, including creaming, sedimentation, flocculation, and coalescence (McClements, 2015) (Chap. 4). Consequently, any plant-based emulsifier used to replace egg in these products must be able to successfully inhibit these processes. Mayonnaise is a highly concentrated oil-in-water emulsion that exhibits semi-solid characteristics because the fat droplets are packed so tightly together that they cannot easily move past each other when an external force is applied. Consequently, the fat droplets are stable to creaming because they cannot move upwards, even though they are relatively large (typically a few to tens of micrometers in diameter). However, the fat droplets in mayonnaise do have a propensity to coalesce because they are packed so tightly together for extended periods. Droplet coalescence in egg-based mayonnaise is usually inhibited through a Pickering mechanism due to the presence of colloidal particles from egg, such as LDL, HDL, granules, and spices, that adsorb to the fat droplet surfaces (Wang et al., 2020). Consequently, it may be important to utilize plant-based colloidal particles to stabilize mayonnaise analogs, since molecular-based emulsifiers are typically much less effective at inhibiting coalescence. Various kinds of plant-based colloidal particles can be used for this purpose, which are typically assembled

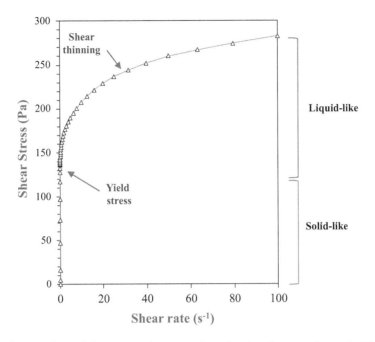

Fig. 7.16 Dependence of shear rate on shear stress for a plant-based mayonnaise product, indicating plastic behavior characterized by a yield stress and shear-thinning. (Data kindly supplied by Hualu Zhou (UMass))

from proteins, polysaccharides, and/or polyphenols (Sarkar & Dickinson, 2020; Schroder et al., 2021). In salad dressings, where the fat droplets are not so close together, molecular or particulate plant-based emulsifiers can be used to stabilize the emulsions, since these systems are far less susceptible to coalescence. However, the lower fat droplet concentration in these products (especially in low or reduced fat versions) means that they are susceptible to creaming. For this reason, these products usually contain thickening agents like starch or xanthan gum that increase the viscosity of the aqueous phase, thereby inhibiting fat droplet movement (as well as other particulate matter, such as herbs and spices). A similar approach is therefore required to prevent creaming in plant-based salad dressings. However, it is also important to match the textural attributes and mouthfeel of the final product to that of egg-based ones.

Chemical Stability Egg-based emulsified products are susceptible to deterioration through chemical reactions that lead to a reduction in product quality, *e.g.*, oxidation or hydrolysis (Jacobsen et al., 2013a). When formulating plant-based analogs it may therefore be important to inhibit these kinds of undesirable reactions. In particular, plant-based emulsified foods containing high levels of polyunsaturated fats (like algal or flaxseed oils) are highly prone to lipid oxidation, which leads to off-flavors (rancidity). The shelf life of these products can be improved by using similar strategies as used for conventional egg-based products, *e.g.*, by adding antioxidants or chelating agents, reducing exposure to light, heat, or oxygen, removing or deactivating pro-oxidants, and controlling the structural organization of the food matrix (Jacobsen et al., 2008; McClements & Decker, 2000). In general, it is important to understand the ingredients used to formulate the product and identify any chemical reactions that may occur. This knowledge can then be used to identify effective strategies to inhibit undesirable chemical reactions.

Microbial Stability Egg-based mayonnaise and dressings are usually relatively resistant to microbial growth because they have acidic aqueous phases (pH < 4.0) and common foodborne pathogens do not tend to grow when the pH is below about 4.4 (Smittle, 2000). Moreover, they may contain antimicrobial preservatives like sodium benzoate, benzoic acid, or citric acid that inhibit microbial growth. As a result, they have relatively long shelf lives – often around a year or more. Again, similar strategies may be required when formulating plant-based emulsified foods. In particular, the aqueous phase of the final product should be less than pH 4.0 and suitable antimicrobial agents may need to be added.

7.5.1.3 Flavor Profile and Sensory Properties

The overall flavor profile of emulsified foods is the result of interactions of different kinds of molecules from the food with natural biosensors in the human body, such as volatile aroma molecules with the receptors in the nose (smell), non-volatile taste molecules with the receptors on the tongue (taste), and polymers and colloidal particles with the pressure sensors on the tongue (mouthfeel) (Ma & Boye, 2013). The smell of these products is mainly the result of small volatile molecules that are released from various kinds of ingredients originally used to formulate the product, such as oils, vinegar, lemon juice, herbs, or spices. In addition, aroma molecules may arise due to chemical degradation reactions that occur during storage, which may lead to off flavors, such as the rancidity caused by lipid oxidation (Jacobsen et al., 2013a). The taste is mainly a result of water-soluble molecules that interact with the tongue such as organic acids (in vinegar and lemon juice), sweeteners (sugars), and seasonings (salt, pepper, mustard, paprika, onion, and garlic). The creamy mouthfeel of emulsified products is primarily due to the fat droplets and thickening agents that lubricate the tongue and palate during mastication (Rudra et al., 2020). Typically, as the fat content of emulsified products is reduced, for example when developing reduced calorie versions, it is necessary to incorporate hydrocolloids (such as starch or gums) to replace the textural and mouthfeel attributes normally provided by the fat droplets (Chung et al., 2016; Ma & Boye, 2013). In addition, it may be necessary to reformulate the product to ensure the flavor profile remains the same, since removing some of the fat droplets alters the equilibrium partitioning and release kinetics of the flavor molecules in oil-in-water emulsions (McClements, 2015).

7.5.1.4 Commercial Products

A number of plant-based mayonnaise products have been introduced onto the market. *Eat Just* (formerly Hampton Creek) from San Francisco (CA) produced JUST Mayo, which is a semi-solid creamy yellow oil-in-water emulsion containing around 72% of emulsified canola oil. The ingredient list of this product also includes water, lemon juice, white vinegar, organic sugar, salt, apple cider vinegar, pea protein, spices, modified food starch, beta-carotene, and calcium disodium EDTA. Presumably, the plant-based emulsifier used to stabilize the fat droplets is this product is the pea protein (which is present at a concentration less than 2%). The various kinds of vinegar and lemon juice lead to an acidic aqueous phase that contributes to the sour taste and provides good antimicrobial stability. The beta-carotene contributes to the yellowish color, while the sugar, salt, and spices contribute to the flavor profile. Presumably, the calcium disodium EDTA is added to chelate any transition metal ions, thereby inhibiting lipid oxidation reactions. Traditional manufacturers of egg-based mayonnaises and dressings have also introduced vegan options in response to changes in consumer demand for more plant-based foods. For instance, *Unilever* (London, UK) introduced a Vegan Dressing and Spread™ in their popular Hellmann's brand. This product contains around 57% emulsified sunflower oil and 7% carbohydrates (mainly modified corn or potato starch), which contribute to the high viscosity and creamy appearance. The ingredient list reports that this product also contained distilled vinegar, sugar, salt, lemon juice concentrate, sorbic acid, calcium disodium EDTA, natural flavor, and paprika extract. These ingredients are therefore quite similar to the ones found in the *Just Mayo* product and play a similar function.

7.5.2 Thickened and Gelled Products: Custards, Flans, and Quiches

Egg proteins are also used as ingredients in some food products because of their thickening, gelling and/or binding properties. For instance, custards are thickened due to the presence of egg proteins. These products are produced by mixing eggs, milk, and sugar together and then slowly heating the mixture while stirring (McGee, 2004). This causes the globular egg proteins to unfold and aggregate with each other, thereby increasing their effective volume, which raises the viscosity (Chap. 4). Because the eggs are diluted with milk and the system is stirred, a 3D protein network that extends throughout the entire volume of the food is not formed, which prevents gelation. However, gels can be formed by using a higher egg protein concentration (less dilution) and by not stirring during cooking, as in semi-solid egg custards, flans, or quiches (McGee, 2004). The unfolded egg proteins may also help to bind other ingredients into the food matrix in these products, such as the meats, vegetables, or spices used in quiches.

Plant-based ingredients may be used to replace the egg (and milk) in these kinds of products. Typically, globular plant proteins are the most suitable choice for mimicking the thickening, gelling, and binding properties of eggs (McClements & Grossmann, 2021). As discussed earlier, it is important to select one or more plant proteins that can unfold at fairly similar temperatures to egg proteins when heated, and then form thickened solutions or gels with rheological characteristics similar to those produced by eggs, such as shear viscosity *versus* shear rate, shear viscosity *versus* temperature, shear modulus *versus* temperature, and/or texture profile analysis. In addition, it is important to select plant-based ingredients that produce the required appearance, which is usually a creamy yellow color. The opacity may be provided by the aggregated proteins but may also require the addition of other types of colloidal particles, such as fat droplets that simulate the granules, LDL, and HDL in eggs. The yellow color may be achieved by incorporating natural colors, such as beta-carotene, curcumin, or amaranth.

7.5.3 Foamed Products: Meringues, Mousses, and Soufflé

The formation of some kinds of traditional food products depends on the ability of the components in eggs to form and stabilize foams, which consist of gas bubbles suspended in a liquid or solid matrix. For instance, egg whites are typically whipped to create a relatively stable foam that is then baked into a meringue (McGee, 2004). During the whipping process, some of the globular proteins in egg white adsorb to the surfaces of the air bubbles formed, where they partially unfold (surface denaturation) and aggregate with their neighbors. This leads to the formation of a viscoelastic shell around the gas bubbles that helps to stabilize them. Nevertheless, they will eventually break down due to coalescence and coarsening phenomena, which involve gas bubbles merging together or gas molecules moving from small to large bubbles, respectively. Cooking the egg white foam leads to the formation of a more stable solid foam (meringue) because of further crosslinking of the proteins due to protein thermal denaturation caused by heat and concentration of the proteins caused by water evaporation. In some products, such as mousses and soufflé, the air bubbles may also be stabilized by other components in the original aqueous solution that form a viscoelastic matrix between them, such as gelatin or cocoa butter (in chocolate mousse) (McGee, 2004).

Research has shown that plant-based foamed foods can be produced that have similar physicochemical characteristics to traditional egg-based ones. For instance, a vegetarian mousse has been produced as an alternative to a traditional animal-based one (Schäfer et al., 2011). The vegetarian mouse was formulated using soy protein isolate as a foaming and thickening agent and was ranked as being acceptable to consumers in sensory trials. Researchers have also shown that lupin protein isolates can be used to form foams that have somewhat similar physicochemical attributes as egg white foams (Raymundo et al., 1998; Volp et al., 2021). The properties of the foams could be manipulated by thermal pretreatment of the lupin proteins, inclusion of thickeners (such as xanthan gum), and/or control of the pH and ionic strength of the aqueous phase, thereby allowing their quality attributes to be more closely tailored to those of egg white foams. Overall, these studies suggest that egg analogs do have potential to be used to formulate high-quality foamed food products, but more research is clearly needed in this area.

7.5.4 Baked Products: Cakes, Cookies, and Pastries

Eggs are also widely used as functional ingredients in a wide variety of baked products, such as cakes, cookies, and pastries (McGee, 2004). In this case, the egg proteins play an important role in determining the microstructure, appearance, texture, and sensory attributes of these products (Deleu et al., 2017; Marcet et al., 2016). These products are usually produced from flour, eggs, sugar, and butter (or another source of semi-solid fat, such as margarine or shortening). Typically, the ingredients are mixed together (either sequentially or all at once) and then beaten to aerate them (McGee, 2004). The aerated mixture is then placed in an oven and baked to create the final product, which can be considered to be a solid foam. The physicochemical and sensory attributes of the baked product, such as its composition, volume, shape, color, stability, texture, and mouthfeel depend on the nature of the gas bubbles formed (*e.g.*, size, number, and location), as well as the properties of the semi-solid matrix that surrounds them. The different functional ingredients used in the formulation all play an important role. Some of the egg proteins adsorb to the surfaces of the gas bubbles and form a protective coating around them, while the rest remain in the surrounding matrix. During cooking, the globular egg proteins unfold and aggregate with each other, which helps to stabilize the gas bubbles and provide mechanical strength to the surrounding matrix. The egg proteins may also react with the sugars through the Maillard reaction, which contributes to the desirable brown crust on the surface of some cakes. They can also help to bind other ingredients into the end product. The cereal proteins in the flour also unfold and aggregate with each other, which also makes an important contribution to the mechanical strength of the final system. The fat and sugar help to modulate the interactions between the wheat protein molecules, thereby altering the texture of the system, as well as contributing to the desirable sensory attributes, such as sweetness and creaminess. The formation of a network of fat crystals when the cake is cooled increases the mechanical strength of the product, as well as contributes to the desirable mouthfeel when these crystals melt during mastication. The starch granules absorb water, swell and burst during heating, which contributes to the mechanical properties and mouthfeel of the final product.

Vegan or vegetarian baked products can be produced by replacing the eggs and butter with plant-based alternatives. For instance, plant proteins can be used to replace the egg proteins, whereas crystalline plant fats (such as coconut oil or cocoa butter) can be used to replace the butter. However, it is important that they accurately simulate the functional attributes of the animal-based ingredient they are designed to replace.

Researchers have examined the possibility of forming plant-based cakes by replacing egg proteins with lupin proteins (Arozarena et al., 2001). This study found that cakes with good volume and textural attributes could be formed by combining the plant proteins with mono-/di-glyceride emulsifiers and xanthan gum. In another related study, researchers examined the impact of replacing egg with soy protein isolate on the properties of batter and cakes (Lin et al., 2017). They examined the impact of the plant protein on the specific gravity and viscosity of the batter, as well as the appearance, microstructure, volume, specific gravity, moisture content, and textural attributes of the cakes. The authors concluded that a mixture of soy proteins and mono-/di-glycerides (1%) could successfully be used to replace egg in the preparation of the cakes. For instance, the egg-free cakes and traditional cakes had somewhat similar specific volumes (1.92 *vs.* 2.08 cm^3/g), specific gravities (0.95 *vs.* 1.03), firmness (320 *vs.* 376 g), and moisture contents (28.0% *vs.* 29.0%) and only slightly different springiness values (77 *vs.* 98%), respectively.

Another study examined the effects of replacing egg white with a plant protein-rich ingredient (lima bean aquafaba, LBA) on the quality attributes of cupcakes (Nguyen et al., 2020). There was an increase in the number of large holes in the cupcakes as more of the egg white was replaced with LBA, leading to a more porous structure. This effect was mainly attributed to the ability of the lima bean proteins (and saponins) to adsorb to the air-water interfaces and stabilize the gas bubbles as well as their ability to form a gel network in the surrounding matrix during heating that provided mechanical strength to the end product. Nevertheless, the overall volume of the cakes containing only plant proteins was quite similar to that produced using egg white proteins. The hardness, gumminess, chewiness, and cohesiveness of the cupcakes decreased with increasing LBA concentration (Table 7.10). The optical properties of the cupcakes were quite similar for both protein sources, with the $L^*a^*b^*$ values of the crumb being around 85.8, −1.4, and +27.3 when egg white was used and 79.5, −1.2 and +26.1 when LBA was used. In future studies, it would be useful to carry out a sensory analysis to determine whether consumers also liked the plant-based versions.

7.5.5 Advantages of Egg Analogs to Food Manufacturer

Replacing eggs with egg analogs also has advantages to food manufacturers who normally use eggs as functional ingredients in their products (Grizio & Specht, 2021). The price and availability of eggs are highly susceptible to disruptions in the supply chain. For example, both Avian flu and Covid-19 caused disturbances in the egg supply chain due to a reduction in the total number of egg-laying chickens available or due to disruptions in the processing and transportation of the eggs. As a result, the replacement of egg-based ingredients with plant-based ones may help reduce fluctuations in the supply and cost of the functional ingredients used to

7.6 Conclusions and Future Directions

Table 7.10 Impact of replacing egg white (EW) with lima bean aquafaba (LBA) in model cupcakes

Sample	Hardness Peak 1 (g)	Hardness Peak 2 (g)	Cohesiveness	Springiness (mm)	Gumminess (g)	Chewiness (mJ)
100% LBA	410^a	340^a	0.61^a	8.8^a	250^a	22^a
25% EW + 75% LBA	660^b	540^b	0.62^{ab}	9.0^{ab}	410^b	36^b
50% EW + 50% LBA	660^b	544^b	0.65^{ab}	8.9^{ab}	430^b	38^b
75% EW + 25% LBA	1000^c	860^c	0.72^{ab}	9.2^b	720^c	65^c
100% EW	1500^d	1300^d	0.71^b	9.1^{ab}	1060^c	95^c

Data adopted from the article by Nguyen and co-workers (2020). Different superscript lowercase letters in the same column denote significant differences (p < .05).

formulate foods. Ingredients isolated from plants are often less expensive than those derived from eggs, which could help food manufacturers to reduce their costs and increase their profits. Moreover, it is often easier to store and handle plant-derived ingredients than egg-derived ones because eggs are highly susceptible to microbial deterioration. For the same reason, plant-based ingredients often pose less food safety risks than plant-based ones. Finally, the utilization of plant-based ingredients may reduce concerns with cholesterol and allergies that are linked to the consumption of egg products, which may be a benefit when labeling and marketing foods. For instance, if a plant-based ingredient is used that does not have allergy concerns, then the food manufacturer does not have to consider how to separate different kinds of ingredients within a factory and clean to remove any potential allergen contaminants.

7.6 Conclusions and Future Directions

Like other areas of the plant-based foods sector, there has been a large amount of research on the development of plant-based alternatives to conventional hen's eggs. This has already led to the introduction of a number of successful egg analogs on the market. However, there is still room for further research in this area. It is possible to improve the quality of existing products so that they more closely mimic the properties and functional versatility of real eggs. Moreover, there is a need for improving the health profile of egg analogs. This may involve fortifying them with beneficial ingredients, such as omega-3 fatty acids, essential amino acids, dietary fibers, vitamins, minerals, and nutraceuticals, as well as reducing the levels of saturated fats, sugars, and salt.

References

Alu'datt, M. H., Rababah, T., Alhamad, M. N., Ereifej, K., Gammoh, S., Kubow, S., & Tawalbeh, D. (2017). Preparation of mayonnaise from extracted plant protein isolates of chickpea, broad bean and lupin flour: Chemical, physiochemical, nutritional and therapeutic properties. *Journal of Food Science and Technology-Mysore, 54*(6), 1395–1405. https://doi.org/10.1007/s13197-017-2551-6

Amagliani, L., & Schmitt, C. (2017). Globular plant protein aggregates for stabilization of food foams and emulsions. *Trends in Food Science & Technology, 67*, 248–259. https://doi.org/10.1016/j.tifs.2017.07.013

Anton, M. (2013). Egg yolk: Structures, functionalities and processes. *Journal of the Science of Food and Agriculture, 93*(12), 2871–2880. https://doi.org/10.1002/jsfa.6247

Arozarena, I., Bertholo, H., Empis, J., Bunger, A., & de Sousa, I. (2001). Study of the total replacement of egg by white lupine protein, emulsifiers and xanthan gum in yellow cakes. *European Food Research and Technology, 213*(4–5), 312–316. https://doi.org/10.1007/s002170100391

Atilgan, M. R., & Unluturk, S. (2008). Rheological properties of liquid egg products (LEPS). *International Journal of Food Properties, 11*(2), 296–309. https://doi.org/10.1080/10942910701329658

Baron, F., & Jan, S. (2011). Egg and egg product microbiology. In Y. Nys, M. Bain, & F. VanImmerseel (Eds.), *Improving the safety and quality of eggs and egg products, Vol 1: Egg chemistry, production and consumption* (pp. 330–350).

Boon, C. S., McClements, D. J., Weiss, J., & Decker, E. A. (2010). Factors influencing the chemical stability of carotenoids in foods. *Critical Reviews in Food Science and Nutrition, 50*(6), 515–532. https://doi.org/10.1080/10408390802565889

Brady, J. W. (2013). *Introductory food chemistry*. Cornell University Press.

Chung, C., Smith, G., Degner, B., & McClements, D. J. (2016). Reduced fat food emulsions: Physicochemical, sensory, and biological aspects. *Critical Reviews in Food Science and Nutrition, 56*(4), 650–685. https://doi.org/10.1080/10408398.2013.792236

Clark, C. (2012). *The science of ice cream*. Royal Society of Chemistry.

Clauer, P. (2021). *Modern egg industry*. Retrieved from https://extension.psu.edu/modern-egg-industry

Cordobes, F., Partal, P., & Guerrero, A. (2004). Rheology and microstructure of heat-induced egg yolk gels. *Rheologica Acta, 43*(2), 184–195. https://doi.org/10.1007/s00397-003-0338-3

Cortinas, L., Galobart, J., Barroeta, A. C., Baucells, M. D., & Grashorn, M. A. (2003). Change in alpha-tocopherol contents, lipid oxidation and fatty acid profile in eggs enriched with linolenic acid or very long-chain omega 3 polyunsaturated fatty acids after different processing methods. *Journal of the Science of Food and Agriculture, 83*(8), 820–829. https://doi.org/10.1002/jsfa.1418

Croguennec, T., Nau, F., & Brule, G. (2002). Influence of pH and salts on egg white gelation. *Journal of Food Science, 67*(2), 608–614. https://doi.org/10.1111/j.1365-2621.2002.tb10646.x

Deleu, L. J., Melis, S., Wilderjans, E., Van Haesendonck, I., Brijs, K., & Delcour, J. A. (2017). Protein network formation during pound cake baking: The role of egg yolk and its fractions. *Food Hydrocolloids, 63*, 226–232. https://doi.org/10.1016/j.foodhyd.2016.07.036

Eke, M. O., Olaitan, N. I., & Ochefu, J. H. (2013). Effect of storage conditions on the quality attributes of shell (table) eggs. *Nigerian Food Journal, 31*(2), 18–24. https://doi.org/10.1016/S0189-7241(15)30072-2

Galobart, J., Barroeta, A. C., Baucells, M. D., & Guardiola, F. (2001). Lipid oxidation in fresh and spray-dried eggs enriched with omega 3 and omega 6 polyunsaturated fatty acids during storage as affected by dietary vitamin E and canthaxanthin supplementation. *Poultry Science, 80*(3), 327–337. https://doi.org/10.1093/ps/80.3.327

GFI. (2021). *State of the industry report: Plant-based meat, eggs, and dairy*. Retrieved from Washington, DC.

Gharibzahedi, S. M. T., Roohinejad, S., George, S., Barba, F. J., Greiner, R., Barbosa-Canovas, G. V., & Mallikarjunan, K. (2018). Innovative food processing technologies on the transglutaminase functionality in protein-based food products: Trends, opportunities and drawbacks. *Trends in Food Science & Technology, 75*, 194–205. https://doi.org/10.1016/j.tifs.2018.03.014

Goldberg, E. M., Gakhar, N., Ryland, D., Aliani, M., Gibson, R. A., & House, J. D. (2012). Fatty acid profile and sensory characteristics of table eggs from laying hens fed hempseed and hempseed oil. *Journal of Food Science, 77*(4), S153–S160. https://doi.org/10.1111/j.1750-3841.2012.02626.x

Grizio, M., & Specht, L. (2021). *Plant-based egg alternatives: Optimizing for functional properties and applications*. Retrieved from Washington, DC.

Hellwig, M. (2019). The chemistry of protein oxidation in food. *Angewandte Chemie-International Edition, 58*(47), 16742–16763. https://doi.org/10.1002/anie.201814144

Hernandez, M. J., Dolz, J., Delegido, J., Cabeza, C., & Dolz, M. (2008). Thixotropic behavior of salad dressings stabilized with modified starch, pectin, and gellan gum. Influence of temperature. *Journal of Dispersion Science and Technology, 29*(2), 213–219. https://doi.org/10.1080/01932690701707191

Hettiarachchy, N., Kannan, A., Schäfer, C., & Wagner, G. (2013). Gelling of plant based proteins. In U. Bröckel, W. Meier, & G. Wagner (Eds.), *Product design and engineering: Formulation of gels and pastes* (pp. 221–246). Wiley.

Huang, L. Y., Wang, T., Han, Z. P., Meng, Y. L., & Lu, X. M. (2016). Effect of egg yolk freezing on properties of mayonnaise. *Food Hydrocolloids, 56*, 311–317. https://doi.org/10.1016/j.foodhyd.2015.12.027

Huang, X., & Ahn, D. U. (2019). How can the value and use of egg yolk be increased? *Journal of Food Science, 84*(2), 205–212. https://doi.org/10.1111/1750-3841.14430

Jacobsen, C. (2015). Some strategies for the stabilization of long chain n-3 PUFA-enriched foods: A review. *European Journal of Lipid Science and Technology, 117*(11), 1853–1866. https://doi.org/10.1002/ejlt.201500137

Jacobsen, C., Horn, A. F., & Nielsen, N. S. (2013a). Enrichment of emulsified foods with omega-3 fatty acids. In C. Jacobsen, N. S. Nielsen, A. F. Horn, & A. D. M. Sorensen (Eds.), *Food enrichment with omega-3 fatty acids* (Vol. 252, pp. 336–352).

Jacobsen, C., Let, M. B., Nielsen, N. S., & Meyer, A. S. (2008). Antioxidant strategies for preventing oxidative flavour deterioration of foods enriched with n-3 polyunsaturated lipids: A comparative evaluation. *Trends in Food Science & Technology, 19*(2), 76–93. https://doi.org/10.1016/j.tifs.2007.08.001

Jacobsen, C., Sorensen, A. D. M., & Nielsen, N. S. (2013b). Stabilization of omega-3 oils and enriched foods using antioxidants. In C. Jacobsen, N. S. Nielsen, A. F. Horn, & A. D. M. Sorensen (Eds.), *Food enrichment with omega-3 fatty acids* (Vol. 252, pp. 130–149).

Javed, A., King, A. J., Imran, M., Jeoh, T., & Naseem, S. (2019). Omega-3 supplementation for enhancement of egg functional properties. *Journal of Food Processing and Preservation, 43*(8). https://doi.org/10.1111/jfpp.14052

Jo, S. H., Kim, K. H., Kim, Y. H., Lee, M. H., Ahn, J. H., Szulejko, J. E., … Kim, A. Y. H. (2013). Study of odor from boiled eggs over time using gas chromatography. *Microchemical Journal, 110*, 517–529. https://doi.org/10.1016/j.microc.2013.05.011

Kaltsa, O., Yanniotis, S., Polissiou, M., & Mandala, I. (2018). Stability, physical properties and acceptance of salad dressings containing saffron (Crocus sativus) or pomegranate juice powder as affected by high shear (HS) and ultrasonication (US) process. *LWT – Food Science and Technology, 97*, 404–413. https://doi.org/10.1016/j.lwt.2018.07.015

Kassis, N., Drake, S. R., Beamer, S. K., Matak, K. E., & Jaczynski, J. (2010). Development of nutraceutical egg products with omega-3-rich oils. *LWT – Food Science and Technology, 43*(5), 777–783. https://doi.org/10.1016/j.lwt.2009.12.014

Katsaros, G., Tsoukala, M., Giannoglou, M., & Taoukis, P. (2020). Effect of storage on the rheological and viscoelastic properties of mayonnaise emulsions of different oil droplet size. *Heliyon, 6*(12). https://doi.org/10.1016/j.heliyon.2020.e05788

Kiosseoglou, V., & Paraskevopoulou, A. (2005). Molecular interactions in gels prepared with egg yolk and its fractions. *Food Hydrocolloids, 19*(3), 527–532. https://doi.org/10.1016/j.foodhyd.2004.10.027

Kovacs-Nolan, J., Phillips, M., & Mine, Y. (2005). Advances in the value of eggs and egg components for human health. *Journal of Agricultural and Food Chemistry, 53*(22), 8421–8431. https://doi.org/10.1021/jf050964f

Le Denmat, M., Anton, M., & Beaumal, V. (2000). Characterisation of emulsion properties and of interface composition in O/W emulsions prepared with hen egg yolk, plasma and granules. *Food Hydrocolloids, 14*(6), 539–549. https://doi.org/10.1016/s0268-005x(00)00034-5

Le Denmat, M., Anton, M., & Gandemer, G. (1999). Protein denaturation and emulsifying properties of plasma and granules of egg yolk as related to heat treatment. *Journal of Food Science, 64*(2), 194–197.

Li, J. H., Wang, C. Y., Li, X., Su, Y. J., Yang, Y. J., & Yu, X. B. (2018a). Effects of pH and NaCl on the physicochemical and interfacial properties of egg white/yolk. *Food Bioscience, 23*, 115–120. https://doi.org/10.1016/j.fbio.2017.12.004

Li, J. H., Wang, C. Y., Zhang, M. Q., Zhai, Y. H., Zhou, B., Su, Y. J., & Yang, Y. J. (2018b). Effects of selected phosphate salts on gelling properties and water state of whole egg gel. *Food Hydrocolloids, 77*, 1–7. https://doi.org/10.1016/j.foodhyd.2017.08.030

Lin, M. Y., Tay, S. H., Yang, H. S., Yang, B., & Li, H. L. (2017). Replacement of eggs with soybean protein isolates and polysaccharides to prepare yellow cakes suitable for vegetarians. *Food Chemistry, 229*, 663–673. https://doi.org/10.1016/j.foodchem.2017.02.132

Ma, Z., & Boye, J. I. (2013). Advances in the design and production of reduced-fat and reduced-cholesterol salad dressing and mayonnaise: A review. *Food and Bioprocess Technology, 6*(3), 648–670. https://doi.org/10.1007/s11947-012-1000-9

MacLeod, A. J., & Cave, S. J. (1975). Volatile flavor components of eggs. *Journal of the Science of Food and Agriculture, 26*, 351–360.

Marcet, I., Collado, S., Paredes, B., & Diaz, M. (2016). Rheological and textural properties in a bakery product as a function of the proportions of the egg yolk fractions: Discussion and modelling. *Food Hydrocolloids, 54*, 119–129. https://doi.org/10.1016/j.foodhyd.2015.09.023

McClements, D. J. (2015). *Food emulsions: Principles, practice, and techniques* (2nd ed.). CRC Press.

McClements, D. J., Bai, L., & Chung, C. (2017). Recent advances in the utilization of natural emulsifiers to form and stabilize emulsions. In M. P. Doyle & T. R. Klaenhammer (Eds.), *Annual review of food science and technology* (Vol. 8, pp. 205–236).

McClements, D. J., & Decker, E. A. (2000). Lipid oxidation in oil-in-water emulsions: Impact of molecular environment on chemical reactions in heterogeneous food systems. *Journal of Food Science, 65*(8), 1270–1282. https://doi.org/10.1111/j.1365-2621.2000.tb10596.x

McClements, D. J., & Grossmann, L. (2021). The science of plant-based foods: Constructing next-generation meat, fish, milk, and egg analogs. *Comprehensive Reviews in Food Science and Food Safety, 20*(4), 4049–4100. https://doi.org/10.1111/1541-4337.12771

McGee, H. (2004). *On food and cooking: The science and lore of the kitchen*. Scribner.

Nguyen, T. M. N., Nguyen, T. P., Tran, G. B., & Le, P. T. Q. (2020). Effect of processing methods on foam properties and application of lima bean (Phaseolus lunatus L.) aquafaba in eggless cupcakes. *Journal of Food Processing and Preservation, 44*(11). https://doi.org/10.1111/jfpp.14886

NRC. (1976). *Fat content and composition of animal products: Proceedings of a symposium*. The National Academies Press.

Panaite, T. D., Mironeasa, S., Iuga, M., & Vlaicu, P. A. (2019). Liquid egg products characterization during storage as a response of novel phyto-additives added in hens diet. *Emirates Journal of Food and Agriculture, 31*(4), 304–314. https://doi.org/10.9755/ejfa.2019.v31.i4.1937

Parker, A., Gunning, P. A., Ng, K., & Robins, M. M. (1995). How does xanthan stabilise salad dressing? *Food Hydrocolloids, 9*(4), 333–342. https://doi.org/10.1016/s0268-005x(09)80263-4

Plagemann, I., Zelena, K., Krings, U., & Berger, R. G. (2011). Volatile flavours in raw egg yolk of hens fed on different diets. *Journal of the Science of Food and Agriculture, 91*(11), 2061–2065. https://doi.org/10.1002/jsfa.4420

Poore, J., & Nemecek, T. (2018). Reducing food's environmental impacts through producers and consumers. *Science, 360*(6392), 987. https://doi.org/10.1126/science.aaq0216

Raikos, V., Campbell, L., & Euston, S. R. (2007). Rheology and texture of hen's egg protein heat-set gels as affected by pH and the addition of sugar and/or salt. *Food Hydrocolloids, 21*(2), 237–244. https://doi.org/10.1016/j.foodhyd.2006.03.015

Raymundo, A., Empis, J., & Sousa, I. (1998). White lupin protein isolate as a foaming agent. *Zeitschrift Fur Lebensmittel-Untersuchung Und-Forschung a-Food Research and Technology, 207*(2), 91–96. https://doi.org/10.1007/s002170050300

Rondoni, A., Asioli, D., & Millan, E. (2020). Consumer behaviour, perceptions, and preferences towards eggs: A review of the literature and discussion of industry implications. *Trends in Food Science & Technology, 106*, 391–401. https://doi.org/10.1016/j.tifs.2020.10.038

Rudra, S. G., Hanan, E., Sagar, V. R., Bhardwaj, R., Basu, S., & Sharma, V. (2020). Manufacturing of mayonnaise with pea pod powder as a functional ingredient. *Journal of Food Measurement and Characterization, 14*(5), 2402–2413. https://doi.org/10.1007/s11694-020-00487-0

Sarkar, A., & Dickinson, E. (2020). Sustainable food-grade Pickering emulsions stabilized by plant-based particles. *Current Opinion in Colloid & Interface Science, 49*, 69–81. https://doi.org/10.1016/j.cocis.2020.04.004

Schäfer, C., Neidhart, S., & Carle, R. (2011). Application and sensory evaluation of enzymatically texturised vegetable proteins in food models. *European Food Research and Technology, 232*(6), 1043–1056. https://doi.org/10.1007/s00217-011-1474-0

Schroder, A., Laguerre, M., Tenon, M., Schroen, K., & Berton-Carabin, C. C. (2021). Natural particles can armor emulsions against lipid oxidation and coalescence. *Food Chemistry, 347*. https://doi.org/10.1016/j.foodchem.2021.129003

Sheldon, B. W., & Kimsey, H. R. (1985). The effects of cooking methods on the chemical, physical, and sensory properties of hard-cooked eggs. *Poultry Science, 64*(1), 84–92. https://doi.org/10.3382/ps.0640084

Silva, F. V. M., & Gibbs, P. A. (2012). Thermal pasteurization requirements for the inactivation of Salmonella in foods. *Food Research International, 45*(2), 695–699. https://doi.org/10.1016/j.foodres.2011.06.018

Smittle, R. B. (2000). Microbiological safety of mayonnaise, salad dressings, and sauces produced in the United States: A review. *Journal of Food Protection, 63*(8), 1144–1153.

Song, H. Y., & McClements, D. J. (2021). Nano-enabled-fortification of salad dressings with curcumin: Impact of nanoemulsion-based delivery systems on physicochemical properties. *LWT – Food Science and Technology*, 145. https://doi.org/10.1016/j.lwt.2021.111299

Stadelman, W. J., Newkirk, D., & Newby, L. (2017). *Egg science and technology* (4th ed.). CRC Press.

Strixner, T., & Kulozik, U. (2013). Continuous centrifugal fractionation of egg yolk granules and plasma constituents influenced by process conditions and product characteristics. *Journal of Food Engineering, 117*(1), 89–98. https://doi.org/10.1016/j.jfoodeng.2013.02.009

Strixner, T., Sterr, J., Kulozik, U., & Gebhardt, R. (2014). Structural study on hen-egg yolk high density lipoprotein (HDL) granules. *Food Biophysics, 9*(4), 314–321. https://doi.org/10.1007/s11483-014-9359-y

Tamanna, N., & Mahmood, N. (2015). Food processing and Maillard reaction products: Effect on human health and nutrition. *International Journal of Food Science, 2015*, 526762–526762. https://doi.org/10.1155/2015/526762

Tang, C. H., & Sun, X. (2010). Physicochemical and structural properties of 8S and/or 11S globulins from Mungbean Vigna radiata (L.) Wilczek with various polypeptide constituents. *Journal of Agricultural and Food Chemistry, 58*(10), 6395–6402. https://doi.org/10.1021/jf904254f

Taslikh, M., Mollakhalili-Meybodi, N., Alizadeh, A. M., Mousavi, M. M., Nayebzadeh, K., & Mortazavian, A. M. (2021). Mayonnaise main ingredients influence on its structure as an emulsion. *Journal of Food Science and Technology-Mysore*. https://doi.org/10.1007/s13197-021-05133-1

Tetrick, J., Boot, J. H. A., Jones, C. M., Clements, M. A., Oliveira, M. A., & Albanello, L. (2019). USA Patent No. US Patent Office.

Tsutsui, T. (1988). Functional-properties of heat-treated egg-yolk low-density lipoprotein. *Journal of Food Science, 53*(4), 1103–1106. https://doi.org/10.1111/j.1365-2621.1988.tb13539.x

USDA. (2020). *FSIS food safety guideline for egg products*. Retrieved from https://www.fsis.usda.gov/sites/default/files/media_file/2021-05/FSIS-GD-2020-0005.pdf

Vega, C., & Mercade-Prieto, R. (2011). Culinary biophysics: On the nature of the 6XA degrees C egg. *Food Biophysics, 6*(1), 152–159. https://doi.org/10.1007/s11483-010-9200-1

Volp, A. R., Seitz, J., & Willenbacher, N. (2021). Structure and rheology of foams stabilized by lupin protein isolate of Lupinus angustifolius. *Food Hydrocolloids, 120*. https://doi.org/10.1016/j.foodhyd.2021.106919

Wang, A. H., Xiao, Z. G., Wang, J. J., Li, G. Q., & Wang, L. J. (2020). Fabrication and characterization of emulsion stabilized by table egg-yolk granules at different pH levels. *Journal of the Science of Food and Agriculture, 100*(4), 1470–1478. https://doi.org/10.1002/jsfa.10154

Wang, Q. L., Jin, G. F., Wang, N., Guo, X., Jin, Y. G., & Ma, M. H. (2017). Lipolysis and oxidation of lipids during egg storage at different temperatures. *Czech Journal of Food Sciences, 35*(3), 229–235. https://doi.org/10.17221/174/2016-cjfs

Waraho, T., McClements, D. J., & Decker, E. A. (2011). Mechanisms of lipid oxidation in food dispersions. *Trends in Food Science & Technology, 22*(1), 3–13. https://doi.org/10.1016/j.tifs.2010.11.003

Warren, M. W., & Ball, H. R. (1991). Effect of concentration of egg-yolk and white on fresh scrambled egg flavor. *Poultry Science, 70*(10), 2186–2190. https://doi.org/10.3382/ps.0702186

Warren, M. W., Larick, D. K., & Ball, H. R. (1995). Volatiles and sensory characteristics of cooked egg-yolk, white and their combinations. *Journal of Food Science, 60*(1), 79. https://doi.org/10.1111/j.1365-2621.1995.tb05611.x

Whiley, H., & Ross, K. (2015). Salmonella and eggs: From production to plate. *International Journal of Environmental Research and Public Health, 12*(3), 2543–2556. https://doi.org/10.3390/ijerph120302543

Xiang, X. L., Jin, G. F., Gouda, M., Jin, Y. G., & Ma, M. H. (2019). Characterization and classification of volatiles from different breeds of eggs by SPME-GC-MS and chemometrics. *Food Research International, 116*, 767–777. https://doi.org/10.1016/j.foodres.2018.09.010

Yuceer, M., Ilyasoglu, H., & Ozcelik, B. (2016). Comparison of flow behavior and physicochemical characteristics of low-cholesterol mayonnaises produced with cholesterol-reduced egg yolk. *Journal of Applied Poultry Research, 25*(4), 518–527. https://doi.org/10.3382/japr/pfw033

Chapter 8
Plant-Based Milk and Cream Analogs

8.1 Introduction

Plant-based milks are currently the largest segment of the overall plant-based food segment, with around $2.5 billion sales in 2020, which made up around 15% of the total milk market (GFI, 2021). Moreover, there has been a much larger increase in the sales of these products than in conventional bovine milk, which has led to many food companies actively developing new and improved products in this category. A variety of plant-based milk products have already been successfully launched on the market, including oat, soy, coconut, almond, cashew, rice, and hemp milks (Chalupa-Krebzdak et al., 2018; Sethi et al., 2016). These products differ in their physicochemical properties and sensory attributes, such as appearance, texture, mouthfeel, and flavor profile (Reyes-Jurado et al., 2021). Despite their popularity, many consumers are still reluctant to adopt these products because they do not like their sensory attributes or because they do not function in the same way as regular milk (such as whitening coffee, creating whipped creams, cooking, or baking). These challenges may be overcome by improving our understanding of the factors that determine the desirable physicochemical and sensory attributes of plant-based milks, and then using this knowledge to create new formulations or processes to fabricate them.

In this chapter, we begin by briefly reviewing the composition, microstructure, and physicochemical attributes of traditional cow's milk since plant-based milk analogs are typically designed to mimic these attributes. We then review the physical and chemical factors that influence the desirable attributes of plant-based milks. The physicochemical, sensory and gastrointestinal properties of cow's milk and milk analogs are then compared. Finally, we highlight a number of areas where further research is required to make high quality plant-based milk analogs with improved nutritional attributes.

Throughout this chapter we use the expression "milk" to refer to plant-based analogs of conventional cow's milk. However, we note that in many countries there have been concerted efforts by the dairy industry and regulators to reserve this expression for edible fluids that come from the mammary gland of some animals, such as cows, goats, and sheep. In these countries, other terms have to be utilized when labeling and marketing these products, such as plant-based beverages. However, we believe that the term "milk" is appropriate for plant-based products because their compositions, structure, properties, and intended use are so similar to those of cow's milk.

8.2 Attributes of Cow's Milk

Plant-based milks are often designed to simulate the composition, microstructure, physicochemical properties, functional performance, and sensory attributes of cow's milk. For this reason, it is important to understand cow's milk when designing plant-based alternatives. Cow's milk is a natural colloidal suspension consisting of fat globules and casein micelles dispersed within an aqueous solution containing lactose, oligosaccharides, whey proteins, and salts (Jukkola & Rojas, 2017). Bovine milk has been designed through evolutionary pressures to deliver nutrients and other bioactive substances from the cow to the calf to promote their growth and stimulate their immune systems. It therefore includes macronutrients and micronutrients that the calve needs to survive and grow, including fats, carbohydrates, proteins, vitamins, and minerals (Chalupa-Krebzdak et al., 2018). In developed countries, raw cow's milk is subjected to various kinds of processing operations prior to being sold, including homogenization, pasteurization, separation, and/or sterilization (Campbell & Marshall, 2016). Raw or processed milk is a highly versatile food ingredient that can be utilized to form a variety of dairy products, including cream, whipped cream, ice cream, butter, yogurt, and cheese. The ability of cow's milk to create these products is a result of its unique composition and microstructure, which is difficult to simulate using plant-based milk analogs.

8.2.1 Composition and Microstructure

Natural (non-processed) cow's milk has a composition and microstructure that is influenced by a variety of factors, including the species, maturity, health status, habitat, and diet of the animals (Pereira, 2014). Even so, the overall composition of cow's milk is fairly similar, with about 87% water, 4.5% lactose, 3.5% fat, 3% protein, 0.8% minerals, and 0.1% vitamins (Pereira, 2014). The most important structural components in raw cow's milk are milk fat globules and casein micelles (Jukkola & Rojas, 2017). The milk fat globules in raw milk have an average diameter of about 4.5 μm, with the majority of the globules having diameters between about 1 to 10 μm. Structurally they are comprised of a hydrophobic core that is mainly made up of triacylglycerols and an amphiphilic shell, known as the milk fat globule membrane (MFGM), that is made up of phospholipids, proteins, glycoproteins, cholesterol, sphingomyelin, and other materials arranged into a complex hierarchical structure (Lopez et al., 2015). The main structural constituents of the MFGM are the phospholipids, which are arranged into a tri-layer structure that has a thickness of around 10 to 20 nm. The other constituents are typically embedded within or between the layers formed by the phospholipid molecules. Other kinds of non-polar substances, such as oil-soluble vitamins and carotenoids, may be dissolved within the hydrophobic triacylglycerol core, which provide an important source of nutrients to the growing calf.

The other major form of colloidal particles in cow's milk are the casein micelles. These particles are assembled from a mixture of various kinds of casein molecules (α_{S1}, α_{S2}, β, and κ) and minerals (colloidal calcium phosphate), which are held together predominantly *via* electrostatic and hydrophobic interactions (Lucey & Horne, 2018). Casein micelles have a mean diameter of about 150 nm, with the majority of the micelles having diameters between about 50 to 500 nm (Broyard & Gaucheron, 2015). The casein molecules have unique molecular characteristics that lead to many of the unique functional attributes of milk. For instance, they tend to be relatively small flexible amphiphilic molecules that may contain numerous phosphate and sugar groups. As such, they are difficult to mimic using plant proteins, which tend to be globular in structure. The aqueous phase in cow's milk is a slightly acidic solution (pH 6.5 to 6.7) that contains whey proteins (β-lactoglobulin, α-lactalbumin, bovine serum albumin and immunoglobulins), lactose, oligosaccharides, and minerals. The whey proteins have globular structures and vary somewhat in their molecular weights, isoelectric points, and thermal denaturation temperatures (Table 8.1).

Table 8.1 Molecular and physicochemical characteristics of casein and whey proteins from cow's milk: T_m = thermal denaturation temperature

	Fraction of total protein (%)	Molar mass (kDa)	Isoelectric point	T_m (°C)
Casein			4.6	–
α_{S1}-casein	39	23.6		–
α_{S2}-casein	10	25.2		–
β-casein	36	24.0		–
κ-casein	13	19.0		–
Whey			5.2	
β-lactoglobulin	51	18.4	5.4	72
α-lactalbumin	19	14.2	4.4	35 and 64[a]
BSA	6	66.3	4.9	64
Immunoglobulins	12	Range	Range	Range
Lactoferrin	1–2	78	8–9	70 and 90[a]

Adapted from McClements (2020) with permission.
[a]The lower and higher temperatures for α-lactalbumin and lactoferrin are for the apo- (calcium or iron free) and holo- (calcium or iron bound) forms, respectively.

8.2.2 Processing

Knowledge of the processing operations used to produce commercial cow's milk, such as homogenization, thermal treatments, and separation, is useful for the design and creation of safe and shelf stable plant-based milk analogs. In non-processed (raw) milk, the native fat globules are relatively large (4.5 μm) and have a density (920 kg m^{-3}) that is appreciably lower than the density of the surrounding aqueous solution (1036 kg m^{-3}). As a result, they have a tendency to move upwards (cream) due to gravitational forces (Lopez et al., 2015). The rate at which particles cream in a colloidal dispersion is proportional to their diameter squared. Consequently, creaming can be reduced by decreasing the size of the fat globules using homogenization. Typically, the fat globules in homogenized milk have dimeters below 0.5 μm, which reduces their creaming rate by almost hundred-fold compared to raw milk. The size of the native fat globules in raw milk is usually reduced by passing it through a high-pressure homogenizer, which generates intense disruptive forces that break them down into smaller fat droplets (Campbell & Marshall, 2016). The composition and structure of the interfacial layer are altered appreciably after milk has been homogenized. Instead of being coated by the MFGM, the fat droplets in homogenized milk are also covered by a layer of caseins (molecules and micelles) and whey proteins. As a result, the physicochemical properties and functional performance of the milk is altered after homogenization.

Raw milk contains enzymes and bacteria that can degrade or utilize the nutrients present to cause a loss in product quality and they can cause food safety concerns. Consequently, raw milk typically undergoes a thermal treatment, such as pasteurization or sterilization, to deactivate enzymes and bacteria, thereby improving its quality, shelf life, and safety (Campbell & Marshall, 2016). These heat treatments can lead to appreciable changes in the molecular structure and interactions of milk proteins, which alters their functionality (Livney et al., 2003). In addition, they can cause changes in milk's color and flavor profile to an extent that depends on the severity of the heat treatment (Deeth, 2017). For instance, the milk may become slightly brownish and have a cooked flavor if heated too much.

Cow's milk typically has a fat concentration of about 3% to 4%. Milk is usually fractionated using centrifugation to create a high-fat cream fraction and a low-fat skim fraction. These fractions can then be used as is or they can be combined in different ratios to create a range of fluid dairy products with different fat contents: skim milk (<0.5%), low-fat milk (1.0%), reduced-fat milk (2.0)%, whole milks (3.3%), half-and-half (10–18%), light cream (18–30%), and heavy cream (36–40%). The appearance, viscosity, and mouthfeel of these products depend on their fat content.

8.2.3 Physicochemical and Sensory Properties

Knowledge of the physicochemical and sensory attributes of cow's milk is useful for designing plant-based alternatives that mimic its properties. Whole milk is a creamy white fluid with a relatively low viscosity and a mild flavor (Schiano et al.,

Table 8.2 Measured lightness and shear viscosity of fluid dairy products with different fat contents. The lightness values were measured with a colorimeter (Kneifel et al., 1992). The values were predicted (20 °C) using an equation fit to experimental measurements (Flauzino et al., 2010)

Product	Fat content	L^*	Viscosity (mPa s)
Skim milk	0.1%	81.7	2.2
Full fat milk	3.6%	86.1	2.7
Creamer	10.0%	86.9	3.9
Cream	36.0%	88.1	17.6

2017). In general, however, the properties of milk depend on its fat content, with the perception of creaminess rising with increasing fat content (McCarthy et al., 2017a). Instrumental colorimeter and rheology measurements have shown that the lightness and viscosity of milks and creams increase with their fat content (Table 8.2), which can be attributed to the impact of the fat droplets on light scattering and fluid flow (Chap. 4). The flavor profile of milk is influenced by the fat content and the type of feed the cow's eat, as well as any chemical changes that occur due to enzymatic, microbial, and physical processes during processing, transport and storage (Schiano et al., 2017). Studies of consumer preferences indicate that flavor plays a major role in determining the overall liking of cow's milk (McCarthy et al., 2017b). The light color and bland flavor of cow's milk mean that the presence of any off-flavors or off-colors are readily perceived by consumers. In contrast, plant-based milk analogs have distinct flavor profiles and mouthfeels that are often quite different from those of cow's milk, which is one of the reasons they are not readily accepted by some consumers (Jeske et al., 2018).

Like other kinds of colloidal dispersion, cow's milk is thermodynamically unstable and has a tendency to break down over time due to processes such as creaming, flocculation, coalescence, and partial coalescence (Dickinson, 1992; McClements, 2015). Milk fat globules have a tendency to move upwards due to gravitational forces because they are less dense than the surrounding aqueous phase. The rate of this process decreases as the droplet size decreases, which is why homogenized milk is much more resistant to creaming than raw milk. Under the normal pH conditions found in homogenized cow's milk (pH 6.5–6.7), the protein-coated fat droplets are fairly resistant to aggregation because the thick anionic interfacial layers generate strong electrostatic and steric repulsive forces. However, they will aggregate if the pH is adjusted towards the isoelectric point of the protein-coated fat droplets, or if mineral ions are added to the aqueous phase, because these changes reduce the electrostatic repulsion between the droplets. Indeed, the controlled aggregation of the fat droplets and proteins in cow's milk play a critical role in the formation of yogurt and cheese (Chap. 9).

As mentioned earlier, cow's milk is usually thermally processed (pasteurization or sterilization) to deactivate enzymes and microbes, thereby increasing its shelf life and safety (Deeth, 2017). Conventional pasteurized milk has a shelf-life of around a few weeks when the container is not opened and it is kept under refrigerated conditions and around a week when it is opened. The shelf life of cow's milk can be extended considerably by using a more extreme heat treatment, such as ultrahigh temperature (UHT) treatment. However, the milk should still be consumed about a week or so after the container is opened. Due to their familiarity with cow's milk, consumers typically expect that plant-based milk analogs should have a similar or better shelf life.

8.2.4 Functional Versatility

One of the unique characteristics of cow's milk is its versatility as a food ingredient. As well as being consumed as a cold drink, milk can also be consumed with breakfast cereals, added to tea or coffee, utilized as an ingredient in cooking, or used to create a diverse range of dairy products like yogurt, cheese, ice cream, and whipped cream (Walstra et al., 2005). It is difficult to use the current generation of plant-based milk analogs in many of these applications. For instance, some milk analogs have a tendency to aggregate when added to tea or coffee, leading to the formation of an unpleasant layer on their surface. Moreover, many milk analogs cannot be used to create yogurt, cheese, or ice cream using the same processes normally used for conventional cow's milk.

The functional versatility of cow's milk is linked to the unique nature of the structural components it contains (Walstra et al., 2005). The production of cheese and yogurt depends on the formation of a 3D network of casein molecules when milk is acidified or when rennet (a hydrolytic enzyme) is added. Caseins and whey proteins are amphiphilic molecules that can adsorb to the surfaces of the air bubbles formed in milk during mechanical agitation or aeration, which means that they can be used to form whipped creams and other foams. The crystallization and melting behavior of the lipid phase inside the fat globules of milk also plays a major role in determining the functional versatility of milk. Milk fat is completely liquid at temperatures above about 37 °C but it becomes partially crystalline when it is cooled to lower temperatures. This phenomenon is important for the production of dairy products like whipped cream, ice cream, and butter. In particular, the fat globules tend to undergo partial coalescence when they are cooled to refrigerator temperatures and sheared or churned. A fat crystal from one partially crystalline fat globule penetrates into a fluid region in another one, which causes them to clump together but retain some of their original shape. For example, in whipped cream and ice cream, clumps of fat globules form shells around the air bubbles, which increases their resistance to collapse. Moreover, a network of aggregated fat globules is formed in the aqueous phase that provides mechanical strength to the system, leading to a semi-solid material being formed. In butter, the partial coalescence of the partly crystalline milk fat globules during chilling and churning leads to phase inversion, with the system transforming from an oil-in-water (milk) to a water-in-oil (butter) emulsion. The fat crystals in the oil phase then form a 3D network that leads to semi-solid behavior, which is important as it provides the plastic-like textural characteristics needed for good spreadability. Again, it is difficult to simulate this kind of behavior using plant-based milks due to the different phase behavior of the lipid phase and different nature of the interfacial coatings around the fat droplets.

Cow's milk is comprised of a particular blend of lipids, proteins, carbohydrates, vitamins, and minerals that contribute to its characteristic mouthfeel, flavor profile, and nutritional effects. This is particularly important for the production of dairy products, such as cheese that rely on specific chemical or biochemical reactions that convert the original constituents in milk into new substances that have their own unique characteristics. For instance, the characteristic aroma and taste of many cheeses are the result of the chemical breakdown of proteins and lipids by the enzymes secreted by natural or added microbes (such as yeasts, molds, or bacteria). Plant-based milk analogs contain different kinds of proteins and lipids, which will result in different kinds of reaction products being generated by the microbes (if the same ones are used). As a result, plant-based cheeses have different flavor profiles than conventional dairy ones, which consumers may not be familiar with or do not find appealing. The science behind the production of plant-based cheeses and other dairy products is covered in more detail in the following chapter.

8.2.5 Nutritional Profile

As mentioned earlier, cow's milk has been designed through evolution to be a complete source of energy and nutrients to the growing calf, containing lipids, proteins, carbohydrates, vitamins, minerals and other important constituents (Chalupa-Krebzdak et al., 2018). The consumption of milks by humans has been reported to meet their nutritional needs and may also help prevent some chronic diseases (Thorning et al., 2016). There is therefore an interest in designing plant-based milks to at least match the nutritional profile of cow's milk. However, it is also possible to create milk analogs that have additional nutritional benefits, such as fortification with nutrients that might be deficient in a fully plant-based diet, such as vitamin B_{12}, vitamin D, or calcium.

8.3 Production of Plant-Based Milk Analogs

Plant-based milks are often designed to have similar physicochemical attributes as cow's milk, such as their appearance, texture, mouthfeel, and flavor profile. Consequently, it is often necessary to simulate the colloidal attributes of cow's milk, since the milk fat globules and casein micelles contribute many of the desirable traits to these products. However, the other constituents in the aqueous solution surrounding these colloidal particles are also important, such as sugars, salts, and whey proteins. In general, there are two main approaches for creating milk-like colloidal dispersions from plant materials: (*i*) disruption of plant tissues (Sethi et al., 2016), and (*ii*) emulsification of plant-based oils, emulsifiers, water, and other ingredients (Do et al., 2018; McClements, 2020).

8.3.1 Plant Tissue Disruption Approaches

For these approaches, a suitable source of plant material, such as oats, soybeans, almonds, coconut flesh, or cashews, is broken down into small fragments using mechanical forces (such as shearing and homogenization) or biochemical reactions (such as controlled enzyme hydrolysis) (McClements, 2020; Nikiforidis et al., 2014; Sethi et al., 2016). The plant materials used for this purpose often contain natural oil bodies, which consist of a triacylglycerol-rich core surrounded by a layer of phospholipids and proteins (Tzen et al., 1993), which is therefore similar to the composition and structure of the fat globules in milk (Michalski, 2009). However, the type and amounts of lipids, phospholipids, and proteins present are different compared to milk fat globules and oil bodies. The most common sources of oil bodies in plants are seeds, such as soybeans (legume seeds), oats (cereal seeds), and almonds (tree seeds). Seeds contain oil bodies because they are a good source of energy during the germination process (Pyc et al., 2017). Oil bodies have diameters ranging from a few hundred to a few thousand nanometers, which is fairly similar to the dimensions of the fat globules in milk. Consequently, the oil bodies in plant-based milks can provide many of the desirable quality attributes provided by the fat globules in cow's milk, such as a creamy appearance and texture. They can also serve as a reservoir for hydrophobic nutrients, such as oil-soluble vitamins and nutraceuticals.

Plant-based milks formed using disruption approaches also contain other types of colloidal particles that can contribute to their physicochemical, sensory, and functional properties. In particular, they contain plant tissue fragments such as starch granules and cell wall materials. These fragments vary in their size, shape, and composition depending on the botanical origin of the plant and the nature of the processing methods used. If the fragments are too large (> 1 μm), they may sediment and form an undesirable layer at the bottom of the product because they are usually denser than water. In addition, large fragments (> 50 μm) will lead to an undesirable rough or gritty mouthfeel. Consequently, it may be important to ensure that the plant tissue fragments have been reduced below a particular size to obtain the required physicochemical and sensory attributes in the end product.

A series of processing steps are typically utilized to disrupt plant seeds and create plant-based milk analogs (Fig. 8.1a) (Campbell et al., 2011; Iwanaga et al., 2007; Nikiforidis et al., 2014). Initially, the seeds are soaked in a suitable aqueous solution (controlled pH and ionic composition) to soften the tissues. The seeds are then ground using a suitable mechanical device to break down the plant tissues and release the oil bodies, starch granules, and cell wall fragments. The resulting slurry may then be fractionated to isolate different fractions, such as oil bodies and cell wall fragments. This can be achieved using gravitational separation, centrifugation, or filtration processes. Any unwanted components can then be removed at this point. In some cases, any large particles present in the system can be broken down further using chemical or enzymatic methods. For instance, specific enzymes that

8.3 Production of Plant-Based Milk Analogs

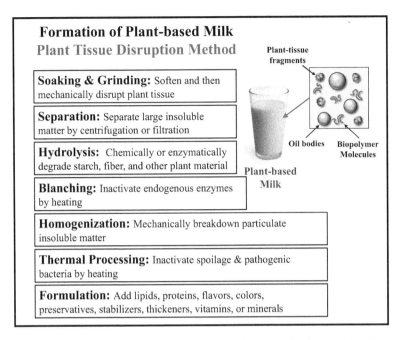

Fig. 8.1a Some commonly used processing operations that are utilized to produce plant-based milks. These processes need not be carried out in the order shown

hydrolyze starch granules or cell wall fragments can be added, which often produces an end product that is more resistant to gravitational separation and has a smoother mouthfeel. In addition, the material used to form the milk analog is typically blanched to inactivate most natural enzymes present that may cause undesirable changes in product quality. It is also typically thermally processed (pasteurized or sterilized) to inactivate any microbes that could lead to spoilage or food safety issues. Moreover, the processing operations may be designed to remove or deactivate any antinutrient factors that may be present. If required, a more intense mechanical disruption process may be used to reduce the size and increase the uniformity of the colloidal particles present in the system. For instance, the slurry can be passed through a high-pressure valve homogenizer that generates intense turbulent flow, shear, and cavitation forces that further break down the particles. In such a product, the aqueous phase in the final product contains a mixture of solutes, such as sugars, minerals, and proteins, that simulate some of the desirable flavor and nutritional characteristics of cow's milk. However, other ingredients may also be added to the product at the end, such as colors, flavors, vitamins, minerals, or preservatives to ensure it has the required shelf life, sensory attributes, and nutritional profile. Finally, the milk analog will be packaged into a suitable container prior to storage and transport.

8.3.2 Emulsification Approaches

An alternative approach of creating milk analogs is to create oil-in-water emulsions that contain fat droplets dispersed within an aqueous medium (McClements, 2020) (Fig. 8.1b). In this case, the fat droplets are designed to mimic the properties of the fat globules found in cow's milk. In addition, other ingredients can be added to simulate the properties of the casein micelles and other functional components in cow's milk, such as the sugars, salts, oligosaccharides, and soluble proteins. This type of emulsion can be prepared by homogenizing plant-based oils, emulsifiers, water, and other ingredients using a suitable mechanical device. A schematic diagram of the various steps needed to form this kind of milk analog is shown in Fig. 8.1b. An oil phase is formed by dissolving all the oil-soluble components (such as hydrophobic vitamins, colors, flavors, and preservatives) into a suitable plant-based oil (such as corn, sunflower, flaxseed, olive oil, or coconut oil). If necessary, the oil phase should be warmed to melt any crystals because these can block the homogenizer. However, it should not be held at elevated temperatures for too long so as to avoid lipid oxidation and other undesirable chemical reactions. A water phase is formed by dispersing a suitable hydrophilic emulsifier (such as a plant-based protein, polysaccharide, phospholipid, or saponin) into an aqueous solution.

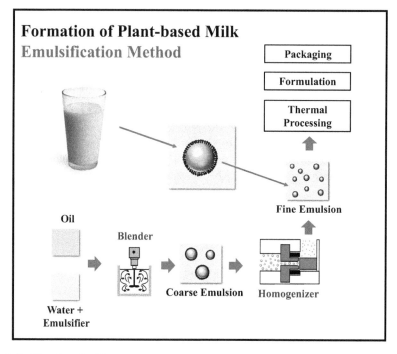

Fig. 8.1b Plant-based milks can also be produced by homogenizing plant-based oil and emulsifiers with water. Typically, a coarse emulsion is formed first, which is then passed through a homogenizer

In some cases, the emulsifiers are fully soluble, but in other cases, they only partially dissolve (especially plant-based proteins). The pH of this solution should usually be controlled by using acids, bases, and/or buffers. The water phase may also contain other water-soluble components, such as flavors, colors, sugars, salts, preservatives, and thickening agents but these could also be added after homogenization. If the oil phase does contain any crystals, then the water phase should be heated prior to combining it with the oil phase. The oil and water phases are then blended together using a high-shear mixer to form a coarse emulsion. This emulsion can then be passed one or more times through a high-energy homogenizer to reduce the fat droplet size further, and therefore improve the stability of the end product. A number of different homogenization devices can be used for this purpose with colloid mills, high-pressure valve homogenizers, sonicators, and microfluidizers being the most common. More detailed information about the ingredients and processing operations that can be used to create plant-based milks using this approach are given in the remainder of this section. Additional information about the functional performance of different ingredients that can be used is also provided in Chap. 2.

8.3.2.1 Ingredients

The type of ingredients used to formulate a plant-based milk analog will play an important role in determining its stability, physicochemical properties, functionality, and sensory attributes. A number of different ingredients are used for this purpose, including plant-based oils, emulsifiers, additives, and water (Do et al., 2018).

Plant-Based Oils Numerous kinds of edible plant-based oils may be utilized to formulate oil-in-water emulsions, such as corn, sunflower, canola, soybean, vegetable, algal, flaxseed, olive, palm, and coconut oils. These oils are mainly comprised of triacylglycerol molecules that vary in the position, chain length, and unsaturation of the fatty acids. The fatty acid profiles of these oils play an important role in determining the properties of the final emulsion. Coconut oils are mainly composed of saturated fatty acids (C_8 to C_{12}), which makes them highly resistant to oxidation. Moreover, they tend to be partially crystalline at low temperatures ($< 25\ °C$), which may be important for simulating some of the desirable attributes of milk fat associated with its melting/crystallization behavior. However, nutritional studies suggest that saturated fats may have undesirable health impacts, such as increasing the risk of coronary heart disease (Ludwig et al., 2018). In contrast, algal and flaxseed oils contain relatively high amounts of polyunsaturated (omega-3) fatty acids, which makes them highly prone to oxidation and means they tend to be fluid at most practical temperatures. In this case, nutritional studies suggest that these omega-3 fatty acids are beneficial to human health (Goyal et al., 2014; Kaur et al., 2018). Moreover, the interfacial tension, viscosity, and melting point of a plant-based oil also influence its ability to form and stabilize emulsions (McClements, 2015). As an example, the droplet size produced by homogenization typically decreases with decreasing interfacial tension and viscosity of the oil phase because these factors facilitate droplet disruption (McClements, 2015). In addition, lipids containing crystals usually have to be heated prior to homogenization to avoid blocking the small tubes and valves inside the homogenizer.

Water The tap water obtained from the general water supply varies in its pH, mineral composition, and organic load from location-to-location and from time-to-time, which can affect the formation and stability of emulsions (Navarini & Rivetti, 2010). Consequently, water should usually be treated prior to use to ensure that it has suitable properties for creating stable plant-based milks. For example, the water may be passed through a carbon filter, reverse osmosis unit, or ion exchange column to remove undesirable organic matter and minerals. It may also be treated with ultraviolet light or heat to deactivate undesirable microorganisms that may be present. Acids, bases, buffers or minerals can then be added to the treated water if required to obtain a specified pH, flavor profile, or functionality.

Plant-Based Emulsifiers Emulsifier selection is one of the most important factors impacting the successful formation of a plant-based milk product with the required properties (McClements, 2015). A number of plant-based emulsifiers can be used to form and stabilize emulsions, such as amphiphilic proteins, phospholipids, saponins, and polysaccharides (McClements et al., 2017; McClements & Gumus, 2016). These emulsifiers may be isolated from natural sources or produced by cellular agriculture methods (Chap. 2). The functional performance of these emulsifiers depends on their molecular characteristics, as well as the environment they are used in (such as pH, ionic strength, temperature, and ingredient interactions). The effectiveness of an emulsifier depends on two main aspects, its ability to form emulsions and its ability to stabilize emulsions:

- *Formation*: The emulsifier must be capable of rapidly adsorbing to the surfaces of the oil droplets produced during homogenization, attaching to the surfaces, and then reducing the interfacial tension. The size of the droplets produced during homogenization tends to be lower for emulsifiers that can rapidly adsorb to the droplet surfaces (faster adsorption kinetics), and for emulsifiers that pack at the oil-water interface in a manner that effectively reduces the contact between the oil and water molecules (lower interfacial tension). Consequently, small surfactants (like saponins) tend to be better at forming small droplets than large amphiphilic polysaccharides (like gum arabic or modified starch). The effectiveness of an emulsifier at forming emulsions can be characterized by two parameters that are determined by measuring the change in mean particle diameter (d_{32}) with increasing emulsifier concentration under standardized conditions, *i.e.*, homogenizer type, homogenization pressure, number of passes, and oil-water ratio (Fig. 8.2). These two parameters are the minimum emulsifier concentration required to form small oil droplets (c_{MIN}) and the minimum oil droplet size that can be produced (d_{MIN}). The amount of emulsifier required to cover a given surface area, which is known as the surface load (Γ), can then be calculated from these parameters: $\Gamma = (d_{MIN} \times c_{MIN})/(6\phi)$, where ϕ is the dispersed phase volume

8.3 Production of Plant-Based Milk Analogs

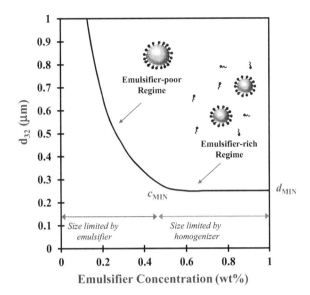

Fig. 8.2 The ability of a plant-based emulsifier to form an emulsion can be characterized by the minimum amount (c_{MIN}) required to form small droplets under standardized homogenization conditions, as well as the smallest droplet size that can be achieved under those conditions (d_{MIN})

fraction. The advantage of using the surface load to characterize an emulsifier is that it should be largely independent of the homogenization conditions. The higher the surface load, the more emulsifier is required to form a given emulsion.

- *Stability*: Once an emulsion has been successfully formed it is important that it remains stable over time or after it is exposed to specific conditions (such as shaking, cooking, or adding to coffee). The stability of an emulsion is mainly determined by the ability of the layer of emulsifier molecules to generate electrostatic and/or steric repulsive forces between the emulsifier-coated fat droplets. However, it may also be influenced by the resistance of the emulsifier coating to disruption, which depends on the nature of any crosslinks formed between the adsorbed emulsifier molecules.

The ability of different kinds of plant-based emulsifiers to form and stabilize emulsions varies considerably depending on their molecular characteristics. For example, studies in our laboratory have shown that the amount of plant-based emulsifier required to form small droplets during homogenization increases in the following order: saponins < phospholipids < proteins < polysaccharides (Fig. 8.3). In practice, the amount of an emulsifier required to form small droplets actually depends on its precise molecular properties (such as molecular weight and surface hydrophobicity). Consequently, different plant-based proteins (or other emulsifiers) may be more or less effective depending on their biological origin and processing history. There are also major differences in the stability of emulsions formed by different plant-based emulsifiers (Fig. 8.4). The fat droplets coated by plant proteins are highly susceptible to aggregation at pH values near their isoelectric points, at high salt concentrations, and at high temperatures, whereas those coated by plant polysaccharides are much more resistant to these factors. This is because plant proteins mainly

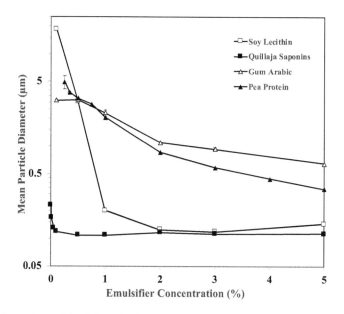

Fig. 8.3 Comparison of the ability of different plant-based emulsifiers to form oil-in-water emulsions by homogenization. (Data taken from authors laboratory)

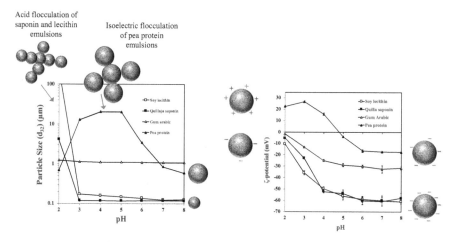

Fig. 8.4 Impact of pH on the electrical surface potential and mean particle diameter of oil-in-water emulsions stabilized by different kinds of plant-based emulsifiers. (Data taken from various studies carried out in the authors lab)

stabilize fat droplets from aggregation through electrostatic repulsion, whereas plant polysaccharides mainly stabilize them through steric repulsion. Moreover, plant proteins tend to unfold when heated above their thermal denaturation temperature (T_m), which increases the surface hydrophobicity of the coated fat droplets, thereby increasing the hydrophobic attraction between them. For instance, emulsions stabilized by lentil, pea, and fava bean proteins exhibit extensive flocculation at pH values near their isoelectric points (pH 5) and at high salt contents (> 50 mM NaCl) because they are stabilized by electrostatic repulsion (Gumus et al., 2017a). Conversely, emulsions stabilized by gum arabic are highly resistant to changes in pH and salt content because they are stabilized by steric repulsion (Ozturk et al., 2015a). For example, the influence of pH on the particle size and charge of emulsions stabilized by various plant-based proteins, polysaccharides, phospholipids, and saponins are compared in Fig. 8.4. These measurements show that the stability of the fat droplets to aggregation is highly dependent on the nature of the emulsifier used to coat them. For instance, protein-coated fat droplets are susceptible to aggregation at intermediate pH values (near their isoelectric point) because they lose their surface charge, which reduces the electrostatic repulsion between them. Phospholipid- and saponin-coated fat droplets are stable from mildly acidic to mildly basic conditions because they have a strong negative charge but they aggregate under strongly acidic conditions because they lose their surface charge when the anionic groups are protonated. Polysaccharide-coated fat droplets are stable to aggregation across the entire pH range because they are primarily stabilized by steric repulsion (but the individual droplets are relatively large, so they may still be prone to creaming). It is therefore important to select an emulsifier that will produce stable milk analogs under the intended range of conditions that will be used.

The effectiveness of plant-based emulsifiers can be increased by using them in combination, rather than individually (McClements & Jafari, 2018). For instance, proteins and carbohydrates can be covalently linked together to form conjugates (Gu et al., 2017; McClements & Decker, 2018; Zha et al., 2019). The protein moiety has good surface-activity, which promotes the attachment of the conjugates to the interface, whereas the carbohydrate moiety generates a strong steric repulsion, which increases the resistance of the emulsions to aggregation. Alternatively, proteins and polysaccharides can be linked together using non-covalent interactions, especially electrostatic attraction, leading to the formation of complexes with improved performance (Li & de Vries, 2018). These complexes can be created before homogenization by mixing the protein and polysaccharide together before emulsion formation. Alternatively, they can be created after homogenization by adding the protein before emulsification and the polysaccharide after (Guzey & McClements, 2006). It is also possible to use mixtures of two or more emulsifiers that do not directly interact with each other to form and stabilize emulsifiers. In this case, the different kinds of emulsifier adsorb to the droplet surfaces and the overall interfacial composition depends on their relative concentrations and surface activities (Reichert et al., 2019). In some situations, using mixed emulsifiers leads to better performance than using either of the individual ones but in other cases it may actually lead to a worse performance. Consequently, the correct type and ratio of emulsifiers used has to be determined empirically. Rather than using conventional molecular emulsifiers, it is also possible to use colloidal emulsifiers, such as plant-based protein, polysaccharide, and/or polyphenol particles, to form Pickering emulsions (Sarkar & Dickinson, 2020). These plant-based particles usually lead to the formation of much larger fat droplets than molecular emulsifiers but are much more resistant to coalescence. The relatively large size of the fat droplets formed by Pickering emulsifiers means that they are likely to be unsuitable for the use in plant-based milks because these products have a relatively low viscosity and so the droplets would rapidly cream.

Additives A number of other additives may be utilized in plant-based milks to improve their quality attributes, extend their shelf life, enhance their nutritional attributes, or increase their safety (McClements, 2020; McClements et al., 2017; McClements & Gumus, 2016). Hydrophobic additives are usually added to the oil phase prior to homogenization, whereas hydrophilic ones are added to the aqueous phase either before or after homogenization. Plant-based colors or flavors may be added to improve the appearance or flavor profile of the products. Plant-based antioxidants or antimicrobials may be added to increase the storage stability and safety of the product by inhibiting undesirable chemical reactions or microbial growth. Thickening agents may be incorporated into the aqueous phase to obtain a desired texture and mouthfeel, as well as to inhibit the creaming or sedimentation of colloidal particles. Numerous kinds of plant-based thickening agents are available for this purpose, with most of them being polysaccharides (Chap. 2). These polysaccharides may be isolated from terrestrial plants (*e.g.,* starch, cellulose, guar gum, locust bean gum, or pectin), marine plants (*e.g.,* alginate and carrageenan), or produced by microbial fermentation (*e.g.,* xanthan gum). Polysaccharides can also be used as stabilizers that inhibit the tendency for fat droplets to aggregate with each other by creating a protective coating around them. As an example, negatively charged pectin molecules are able to adsorb to cationic groups located on the surfaces of protein-coated fat droplets, which reduces isoelectric flocculation by increasing the electrostatic and steric repulsion between the droplets (Fig. 8.5) (Guzey & McClements, 2006). This phenomenon may be used in emulsions containing plant protein-coated fat droplets to improve their resistance to flocculation, especially when added to acidic coffees. It is therefore important to select an appropriate combination of plant-based additives to ensure the final product has the desired properties.

The nutrient composition of some common plant-based milks is compared to that of cow's milk in Table 8.3. It can be seen that there are appreciable differences in both the macronutrient and micronutrient levels in the different kinds of milks, which may have some nutritional and health consequences, which are discussed in Sect. 8.6.

8.3.2.2 Processing Operations

As discussed earlier, a number of different processing operations are used to produce plant-based milk analogs using the emulsification approach (Vogelsang-O'Dwyer et al., 2021). However, the most important step is homogenization, which involves reducing the size of the fat droplets to a level where they provide the required physicochemical and stability attributes (Hakansson, 2019). Many

8.3 Production of Plant-Based Milk Analogs

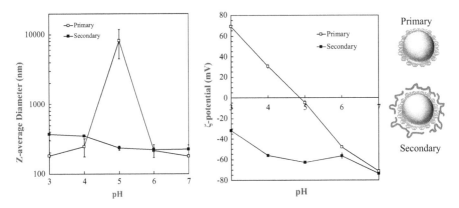

Fig. 8.5 Impact of pH on the electrical surface potential and mean particle diameter of oil-in-water emulsions stabilized by either a protein (β-lactoglobulin) or a protein/polysaccharide (β-lactoglobulin/pectin), which are referred to as primary and secondary emulsions, respectively. A similar phenomenon should occur for plant-protein coated fat droplets

different types of homogenizers are available for producing emulsions in the food industry, such as high-shear mixers, colloid mills, high-pressure valve homogenizers (HPVHs), ultrasonic homogenizers, and microfluidizers. It should be noted that HPVHs and ultrasonic homogenizers are also widely utilized to create homogenized cow's milk. Each homogenization technology has specific benefits and drawbacks and it is important to select the most suitable one for the product being prepared. In particular, homogenizers vary in terms of their initial capital costs, running costs, robustness, versatility, production rates, and ability to generate fine uniform fat droplets (McClements, 2015). HPVHs are currently the most commonly used type of homogenizer in the food industry at present, but there is increasing interest in the utilization of sonicators and microfluidizers because they have been shown to be highly effective at producing fine droplets ($d_{32} < 300$ nm).

Milk analogs are typically subjected to some kind of thermal processing (such as pasteurization or sterilization) after they have been prepared to inactive enzymes or microorganisms that may reduce their shelf lives or pose a safety risk. The resistance of milk analogs to thermal processing is strongly influenced by the type of emulsifier used to coat the fat droplets, as well as the time-temperature combination used (McClements et al., 2017). Some emulsifiers are much more resistant to heating than others. For instance, polysaccharides and saponins tend to produce more heat-stable emulsions than proteins. However, the heat-stability of protein-coated fat droplets is highly dependent on the solution conditions. Typically, they are much more resistant to heating at low salt contents and at pH values sufficiently far from the isoelectric point (Gumus et al., 2017a; Qamar et al., 2019). Consequently, it is important to select a plant-based emulsifier that is resistant to any thermal processing operations used. It is also important to ensure that the emulsifier-coated droplets do not undergo any adverse interactions with other additives in the systems. For instance, polyphenols and some other phytochemicals may interact with proteins and influence their performance as emulsifiers.

Table 8.3 Nutrient levels in cow's milk and a number of representative plant-based milks (per 100 g). It should be noted that the nutrient levels may vary within a particular product category depending on its formulation (*e.g.*, reduced fat, sweetened, or sugar free)

Value	Unit	Cows' milk	Soy milk	Almond milk	Coconut milk	Rice milk
Energy	kcal	67	42	15	80	113
Macronutrients						
Protein	g	3.3	2.9	0.4	0.7	0.7
Fat	g	3.3	1.7	1.0	2.7	2.3
Carbohydrate	g	5.4	3.3	1.3	14.0	22.0
Dietary fiber	g	0	0.4	0.2	1.3	0.7
Sugar	g	5	2.5	0.8	10.0	12.7
Minerals						
Calcium, Ca	mg	130	130	184	130	280
Iron, Fe	mg	0	0.45	0.28	0.48	0.48
Magnesium, Mg	mg	–	–	6	–	26
Potassium, K	mg	–	130	–	–	65
Sodium, Na	mg	52	42	72	20	94
Vitamins						
Vitamin C	mg	0.50	0	0	3.2	0
Riboflavin	mg	–	0.21	0.01	–	0.34
Vitamin B-12	µg	–	1.3	0	0.80	1.5
Vitamin A	IU	210	210	0	0	500
Vitamin D	IU	42	50	1	53	100
Lipids						
Fatty acids (saturated)	g	2.1	0.21	0.1	2.3	0
Fatty acids (trans)	g	0	0	0	0	0
Cholesterol	mg	15	0	0	0	0

The data taken was obtained from the USDA Food Database (https://fdc.nal.usda.gov).

8.4 Physicochemical Attributes

Milk analogs are usually designed to have physicochemical and functional attributes mimicking those of the cow's milks they are designed to replace. In this section, we give an overview of the main physicochemical properties of milk analogs and highlight the main factors that influence them, with a special emphasis on presenting mathematical models that can be used to quantify the impact of these factors. More general information about the optical, rheological, and stability properties of plant-based foods is given in Chap. 4.

8.4.1 Appearance

The optical properties of milk analogs are usually designed to mimic those of cow's milk, *i.e.,* they are expected to be whitish opaque fluids with uniform appearances (McClements, 2020). In practice, the appearance of milk analogs depend on the type of ingredients and processing methods used to produce them since this affects the absorption and scattering of light waves (Chap. 4). In some cases, a milk analog may have an appearance that significantly deviates from that of cow's milk but is still acceptable to consumers. For instance, hazelnut milk may have a light brown color that consumers find appealing because they associate it with a "nutty" flavor.

8.4.1.1 Physical Basis of Appearance

The physical basis of the optical properties of plant-based foods was discussed in detail in Chap. 4 and so only a brief overview is given here. The creamy appearance of milk analogs is primarily a result of light scattering by the fat droplets, oil bodies, or other colloidal particles within them. The degree of light scattering depends on the size, concentration, and refractive index contrast of the particles (Griffin & Griffin, 1985; Stocker et al., 2017). It is difficult to control the refractive index of the colloidal particles but the size and concentration can be controlled. Typically, the lightness of a milk analog increases as the particle concentration increases and has a maximum value at an intermediate particle diameter of around 200 nm (Fig. 8.6). Consequently, it is important to control the particle size and concentration to obtain a lightness in a milk analog that is similar to that found in cow's milk.

The color of milk analogs is governed by selective absorption of light waves over particular wavelengths in the visible spectrum due to the presence of chromophores (Chap. 4). Cow's milk often has a slightly yellowish-green tinge (positive $b*$ and negative $a*$) due to the presence of low levels of natural pigments in the animal feed, such as carotenoids or riboflavin (McClements, 2015; Schiano et al., 2017). Thus, the color of cow's milk may vary depending on the nature of the food in the animal's diet, which varies depending on the farming method, geographical location, and season (Agabriel et al., 2007; Scarso et al., 2017). Milk analogs may also contain pigments that influence their color, which may be naturally present in the ingredients used to formulate them or generated during processing or storage (*e.g.,* non-enzymatic browning products). Moreover, pigments may be intentionally incorporated into some milk analogs to provide the expected color for a particular product type. For instance, brown, red, or yellow pigments may be added to chocolate, strawberry, or banana flavor milk analogs. For labeling purposes, it is advantageous to utilize natural plant-derived pigments for this purpose, such as those discussed in Chap. 2.

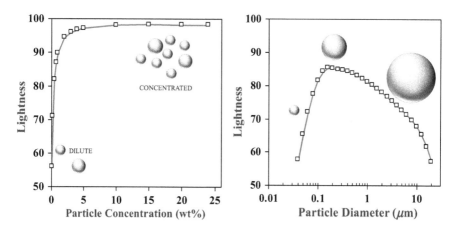

Fig. 8.6 The lightness of a plant-based milk is dependent on the size and concentration of any particulate matter that scatters light. The theoretical predictions of lightness were made by the author using light scattering theory for oil droplets dispersed in water. For the left figure it was assumed the particle size was constant, while for the right figure it was assumed the particle concentration was constant

8.4.1.2 Measurement of Appearance

The overall appearance of milk analogs can be characterized by digital photography, which can help to establish whether they are uniform in appearance or have separated due to creaming (layer at the top) and/or sedimentation (layer at the bottom). The optical properties, such as lightness and color, can be conveniently quantified using instrumental colorimeters or spectrometers (Chap. 4). Typically, the optical properties are reported by these instruments in terms of tristimulus color coordinates, such as the $L^*a^*b^*$ system, where L^* represents the lightness (black to white, 0 to 100), a^* represents the redness-greenness (+/−), and b^* represents the yellowness/blueness (+/−) (Hutchings, 1999). These values are commonly measured by pouring the milk analog into a transparent container and a standardized light beam is reflected from its surface. The measured reflectance spectrum is then converted into tristimulus color coordinates using a suitable mathematical model (McClements, 2002b).

8.4.1.3 Comparison to Cow's Milk

Cow's Milk Whole milk from cows has a high lightness and slightly yellowish-green tinge, as indicated by the high L^* values, slightly negative a^*-values, and slightly positive b^*-values (Table 8.4). The high lightness is due to the strong scattering of light waves by the milk fat globules and casein micelles. The slight tinge is due to the selective absorption of light waves by dietary pigments (such as carotenoids and riboflavin) that accumulate inside the milk fat globules (Schiano et al.,

2017). The appearance of cow's milk is also influenced by the level of fat that is present (Kneifel et al., 1992). As the fat content is raised, the lightness (L^*) increases, the greenness ($-a^*$) decreases, and the yellowness ($+b^*$) increases (Table 8.4). These changes can be attributed to the increase in light scattering that occurs when the fat globule concentration increases, as well as the change in selective light absorption that occurs as the pigment concentration increases. These results suggest that it may be necessary to design milk analogs with different optical properties depending upon whether one is trying to mimic different kinds of cow's milk or creams (*e.g.,* skim, low-fat, whole milk, or half-and-half).

Milk Analogs Milk analogs also tend to be creamy looking fluids but their lightness and color coordinates can differ appreciably from cow's milk. In general, the raw materials and processing operations used to produce milk analogs influences their appearance. The tristimulus color coordinates of a number of milk analogs are compared to that of cow's milk products in Table 8.4. The milk analogs tend to have a lower lightness than cow's milk, which means they look waterier. In addition, they have appreciably different color coordinates (a^* and b^*) than cow's milk. Moreover, the optical properties of a particular type of plant-based milk can vary considerably depending on the ingredients and manufacturing operations used. For instance, colorimetry analysis of commercial soy milks has shown that their optical properties can vary significantly (Liu & Chang, 2013): L^* from around 62 to 81 (increasing lightness); a^* from around -3 to $+3$ (reddish to greenish tinge); and b^* from around $+7$ to $+21$ (increasing yellowish). The origin of this effect is probably due to differences in the composition and structure of the milks, in particular the type and concentration of pigments and colloidal particles they contain. However, some kinds of milk analogs are expected to have their own characteristic appearances that consumers find desirable. For instance, nut and oat milks are expected to have a light brownish tinge, which is the result of specific chromophores in the plant materials used to prepare them. However, some consumers may not find milk analogs that do not look like cow's milk appealing. For this reason, milk analog manufacturers may use specific processing operations to remove pigments or bleach them. The lightness of milk analogs depends on the number, type, and size of the colloidal particles

Table 8.4 Reported $L^*a^*b^*$ values for several animal- and plant-based milk products. The fat content is shown in brackets for the cow's milk products

Product	L^*	a^*	b^*	References
Cow's skim milk (0.1%)	81.7	−4.8	+4.1	Kneifel et al. (1992)
Cow's full fat milk (3.6%)	86.1	−2.1	+7.8	Kneifel et al. (1992)
Cow's coffee creamer (10%)	86.9	−0.5	+8.6	Kneifel et al. (1992)
Cow's cream (36%)	88.1	−0.2	+8.8	Kneifel et al. (1992)
Almond milk	71.4	+3.3	+16.0	Zheng et al. (2021)
Cashew milk	72.7	+2.9	+15.1	Zheng et al. (2021)
Coconut milk	85.9	−1.0	+5.1	Zheng et al. (2021)
Oat milk	67.8	+4.2	+13.8	Zheng et al. (2021)
Soy milk	73.1	+12.1	+2.1	Durazzo et al. (2015)

they contain. Manufacturers can therefore manipulate the lightness of their products by controlling these parameters so as to more closely resemble the appearance of cow's milk (McClements, 2002a, b).

8.4.2 Texture

The rheological properties of fluid dairy products are mainly governed by their fat content, with the shear viscosity increasing with increasing milk fat globule content (Table 8.2). Skim milk is a relatively low viscous ideal fluid, whereas heavy cream is a high viscosity shear-thinning fluid. Milk analogs are usually formulated to mimic the textural characteristics of the dairy products they are designed to simulate. As discussed in Chap. 4, the rheological properties of colloidal dispersions like milk and milk analogs are mainly determined by the nature of the colloidal particles and polymers they contain, including milk fat globules, fat droplets, plant cell fragments, protein aggregates, and hydrocolloids (such as thickening agents).

8.4.2.1 Physical Basis of Rheology

The viscosity of milk and milk analogs can be described utilizing the theories developed to describe the viscosity of colloidal dispersions (Chap. 4). For dilute systems ($\phi < 5\%$) containing non-interacting spherical particles, the shear viscosity (η) can be described by Einstein's equation:

$$\eta = \eta_1 (1 + 2.5\phi) \tag{8.1}$$

Here, η_1 is the shear viscosity of the continuous phase and ϕ is the volume fraction of the particles. This equation predicts that the viscosity of a milk-like product should increase linearly with increasing fat droplet concentration. It provides a relatively good description of the shear viscosity of cow's milks (skim to whole) due to their fairly low particle content. However, it is unsuitable for creams because the fluid flow around one milk fat globule influences that around the neighboring ones. In this case, an effective medium theory (EMT) can be used to describe the dependence of the shear viscosity on particle concentration (Genovese et al., 2007; McClements, 2015):

$$\eta = \eta_1 \left(1 - \frac{\phi}{\phi_c}\right)^{-2} \tag{8.2}$$

Here, ϕ_c is known as the critical packing parameter, which has a numerical value of around 0.65. This parameter represents the volume fraction at which the colloidal particles are packed so tightly together that they cannot move past each other and so the whole system gains solid-like behavior (elasticity). Predictions made using the EMT show that the viscosity of milk-like colloidal dispersions should increase with

8.4 Physicochemical Attributes

increasing fat droplet concentration (Fig. 8.7). In practice, the viscosities may be higher than these predicted values because of the presence of other kinds of colloidal particles, such as casein micelles in milk or hydrocolloids in milk analogs.

The shear viscosity increases relatively slowly with increasing particle concentration from around 0% to 20% but then increases much more rapidly from around 20% to 50% (Fig. 8.7). Cow's milk products (skim to whole) therefore have relatively low viscosities because they have low fat contents (0.1–3.6%). In contrast, cream products (light to heavy) have relatively high viscosities because they have much higher fat contents (18% to over 36%). As mentioned earlier, the casein micelles will also make some contribution to the shear viscosity of milk and cream products.

Typically, dilute colloidal dispersions (like milk) tend to be ideal fluids, *i.e.,* their shear viscosity is independent of the applied shear rate. In contrast, concentrated dispersions (like cream) tend to be shear thinning fluids, *i.e.,* their viscosity decreases with increasing shear rate. This effect is due to a balance between Brownian motion effects (which favor a random distribution of particles) and shear effects (which favor alignment of the particles with the shear field) (McClements, 2015). The viscosity of fluid dairy products, especially creams, also depends on the aggregation state of the colloidal particles. Fat globules and casein micelles may flocculate when a dairy product is acidified, since this causes the pH to move towards the isoelectric point of the proteins (around pH 5), thereby reducing the electrostatic repulsion between them. Partially crystalline milk fat globules may undergo partial coalescence, especially when the product is sheared, which can lead to extensive

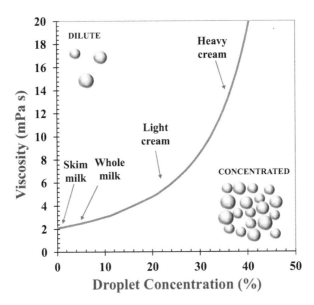

Fig. 8.7 The shear viscosity of milk-type products increases with increasing fat content. In reality, the viscosities may be higher than these predicted values because other kinds of colloidal particles (such as casein micelles in milk or hydrocolloids in milk analogs) also contribute to the viscosity

clumping. Flocculation and clumping of fat globules lead to an increase in the effective volume fraction of the disperse phase due to the presence of aqueous phase trapped within the aggregates. As a result, these effects lead to an increase in the viscosity (Eq. 8.2). For plant-based milks, the addition of thickening agents (like polysaccharides) may also contribute to the overall viscosity of the system. As a result, they may exhibit strong shear thinning behavior, leading to flow index values considerably below 1 (Table 8.5).

8.4.2.2 Measurement of Rheology

In research and development laboratories, the rheology of milks and creams is usually characterized using sophisticated instrumental rotational viscometers or shear rheometers (Rao, 2013). In quality assurance laboratories, simpler and cheaper versions of these instruments are typically used to provide a rapid assessment of whether a product meets some specified textural criteria. The most commonly utilized measurement cells for fluid samples are the cup-and-bob and cone-and-plate types (Fig. 8.8). When using the cup-and-bob type, the sample is poured into a cup (outer cylinder) and the bob (inner cylinder) is lowered into it. Prior to analysis, the sample is allowed to reach the required measurement temperature since the viscosity often depends on this parameter. A known force (shear stress) is applied to the inner cylinder and its rotational speed (shear rate) is measured using a suitable device (or *vice versa*). The shear stress (τ) is then plotted as a function of the shear rate ($d\gamma/dt$ or $\dot{\gamma}$). The apparent shear viscosity can then be calculated as a function of the shear rate from the slope of this graph since $\tau = \eta \times \dot{\gamma}$ (McClements, 2015).

Table 8.5 Rheological, optical, stability, and particle size (D_{43}) properties of a number of plant-based milks. The viscosity is reported at a shear rate of 10 s^{-1}. Ranges are given for examples where a number of different products in the same category were analyzed. Data adopted from Jeske et al. 2017

Milk type	Viscosity [mPa·s]	Flow index n	D_{43} [µm]	Separation rate (%/h)	Whiteness Index
Cow's	3.2	1.00	0.60	3.9	81.9
Almond	3.9–26.3	0.56–0.98	0.9–6.0	1.4–52	52–76
Cashew	5.6	0.97	29	28	66
Coconut	48	0.40	1.7	37	68
Hazelnut	25	0.67	2.2	1.3	56
Hemp	25	0.73	1.5	4.4	69
Macadamia	2.2	1.00	3.4	54	52
Oat	6.8	0.89	3.8	40	60
Quinoa	13	0.76	82	32	71
Rice	2.8	0.97	11	43	67
Brown rice	2.2	1.00	0.72	51	64
Soy	2.6–7.6	0.90–1.00	1.0–1.3	8.6–23	69–75

The data was adapted from Jeske and others (2017).

8.4 Physicochemical Attributes

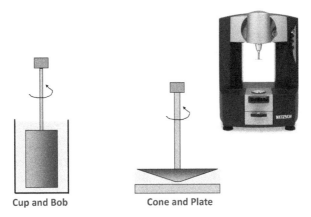

Fig. 8.8 The shear viscosity of milk-type products is usually measured using an instrumental rheometer or viscometer. The most common measurement cells are shown here. Image of dynamic shear rheometer kindly supplied by Philip Rolfe (Netzsch in Selb, Germany)

As discussed in Chap. 4, for an ideal liquid, the apparent shear viscosity is independent of the shear rate, which is the case for cow's milk products because they have a low concentration of colloidal particles. However, for a shear thinning liquid, the apparent shear viscosity may decrease appreciably with increasing shear rate, which is the case for dairy creams because they have a high fat content, as well as some milk analogs because they contain hydrocolloid-based thickening agents. The apparent shear viscosity of these shear-thinning fluids can be modelled using a power-law equation:

$$\eta = K(\dot{\gamma})^{n-1} \tag{8.3}$$

Here, K and n are known as the consistency index and power index, respectively. The power index provides information about the ideality of fluid flow: $n = 1$ for an ideal fluid; $n < 1$ for a shear thinning fluid; and $n > 1$ for a shear thickening fluid. For a shear-thinning fluid, like many dairy creams and milk analogs, the smaller the value of n the more pronounced the decrease in the viscosity with increasing shear rate. This kind of behavior is important because it impacts the stability, functionality, and mouthfeel of the final product.

8.4.2.3 Comparison to Cow's Milk

Cow's Milk Based on the equations describing the rheology of colloidal dispersions, the viscosity of cow's milk is expected to mainly depend on the dispersed phase volume fraction of the colloidal particles it contains (milk fat globules and casein micelles), as well as the shear viscosity of the surrounding aqueous phase (Eqs. 8.1 and 8.2). The aqueous phase of milk contains a mixture of soluble sugars (mainly lactose), globular proteins (mainly whey proteins), and minerals (mainly potassium, sodium, and chloride). It tends to have a viscosity that is fairly similar to that of water. The volume fraction of milk fat globules depends on milk type, ranging from around 0.1% (ϕ = 0.001) for skim milk to 3.6% (ϕ = 0.036) for whole milk. The volume fraction of casein micelles in milk typically varies from around 6 to 12% (ϕ = 0.06 to 0.12) (Goff, 2019). The reason that the volume fraction of casein micelles is considerably higher than the protein content (around 4%) is because they trap a considerable amount of water. The total particle volume fraction (milk fat globules + casein micelles) found in cow's milk products (skim to whole) would therefore be expected to range from around 6% to 16%, which would lead to a predicted viscosity of around 1.2- to 1.8-fold larger than the surrounding aqueous solution that contains only soluble solutes, such as sugars, salts, and globular proteins (Eq. 8.2). Experimental measurements support the predictions of the effective medium theory. For instance, the measured shear viscosities of cow's milk have been reported to range from about 1.9 mPa s (skim milk) to 2.2 mPa s (whole milk) (Li et al., 2018). Experiments have also shown that milk products exhibit ideal fluid behavior (n = 1), with their shear viscosities increasing with fat content and decreasing with temperature (Flauzino et al., 2010). Nevertheless, there may be an increase in viscosity and shear-thinning behavior may occur at temperatures where the fat globules in milk tend to aggregate with each other.

An empirical equation has been developed to estimate the impacts of fat content and temperature on the shear viscosity of cow's milk products (Flauzino et al., 2010):

$$\eta(mPas) = 0.00847 e^{\frac{E_a}{RT}} e^{5.83X} \tag{8.4}$$

In this equation, E_a is the activation energy (13.5 kJ/mol), R is the gas constant (8.31 J/K/mol), T is the absolute temperature (K), and X is the fraction of fat globules present, which varies from around 0.005 for skim milk to 0.4 for whole milk. Predictions made using this equation are shown in Table 8.2, which highlight the increase in shear viscosity with increasing fat content.

The perceived mouthfeel of milk-like products is believed to depend on their ability to lubricate the mouth, *i.e.,* their ability to reduce the friction generated when the tongue moves against the palate. This kind of rheological behavior is described as tribology. Measurements of the tribology of cow's milk have shown that the friction coefficient decreases as the milk fat globule content decreases, which may play a role in determining the perceived creaminess of high fat products (Li et al., 2018).

Milk Analogs Like cow's milk, plant-based milks are also colloidal dispersions but they contain different kinds of particles (such as oil bodies, fat droplets, plant cell fragments, and protein aggregates), as well as polymers (such as hydrocolloid thickeners). Their rheological properties can therefore be described using similar mathematical models as used to describe the rheology of cow's milk (Eqs. 8.1 and 8.2). In this case, however, it is necessary to know the effective volume fraction of the different kinds of colloidal particles and polymers present, which is often more dif-

8.4 Physicochemical Attributes

ficult because of their heterogeneous nature. In particular, it is difficult to know the total concentration and structure of the plant tissue fragments present. Moreover, many milk analogs contain hydrocolloid thickening agents, like locust bean and guar gums, to modify their textures and mouthfeels, as well as to retard creaming and sedimentation of particulates. The thickening power of these hydrocolloids depends on their molecular conformation, aggregation state, and concentration, which is often difficult to ascertain. Typically, milk analogs containing thickening agents display shear-thinning behavior ($n < 1$). In this case, the decrease in apparent viscosity with increasing shear rate may have to be controlled to obtain the desired functional and sensory attributes required in the end product. In general, the viscosity of milk analogs increases as the concentration of fat droplets, oil bodies, cell wall fragments, and thickening agents increases, with the degree of the increase depending on the conformation of the thickening agents.

As discussed in Chap. 4, the effective volume fraction of polymer molecules (ϕ_{eff}) should be utilized in the equations used to predict the viscosity of polymer solutions rather than the volume fraction of the polymer chain (ϕ): $\phi_{eff} = R_V \phi$. Here, R_V is the volume ratio of the polymer, which is the effective volume of the polymer in solution (polymer chain and solvent) divided by the volume of only the polymer chain. The effective volume fraction can be much greater than the actual volume fraction when a polymer molecule has a high molecular weight and a highly extended structure, since it then encompasses a large amount of water.

Consequently, polysaccharides with high molecular weights and highly extended molecular conformations (like locust bean, guar, or xanthan gum) are much more effective thickening agents than compact globular proteins (like pea or soy protein). It should be noted that food grade-hydrocolloids vary in their molecular weights and conformations from batch-to-batch and from supplier-to-supplier, which should be considered when formulating plant-based milk analogs.

Researchers have measured the rheological properties of a number of commercial plant-based milk analogs (Table 8.5) (Jeske et al., 2018). The apparent shear viscosity was measured as a function of shear rate in these studies using a shear viscometer. There is a wide variation in the viscosities of these products, with values ranging from around 2 to 48 mPa s, depending on product type. Moreover, there is a considerable variation in the shear thinning behavior of the milk analogs, with the power index (n) varying from around 0.4 (strong shear thinning) to 1.0 (ideal). Even within the same product category there are appreciable variations. For instance, the viscosity has been reported to range from around 1.2 to 9.9 mPa s for different commercial soy milk products (Liu & Chang, 2013). Differences in the shear viscosities of milk analogs are mainly a result of differences in the types and concentrations of thickening agents they contain. As mentioned earlier, these thickening agents are often added to inhibit creaming or sedimentation of particulate matter during product storage but they also influence the texture and mouthfeel of the end product.

In conclusion, the rheological attributes of some milk analogs are appreciably different from those of cow's milk. For instance, cow's milk is an ideal fluid ($n = 1$) with a relatively low viscosity (2 to 3 mPa s), whereas many milk analogs are shear thinning fluids ($n < 1$) with relatively high viscosities (Table 8.5). However, some milk analogs do have rheological attributes that closely simulate those of cow's milk, such as certain kinds of soy and rice milk.

8.4.3 Stability

Ideally, the stability of milk analogs should be similar or better than that of cow's milk. The products should remain stable during storage, which means that they should be resistant to physical, chemical or biological changes. In addition, they may also have to be resistant to specific processing operations (such as pasteurization or sterilization) or food preparation procedures (such as addition to coffee) (Chung et al., 2017). For this reason, it is necessary to have knowledge of the main major factors influencing the stability of plant-based milks.

Milk analogs may break down as a result of physical instability mechanisms, such as sedimentation, creaming, flocculation, coalescence, or oiling off (McClements, 2015). Alternatively, they may break down as a result of chemical instability mechanisms, such as hydrolysis or oxidation. In addition, they may break down as a result of microbial contamination, such as the growth of molds, yeast, or bacteria. An improved understanding of the origin of these instability mechanisms can be used to extend the shelf life and improve the safety of milk analogs.

8.4.3.1 Physical Basis of Instability

Gravitational Separation One of the most common instability mechanisms in milk analogs that lead to a reduction in product quality is gravitational separation, which arises because the colloidal particles they contain have different densities compared to the aqueous solution surrounding them (McClements, 2015, 2020). Particles with a lower density than the aqueous, such as oil bodies or fat droplets, tend to move upwards due to creaming. In contrast, particles with a higher density, such as plant cell fragments or protein aggregates, tend to move downwards due to sedimentation. These processes lead to the formation of an unsightly ring at the top and/or a sediment at the bottom of the product.

As discussed in Chap. 4, the creaming velocity (v) of a particle in a dilute colloidal dispersion is given by Stokes' law:

$$v = -\frac{gd^2 (\rho_2 - \rho_1)}{18\eta_1} \tag{8.5}$$

Here, v is the velocity the particle moves upwards, d is the particle diameter, ρ_1 is the density of the aqueous phase, ρ_2 is the density of the particles, g is the gravitational constant, and η_1 is the viscosity of the aqueous phase. This equation assumes that the particles are rigid and spherical, do not interact with each other, and are

8.4 Physicochemical Attributes

dispersed in an ideal fluid. Stokes' law predicts that the creaming rate decreases with decreasing particle size and density contrast and with increasing aqueous phase viscosity. The particles move upwards when the density contrast is negative ($\rho_2 < \rho_1$) and downwards when it is positive ($\rho_2 > \rho_1$).

Stokes's law was used to predict the influence of particle size on the creaming rate of fat droplets ($\rho_2 = 930$ kg m^{-3}) and plant-tissue fragments ($\rho_2 = 1350$ kg m^{-3}) in model plant-based milks (Table 8.6). These predictions indicate that gravitational separation may occur quite quickly when the diameter of the colloidal particles exceeds about 500 nm. This is one of the most important reasons for controlling the size of the colloidal particles in milk analogs. If the particles are not sufficiently small, then the product may separate during storage due to gravitational effects, leading to an unappealing appearance. In some cases, the particle size can be reduced using mechanical (homogenization) or chemical (enzyme hydrolysis) methods. If it is not possible to reduce the particle size, then gravitational separation may be reduced by adding thickening agents to increase the viscosity of the aqueous phase. For this reason, some commercial milk analogs contain thickening agents, such as gellan or locust bean gums. However, addition of these thickening agents changes their texture and mouthfeel, which needs to be considered when ensuring they are acceptable to consumers.

Stokes' law highlights the various approaches that can be utilized to reduce the adverse effects of gravitational separation on product quality:

- *Particle size reduction:* The creaming velocity is proportional to the square of the diameter of the colloidal particles ($v \propto d^2$). Consequently, an efficacious mean of retarding gravitational separation in milk analogs is to create a product where the majority of colloidal particles are relatively small ($d < 300$ nm). For milk analogs formulated from seeds containing oil bodies (like soy and nut milks), it may be possible to select breeds or growing conditions that lead to the formation of small oil bodies (Tzen et al., 1993; Tzen & Huang, 1992). Alternatively, it may be possible to reduce the size of the oil bodies by homogenizing the slurry of plant material generated during their processing (Al Loman et al., 2018; Preece et al., 2015). For milk analogs produced by emulsifying oil and water phases together, the fat droplet dimensions can be decreased by optimizing the homogenization process, *i.e.*, such as homogenizer type, operating pressure, number of passes, and emulsifier characteristics. The fat droplet size typically decreases as the operating pressure and number of passes increases, as well as when the emulsifier concentration increases (McClements, 2015). Once small colloidal particles have been formed in a milk analog, it is necessary to prevent their aggregation during storage or processing, otherwise there will be an increase in particle size that leads to faster gravitational separation. Strategies for preventing particle aggregation are considered in the next section.
- *Viscosity enhancement:* As mentioned earlier, the shear viscosity of the aqueous phase that surrounds the colloidal particles in milk analogs can be increased by incorporating thickening agents, which are usually polysaccharide-based hydrocolloids such as guar, locust bean, or xanthan gums. The effectiveness of these

Table 8.6 Calculations of the impact of particle size on the creaming velocity of fat droplets ($\rho_2 = 930$ kg m^{-3}) and plant-tissue fragments ($\rho_2 = 1350$ kg m^{-3}) suspended in aqueous solutions ($\rho_1 = 1050$ kg m^{-3}) with different viscosities (1-500 mPa s) made using Stokes Law

Diameter (μm)	Creaming velocity (mm/day) Viscosity (mPa s)					
Fat globules						
	1	5	10	50	100	500
0.1	0.1	0.0	0.0	0.0	0.0	0.0
0.2	0.2	0.0	0.0	0.0	0.0	0.0
0.5	1.4	0.3	0.1	0.0	0.0	0.0
1	5.8	1.2	0.6	0.1	0.1	0.0
2	23.0	4.6	2.3	0.5	0.2	0.0
5	144	28.8	14.4	2.9	1.4	0.3
10	576	115	57.6	11.5	5.8	1.2
20	2304	461	230	46.1	23.0	4.6
50	14,400	2880	1440	288	144	28.8
Plant-fragments						
	1	5	10	50	100	500
0.1	−0.1	0.0	0.0	0.0	0.0	0.0
0.2	−0.6	−0.1	−0.1	0.0	0.0	0.0
0.5	−3.6	−0.7	−0.4	−0.1	0.0	0.0
1	−14.4	−2.9	−1.4	−0.3	−0.1	0.0
2	−57.6	−11.5	−5.8	−1.2	−0.6	−0.1
5	−360	−72	−36.0	−7.2	−3.6	−0.7
10	−1440	−288	−144	−28.8	−14.4	−2.9
20	−5760	−1152	−576	−115	−57.6	−11.5
50	−36,000	−7200	−3600	−720	−360	−72.0

substances at thickening solutions depends on their molecular characteristics, such as their molecular weight and conformation. Typically, the higher the molecular weight and the more extended the conformation, the greater their thickening power, which means that a lower concentration is required to reach the same viscosity (Chap. 4). Aqueous solutions containing thickening agents typically exhibit shear thinning behavior due to alignment and disentanglement of the polymer molecules as the shear rate increases, which may be an important textural attribute of some products. When using thickening agents it is important to be aware that they can promote aggregation of certain kinds of colloidal particles through a bridging or depletion mechanism (McClements, 2015). It is also important to select a thickening agent that provides textural and mouthfeel properties that are acceptable in the end product.

- *Reduce density contrast:* The rate of gravitational separation is directly proportional to the density contrast ($\Delta\rho = \rho_2 - \rho_1$) between the colloidal particles and the aqueous phase (McClements, 2015, 2020). Consequently, it is possible to slow down creaming or sedimentation by reducing this density contrast. Practically, this is challenging since plant-based oils have densities that fall within a fairly narrow range, *e.g.,* 910 to 930 kg m^{-3} for fully liquid oils at room temperature. It is possible to increase the density of the oil phase by using a fat that is partially crystalline (such as coconut oil, cocoa butter, or palm oil) at the required temperature. However, there is a possibility that the fat droplets will then undergo partial coalescence (Fredrick et al., 2010). Other techniques to decrease the density contrast are discussed in Chapter 4.

Particle Aggregation A common cause of instability in milk analogs is the tendency for the colloidal particles, such as fat droplets, oil bodies, or cell wall fragments, to aggregate during food processing, storage, or preparation (McClements, 2015, 2020). Particle aggregation may negatively influence the quality attributes of milk analogs in various ways: (a) aggregates large enough to be discerned by the human eye (> 100 μm) may lead to an undesirable appearance; (b) aggregates large enough to be discerned as discrete entities by the tongue (> 100 μm) may lead to an undesirable mouthfeel (*e.g.,* grittiness); (c) aggregates large enough to undergo gravitational separation (> 500 nm) may lead to an unsightly cream layer at the top of a product and/or a sediment layer at the bottom; and (d) extensive aggregation may lead to thickening or gelling of the product that is undesirable.

The propensity of colloidal particles to aggregate depends on the balance of attractive and repulsive interactions them between (McClements, 2015, 2020). The most important attractive interactions operating in milk analogs are van der Waals, hydrophobic, depletion, and bridging interactions, whereas the most important repulsive interactions are steric and electrostatic interactions (Fig. 8.9). The particles tend to aggregate with each other when the attractive forces dominate but remain as individual entities when the repulsive forces dominate (McClements, 2015). In this section, different strategies that can be used to improve the stability of milk analogs by controlling the colloidal interactions are highlighted:

- *Increase steric repulsion:* Steric repulsive forces are one of the most effective methods of retarding particle aggregation. This is usually achieved by ensuring the colloidal particles are coated by layers of hydrophilic polymers, such as polysaccharides and/or proteins. When two colloidal particles approach close enough for these polymeric layers to overlap, an extremely strong but short range repulsion is generated. The origin of this effect is mainly due to the decrease in the configurational entropy of the polymer chains in the interfacial layers when they overlap with each other. The range of the steric repulsion increases as the thickness of the polymer coating increases, which usually leads to more effective steric stabilization. Colloidal particles stabilized by steric repulsion are typically resistant to changes in solution and environmental conditions that a milk analog may experience, such as alterations in its pH, salt composition, or temperature.
- *Increase electrostatic repulsion*: Colloidal particles may also be prevented from aggregating with each other by generating a sufficiently strong electrostatic repulsion between them. This electrostatic repulsion arises from the presence of charged groups on the surfaces of the colloidal particles, such as carboxyl (–CO_2^-), amino (–NH_3^+), or phosphate (–PO_4^-) groups. Colloidal particles with plant phospholipids, proteins or polysaccharides at their surfaces typically have

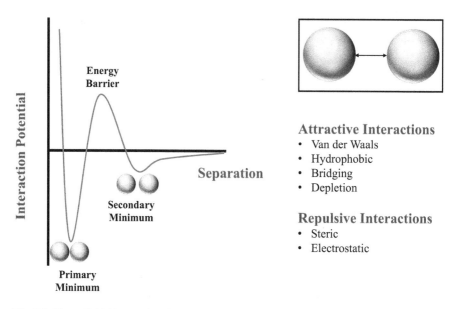

Fig. 8.9 The colloidal interactions between two particles can be modeled by calculating the interaction potential versus particle separation. The overall interaction depends on the combination of attractive and repulsive interactions, which is system specific. For electrostatically stabilized systems, the profile often has a primary minimum, secondary minimum, and an energy barrier

a net charge whose magnitude depends on the pH of the surrounding aqueous phase (Fig. 8.4). For instance, phospholipid- or saponin-coated particles have a high negative charge under neutral conditions, which decreases in magnitude once the solution is reduced below about pH 5. As a result, they tend to aggregate at highly acidic pH values where they lose their negative charge, due to weakening of the electrostatic repulsion. Plant protein-coated particles are strongly negatively charged under neutral conditions but become strongly positively charged under acidic conditions because the pH is reduced below their isoelectric point. As a result, they tend to aggregate at intermediate pH values near the protein's isoelectric point, which is again due to a weakening of the electrostatic repulsion. In contrast, polysaccharide-coated particles (like gum arabic) tend to have a moderate negative charge at neutral pH, which decreases in magnitude as the pH is lowered. However, because they are mainly stabilized by steric repulsion their aggregation stability does not strongly depend on pH. The ionic composition of the aqueous phase surrounding the colloidal particles in milk analogs also influences the magnitude of the electrostatic repulsive forces. As the mineral ion concentration increases, the magnitude and range of the electrostatic repulsive forces decrease as a result of electrostatic screening effects, *i.e.,* the accumulation of counter-ions around oppositely charged groups on the particle surfaces. This effect is particularly important for multivalent counter-ions, such as cationic calcium ions in the presence of anionic colloidal particles like those found in milk analogs around neutral pH.

8.4 Physicochemical Attributes

- *Reduce hydrophobic attraction*: The colloidal particles in milk analogs may have non-polar patches on their surfaces that are exposed to the surrounding aqueous phase. For example, many globular proteins (like soy or pea protein) unfold when they are heated above their thermal denaturation temperature, which exposes non-polar amino acid side groups at their surfaces. The exposure of these non-polar groups generates a strong long-range hydrophobic attraction that favors particle aggregation by reducing the unfavorable contact area between non-polar groups and water. The hydrophobic attraction between colloidal particles in milk analogs may be reduced by ensuring that any proteins are not heated above their thermal denaturation temperatures, or by adding amphiphilic ingredients that adsorb to the exposed non-polar regions and cover them.
- *Avoid depletion flocculation:* Milk analogs often contain appreciable levels of non-adsorbed polymers in the aqueous phase that surrounds the colloidal particles, such as polysaccharides arising from the plant cell walls (like pectin) or that are added as thickening agents (like guar or locust bean gum). These polymers can generate an attraction between the colloidal particles due to an osmotic pressure effect. The polymers cannot get closer to the surfaces of the colloidal particles than their radius of hydration. Consequently, there is a narrow layer of aqueous phase around each colloidal particle where the polymer concentration is effectively zero. As a result, there is a difference in polymer concentration between the bulk of the aqueous phase and this exclusion zone, which generates an osmotic pressure. This osmotic pressure favors particle aggregation, since this would reduce the total volume of the exclusion zone in the system. The strength of this depletion attraction is influenced by the molar mass and radius of hydration of the polymer molecules. It also tends to become stronger as the polymer concentration in the aqueous phase increases. Once the polymer concentration exceeds a critical level, the attractive forces may outweigh the repulsive forces, leading to particle aggregation. Consequently, it is possible to prevent this kind of aggregation by designing the system in a way that the concentration of non-adsorbed polymers in the aqueous phase is not too high.
- *Avoid bridging flocculation*: Milk analogs may also contain other kinds of polymers that can become attached to the surfaces of the colloidal particles through attractive interactions. For instance, they may contain anionic polysaccharides (like pectin, alginate, or carrageenan) that can bind to cationic regions on the surfaces of protein-coated particles. In this case, an individual polymer molecule can become attached to the surfaces of two or more colloidal particles, thereby linking them together and promoting aggregation through a bridging flocculation mechanism. The most common forms of attractive interaction between polymers and particle surfaces are usually either electrostatic or hydrophobic in origin. Aggregation due to bridging flocculation can be avoided by controlling the composition of the system so that there are no polymers that can become attached to the surfaces of the colloidal particles.

In general, the most important parameters that can be adjusted to control the colloidal interactions in milk analogs are: (a) the homogenizer type and operating conditions, as this influences the mean particle size and polydispersity; (b) the type of any emulsifiers used to coat the colloidal particles, as this influences their interfacial thickness, charge, and hydrophobicity; (c) the pH and mineral composition of the aqueous phase as this influences the magnitude and range of the electrostatic interactions; and (d) polymer type and concentration because this influences depletion and bridging flocculation.

Aggregation Type The colloidal particles in milk analogs may aggregate with each other through a number of mechanisms, including flocculation, coalescence, and partial coalescence (McClements, 2020). Flocculation occurs when a number of particles come together and form a clump, but each individual particle retains its original size. The size and shape of the clumps formed, as well as their resistance to disruption when stirred, depends on the strength of the attractive forces holding them together. Flocculation often leads to rapid gravitational separation, as well as to an increase in the viscosity of a product. Coalescence occurs when a number of particles come together and merge into a single larger particle. It tends to occur when the attractive interactions between the particles are relatively strong and the interfacial coatings surrounding the particles are not sufficiently robust to prevent them from being disrupted. Coalescence also leads to an increase in gravitational separation and sometimes to oiling off (the formation of an oil layer on the top of the product), but it does not typically lead to an increase in viscosity. Partial coalescence can occur when a number of partially crystalline fat droplets approach each other and partly merge together to form an irregular clump. These clumps are held together by crystals that range from one droplet penetrating through the interfacial layer of another droplet and into a liquid oil region. The formation of these clumps usually leads to rapid creaming and to an increase in the viscosity of the product. If the sample is sheared at a high rate, the clumping may become extensive and eventually lead to phase inversion, *i.e.,* a change from an O/W to W/O system. Partial coalescence plays an important role in the formation of dairy products like butter, whipped cream, and ice cream. Consequently, it may be important to mimic this phenomenon when designing dairy analogs.

8.4.3.2 Quantification of Milk Stability

The deterioration in the quality of milk analogs during storage or after subjection to specific conditions can be assessed using a number of analytical methods (McClements, 2015). Typically, a combination of methods is required to obtain a full understanding of the origin and nature of the instability mechanisms involved. For instance, it may be important to measure changes in the overall appearance, shear viscosity, microstructure, particle size, and particle charge to obtain insights into the origin of the failure of a product during processing or storage. A detailed discussion of the different analytical instruments and testing protocols for measuring the stability of colloidal dispersions has been given elsewhere (McClements, 2015). Most of these methods are appliable for measuring the properties of milk analogs. For this reason, we only give a brief overview of the most important methods that are suitable for characterizing the stability of milk analogs in this section.

Gravitational Separation The tendency for milk analogs to separate due to sedimentation or creaming is typically monitored by recording changes in the appearance of test samples kept in clear containers under controlled conditions, which can simply be carried out using a digital camera. The rate and amount of separation that has occurred in a sample by a particular time can be quantified by measuring the heights of any visible layers of cream, serum, or sediment. However, it is often difficult to precisely establish the position of the different layers in optically opaque samples. Consequently, alterations in the vertical concentration profile of the particles in the sample are often determined using more sensitive instrumental methods specially designed for the purpose. For instance, creaming and/or sedimentation can be monitored using analytical instruments that direct a beam of laser light at the sample that can be moved upward and downwards in a vertical direction. The instrument can measure reflectance-height and transmittance-height profiles of the test sample over time (Fig. 8.10). As the particle concentration increases at a particular height, the reflectance increases (the fraction of light reflected from the surface of the sample) while the transmittance decreases (the fraction of light transmitted through the sample). Detailed information about the kinetics of gravitational separation can be obtained by measuring the reflectance and transmittance profiles over time. A limitation of this type of instrument is that one often has to wait for an extended period to obtain information about the separation of the system, *e.g.,* days, weeks or months. Another instrument, that is based on a fairly similar principle, accelerates the separation process by centrifuging the sample. The reflectance-height and transmittance-height profiles are measured during this process to monitor the kinetics of the separation process. Although this instrument speeds up the process, it should be noted that it may not capture events that occur over extended storage periods that can lead to separation, such as flocculation or coalescence.

Particle Size It is important to have information about the size of the colloidal particles in milk analogs because this influences their susceptibility to aggregation and gravitational separation, as well as various other important physicochemical attributes, such as the appearance and mouthfeel of the end product. The most commonly used analytical instruments for measuring the size of colloidal particles are static light scattering (SLS) and dynamic light scattering (DLS). Instruments based on SLS direct a beam of laser light through a diluted colloidal dispersion and then have a series of detectors that measures the light scattering pattern, *i.e.,* the change in the light intensity with scattering angle. An appropriate mathematical model, such as the Mie theory, is then utilized by the instrument software to convert the measured scattering pattern into a particle size distribution (PSD). The user of the instrument has to input the refractive index values of the dispersed phase (particles) and continuous phase (water) since the mathematical model requires these parameters. Instruments based on DLS typically direct a beam of laser light at the sample and then measure changes in the intensity of the reflected light over time. The rate

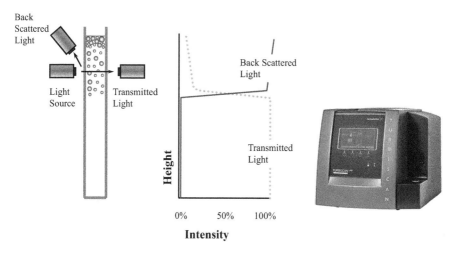

Fig. 8.10 The stability of milk analogs to sedimentation or creaming can be conveniently monitored using instruments that measure the change in transmitted and backscattered light with sample height over time. (Image of instrument kindly provided by Formulaction (Toulouse, France) and used with permission)

of the fluctuations in the intensity-time profile provides information about the particle size: the faster the fluctuations, the smaller the size. Again, a suitable mathematical model is required to calculate the particle size distribution from the intensity-time profile. A number of the DLS instruments commonly used to measure particle size can also measure particle charge (ζ-potential). They do this by using a laser beam to measure the velocity and direction of movement of the colloidal particles in a well-defined electrical field. Information about the size and aggregation state of the colloidal particles within milk analogs is also often obtained using different types of microscopes, such as conventional optical microscopy, confocal laser scanning fluorescence microscopy, scanning electron microscopy (SEM), transmission electron microscopy (TEM), and atomic force microscopy (AFM). Information about the stability of milk analogs to aggregation can sometimes be obtained by measuring changes in their viscosity using a shear viscometer or rheometer – the viscosity typically increases with increasing aggregation. In addition, aggregation may lead to an increase in the rate of creaming or sedimentation, which can be monitored using the devices described earlier.

8.4.3.3 Comparison to Cow's Milk

Raw (non-homogenized) cow's milk tends to be unstable to creaming because of the relatively large size ($d \approx 3.5$ μm) and high density contrast ($\Delta\rho = -110$ kg m^{-3}) of the milk fat globules, as well as the relatively low viscosity of the aqueous phase ($\approx$ 2 mPa s). The creaming stability of milk is greatly increased after it is homogenized due to the reduction in the size of the milk fat globules ($d \approx 0.5$ μm). Instability due to gravitational separation is also a problem in milk analogs. Fat droplets or oil bodies (which have a lower density than water) have a propensity to cream, whereas cell wall fragments and protein aggregates (which have a higher density than water) have a tendency to sediment. As shown earlier, the creaming or sedimentation rate decreases as the size of the particles decreases (Table 8.6). Consequently, it is usually important to design the processing operations used to fabricate milk analogs in a way that the particle diameter is below some critical level (<0.5 µm) and that the particles do not aggregate with each other during storage.

The colloidal particles in milk analogs may aggregate with each other during processing, storage, transport, or utilization, which is usually undesirable because it leads to an increase in the creaming rate and viscosity. As highlighted earlier, this can be avoided by ensuring that there is a relatively strong repulsive interaction between the particles, such as steric and electrostatic repulsion (Fig. 8.9). Moreover, it may be important to avoid hydrophobic, depletion, or bridging flocculation from occurring. The resistance of milk analogs to aggregation can be controlled by manipulating their composition, microstructure, and processing, particularly by controlling the pH, mineral composition, hydrocolloid content, thermal history, and homogenization conditions used. The susceptibility of fat droplets coated by different kinds of plant-based emulsifiers to pH-induced aggregation is highlighted in Fig. 8.4. As discussed earlier, the protein-coated fat droplets tend to aggregate around the isoelectric point of the adsorbed proteins due to the decrease in droplet charge and electrostatic repulsion over this pH range. The phospholipid- and the saponin-coated fat droplets tend to aggregate at low pH values because of the decrease in their negative charge, which again reduces the electrostatic repulsion. In contrast, the polysaccharide-coated fat droplets are stable to aggregation across the whole pH range because they are mainly stabilized by steric repulsion. It is therefore critical to choose a suitable plant-based emulsifier when formulating milk analogs produced by emulsification.

8.5 Sensory Attributes

Milk analogs must be acceptable and desirable to consumers, otherwise people will not incorporate them into their diets (Jeske et al., 2018; Makinen et al., 2016). Some consumers do not find milk analogs appealing because they do not have sensory attributes that are sufficiently similar to those of cow's milk, such as their look, feel, mouthfeel, and flavor profile. Consequently, food scientists are attempting to understand the major factors impacting the sensory attributes of milk analogs and (in some cases) use this knowledge to better match their sensory profiles to those of cow's milk. Improving the sensory attributes of milk analogs therefore depends on the knowledge of the interaction of plant-based milks with the human senses (especially sight, touch, taste, and smell). In addition, information about the sensory attributes of cow's milk products is useful for designing better milk analogs.

Cow's milk has a creamy appearance, creamy mouthfeel, and subtle flavor profile that are influenced by the concentration of fat globules present (McCarthy et al.,

2017a). Typically, the perceived creaminess increases as the fat content increases. Consumers have individual preferences for particular types of milk, such as skim, low-fat, semi-skimmed or whole milk, which has implications for the development of milk analogs. The unique sensory properties of cow's milk are strongly influenced by the milk fat globules, since they contribute to their creamy appearance (by scattering light), texture (by altering fluid flow), mouthfeel (by lubricating the tongue), and flavor profile (by solubilizing non-polar flavors) (Schiano et al., 2017). In addition, the presence of the casein micelles contributes to their desirable appearance through light scattering effects. The subtle look, feel, and flavor of cow's milk make it difficult to accurately mimic its sensory qualities.

Milk analogs are also creamy colloidal dispersions but their appearance, texture, and flavor profile are often appreciably different from those of cow's milk. As discussed earlier, many milk analogs have a different lightness and color ($L*a*b*$) than cow's milk due to differences in their composition and structure, which leads to differences in their perceived appearance. Moreover, many milk analogs have appreciably higher viscosities than cow's milk because they contain thickening agents, which alters their perceived flow characteristics and mouthfeel. Milk analogs may also contain taste and aroma molecules that come from the original plants or are generated during processing that give them particular flavor profiles, *e.g.,* "almond", "beany", "grassy" "oat", "painty", or "nutty" (Jeske et al., 2018). In some cases, these flavors are desirable but in other cases they are undesirable. Milk analogs may also contain particles, like cell tissue fragments or protein aggregates, which lead to unpleasant chalky or gritty feelings in the mouth. The mouthfeel characteristics of milk analogs can often be improved by homogenizing or enzymatically treating them to reduce the size of the particulate matter. Ideally, the particle dimensions should be reduced below around 50 μm for the product to appear smooth in the mouth. For example, researchers have carried out sensory studies to rank the overall liking of consumers for different kinds of milk analogs and found that liking decreased in the following order oat > rice > almond > soy > lentil > hemp (Jeske et al., 2019). A comparison of the sensory attributes of milk analogs with cow's milk found that consumers liked almond milk as much as cow's milk but liked soy milk much less than cow's milk (Kundu et al., 2018). The poor liking of the soy milk was mainly attributed to its undesirable color, flavor, and mouthfeel rankings. However, as consumers get more and more accustomed to such new food products it is likely that the acceptance will change in the future.

8.6 Nutritional Attributes

Ideally, milk analogs should match or exceed the nutritional profile of cow's milk so that there are no adverse health consequences from switching from cow's to plant-based milks. The nutrient and calorie content of milk analogs depend on the ingredients and methods used to produce them. The nutrition profiles of milk analogs are therefore different from each other, as well as from cow's milk (Table 8.3). Even within the same product category (such as soy milk) there may be appreciable differences in their nutrition profiles depending on the amount of sugars and other additives used. Of the commercial products shown in Table 8.3, the coconut and rice milks have greater calorie levels than cow's milk, while the almond and soy milk have a lower calorie level. In terms of protein content, the soy milk has a similar level to cow's milk, but the almond, coconut and rice milks have appreciably lower levels. In terms of the sugar content, the coconut and rice milks have much higher levels than the cow's milk, which may be undesirable from a nutritional viewpoint. Interestingly, a study by scientists in the United Kingdom concluded that there may be adverse effects on dental health associated with increasing consumption of certain milk analogs due to their relatively high sugar contents (Sumner & Burbridge, 2021).

The quality of the proteins in cow's milk and milk analogs also differs, such as the amino acid scores (Table 8.7) and amino acid compositions (Table 8.8). Some of the proteins used to formulate milk analogs are lacking essential amino acids, which could have nutritional consequences if they are a major source of proteins in a consumer's diet. Moreover, cow's milk and milk analogs differ in the amounts of vitamins and minerals they contain (Table 8.3), which could also have a health impact if these micronutrients are limiting in a person's diet. There may therefore be a need to supplement milk analogs with micronutrients so as to improve their nutritional profiles, which is discussed in a following section. Indeed, one recent study in the United Kingdom reported that consumers who only drink plant-based milk analogs have a risk of suffering from iodine deficiency if they are not fortified, which was attributed to the fact that cow's milk was the main source of iodine in the diet in the UK, whereas milk analogs typically have low iodine levels (Dineva et al., 2021).

There have been few systematic studies examining the long-term impact of substituting milk analogs for cow's milk on human nutrition and health (Vanga & Raghavan, 2018). As discussed in Chap. 5, it is possible that individuals may lack some essential vitamins and minerals in their diet if they adopt a fully plant-based diet and do not take supplements, *e.g.,* vitamin B_{12}, vitamin D, calcium, and iodine. On the other hand, individuals on a healthy plant-based diet tend to be healthier than those on a conventional Western diet. However, the calorie content and nutrient profile of plant-based milks vary widely within and between types (Table 8.3). Consequently, the type of milk analog consumed will have an important impact on their potential nutritional effects. Consuming high-calorie sugar-sweetened milk analogs in place of cow's milk is likely to have negative nutritional consequences, whereas consuming low-calorie unsweetened ones is likely to have positive or neutral effects.

Knowledge of the behavior of foods within the gastrointestinal tract (GIT) is useful for controlling nutrient bioavailability, pharmacokinetics, and physiological responses, which can be used to design healthier products (Dupont et al., 2018). For this reason, many researchers are studying the digestion, solubilization, metabolism, and absorption of the various nutrients in foods as they pass through the GIT. Recently, there has been an interest in understanding the behavior of milk analogs in the human gut, as this may impact their nutritional and health effects (Zheng et al., 2019a, 2021). Again, it is important to understand how the behavior of milk analogs differs from that of cow's milk so as to better understand the potential impacts of adopting a more plant-based diet on human health.

Table 8.7 Digestible indispensable amino acid scores (DIAAS) for milk and selected plant proteins used to formulate plant-based milks

Protein source	Histidine	Isoleucine	Leucine	Lysine	Met + Cys	Phe + Tyr	Threonine	Tryptophan	Valine	DIAAS	Limiting AA
Corn	137 ± 37	95 ± 16	176 ± 63	43 ± 18	148 ± 26	178 ± 54	106 ± 13	66 ± 46	96 ± 16	43	Lys
Rice	116 ± 9	95 ± 19	87 ± 13	56 ± 3	122 ± 13	151 ± 38	93 ± 5	146 ± 37	102 ± 19	56	Lys
Wheat	148 ± 28	98 ± 11	95 ± 12	56 ± 14	150 ± 22	138 ± 26	97 ± 11	162 ± 25	98 ± 11	56	Lys
Hemp	116 ± 9	95 ± 18	87 ± 13	56 ± 23	122 ± 13	151 ± 38	93 ± 5	146 ± 37	102 ± 19	56	Lys
Fava bean	135 ± 5	114 ± 2	103 ± 6	113 ± 5	64 ± 6	150 ± 4	113 ± 8	87 ± 10	90 ± 2.4	64	M + C
Oat	114 ± 14	106 ± 4	102 ± 5	68 ± 7	177 ± 62	171 ± 12	106 ± 7	142 ± 22	110 ± 4	68	Lys
Rapeseed	133 ± 7	95 ± 5	85 ± 5	79 ± 11	145 ± 16	116 ± 14	119 ± 7	136 ± 10	98 ± 5	79	Lys
Lupin	153 ± 20	124 ± 11	104 ± 10	92 ± 10	83 ± 10	170 ± 13	129 ± 17	105 ± 5	88 ± 10	83	M + C
Pea	124 ± 12	108 ± 14	94 ± 12	130 ± 13	83 ± 14	148 ± 21	117 ± 10	100 ± 9	89 ± 11	83	M + C
Canola	131 ± 9	99 ± 11	85.4 ± 9	85.1 ± 11	143 ± 12	123 ± 8	119 ± 12	144 ± 25	93 ± 11	85	Lys
Soy	149 ± 12	133 ± 9	110 ± 7	114 ± 11	106 ± 14	186 ± 17	130 ± 8	170 ± 27	103 ± 8	103	NA
Potato	125 ± 9	167 ± 10	155 ± 12	145 ± 6	135 ± 7	266 ± 23	204 ± 15	165 ± 18	149 ± 6	125	NA
Whey	106 ± 14	173 ± 24	145 ± 25	151 ± 29	152 ± 26	124 ± 20	213 ± 29	220 ± 58	122 ± 15	106	NA
Casein	183 ± 12	163 ± 5	152 ± 7	160 ± 5	137 ± 6	255 ± 10	161 ± 5	205 ± 17	159 ± 3	137	NA

The data is applicable to individuals of 3 years and older (Herreman et al., 2020). More information about protein quality can be found in Chapter 5.

8.6 Nutritional Attributes

Table 8.8 Amino acid compositions of selected milk and plant proteins used to formulate milk-like products

Protein source	Oat	Lupin	Hemp	Soy	Brown rice	Pea	Whey	Milk	Caseinate	Casein
Essential amino acid										
Threonine	1.5	1.6	1.3	2.3	2.3	2.5	5.4	3.5	3.5	2.6
Methionine	0.1	0.2	1	0.3	2	0.3	1.8	2.1	2.2	1.6
Phenylalanine	2.7	1.8	1.8	3.2	3.7	3.7	2.5	3.5	4.2	3.1
Histidine	0.9	1.2	1.1	1.5	1.5	1.6	1.4	1.9	2.2	1.7
Lysine	1.3	2.1	1.4	3.4	1.9	4.7	7.1	5.9	5.9	4.6
Valine	2	1.4	1.3	2.2	2.8	2.7	3.5	3.6	3.8	3
Isoleucine	1.3	1.5	1	1.9	2	2.3	3.8	2.9	3	2.3
Leucine	3.8	3.2	2.6	5	5.8	5.7	8.6	7	7.8	5.8
Σ essential	13.7	13.1	11.6	19.9	22.1	23.6	34.1	30.3	32.8	24.8
Non-essential amino acid										
Serine	2.2	2.5	2.3	3.4	3.4	3.6	4	4	4.2	3.4
Glycine	1.7	2.1	2.1	2.7	3.4	2.8	1.5	1.5	1.5	1.2
Glutamic acid	11	12.4	7.4	12.4	12.7	12.9	15.5	16.7	16	13.9
Proline	2.5	2	1.8	3.3	3.4	3.1	4.8	7.3	8.7	6.5
Cysteine	0.4	0.2	0.2	0.2	0.6	0.2	0.8	0.2	0.1	0.1
Alanine	2.2	1.7	1.9	2.8	4.3	3.2	4.2	2.6	2.6	2
Tyrosine	1.5	1.9	1.3	2.2	3.5	2.6	2.4	3.8	4.4	3.4
Arginine	3.1	5.5	5.3	4.8	5.4	5.9	1.7	2.6	2.9	2.1
Σ non-essential	24.7	28.2	22.4	31.9	36.8	34.4	34.9	38.6	40.4	32.5

Data derived from previous studies (Gorissen et al., 2018). The aspartic acid, asparagine, glutamine and tryptophan concentrations were not measured. The data are given as grams per 100 grams of raw material.

Both *in vitro* and *in vivo* studies have been performed to better understand the behavior of cow's milk in the human gut (Egger et al., 2017, 2018, 2019;). During its passage through the upper intestine, the lipids and proteins in cow's milk are hydrolyzed by lipases and proteases, releasing fatty acids, monoglycerides, peptides, and amino acids. In addition, various kinds of small molecules from the milk are dissolved in the gastrointestinal fluids, such as lactose, oligosaccharides, and minerals. Small hydrophilic molecules diffuse through the GIT fluids to the surfaces of the epithelium cell walls where they are absorbed by passive or active transport mechanisms. Hydrophobic molecules resulting from lipid digestion, like fatty acids and monoglycerides, combine with bile salts and phospholipids secreted by the human gut to form mixed micelles. These mixed micelles may also solubilize other hydrophobic substances originally located in the milk fat globules, such as carotenoids, vitamin A, and vitamin D. The mixed micelles then travel through the GIT fluids to the surfaces of the epithelium cells. The constituents contained within them can then be absorbed by the epithelium cells. After absorption, hydrophilic substances tend to move into the portal vein where they are transported to the liver (where they may be metabolized) prior to entering the bloodstream. In contrast, hydrophobic substances are packaged into chylomicrons, move through the lymphatic system, and directly enter the bloodstream. The behavior of cow's milk within the human gut has been studied using *in vitro* digestion models (Van Hekken et al., 2017; Ye et al., 2019). The casein micelles and protein-coated fat globules in the milk underwent appreciable aggregation in the simulated stomach environment, which was attributed to the change in pH and ionic strength, as well as to bridging flocculation of the cationic proteins by anionic mucin molecules. These aggregates were largely broken down after entering the simulated small intestine, which was mainly attributed to the change to a neutral pH (where the proteins and mucin are both negative), the presence of amphiphilic bile acids, and the hydrolysis of the lipids and proteins by digestive enzymes. The rate of lipid digestion in the small intestine increased after homogenization of the cow's milk as a result of the increase in the surface area of lipids exposed to the lipases (Van Hekken et al., 2017). The rate of lipid and protein digestion increased after ultra high temperature processing, which was ascribed to changes in the structure of the aggregates formed in the gastric fluids (Ye et al., 2019). Overall, these results show that cow's milks are effectively digested under GIT conditions, thereby releasing their nutrients, but the rate of this process depends on their processing.

Cow's milk has been designed through evolution to provide the growing calf with all the nutrients it requires in a bioaccessible form (Bourlieu et al., 2018; Le Huerou-Luron et al., 2018; Lee et al., 2018). Cow's milk has some compositional and structural similarities to human milk, which means that it can also serve as a valuable source of nutrients to growing infants. As well as containing nutrients, cow's milk also contains other substances that can promote the formation of a healthy gut microbiome and strengthen the infant's immune system (Bourlieu et al., 2017). In particular, cow's milk contains various kinds of oligosaccharides that act as prebiotics that stimulate the formation of a healthy gut microbiome (Oliveira et al., 2015; Robinson, 2019). The digestion of the proteins in cow's milk within the human gut also leads to the generation of a specific profile of bioactive peptides that can have health benefits (Park & Nam, 2015; Sah et al., 2015). For instance, these milk peptides may exhibit antimicrobial, blood pressure lowering, and prebiotic activities. Cow's milk is a good source of bioavailable calcium in the human diet, which has been attributed to the fact that it is contained within casein micelles that are rapidly digested in the human gut and release the calcium in a form that can easily be absorbed (Gueguen & Pointillart, 2000). When designing milk analogs, it may therefore be useful to simulate the gastrointestinal fate of cow's milk products so as to obtain similar nutritional profiles and physiological effects.

A number of researchers have also examined the gastrointestinal fate of milk analogs using simulated GIT models (Capuano & Pellegrini, 2019). The results of these studies suggest that milk analogs may behave differently than cow's milk, which can be attributed to differences in the compositions and structures of these two types of colloidal dispersions (Do et al., 2018). Studies using simulated GIT models have shown that oil bodies from different plants are digested differently. As an example, oil bodies extracted from oats are fairly resistant to hydrolysis for a number of reasons: (a) they tend to aggregate when exposed to the stomach and small intestine; (b) they are coated by interfacial layers that are resistant to digestion; and (c) they contain minor constituents that can inhibit digestion, like dietary fibers and phytochemicals (Wilde et al., 2019). In contrast, oil bodies extracted from almonds are digested more rapidly because they are less susceptible to aggregation in the small intestine and their interfacial layers are more prone to digestion (Gallier & Singh, 2012). *In vitro* digestion models have also been used to study the digestibility of oil bodies from other plant materials, such as from soybeans (Ding et al., 2019), hazelnuts (Capuano et al., 2018), and sunflower seeds (Makkhun et al., 2015). These models have also been used to examine the digestibility of plant protein-coated fat droplets in milk analogs created by emulsification, which showed that these kinds of milk analogs tend to be rapidly and fully digested (Gumus et al., 2017b). These *in vitro* studies indicate that the digestibility of milk analogs depends on the characteristics of the oil bodies or fat droplets they contain (particularly their size, composition, and structure), as well as the properties of the surrounding food matrix, and the nature of any processing operations they are exposed to. These differences in digestibility may lead to alterations in the bioavailability of nutrients, the change in blood nutrient levels over time, and hormonal responses (such as hunger, satiety, and satiation) that could influence their nutritional consequences and health effects.

The remnants of milk analogs that reach the colon after passing through the upper gastrointestinal tract will influence the composition and function of the gut microbiome (Do et al., 2018). For instance, certain kinds of dietary fibers or phytochemicals in milk analogs may act as prebiotics that can stimulate the growth of beneficial microbes in the colon. However, long-term randomized controlled trials are required to compare the impact of milk analogs and cow's milk on human nutrition and health.

8.7 Nutritional Fortification

Some scientists have expressed concerns about moving from an omnivore to a plant-based diet (Vanga & Raghavan, 2018). Diets that do not include any animal products may lack some nutrients that are essential for human health or improve wellbeing, such as essential amino acids, omega-3 fatty acids, vitamin B_{12}, vitamin D, calcium, iron, and iodine (Obeid et al., 2019; Sebastiani et al., 2019). Long-term deficiencies in essential nutrients may cause adverse health effects, especially in infants and the elderly (Hunt, 2019; Sebastiani et al., 2019). For this reason, there is interest in fortifying plant-based foods with nutrients that may be lacking from a plant-based diet. In addition, there is also interest in fortifying them with nutraceuticals that could further enhance their nutritional profile, such as β-carotene, lutein, zeaxanthin, lycopene, curcumin, resveratrol, quercetin, and various other health-promoting phytochemicals (Abuajah et al., 2015; Assadpour & Jafari, 2019; McClements, 2020). Milk analogs are particularly suitable for fortification with these kinds of bioactive components. First, they can form a regular part of a person's daily diet because they can be drunk as a beverage, used as a creamer in coffee or tea, or poured over

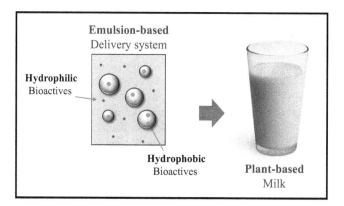

Fig. 8.11 Plant-based milks can be fortified with hydrophobic or hydrophilic bioactive components because they contain water, oil, and oil-water regions

breakfast cereals. Second, milk analogs have both non-polar and polar domains within them so they can solubilize both hydrophilic and hydrophobic bioactive substances (Fig. 8.11). Third, structural design principles can be utilized to enhance the bioavailability and bioactivity of the bioactive substances, for instance the droplet size, composition, or interfacial properties can be modified. Nonetheless, when designing these fortified milk analogs it is important to make sure that the introduction of the bioactive substances does not adversely affect product attributes, like the appearance, mouthfeel, or taste. Moreover, the milk analogs should be designed to protect the bioactive substances from degradation during storage, as well as to ensure they are in a bioavailable form after they have been consumed.

Hydrophilic bioactives (like vitamin B_{12}) can often be simply dissolved within the aqueous phase of a milk analog. Insoluble mineral salts, like calcium carbonate, may be dispersed in the aqueous phase in the form of colloidal particles. Hydrophobic bioactives usually have to be incorporated into the oil phase. This can be achieved in a number of ways depending on the type of plant-based milk. First, for milk analogs produced by emulsification, the hydrophobic bioactives can simply be mixed with the oil phase prior to homogenization (Fig. 8.11). Second, an emulsion-based delivery system containing bioactive-loaded fat droplets can mixed with the milk analog (Fig. 8.11). Third, a milk analog may be homogenized with an oil phase containing the dissolved bioactive substances, which leads to the creation of bioactive-loaded fat droplets provided there are sufficient natural emulsifiers in the original milk analog, *e.g.,* phospholipids or proteins. Fourth, certain kinds of hydrophobic bioactives (usually phenolic ones like curcumin, resveratrol, or quercetin) can be introduced into the fat droplets or oil bodies after the formation of the milk analog, *e.g.,* by using pH shift methods (Zhang et al., 2020). In this latter case, the bioactives are first dissolved in a concentrated alkaline solution, which is then mixed with the near neutral milk analog, which causes the bioactives to move into the hydrophobic core of the fat droplets or oil bodies (Zheng et al., 2019a, b).

In the remainder of this section, we focus on the encapsulation of oil-soluble bioactive substances since they are usually the most difficult to incorporate. Vitamin

D has been incorporated into a number of different model milk analogs, including a dispersion of pea protein-coated fat droplets in water (Walia & Chen, 2020). The uptake of this vitamin by model epithelium cells (Caco-2 cells) was enhanced almost 2.5-times when the average droplet diameter was reduced from 350 to 233 nm, which suggests that smaller fat droplets may be more efficacious for delivering this oil-soluble vitamin. The type of oil used to form pea-protein coated fat droplets has also been found to influence the bioaccessibility of vitamin D (Schoener et al., 2019). In this study, the vitamin bioaccessibility was higher when it was encapsulated in corn oil droplets (rich in monounsaturated fatty acids) than in flaxseed oil droplets (rich in polyunsaturated fatty acids), which was attributed to differences in the solubilization capacity of the mixed micelles formed after digestion. Other studies indicated that oil-soluble vitamins can be encapsulated into model milk analogs prepared using various other types of plant-based emulsifiers, including soy protein (Zhang et al., 2020), quillaja saponin (Lv et al., 2019; Ozturk et al., 2015b; Tan et al., 2021), gum arabic (Lv et al., 2019) and soy lecithin (Mehmood et al., 2019). A recent study showed that incorporating calcium into a model milk analog reduced the bioaccessibility of vitamin D encapsulated within the fat droplets (Zhou et al., 2021), which may have been due to the ability of the calcium ions to precipitate the vitamin-loaded mixed micelles. This study highlights the importance of carefully designing milk analogs containing multiple bioactive substances to ensure they are all bioavailable.

The composition and microstructure of any delivery systems used to introduce hydrophobic bioactive substances into milk analogs must be optimized to ensure that they have a high bioavailability (McClements, 2018). The size, composition, and coating of the oil droplets used influence the bioaccessibility, stability and absorption of hydrophobic bioactives inside the GIT, which determines their overall bioavailability (McClements, 2018; McClements et al., 2015). These delivery systems should therefore be designed to create fortified milk analogs that remain stable throughout storage but efficiently release the nutrients or nutraceuticals in a bioavailable form in the GIT. The importance of the various factors that impact the overall bioavailability of hydrophobic bioactives in colloidal delivery systems is discussed in detail elsewhere (McClements, 2018).

8.8 Environmental Impact: Life Cycle Analysis

Many consumers are adopting milk analogs as part of a plant-based diet because this type of eating pattern has been shown to have a lower environmental impact than one containing appreciable quantities of animal-derived foods. Life cycle analysis (LCA) is a systematic approach that has been utilized to compare the environmental impacts of milk analogs with cow's milk. LCA quantifies the environmental impacts associated with producing, transporting, storing, and selling a product, such as greenhouse gas emissions, pollution, water use, land use, fossil fuel use, and biodiversity loss. LCA has been carried out to compare the environmental impacts of producing cow's milk to those of producing almond and soy milks (Grant & Hicks, 2018). It was reported that the milk analogs had fewer negative effects on the environment than cow's milk, which was primarily attributed to the relatively high energy use, fossil fuel use, and pollution associated with rearing dairy cows and transporting their milk. Even so, almond milk did score relatively poorly on the amount of water required to produce it, which needs to be considered in arid areas. Another LCA study also found that the overall environmental impact of producing milk analogs (almond, oat, rice, or soy) was much lower than that of producing cow's milk (Wenzel & Jungbluth, 2017). A comprehensive LCA study of many different kinds of foods also reported that milk analogs had a much lower environmental impact than cow's milk (Table 8.9) (Poore & Nemecek, 2018). Production of the milk analogs led to fewer greenhouse gas emissions, caused less pollution, and required less land and water use than the production of cow's milk (Poore & Nemecek, 2018). Notably, the environmental impacts of the milk analogs depended on the plant materials used to produce them. For instance, the amount of water required to produce almond and rice milks was much higher than that required to produce soy and oat milks. In contrast, the opposite was true for the land use. Moreover, the production of rice milk caused more pollution (eutrophication) than the production of other kinds of milk analogs. Another detailed LCA study comparing the environmental impacts of cow's milk and milk analogs also found that the latter have far fewer negative impacts on the environment (Detzel et al., 2021). All of the LCA studies discussed so far have been for plant-based milks produced using the plant-tissue disruption approach. A LCA conducted by Ripple Foods on plant-based milks that are produced using the emulsification approach showed that milk analogs consisting of pea protein stabilized sunflower oil-in-water emulsions had much lower greenhouse gas emissions and water usage than both cow's milk and almond milk (www.ripplefoods.com). Indeed, this analysis showed that there was 100-fold and 25-fold more water required to produce almond milk and cow's milk than the emulsified sunflower product (per unit amount of protein).

Overall, these LCA studies indicate that there are substantial benefits to the environment when replacing cow's milk with milk analogs, but the nature and degree of these benefits depends on the type of milk analog.

It should be noted that the findings of LCA studies are influenced by the nature and quality of the data used, as well as the assumptions underlying their

Table 8.9 Comparison of environmental impacts of plant-based milk analogs and cow's milk (Poore & Nemecek, 2018).

Milk product	Land use (m^2)	GHG (kg CO_2eq)	Eutrophication (g PO_4^{3-}eq)	Water use (L)
Cow's milk	9.0	3.2	10.7	628
Soy milk	0.7	1.0	1.1	28
Almond milk	0.5	0.7	1.5	371
Oat milk	0.8	0.9	1.6	48
Rice milk	0.3	1.2	4.7	270

Data is given per liter or product. Here GHG is greenhouse gas emissions.

interpretation. As a result, there are often appreciable variations in the calculated impacts of milk analogs and cow's milk depending on factors like the type of raw materials used, the geographical locations of the raw materials, the manufacturing factories, the retail outlets for the end products, the growing conditions, the processing operations, the transportation method, and the storage conditions. Even so, the findings of many different LCA indicate that milk analogs are more environmentally friendly and sustainable than cow's milk.

8.9 Conclusions and Future Work

There has been growing interest amongst many consumers in switching from cow's milk to plant-based milk alternatives, which is mainly driven by the sustainability, health, and ethical concerns associated with animal-derived foods (McClements, 2020; McClements et al., 2019). For this reason, many sections of the modern food industry are creating milk analogs to replace cow's milk, including those derived from soybeans, almonds, oats, coconut, and rice. The sensory attributes, cost, and convenience of these milk analogs will determine their commercial success. In this chapter, we have reviewed different methods of producing milk analogs and have examined the main factors that influence their physicochemical, functional, sensory, and nutritional attributes. In addition, we have highlighted differences and similarities between milk analogs and the cow's milk products they are designed to replace. Many consumers do not currently incorporate milk analogs into their diets because they do not like the taste or they do not perform as expected (*e.g.,* in coffee or tea). In the future, it is likely that more people will adopt milk analogs as their quality attributes and functional performances are improved.

In the future, research should be directed towards creating milk analogs with improved flavor profiles and a greater functional versatility, which will depend on developing a more fundamental understanding of the relationships between their composition, structure, physicochemical, and sensory attributes. In addition, research is needed to create products that have improved nutritional profiles and lower environmental impacts. In particular, it will be important to develop milk analogs that are fortified with bioavailable forms of nutrients that might be deficient in a plant-based diet, such as iodine, calcium, vitamin B_{12}, and vitamin D.

References

Abuajah, C. I., Ogbonna, A. C., & Osuji, C. M. (2015). Functional components and medicinal properties of food: A review. *Journal of Food Science and Technology-Mysore, 52*(5), 2522–2529. https://doi.org/10.1007/s13197-014-1396-5

Agabriel, C., Cornu, A., Journal, C., Sibra, C., Grolier, P., & Martin, B. (2007). Tanker milk variability according to farm feeding practices: Vitamins A and E, carotenoids, color, and terpenoids. *Journal of Dairy Science, 90*(10), 4884–4896. https://doi.org/10.3168/jds.2007-0171

Al Loman, A., Callow, N. V., Islam, S. M. M., & Ju, L. K. (2018). Single-step enzyme processing of soybeans into intact oil bodies, protein bodies and hydrolyzed carbohydrates. *Process Biochemistry, 68*, 153–164. https://doi.org/10.1016/j.procbio.2018.02.015

Assadpour, E., & Jafari, S. M. (2019). A systematic review on nanoencapsulation of food bioactive ingredients and nutraceuticals by various nanocarriers. *Critical Reviews in Food Science and Nutrition, 59*(19), 3129–3151. https://doi.org/10.1080/10408398.2018.1484687

Bourlieu, C., Deglaire, A., de Oliveira, S. C., Menard, O., Le Gouar, Y., Carriere, F., & Dupont, D. (2017). Towards infant formula biomimetic of human milk structure and digestive behaviour. *OCL - Oilseeds and Fats Crops and Lipids, 24*(2). https://doi.org/10.1051/ocl/2017010

Bourlieu, C., Deglaire, A., de Oliveira, S. C., Menard, O., Le Gouar, Y., Carriere, F., & Dupont, D. (2018). Towards infant formula biomimetic of human milk structure and digestive behaviour. *Cahiers De Nutrition Et De Dietetique, 53*(4), 218–231. https://doi.org/10.1016/j.cnd.2017.12.002

Broyard, C., & Gaucheron, F. (2015). Modifications of structures and functions of caseins: A scientific and technological challenge. *Dairy Science & Technology, 95*(6), 831–862. https://doi.org/10.1007/s13594-015-0220-y

Campbell, J. R., & Marshall, R. T. (2016). *Dairy production and processing: The science of milk and milk products*. Waveland Press.

Campbell, K. A., Glatz, C. E., Johnson, L. A., Jung, S., de Moura, J. M. N., Kapchie, V., & Murphy, P. (2011). Advances in aqueous extraction processing of soybeans. *Journal of the American Oil Chemists Society, 88*(4), 449–465. https://doi.org/10.1007/s11746-010-1724-5

Capuano, E., & Pellegrini, N. (2019). An integrated look at the effect of structure on nutrient bioavailability in plant foods. *Journal of the Science of Food and Agriculture, 99*(2), 493–498. https://doi.org/10.1002/jsfa.9298

Capuano, E., Peltegrini, N., Ntone, E., & Nikiforidis, C. V. (2018). In vitro lipid digestion in raw and roasted hazelnut particles and oil bodies. *Food & Function, 9*(4), 2508–2516. https://doi.org/10.1039/c8fo00389k

Chalupa-Krebzdak, S., Long, C. J., & Bohrer, B. M. (2018). Nutrient density and nutritional value of milk and plant-based milk alternatives. *International Dairy Journal, 87*, 84–92. https://doi.org/10.1016/j.idairyj.2018.07.018

Chung, C., Sher, A., Rousset, P., Decker, E. A., & McClements, D. J. (2017). Formulation of food emulsions using natural emulsifiers: Utilization of quillaja saponin and soy lecithin to fabricate liquid coffee whiteners. *Journal of Food Engineering, 209*, 1–11. https://doi.org/10.1016/j.jfoodeng.2017.04.011

Deeth, H. (2017). Optimum thermal processing for extended shelf-life (ESL) milk. *Food, 6*(11). https://doi.org/10.3390/foods6110102

Detzel, A., Kruger, M., Busch, M., Blanco-Gutierrez, I., Varela, C., Manners, R., Bez, J., & Zannini, E. (2021). Life cycle assessment of animal-based foods and plant-based protein-rich alternatives: An environmental perspective. *Journal of the Science of Food and Agriculture*. https://doi.org/10.1002/jsfa.11417

Dickinson, E. (1992). *Introduction to food colloids*. Oxford University Press.

Dineva, M., Rayman, M. P., & Bath, S. C. (2021). Iodine status of consumers of milk-alternative drinks v. cows' milk: Data from the UK National Diet and Nutrition Survey. *British Journal of Nutrition, 126*(1), 28–36. https://doi.org/10.1017/S0007114520003876

Ding, J., Xu, Z., Qi, B., Cui, S., Wang, T., Jiang, L., Zhang, Y., & Sui, X. (2019). Fabrication and characterization of soybean oil bodies encapsulated in maltodextrin and chitosan-EGCG conjugates: An in vitro digestibility study. *Food Hydrocolloids, 94*, 519–527. https://doi.org/10.1016/j.foodhyd.2019.04.001

Do, D. T., Singh, J., Oey, I., & Singh, H. (2018). Biomimetic plant foods: Structural design and functionality. *Trends in Food Science & Technology, 82*, 46–59. https://doi.org/10.1016/j.tifs.2018.09.010

Dupont, D., Le Feunteun, S., Marze, S., & Souchon, I. (2018). Structuring food to control its disintegration in the gastrointestinal tract and optimize nutrient bioavailability. *Innovative Food Science & Emerging Technologies, 46*, 83–90. https://doi.org/10.1016/j.ifset.2017.10.005

References

Durazzo, A., Gabrielli, P., & Manzi, P. (2015). Qualitative study of functional groups and antioxidant properties of soy-based beverages compared to cow milk. *Antioxidants, 4*(3), 523–532. https://doi.org/10.3390/antiox4030523

Egger, L., Menard, O., Baumann, C., Duerr, D., Schlegel, P., Stoll, P., Vergères, G., Dupont, D., & Portmann, R. (2019). Digestion of milk proteins: Comparing static and dynamic in vitro digestion systems with in vivo data. *Food Research International, 118*, 32–39. https://doi.org/10.1016/j.foodres.2017.12.049

Egger, L., Schlegel, P., Baumann, C., Stoffers, H., Guggisberg, D., Brugger, C., Dürr, D., Stoll, P., Vergères, G., & Portmann, R. (2017). Physiological comparability of the harmonized INFOGEST in vitro digestion method to in vivo pig digestion. *Food Research International, 102*, 567–574. https://doi.org/10.1016/j.foodres.2017.09.047

Egger, L., Schlegel, P., Baumann, C., Stoffers, H., Guggisberg, D., Brugger, C., Dürr, D., Stoll, P., Vergères, G., & Portmann, R. (2018). Mass spectrometry data of in vitro and in vivo pig digestion of skim milk powder. *Data in Brief, 21*, 911–917. https://doi.org/10.1016/j.dib.2018.09.089

Flauzino, R. D., Gut, J. A. W., Tadini, C. C., & Telis-Romero, J. (2010). Flow properties and tube friction factor of milk cream: Influence of temperature and fat content. *Journal of Food Process Engineering, 33*(5), 820–836. https://doi.org/10.1111/j.1745-4530.2008.00307.x

Fredrick, E., Walstra, P., & Dewettinck, K. (2010). Factors governing partial coalescence in oil-in-water emulsions. *Advances in Colloid and Interface Science, 153*(1–2), 30–42. https://doi.org/10.1016/j.cis.2009.10.003

Gallier, S., & Singh, H. (2012). Behavior of almond oil bodies during in vitro gastric and intestinal digestion. *Food & Function, 3*(5), 547–555. https://doi.org/10.1039/c2fo10259e

Genovese, D. B., Lozano, J. E., & Rao, M. A. (2007). The rheology of colloidal and noncolloidal food dispersions. *Journal of Food Science, 72*(2), R11–R20. Retrieved from https://doi.org/10.1111/j.1750-3841.2006.00253.x

GFI. (2021). *State of the industry report: Plant-based meat, eggs, and dairy*. Retrieved from Washington, DC. https://gfi.org/resource/plantbased-meat-eggs-and-dairy-state-of-the-industry-report/

Goff, H. D. (2019). *The dairy science and technology eBook*. University of Guelph.

Gorissen, S. H. M., Crombag, J. J. R., Senden, J. M. G., Waterval, W. A. H., Bierau, J., Verdijk, L. B., & van Loon, L. J. C. (2018). Protein content and amino acid composition of commercially available plant-based protein isolates. *Amino Acids, 50*(12), 1685–1695. https://doi.org/10.1007/s00726-018-2640-5

Goyal, A., Sharma, V., Upadhyay, N., Gill, S., & Sihag, M. (2014). Flax and flaxseed oil: An ancient medicine & modern functional food. *Journal of Food Science and Technology-Mysore, 51*(9), 1633–1653. https://doi.org/10.1007/s13197-013-1247-9

Grant, C. A., & Hicks, A. L. (2018). Comparative life cycle assessment of Milk and plant-based alternatives. *Environmental Engineering Science, 35*(11), 1235–1247. https://doi.org/10.1089/ees.2018.0233

Griffin, M. C. A., & Griffin, W. G. (1985). A simple turbidimetric method for the determination of the refractive-index of large colloidal particles applied to casein micelles. *Journal of Colloid and Interface Science, 104*(2), 409–415. Retrieved from https://doi.org/10.1016/0021-9797(85)90049-9

Gu, L. P., Su, Y. J., Zhang, M. Q., Chang, C. H., Li, J. H., McClements, D. J., & Yang, Y. J. (2017). Protection of beta-carotene from chemical degradation in emulsion-based delivery systems using antioxidant interfacial complexes: Catechin-egg white protein conjugates. *Food Research International, 96*, 84–93. https://doi.org/10.1016/j.foodres.2017.03.015

Gueguen, L., & Pointillart, A. (2000). The bioavailability of dietary calcium. *Journal of the American College of Nutrition, 19*(2), 119S–136S. https://doi.org/10.1080/07315724.2000.10718083

Gumus, C. E., Decker, E. A., & McClements, D. J. (2017a). Formation and stability of omega-3 oil emulsion-based delivery systems using plant proteins as emulsifiers: Lentil, pea, and faba bean proteins. *Food Biophysics, 12*(2), 186–197. https://doi.org/10.1007/s11483-017-9475-6

Gumus, C. E., Decker, E. A., & McClements, D. J. (2017b). Gastrointestinal fate of emulsion-based omega-3 oil delivery systems stabilized by plant proteins: Lentil, pea, and faba bean proteins. *Journal of Food Engineering, 207*, 90–98. https://doi.org/10.1016/j.jfoodeng.2017.03.019

Guzey, D., & McClements, D. J. (2006). Formation, stability and properties of multilayer emulsions for application in the food industry. *Advances in Colloid and Interface Science, 128*, 227–248. https://doi.org/10.1016/j.cis.2006.11.021

Hakansson, A. (2019). Emulsion formation by homogenization: Current understanding and future perspectives. In M. P. Doyle & D. J. McClements (Eds.). *Annual Review of Food Science and Technology, 10*, 239–258.

Herreman, L., Nommensen, P., Pennings, B., & Laus, M. C. (2020). Comprehensive overview of the quality of plant- and animal-sourced proteins based on the digestible indispensable amino acid score. *Food Science & Nutrition, 8*(10), 5379–5391. https://doi.org/10.1002/fsn3.1809

Hunt, M. W. (2019). Veganism and children: Physical and social well-being. *Journal of Agricultural & Environmental Ethics, 32*(2), 269–291. https://doi.org/10.1007/s10806-019-09773-4

Hutchings, J. B. (1999). *Food color and appearance* (2nd ed.). Springer.

Iwanaga, D., Gray, D. A., Fisk, I. D., Decker, E. A., Weiss, J., & McClements, D. J. (2007). Extraction and characterization of oil bodies from soy beans: A natural source of pre-emulsified soybean oil. *Journal of Agricultural and Food Chemistry, 55*(21), 8711–8716. https://doi.org/10.1021/jf071008w

Jeske, S., Bez, J., Arendt, E. K., & Zannini, E. (2019). Formation, stability, and sensory characteristics of a lentil-based milk substitute as affected by homogenisation and pasteurisation. *European Food Research and Technology, 245*(7), 1519–1531. https://doi.org/10.1007/s00217-019-03286-0

Jeske, S., Zannini, E., & Arendt, E. K. (2017). Evaluation of physicochemical and glycaemic properties of commercial plant-based milk substitutes. *Plant Foods for Human Nutrition, 72*(1), 26–33. https://doi.org/10.1007/s11130-016-0583-0

Jeske, S., Zannini, E., & Arendt, E. K. (2018). Past, present and future: The strength of plant-based dairy substitutes based on gluten-free raw materials. *Food Research International, 110*, 42–51. https://doi.org/10.1016/j.foodres.2017.03.045

Jukkola, A., & Rojas, O. J. (2017). Milk fat globules and associated membranes: Colloidal properties and processing effects. *Advances in Colloid and Interface Science, 245*, 92–101. https://doi.org/10.1016/j.cis.2017.04.010

Kaur, P., Waghmare, R., Kumar, V., Rasane, P., Kaur, S., & Gat, Y. (2018). Recent advances in utilization of flaxseed as potential source for value addition. *OCL - Oilseeds and Fats Crops and Lipids, 25*(3). https://doi.org/10.1051/ocl/2018018

Kneifel, W., Ulberth, F., & Schaffer, E. (1992). Tristimulus color reflectance measurement of milk and dairy-products. *Lait, 72*(4), 383–391. Retrieved from https://hal.archives-ouvertes.fr/hal-00929301/document

Kundu, P., Dhankhar, J., & Sharma, A. (2018). Development of non dairy milk alternative using soymilk and almond milk. *Current Research in Nutrition and Food Science, 6*(1), 203–210. https://doi.org/10.12944/crnfsj.6.1.23

Le Huerou-Luron, I., Lemaire, M., & Blat, S. (2018). Health benefits of dairy lipids and MFGM in infant formula. *Ocl - Oilseeds and Fats Crops and Lipids, 25*(3). https://doi.org/10.1051/ocl/2018019

Lee, H., Padhi, E., Hasegawa, Y., Larke, J., Parenti, M., Wang, A. D., Hernell, O., Lönnerdal, B., & Slupsky, C. (2018). Compositional dynamics of the milk fat globule and its role in infant development. *Frontiers in Pediatrics, 6*. https://doi.org/10.3389/fped.2018.00313

Li, X. F., & de Vries, R. (2018). Interfacial stabilization using complexes of plant proteins and polysaccharides. *Current Opinion in Food Science, 21*, 51–56. https://doi.org/10.1016/j.cofs.2018.05.012

Li, Y., Joyner, H. S., Lee, A. P., & Drake, M. A. (2018). Impact of pasteurization method and fat on milk: Relationships among rheological, tribological, and astringency behaviors. *International Dairy Journal, 78*, 28–35. https://doi.org/10.1016/j.idairyj.2017.10.006

Liu, Z. S., & Chang, S. K. C. (2013). Nutritional profile and physicochemical properties of commercial soymilk. *Journal of Food Processing and Preservation, 37*(5), 651–661. https://doi.org/10.1111/j.1745-4549.2012.00696.x

Livney, Y. D., Corredig, M., & Dalgleish, D. G. (2003). Influence of thermal processing on the properties of dairy colloids. *Current Opinion in Colloid & Interface Science, 8*(4–5), 359–364. https://doi.org/10.1016/s1359-0294(03)00092-x

Lopez, C., Cauty, C., & Guyomarc'h, F. (2015). Organization of lipids in milks, infant milk formulas and various dairy products: Role of technological processes and potential impacts. *Dairy Science & Technology, 95*(6), 863–893. https://doi.org/10.1007/s13594-015-0263-0

Lucey, J. A., & Horne, D. S. (2018). Perspectives on casein interactions. *International Dairy Journal, 85*, 56–65. https://doi.org/10.1016/j.idairyj.2018.04.010

Ludwig, D. S., Willett, W. C., Volek, J. S., & Neuhouser, M. L. (2018). Dietary fat: From foe to friend? *Science, 362*(6416), 764. Retrieved from http://science.sciencemag.org/content/362/6416/764.abstract

Lv, S. S., Zhang, Y. H., Tan, H. Y., Zhang, R. J., & McClements, D. J. (2019). Vitamin E encapsulation within oil-in-water emulsions: Impact of emulsifier type on physicochemical stability and bioaccessibility. *Journal of Agricultural and Food Chemistry, 67*(5), 1521–1529. https://doi.org/10.1021/acs.jafc.8b06347

Makinen, O. E., Wanhalinna, V., Zannini, E., & Arendt, E. K. (2016). Foods for special dietary needs: Non-dairy plant-based milk substitutes and fermented dairy-type products. *Critical Reviews in Food Science and Nutrition, 56*(3), 339–349. https://doi.org/10.1080/10408398.2012.761950

Makkhun, S., Khosla, A., Foster, T., McClements, D. J., Grundy, M. M. L., & Gray, D. A. (2015). Impact of extraneous proteins on the gastrointestinal fate of sunflower seed (Helianthus annuus) oil bodies: A simulated gastrointestinal tract study. *Food & Function, 6*(1), 125–134. https://doi.org/10.1039/c4fo00422a

McCarthy, K. S., Lopetcharat, K., & Drake, M. A. (2017a). Milk fat threshold determination and the effect of milk fat content on consumer preference for fluid milk. *Journal of Dairy Science, 100*(3), 1702–1711. https://doi.org/10.3168/jds.2016-11417

McCarthy, K. S., Parker, M., Ameerally, A., Drake, S. L., & Drake, M. A. (2017b). Drivers of choice for fluid milk versus plant-based alternatives: What are consumer perceptions of fluid milk? *Journal of Dairy Science, 100*(8), 6125–6138. https://doi.org/10.3168/jds.2016-12519

McClements, D. J. (2002a). Colloidal basis of emulsion color. *Current Opinion in Colloid & Interface Science, 7*(5–6), 451–455. https://doi.org/10.1016/s1359-0294(02)00075-4

McClements, D. J. (2002b). Theoretical prediction of emulsion color. *Advances in Colloid and Interface Science, 97*(1–3), 63–89. https://doi.org/10.1016/s0001-8686(01)00047-1

McClements, D. J. (2015). *Food emulsions: Principles, practice, and techniques* (2nd ed.). CRC Press.

McClements, D. J. (2018). Enhanced delivery of lipophilic bioactives using emulsions: A review of major factors affecting vitamin, nutraceutical, and lipid bioaccessibility. *Food & Function, 9*(1), 22–41. https://doi.org/10.1039/c7fo01515a

McClements, D. J. (2020). Development of next-generation nutritionally fortified plant-based milk substitutes: Structural design principles. *Food, 9*(4). https://doi.org/10.3390/foods9040421

McClements, D. J., Bai, L., & Chung, C. (2017). Recent advances in the utilization of natural emulsifiers to form and stabilize emulsions. In M. P. Doyle & T. R. Klaenhammer (Eds.). *Annual Review of Food Science and Technology, 8*, 205–236.

McClements, D. J., & Decker, E. (2018). Interfacial antioxidants: A review of natural and synthetic emulsifiers and coemulsifiers that can inhibit lipid oxidation. *Journal of Agricultural and Food Chemistry, 66*(1), 20–35. https://doi.org/10.1021/acs.jafc.7b05066

McClements, D. J., & Gumus, C. E. (2016). Natural emulsifiers – Biosurfactants, phospholipids, biopolymers, and colloidal particles: Molecular and physicochemical basis of functional performance. *Advances in Colloid and Interface Science, 234*, 3–26. https://doi.org/10.1016/j.cis.2016.03.002

McClements, D. J., & Jafari, S. M. (2018). Improving emulsion formation, stability and performance using mixed emulsifiers: A review. *Advances in Colloid and Interface Science, 251*, 55–79. https://doi.org/10.1016/j.cis.2017.12.001

McClements, D. J., Li, F., & Xiao, H. (2015). The nutraceutical bioavailability classification scheme: Classifying nutraceuticals according to factors limiting their oral bioavailability. In M. P. Doyle & T. R. Klaenhammer (Eds.). *Annual Review of Food Science and Technology, 6*, 299–327.

McClements, D. J., Newman, E., & McClements, I. F. (2019). Plant-based milks: A review of the science underpinning their design, fabrication, and performance. *Comprehensive Reviews in Food Science and Food Safety, 18*(6), 2047–2067. https://doi.org/10.1111/1541-4337.12505

Mehmood, T., Ahmed, A., Ahmed, Z., & Ahmad, M. S. (2019). Optimization of soya lecithin and Tween 80 based novel vitamin D nanoemulsions prepared by ultrasonication using response surface methodology. *Food Chemistry, 289*, 664–670. https://doi.org/10.1016/j.foodchem.2019.03.112

Michalski, M. C. (2009). Specific molecular and colloidal structures of milk fat affecting lipolysis, absorption and postprandial lipemia. *European Journal of Lipid Science and Technology, 111*(5), 413–431. https://doi.org/10.1002/ejlt.200800254

Navarini, L., & Rivetti, D. (2010). Water quality for espresso coffee. *Food Chemistry, 122*(2), 424–428. https://doi.org/10.1016/j.foodchem.2009.04.019

Nikiforidis, C. V., Matsakidou, A., & Kiosseoglou, V. (2014). Composition, properties and potential food applications of natural emulsions and cream materials based on oil bodies. *RSC Advances, 4*(48), 25067–25078. https://doi.org/10.1039/c4ra00903g

Obeid, R., Heil, S. G., Verhoeven, M. M. A., van den Heuvel, E., de Groot, L., & Eussen, S. (2019). Vitamin B12 intake from animal foods, biomarkers, and health aspects. *Frontiers in Nutrition, 6*. https://doi.org/10.3389/fnut.2019.00093

Oliveira, D. L., Wilbey, R. A., Grandison, A. S., & Roseiro, L. B. (2015). Milk oligosaccharides: A review. *International Journal of Dairy Technology, 68*(3), 305–321. https://doi.org/10.1111/1471-0307.12209

Ozturk, B., Argin, S., Ozilgen, M., & McClements, D. J. (2015a). Formation and stabilization of nanoemulsion-based vitamin E delivery systems using natural biopolymers: Whey protein isolate and gum arabic. *Food Chemistry, 188*, 256–263. https://doi.org/10.1016/j.foodchem.2015.05.005

Ozturk, B., Argin, S., Ozilgen, M., & McClements, D. J. (2015b). Nanoemulsion delivery systems for oil-soluble vitamins: Influence of carrier oil type on lipid digestion and vitamin D-3 bioaccessibility. *Food Chemistry, 187*, 499–506. https://doi.org/10.1016/j.foodchem.2015.04.065

Park, Y. W., & Nam, M. S. (2015). Bioactive peptides in milk and dairy products: A review. *Korean Journal for Food Science of Animal Resources, 35*(6), 831–840. https://doi.org/10.5851/kosfa.2015.35.6.831

Pereira, P. C. (2014). Milk nutritional composition and its role in human health. *Nutrition, 30*(6), 619–627. https://doi.org/10.1016/j.nut.2013.10.011

Poore, J., & Nemecek, T. (2018). Reducing food's environmental impacts through producers and consumers. *Science, 360*(6392), 987-+. https://doi.org/10.1126/science.aaq0216

Preece, K. E., Drost, E., Hooshyar, N., Krijgsman, A., Cox, P. W., & Zuidam, N. J. (2015). Confocal imaging to reveal the microstructure of soybean processing materials. *Journal of Food Engineering, 147*, 8–13. https://doi.org/10.1016/j.jfoodeng.2014.09.022

Pyc, M., Cai, Y. Q., Greer, M. S., Yurchenko, O., Chapman, K. D., Dyer, J. M., & Mullen, R. T. (2017). Turning over a new leaf in lipid droplet biology. *Trends in Plant Science, 22*(7), 596–609. https://doi.org/10.1016/j.tplants.2017.03.012

Qamar, S., Bhandari, B., & Prakash, S. (2019). Effect of different homogenisation methods and UHT processing on the stability of pea protein emulsion. *Food Research International, 116*, 1374–1385. https://doi.org/10.1016/j.foodres.2018.10.028

Rao, M. A. (2013). *Rheology of fluid, semisolid, and solid foods: Principles and applications* (3rd ed.). Springer Science.

Reichert, C. L., Salminen, H., Bonisch, G. B., Schafer, C., & Weiss, J. (2019). Influence of concentration ratio on emulsifying properties of Quillaja saponin – Protein or lecithin mixed systems. *Colloids and Surfaces A: Physicochemical and Engineering Aspects, 561*, 267–274. https://doi.org/10.1016/j.colsurfa.2018.10.050

Reyes-Jurado, F., Soto-Reyes, N., Dávila-Rodríguez, M., Lorenzo-Leal, A. C., Jiménez-Munguía, M. T., Mani-López, E., & López-Malo, A. (2021). Plant-based milk alternatives: Types, processes, benefits, and characteristics. *Food Reviews International*, 1–32. https://doi.org/10.1080/87559129.2021.1952421

Robinson, R. C. (2019). Structures and metabolic properties of bovine milk oligosaccharides and their potential in the development of novel therapeutics. *Frontiers in Nutrition, 6*. https://doi.org/10.3389/fnut.2019.00050

Sah, B. N. P., Vasiljevic, T., McKechnie, S., & Donkor, O. N. (2015). Identification of anticancer peptides from bovine milk proteins and their potential roles in management of cancer: A critical review. *Comprehensive Reviews in Food Science and Food Safety, 14*(2), 123–138. https://doi.org/10.1111/1541-4337.12126

Sarkar, A., & Dickinson, E. (2020). Sustainable food-grade Pickering emulsions stabilized by plant-based particles. *Current Opinion in Colloid & Interface Science, 49*, 69–81. https://doi.org/10.1016/j.cocis.2020.04.004

Scarso, S., McParland, S., Visentin, G., Berry, D. P., McDermott, A., & De Marchi, M. (2017). Genetic and nongenetic factors associated with milk color in dairy cows. *Journal of Dairy Science, 100*(9), 7345–7361. https://doi.org/10.3168/jds.2016-11683

Schiano, A. N., Harwood, W. S., & Drake, M. A. (2017). A 100-year review: Sensory analysis of milk. *Journal of Dairy Science, 100*(12), 9966–9986. https://doi.org/10.3168/jds.2017-13031

Schoener, A. L., Zhang, R. J., Lv, S. S., Weiss, J., & McClements, D. J. (2019). Fabrication of plant-based vitamin D-3-fortified nanoemulsions: Influence of carrier oil type on vitamin bioaccessibility. *Food & Function, 10*(4), 1826–1835. https://doi.org/10.1039/c9fo00116f

Sebastiani, G., Herranz Barbero, A., Borras-Novell, C., Casanova, M. A., Aldecoa-Bilbao, V., Andreu-Fernandez, V., Tutusaus, M. P., Martínez, S. F., Roig, M. D. G., & Garcia-Algar, O. (2019). The effects of vegetarian and vegan diet during pregnancy on the health of mothers and offspring. *Nutrients, 11*(3). https://doi.org/10.3390/nu11030557

Sethi, S., Tyagi, S. K., & Anurag, R. K. (2016). Plant-based milk alternatives an emerging segment of functional beverages: A review. *Journal of Food Science and Technology-Mysore, 53*(9), 3408–3423. https://doi.org/10.1007/s13197-016-2328-3

Stocker, S., Foschum, F., Krauter, P., Bergmann, F., Hohmann, A., Happ, C. S., & Kienle, A. (2017). Broadband optical properties of milk. *Applied Spectroscopy, 71*(5), 951–962. https://doi.org/10.1177/0003702816666289

Sumner, O., & Burbridge, L. (2021). Plant-based milks: The dental perspective. *British Dental Journal*. https://doi.org/10.1038/s41415-020-2058-9

Tan, Y. B., Zhou, H. L., Zhang, Z. Y., & McClements, D. J. (2021). Bioaccessibility of oil-soluble vitamins (A, D, E) in plant-based emulsions: Impact of oil droplet size. *Food & Function, 12*(9), 3883–3897. https://doi.org/10.1039/d1fo00347j

Thorning, T. K., Raben, A., Tholstrup, T., Soedamah-Muthu, S. S., Givens, I., & Astrup, A. (2016). Milk and dairy products: Good or bad for human health? An assessment of the totality of scientific evidence. *Food & Nutrition Research, 60*. https://doi.org/10.3402/fnr.v60.32527

Tzen, J. T. C., Cao, Y. Z., Laurent, P., Ratnayake, C., & Huang, A. H. C. (1993). Lipids, proteins, and structure of seed oil bodies from diverse species. *Plant Physiology, 101*(1), 267–276. https://doi.org/10.1104/pp.101.1.267

Tzen, J. T. C., & Huang, A. H. C. (1992). Surface-structure and properties of plant seed oil bodies. *Journal of Cell Biology, 117*(2), 327–335. https://doi.org/10.1083/jcb.117.2.327

Van Hekken, D. L., Tunick, M. H., Ren, D. X., & Tomasula, P. M. (2017). Comparing the effect of homogenization and heat processing on the properties and in vitro digestion of milk from organic and conventional dairy herds. *Journal of Dairy Science, 100*(8), 6042–6052. https://doi.org/10.3168/jds.2016-12089

Vanga, S. K., & Raghavan, V. (2018). How well do plant based alternatives fare nutritionally compared to cow's milk? *Journal of Food Science and Technology-Mysore, 55*(1), 10–20. https://doi.org/10.1007/s13197-017-2915-y

Vogelsang-O'Dwyer, M., Zannini, E., & Arendt, E. K. (2021). Production of pulse protein ingredients and their application in plant-based milk alternatives. *Trends in Food Science & Technology, 110*, 364–374. https://doi.org/10.1016/j.tifs.2021.01.090

Walia, N., & Chen, L. Y. (2020). Pea protein based vitamin D nanoemulsions: Fabrication, stability and in vitro study using Caco-2 cells. *Food Chemistry, 305*. https://doi.org/10.1016/j.foodchem.2019.125475

Walstra, P., Wouters, J. T. M., & Geurts, T. J. (2005). *Dairy science and technology*. CRC Press.

Wenzel, P., & Jungbluth, N. (2017). *The environmental impact of vegan drinks compared to whole milk* (p. 1). ESU-Services.

Wilde, P. J., Garcia-Llatas, G., Lagarda, M. J., Haslam, R. P., & Grundy, M. M. L. (2019). Oat and lipolysis: Food matrix effect. *Food Chemistry, 278*, 683–691. https://doi.org/10.1016/j.foodchem.2018.11.113

Ye, A. Q., Liu, W. L., Cui, J., Kong, X. N., Roy, D., Kong, Y. Y., Han, J., & Singh, H. (2019). Coagulation behaviour of milk under gastric digestion: Effect of pasteurization and ultra-high temperature treatment. *Food Chemistry, 286*, 216–225. https://doi.org/10.1016/j.foodchem.2019.02.010

Zha, F. C., Dong, S. Y., Rao, J. J., & Chen, B. C. (2019). The structural modification of pea protein concentrate with gum arabic by controlled Maillard reaction enhances its functional properties and flavor attributes. *Food Hydrocolloids, 92*, 30–40. https://doi.org/10.1016/j.foodhyd.2019.01.046

Zhang, A. Q., Chen, S., Wang, Y. Y., Wang, X. B., Ku, N., & Jiang, L. Z. (2020). Stability and in vitro digestion simulation of soy protein isolate-vitamin D-3 nanocomposites. *LWT - Food Science and Technology, 117*. https://doi.org/10.1016/j.lwt.2019.108647

Zheng, B., Zhang, X., Lin, H., & McClements, D. J. (2019a). Loading natural emulsions with nutraceuticals using the pH-driven method: Formation & stability of curcumin-loaded soybean oil bodies. *Food & Function, 10*(9), 5473–5484. https://doi.org/10.1039/c9fo00752k

Zheng, B., Zhang, X., Peng, S., & McClements, D. J. (2019b). Impact of curcumin delivery system format on bioaccessibility: Nanocrystals, nanoemulsion droplets, and natural oil bodies. *Food & Function, 10*(7), 4339–4349. https://doi.org/10.1039/c8fo02510j

Zheng, B. J., Zhou, H. L., & McClements, D. J. (2021). Nutraceutical-fortified plant-based milk analogs: Bioaccessibility of curcumin-loaded almond, cashew, coconut, and oat milks. *LWT - Food Science and Technology, 147*. https://doi.org/10.1016/j.lwt.2021.111517

Zhou, H. L., Zheng, B. J., Zhang, Z. Y., Zhang, R. J., He, L. L., & McClements, D. J. (2021). Fortification of plant-based milk with calcium may reduce vitamin D bioaccessibility: An in vitro digestion study. *Journal of Agricultural and Food Chemistry, 69*(14), 4223–4233. https://doi.org/10.1021/acs.jafc.1c01525

Chapter 9
Dairy Alternatives – Cheese, Yogurt, Butter, and Ice Cream

9.1 Introduction

Plant-based milks have become an important commodity and a staple food for many consumers (Chap. 8). These milk analogs are typically designed to exhibit similar physicochemical and sensory attributes as milks derived from mammals, which are usually cows but may also be other livestock animals such as buffalo, goats, or sheep. In principle, milk analogs can be used to produce a variety of other dairy analogs, such as cheese, yogurt, ice cream, or whipped cream. Alternatively, dairy analogs can be produced from plant-based ingredients without making a milk analog first. However, in both cases, they must be carefully designed to exhibit functional attributes that are similar to their conventional counterparts.

Originally, dairy products were created to increase the shelf life of milk by changing its physicochemical attributes (such as pH, ionic strength, or water activity) or its biological properties (such as enzyme activity or microbial load), e.g., by heating, acidification, or dehydration. Nowadays, dairy products are familiar commodities that are also consumed for their desirable flavors and nutritional profiles. Plant-based ingredients can be used to create dairy analogs by utilizing appropriate processing operations to convert them into materials with similar optical, textural, and sensorial attributes as their dairy counterparts. This chapter reviews the main ingredients, processes, and properties of some major plant-based dairy products: cheese, yogurt, ice cream, whipping cream, and butter analogs.

9.2 History of Plant-Based Cheeses

Plant-based cheeses have a long tradition but they were not originally considered as cheese alternatives. Most likely, the oldest plant-based 'cheese' is fermented tofu, which has been consumed in China since the seventeenth century. This food is also known as *'furu'*, which can be translated as *'spoiled milk'*, and thus has a similar connotation as animal-based cheese (Shurtleff & Aoyagi, 2011). Interestingly, tofu is not typically regarded as a cheese alternative but as a meat analog, and fermented tofu is rarely eaten in western societies as an alternative to cheese. Cheese analogs are a rather new development and were mostly regarded as cheap imitations of real cheese when they were first released to the market. However, this attitude has changed over the past few years with the release of improved versions of next-generation cheese analogs and the increasingly positive perception of plant-based foods by consumers (GFI, 2020).

Plant-based cheeses are usually designed to mimic the physicochemical, textural, and sensory attributes of specific dairy cheeses, such as cheddar, Roquefort, or Camembert. Conventional dairy cheeses have only one main ingredient: milk. The numerous different kinds of cheeses obtained from animal milk are created by adjusting the milk composition, the processing operations, and the ripening conditions. This is in contrast to plant-based cheeses, which are typically produced using a variety of ingredients and processes (Table 9.1). In this section, we begin by reviewing the production and properties of animal-based cheeses, as this information is critical for the creation of high-quality plant-based cheeses that accurately simulate their desirable characteristics. We then describe the raw materials and processing operations used to obtain plant-based cheeses, as well as their physicochemical attributes. In addition, we compare the environmental sustainability and nutritional properties of plant-based cheeses with animal-based ones.

9.3 Animal-Based Cheeses

The production of cheese from animal milk is believed to be one of the oldest processes used to preserve milk, with examples of these products already being manufactured around 8000 years ago (Fox & McSweeney, 2017). The term 'cheese' includes a broad range of foods derived from milk that exhibit different physicochemical and sensory properties, but they are all prepared using fairly similar processing methods. In general, cheeses are formed by the controlled agglomeration of milk proteins, particularly caseins, which leads to a gelled product with semi-solid characteristics. The agglomeration of the caseins is typically induced by acidification and/or enzyme addition, which leads to the formation of a curd that is then separated from the non-aggregated whey part (Johnson, 2017). Typically, cheese is classified according to its type (rennet, fresh, or processed cheese), its textural attributes (soft, semi-soft, medium-hard, semi-hard, and hard), and its origin (*e.g.,* cow,

9.3 Animal-Based Cheeses

Table 9.1 Selected plant-based cheese analogs and regular cow milk cheeses on the market and their composition in g per 100 g sorted by their main ingredient into two different processing routes

Product	Ingredients	kcal	F (SF)	CHO	P	Fi	S	Others	Type	Company website
Material fractionation route										
Field roast: Creamy original: Chao block	Filtered water, coconut oil, corn and potato starch, modified potato starch, fermented chao tofu (soybeans, water, salt, sesame oil, calcium sulfate), sea salt, natural flavor, olive extract (antioxidant used as a preservative), beta carotene(color), powdered cellulose (to prevent caking).	286	21 (21)	21	0	0	0.93	Calcium: 0% DV	Semi-hard, ripened flavor	https://fieldroast.com/product/chao-block-creamy-original/
Daiya: Medium Cheddar style block	Filtered water, tapioca starch, coconut oil, vegan natural flavors, pea protein, expeller pressed: Canola and/or safflower oil, chicory root fiber, Tricalcium phosphate, salt, xanthan gum, lactic acid (vegan), pea starch, vegan enzyme, cane sugar, annatto (color), yeast extract, coconut cream.	286	21 (16)	25	4	0	0.82	Calcium: 0.47g	Cheddar-style, semi-hard	https://daiyafoods.com/our-foods/blocks/medium-cheddar/
Follow your heart: Dairy-free Mozzarella blocks	Filtered water, coconut oil, modified potato and corn starches, potato starch, sea salt, natural flavor, olive extract, Beta carotene for color.	286	21 (21)	21	0	0	0.93	Calcium: 0% DV	Mozzarella-style, semi-soft	https://followyourheart.com/products/dairy-free-mozzarella-blocks/

(continued)

Table 9.1 (continued)

Product	Ingredients	kcal	F	(SF)	CHO	P	Fi	S	Others	Type	Company website
Follow your heart: Dairy-free parmesan shredded	Filtered water, organic palm fruit oil†, modified potato starch, expeller-pressed canola oil, natural flavors (contains autolyzed yeast), organic vegetable glycerin, less than 2% of: Sea salt, calcium phosphate, bamboo fiber, sodium phosphate, carrageenan, lactic acid, nutritional yeast, organic chickpea miso (organic handmade rice Koji, organic whole chickpeas, sea salt, water, Koji spores), sunflower lecithin, citric acid, annatto.	321	25	(13)	29	0	0	1.64	Calcium: 20% DV	Parmesan-style, hard	https://followyourheart.com/products/dairy-free-parmesan-shredded/
Follow your heart: Dairy-free Mozzarella	Filtered water, organic expeller-pressed soybean oil, organic soymilk powder (organic soybeans), natural flavors (plant sources), inulin (chicory root extract), agar agar, sea salt, organic soy protein, lactic acid (vegetable source)	286	29	(5)	7	4	4	0.34	Calcium: 0% DV	Mozzarella-style, soft	https://followyourheart.com/products/dairy-free-mozzarella-block-soy/
Violife: Just like feta block	Filtered water, coconut oil, potato starch, salt (sea salt), glucono delta lactone, flavor (vegan sources), olive extract, vitamin B12.	321	29	(25)	11	0	0	0.68	Calcium: 0.0g Vit. B12: 30% DV	Feta-style	https://violifefoods.com/us/product/just-like-feta-block/
Violife: Just like cream cheese original	Filtered water, coconut oil, potato starch, salt (sea salt), glucono-delta-lactone, flavor (vegan sources), olive extract, vitamin B12.	233	23	(20)	7	0	0	0.47	Calcium: 0.0g Vit. B12: 30% DV	Cream-cheese style	https://violifefoods.com/us/product/just-like-cream-cheese-original/

9.3 Animal-Based Cheeses

Product	Ingredients	kcal	F (SF)	CHO	P	Fi	S	Others	Type	Company website
Bute Island foods: Sheese block gouda style	Water, coconut oil (21%), modified potato starch, maize starch, gluten free oat fibre, modified maize starch, thickeners (carrageenan, guar gum), salt, natural flavourings, acidity regulators (lactic acid, sodium lactate), yeast extract, colour (carotenes).	292	23 (19)	19	0.5	5.1	1.7 (as NaCl)	Calcium: 0.15g	Gouda-style, semi-hard	https://www.buteisland.com/products/sheese-blocks/gouda-style/
Simply V: Fein cremige Genießerscheiben	53% cashew nut preparation (drinking water, 2% roasted cashews), coconut oil, modified starch, starch, salt, potato protein, flavor, coloring food (carrot and apple concentrate), antioxidant: Sodium ascorbate.	272	19 (18)	24	<0.5		2 (as NaCl)		Semi-hard	https://www.simply-v.de/
OATzarella: Original mini cheese wheel	Water, organic steel cut oats, organic extra virgin olive oil, organic tapioca flour, natural flavors, sea salt, organic agar agar, lactic acid (vegan).	179	16	11	0	0	0.68		Mozzarella-style, soft	https://oatzarella.com/cheeses
Tissue disruption route										
Fermented tofu	Tofu (98%) (soya beans, water, magnesium chloride, calcium sulphate), sea salt, vegan yogurt culture: *Str. Thermophilus*. (depending on manufacturer)	116	8 (1.2)	5.15	8.15		2.87	Calcium: 1.23g		https://fdc.nal.usda.gov/fdc-app.html#/food-details/174305/nutrients
Kite Hill: Almond Milk ricotta alternative	Almond Milk (water, almonds), salt, enzymes, tartaric acid, cultures.	246	21 (2)	9	9	4	0.40	Calcium: 0.1g	Ricotta-style, semisoft	https://www.kite-hill.com/our-food/artisanal-delicacy/

(continued)

Table 9.1 (continued)

Product	Ingredients	kcal	F (SF)	CHO	P	Fi	S	Others	Type	Company website
Parmela creamery: Nut cheese pepper Jack slices	Nutmilk (water, cashews), coconut oil, modified food starch, potato starch, sea salt, peppers (jalapeno, habanero), natural flavor, annatto, yeast extract, cultures	286	25 (18)	21	7	0	0.91	Calcium: 2% DV	Monterey Jack-style, semi-hard	https://www.parmelacreamery.com/products/slices
Catalyst creamery: Muenster style H*mp seed cheese	Hemp milk (filtered water, hemp hearts), organic apple cider vinegar, organic coconut oil, tapioca, nutritional yeast, kappa carrageenan, sea salt, cultured cane sugar, organic mustard, organic onion powder, organic smoked paprika	Not available							Muenster-style, semi-soft	https://www.catalystcreamery.com/cheesecave/muenster
Cheeze & Thank you: Artisanal herbed feta	Organic tofu (water, organic whole soybeans, calcium sulfate, magnesium chloride), organic refined coconut oil, white wine vinegar, sea salt, less than 2% of spice, vegan source lactic acid	214	16 (11)	4	7	0	1.46	Calcium: 15% DV	Feta-style	https://www.cheezeandthankyou.com/cheezes
Cheeze & Thank you: Artisanal Mozzarella Capri	Organic soy milk (soybeans, water), organic refined coconut oil, tapioca flour, less than 2% of garlic, sea salt, spice, Kalamata olives, sundried tomato, kappa carrageenan, vegan source lactic acid	393	36 (29)	11	4	0	0.68	Calcium: 2% DV	Mozzarella-style	https://www.cheezeandthankyou.com/cheezes
Treeline cheese: Classic	Cashew nuts, filtered water, vegan lactic acid, vegan L. acidophilus, hickory smoked sea salt.	500	39 (7)	18	18	4	0.45	Calcium: 0.0g	Soft-ripened	https://www.treelinecheese.com/pages/classic
Treeline cheese: Plain	Cashew nuts, filtered water, sea salt, lemon juice, vegan L. acidophilus.	321	25 (4)	14	7	0	0.46	Calcium: 0.07g	Cream-cheese	https://www.treelinecheese.com/pages/plain

9.3 Animal-Based Cheeses

Product	Ingredients	kcal	F (SF)	CHO	P	Fi	S	Others	Type	Company website
Happy Cheeze: Happy White	Cashews (64%), water, salt, vegan fermentation and edible mold cultures.	359	27.3 (5.9)	13.3	13.9		1.2 (as NaCl)		Camembert-style, soft-ripened	https://happy-cheeze.com/products/vegane-camembert-kaese-alternative-dr-mannahs
The Frauxmagerie: Botanic true blue	Raw cashews, nutritional yeast, vegan probiotics, salt, bacterial culture, filtered water	440	36 (4)	12	16	4	0.26	Calcium: 4% DV	Blue-cheese style, soft-ripened	https://thefrauxmagerie.com/product/botanic-true-blue/
Miyoko's creamery: Aged sharp English farmhouse cashew milk cheese	Organic cashew milk (organic cashews, filtered water), organic chickpea miso (organic rice Koji (organic rice, Koji spores), organic whole chickpeas, sea salt, water), nutritional yeast, sea salt, natural flavors (derived from oregano, plum, flaxseed), cultures	393	29 (5)	25	14	4	0.71	Calcium: 0.04g	Farmstead cheese style	https://miyokos.com/products/vegan-cheese-wheel
Miyoko's creamery: Organic cashew milk Mozzarella	Organic cashew milk (filtered water, organic cashews), organic coconut oil, organic tapioca starch, sea salt, organic agar, mushroom extract, organic Konjac, cultures.	214	18 (11)	4	4	0	0.75	Calcium: 0.04g	Mozzarella-style, soft	https://miyokos.com/products/fresh-vegan-mozzarella-cheese

(continued)

Table 9.1 (continued)

Product	Ingredients	kcal	F (SF)	CHO	P	Fi	S	Others	Type	Company website
Miyoko's creamery: Cultured vegan Cheddar cheese block	Miyoko's cultured vegan milk (oat milk (filtered water, organic oats), navy beans, organic garbanzo beans, cultures), filtered water, organic coconut oil, Faba bean protein, potato starch, organic tapioca starch, contains less than 2% of sea salt, calcium sulfate, natural flavors, organic yeast extract, organic annatto, organic cultured dextrose, Konjac, organic locust bean gum	250	16 (13)	18	11	0	0.96	Calcium: 0.52g	Farmstead-style, semi-hard	https://miyokos.com/products/cultured-vegan-farmhouse-cheddar-chunk
Nuts for cheese: Organic un-brie-Lievable wedge	Organic cashews, organic coconut oil, organic coconut milk, water, organic quinoa rejuvelac (water, organic quinoa), sea salt, nutritional yeast, organic chickpea miso (organic rice, organic chickpeas, sea salt, water, koji spores), fermented organic oregano extract (water, organic oregano, organic raw cane sugar, active cultures)	467	43 (20)	17	10	3	0.67	Calcium: 0.03g	Brie-style, soft-ripened	https://nutsforcheese.com/vegan-cheese-products/
Regular cheese from cow milk										
Camembert	Pasteurized milk, salt, starter cultures, non-animal rennet. (depending on manufacturer)	300	24 (15)	0.5	20	0	0.84	Calcium: 0.39g Vit. A: 241 μg Vit. B12: 1.3 μg	Soft-ripened	USDA (2021)

9.3 Animal-Based Cheeses

Product	Ingredients	kcal	F (SF)	CHO	P	Fi	S	Others	Type	Company website
Cheddar	Milk, salt, cultures, enzyme (rennet). (depending on manufacturer)	408	34 (19)	2.4	23	0	0.65	Calcium: 0.71g; Vit. A: 316 µg; Vit. B12: 1.1 µg	Semi-hard	USDA (2021)
Mozzarella	Pasteurized cows milk, salt, non-animal rennet, starter cultures. (depending on manufacturer)	298	20 (12)	4.4	24	0	0.7	Calcium: 0.69g; Vit. A: 203 µg; Vit. B12: 1.7 µg	Soft	USDA (2021)
Parmigiano Reggiano	Milk, salt, animal rennet. (depending on manufacturer)	393	29 (21)	0	32	0	0.68	Calcium: 1.25g; Vit. A: 1071 IU	Hard	USDA (2021)

The websites were accessed in April 2021. Note that some products are combinations of the fractionation and tissue disruption route. *Key:* Kcal = kilocalories; F = Fat; CHO = Carbohydrates; P = Protein; SF = Saturated fats; Fi = Fiber, and S = Sodium. Reprinted with permission from Elsevier (Grossmann & McClements, 2021) buffalo, or goat). However, much more cheese classifications are known and many of them follow regional traditions and regulations. In the remainder of this section, we briefly describe the processing operations used to produce animal-based cheeses and their physicochemical properties. An understanding of the formation and the properties of dairy cheeses is critical for the development of plant-based analogs with similar physicochemical and sensory attributes.

9.3.1 Raw Materials

Cheeses are produced from milk, which is the nutrient-rich fluid produced by the mammary glands of mammals to feed their young (Fox & McSweeney, 2017). In principal, every milk that is rich in casein can be used to produce cheese but most commonly milk obtained from cows, goats, buffalos, and sheep is used. More rare artisan-style types of cheeses are produced from milk derived from other kinds of animals, such as donkeys, camels, llamas, and moose (Faccia et al., 2019; Holsinger et al., 1995; Konuspayeva et al., 2017). In the following, we will mainly refer to bovine milk because it is the most frequently used milk for cheese manufacturing.

The quality of cow's milk depends on a variety of factors such as the breed, lactation stage, climate, feed, and type of husbandry system used (Franzoi et al., 2019). For example, a study performed on 2800 Holstein-Friesian cows showed that the average casein content in the first 10 days of lactation was 3.05%, which decreased to below 2.5% in the second month before going up to 2.96% in their 11th month. Similar dependencies were found for factors such as age and season (Ng-Kwai-Hang et al., 1982). Moreover, milk quality is influenced by the storage conditions used before it is further processed. In general, the milk used for cheese processing has to fulfill certain quality standards such as the presence of antimicrobial compounds, the microbial load (including spores and relevant pathogens/coliforms), acidity, somatic cell count, and the sensitivity to fermentation with starter cultures and rennet (Metz et al., 2020).

The main constituents of milk are water (87%), lactose (4–5%), fat (3–4%), protein (3%), minerals (0.8%), and vitamins (0.1%) (McClements et al., 2019). The presence of proteins and calcium in milk is particularly important for the creation of cheeses, and the levels of these components influence cheese yield and quality. Typically, the casein fraction, which comprises around 80% of the total proteins in milk, plays the most important role in cheese formation but the whey protein fraction, which makes up the remaining 20% of the total proteins, are also important in some types of cheese (*e.g.*, ricotta). The caseins in milk are assembled into micelles, which can be considered to be natural nanoparticles with a mean diameter of about 200 nm. They are composed of several sub-protein fractions that are present in milk at different concentrations: α_{S1}, β, α_{S2}, and κ caseins (Farrell et al., 2004). These sub-protein fractions are held together in the casein micelles by hydrophobic interactions and salt bridges (calcium phosphate nanoclusters) (Lucey & Horne, 2018). Casein micelles are normally prevented from aggregating with each other due to the

9.3 Animal-Based Cheeses

presence of κ-casein, a glycosylated protein, which is located on their surface. The hydrophilic part of this molecule (the 'glycomacropeptide') protrudes into the surrounding water phase and generates a strong steric and electrostatic repulsion between the casein micelles by forming a charged hydrophilic 'hairy layer' at their surfaces, which increases the hydrodynamic radius of the micelles by about 5–10 nm (Dalgleish, 2011). During cheese making using rennet, κ casein is partly hydrolyzed by this enzyme, resulting in the release of the hydrophilic glycomacropeptide. As a result, the steric and electrostatic repulsion between the casein micelles is reduced, which promotes their aggregation. Thus, cheese making is commonly not possible for milk that is low in κ-casein (Hallén et al., 2010).

9.3.2 Cheese Production

The main aim of cheese production is to obtain a viscoelastic solid material from fluid milk by inducing a sol-gel transition and collecting the curd formed, which usually has around a tenfold higher protein content than the original milk. In this section, the main processing routes that can be used to achieve this goal are described (Kammerlehner, 2009; Kessler, 2002; McSweeney et al., 2017). Typically, cheese manufacturing involves a number of processing and ripening steps (Figs. 9.1 and 9.7):

- *Pasteurization*: The milk used for cheese production is normally pasteurized to inactive pathogens, reduce the overall microbial load, and inactivate indigenous enzymes that might interfere with the cheese ripening process. Typically, this is achieved using a high-temperature/short-time (HTST) pasteurization step, which involves heating at 72–74 °C for 15–30 s. An advantage of this method is that extensive milk protein denaturation does not occur, which would negatively influence their ability to form the gel-like structures required to create cheese (Hougaard et al., 2010).

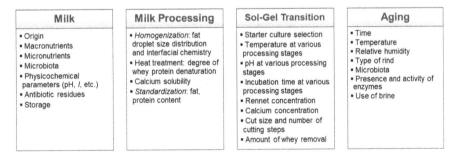

Fig. 9.1 Overview of some important regular cheese production parameters that influence the final texture and flavor of the cheese

- *Standardization*: The composition of milk used for cheese production is typically standardized to achieve the specific proximate composition required for the type of cheese being produced (*e.g.,* fat and protein content). The fat content is standardized by mixing whole milk, cream, and skim milk in different ratios, or by automated processing equipment that mixes the cream and skim milk phases after separation by controlling the volume flow and using continuous fat content measurements (*e.g.,* using density or infrared spectroscopy measurements). The protein content is typically standardized by using filtration techniques such as microfiltration and ultrafiltration.
- *Acidification*: The transformation of lactose to lactic acid by bacteria plays a critical role in most cheese production techniques and strongly influences cheese characteristics. Typically, the pH decreases to around 5.6 to 4.9 (rennet coagulated cheeses) or 4.9 to 4.5 (acid coagulated cheeses) during the first 24/48 h of cheese production, before increasing during ripening. Depending on the cheese variety, the final pH in ripened cheese may range from around 5.1 to 7.0. Different mesophilic and thermophilic lactic acid and propionic acid bacteria are added to the milk to induce this acidification. The reduction in pH influences process parameters such as enzyme activity, syneresis, calcium solubility, and proteolysis.
- *Coagulation*: Depending on the desired cheese type, the sol-gel transition can be achieved by different means. For rennet-coagulated cheese, the pH is lowered to between 6.7 and 6.3, and rennet is added to induce protein coagulation. Rennet is a mixture of proteases, which is mainly comprised of aspartic endopeptidase chymosin. Chymosin cleaves the glycomacropeptide from κ-casein, resulting in the aggregation of the casein micelles and the formation of calcium-bridges. For acid-precipitated cheeses, the pH is adjusted to 4.6 or slightly higher if small amounts of rennet are added. For some cheeses, such as Ricotta, acidification is combined with a thermal treatment to achieve the required textural attributes in the end product.
- *Curd processing*: After sufficient casein aggregation is achieved, the coagulum can be cut into smaller pieces to facilitate the release of the fluid whey from the curd. The size of these pieces influences the hardness of the final product: the smaller the size, the harder the cheese. For some cheeses, however, the curd is not cut, *e.g.* quark. The amount of whey released also depends on the time when the curd is cut. Shorter cutting times are used for soft cheeses that have high moisture contents, whereas longer ones are used for semi-soft and hard cheeses that have lower moisture contents. For some cheeses, syneresis is further promoted by heating the curd to around 36 to 55 °C to increase the hydrophobic interactions between the casein molecules, thereby promoting the expulsion of the fluid whey. Part of the lactic acid and lactose in the cheese can be removed by using an additional washing step so as to prevent a further drop in pH (if required). Subsequently, the curd is transferred into forms, pressed, and (depending on cheese type) incubated in a salt bath.
- *Ripening*: Cheese ripening is typically performed at 10–25 °C for a period that may range from several days to several weeks. Typically, lower temperatures are

used after the initial ripening stage when the ripening is extended for several weeks or months. The water loss during the ripening process induces the formation of a natural rind and increases the hardness of the cheese (if not ripened in plastic bags). The protein-fat globule network in the cheese is partially broken down by proteolysis and lipolysis during the ripening phase, which changes the microstructure, texture, and flavor. Moreover, glycolysis converts lactose and lactate to intermediate compounds in the first few days. The hydrolyzed intermediates resulting from these processes are further converted into a variety of volatile compounds that are responsible for the unique flavor profiles of different cheeses. These processes are a result of indigenous active enzymes, as well as the bacteria still present in the milk or added as starter cultures. The diverse range of flavors exhibited by different kinds of cheeses is mainly due to the different kinds of aromatic molecules produced by the different enzymes and microbes utilized in the cheese ripening process. For the production of some cheeses, the surface of the cheese is smeared with the ripening culture such as yeasts and bacteria (Dugat-Bony et al., 2015). Yeast is usually the first microbial species to grow on the cheese rind. They metabolize the lactic acid produced by the starter cultures and deaminate amino acids, which causes the pH to increase, thereby facilitating subsequent bacterial growth.

The above described operations result in the production of a viscoelastic semi-solid product that has a characteristic appearance, texture, and flavor profile. The majority of commercial dairy cheeses follow this overarching processing scheme but each with some variations (as described in the next section). This process has been developed, improved, and fine-tuned over several centuries. In principle, plant-based cheese products can be produced using an analogous process. However, they may also be created using different processing routes, which are described later.

9.3.3 Production of Cheese Varieties

The overarching cheese production process described in the previous section can be fine-tuned to obtain different cheese varieties. In principle, each of the outlined materials and operations (milk, pasteurization, standardization, acidification, coagulation, curd processing, and ripening) can be altered to obtain a cheese product with different characteristics (Coker et al., 2005). This leads to a great variety of parameters that can be changed, as outlined in Fig. 9.1. As an example, camembert and cheddar, which have distinctly different textures and flavors, will be described here. More details about variations in the properties of these and other cheeses are described by Kammerlehner (2009).

- *Camembert*: Camembert is a soft cheese with a fat in dry matter content of around 45–60% and a NaCl content of below 2%. Initially, the milk is standardized to a fat content of around 3% or higher and then pasteurized at 72 °C for 15 s (if not made from raw milk). After cooling to 30–32 °C, starter cultures (such as *Lactococcus lactis* subsp. *lactis* and *cremoris*) and surface ripening cultures (such as *Penicillium candidum*, *Geotrichum candidum*, and *Kluyveromyces marxianus*) are added. The milk is then fermented until a pH of around 6.5–6.3 is obtained. Rennet is added and the mixture is incubated for 15–20 min. The coagulum formed is cut into pieces of about 15 to 20 mm. Around 10–20% of the initial amount is separated at a pH of around 5–6 as whey before the curd is mechanically treated to foster syneresis and then another 15–30% of whey is separated. The resulting curd is transported into cheese molds and formed for 10–20 h without the addition of pressure. Subsequently, the cheese (pH 4.5–5.2) is incubated for around 100 min in a salt solution at a NaCl concentration of 16–18% and a pH of 4.9 at 16–20 °C. Finally, the cheese is dried and stored for 6–11 days at 12–17 °C and 85–95% relative humidity to initiate and promote the growth of the surface mold. The camembert cheese produced has a final pH of around 7 in the center after production and at the surface after longer storage.

- *Cheddar*: Cheddar is a hard cheese with a fat in dry matter content of >48%. The raw milk is standardized to around 3.2–3.5% fat and is pasteurized at 72 °C for 15 s, followed by cooling to around 30 °C and the addition of starter cultures (typically *Lactococcus lactis subsp. cremoris*). The milk is incubated until a pH of 6.6–6.5 is reached and then rennet is added. After around 45 min, the coagulum is cut into pieces of 4–5 mm and the curd is stirred for 20–25 min. Subsequently, the curd is heated to 38 °C for 20–25 min followed by 50–60% (from initial mass) of whey being removed at a pH of around 6.0. The curd is then set to rest, followed by cutting it into blocks that are turned several times to foster the removal of the whey. Finally, the blocks are cut again into pieces of around 4 mm at a pH of 5.3–5.4. Salt is then added (3%) and the cheese is transferred into forms. The cheese is pressed by its own weight or by the addition of pressure, dried at the surface, and aged in paraffin foil for up to several months, after which it reaches a final pH of around 5.0–5.4.

The great variety of dairy cheeses available to consumers is the result of fine-tuning these processes that convert milk into products with characteristic physicochemical and sensory properties, such as appearance, texture, and flavor.

These processes result in a typical microstructure that makes cheeses a popular food. As an example, the resulting microstructure of two cheese types is shown in Fig. 9.2, with Parmigiano Reggiano having a continuous rennet-induced casein protein network with embedded partially coalesced fat globules, whereas cream cheese consists of a corpuscular structure with small clusters of fat globules that are surrounded by protein aggregates and with pores that are filled with whey (Wolfschoon Pombo, 2021). In principle, some of the same approaches could be used to create a diverse range of plant-based cheese types but further work is still needed in this area.

9.3 Animal-Based Cheeses

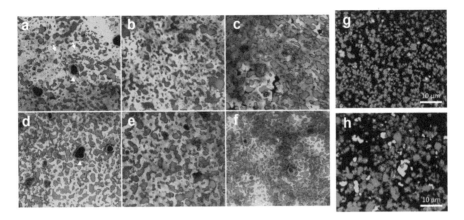

Fig. 9.2 Microstructure of Parmigiano Reggiano during ripening from 12 to 50 months (**a–f**) and cream cheese with different stabilizers (carrageenan, locust bean gum; *g*) and (carrageenan, locust bean gum, gelatin, citrus fiber; (**h**) illustrated by confocal laser scanning microscopy. Arrows show intact fat globules. Green color illustrates protein; red illustrates fat globules. (Modified from Alinovi et al. 2020; D'Incecco et al. 2020, under CC BY 4.0 (http://creativecommons.org/licenses/by/4.0/))

9.3.4 Key Physicochemical Properties

A cheese manufacturing process is designed to convert milk into a specific type of cheese product. Different kinds of cheeses are expected to have different material and functional characteristics. For example, cheeses used as pizza toppings should have appropriate meltability and stretchability. In this section, we review some of the key characteristics of cheeses, especially their texture, appearance, melting properties, flavor characteristics, and sliceability as well as the methods used to measure these parameters.

9.3.4.1 Textural Properties and Appearance

When designing plant-based cheeses it is important to match the color and texture of real cheeses. Representative measurements of the color ($L^*a^*b^*$) of a number of animal- and plant-based cheeses are summarized in Table 9.2. These results show that the color of real cheddar is moderately light ($L^* = 53$–58), slightly reddish ($a^* = +6$ to $+7$), and strongly yellowish ($b^* = +28$ to $+32$), whereas that of the plant-based cheese analogs is somewhat different: $L^* = 36$–51, $a^* = +6$ to $+23$, $b^* = +24$ to $+44$.

In terms of textural attributes, cheeses exhibit both viscous and elastic properties and can often be categorized as viscoelastic solids (Chap. 4). The textural properties are commonly analyzed by uniaxial tension and tensile tests, as well as by other rheological techniques, such as dynamic shear rheometry. Often, texture profile

Table 9.2 Comparison of some color and textural attributes of animal-based and plant-based cheeses

Product	Color values			Texture profile analysis					
	L^*	a^*	b^*	Hardness (g)	Adhesiveness (g s)	Resilence (%)	Cohesiveness	Springiness %	Chewiness
Grafton village 2 year aged cheddar cheese	58.0	5.7	31.7	15,300	−489	3.82	0.14	39.5	845
Kerrygold Skellig sweet natural cheese	53.6	7.4	28.8	15,300	−269	2.78	0.10	23.7	367
Kerrygold reserve cheddar cheese	55.6	6.0	28.0	16,000	−180	3.1	0.10	27.3	439
Miyoko's creamery cultured vegan cheese block	35.6	17.1	32.7	18,000	−245	9.9	0.24	38.8	1680
Violife epic mature cheddar flavor block	51.2	5.8	24.4	19,200	−31.2	24.2	0.43	41.2	3290
Violife just like Cheddar slice 100% vegan	46.3	22.1	41.8	16,100	−23.3	41.3	0.74	42.0	4970
Daiya deliciously dairy-free cheddar style slices	45.0	23.4	44.1	8380	−144.1	4.9	0.09	36.9	282
365 by whole foods market plant-based cheddar cheese alternative	43.5	18.6	32.8	14,000	−56.1	31.4	0.62	36.7	3200

Description of texture profile attributes can be found in Chap. 4. Measurements of texture and color of dairy and plant-based cheeses made in the authors' laboratory. The authors thank Kanokporn Leethanapanich (UMass) for kindly providing the data

9.3 Animal-Based Cheeses

analyses (double compression tests) are carried out to obtain insights into the textural attributes of cheeses, which can range from soft to hard depending on the cheese type (Table 9.2). These results show that the hardness of different plant-based cheeses (8.4–19 kg) is somewhat similar to that of real cheddar cheese (15–16 kg). The hardness provides insights into the resistance to deformation, whereas other important parameters of these measurements are: (*i*) *springiness*, which describes the ability of the material to recover from deformation and spring back after the first deformation; (*ii*) *adhesiveness*, which is the tendency for a material to stick to another material (*e.g.,* tongue, knife); (*iii*) *cohesiveness*, which is the resistance to fracture into pieces and the ability of the material to withstand a second deformation; and (*iv*) *chewiness*, which is resistance to disintegration during chewing (Fox et al., 2017). For example, mozzarella should have a low hardness and high cohesiveness, but cheddar cheese should have opposite properties.

More insights into the viscoelastic properties of cheeses can be obtained by using dynamic shear rheology. Figure 9.3 shows an amplitude sweep carried out for cheddar cheese and two plant-based cheeses. Here, the linear viscoelastic region (LVR) of a cheese product is determined by measuring G' and G'' *versus* the applied strain (or stress). Within the LVR, the cheese structure is not irreversibly damaged by the

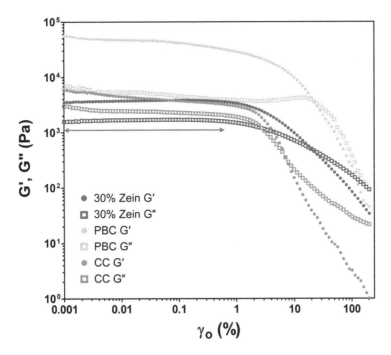

Fig. 9.3 Amplitude sweeps of regular cheddar cheese (CC), plant-based cheddar cheese (PBC), and 30% zein 'cheese'. The horizontal arrow indicates the approximate range of the linear viscoelastic region of the zein cheese. Plant-based cheeses can exhibit quite different textural attributes as shown here with higher solid-like characteristics (G') and different yield stress behavior (limit of the linear viscoelastic range) compared to regular cheese. (Modified and reprinted with permission of Elsevier from Mattice and Marangoni 2020)

applied stress but above this region, it breaks down. The endpoint of the LVR is the yield point, which describes the minimum stress required to break down the material's structure and soften the material. As shown in Fig. 9.3, plant-based cheeses can have considerably different moduli and yield points, which results in different sensations during cutting, chewing, and processing. Moreover, G' is greater than G'' because the cheeses are predominantly elastic-like materials (at relatively low strains). The higher moduli and yield point of the plant-based cheddar cheese indicates that it is more resistant to deformation and that the plastic deformation requires more force to be applied.

Other methods used to describe the viscoelastic behavior of cheeses include creep and stress relaxation tests. These tests can be performed under shear or normal force conditions. In these tests, the material is deformed in the linear viscoelastic range under constant stress (creep) or strain (relaxation) and the resulting strain or stress is measured, respectively (see Chap. 4). To gain more insights into such material behavior, several models have been used to describe the viscoelastic rheological properties of cheeses (Muthukumarappan & Swamy, 2017):

- *Maxwell model*: Describes the rheology of a material using a combination of a spring and a dashpot in serial connection. It is suitable for modeling stress relaxation experiments *i.e.*, measurements of the change in stress over time when a material is deformed to a constant point. The Maxwell model has been used to describe the rheology of regular and reduced-fat cheeses.
- *Voigt-Kelvin model*: Describes the rheology of a material using a combination of numerous springs and dashpots arranged into specific serial and parallel arrangements (Fig. 9.4). It is suitable for modeling creep recovery experiments *i.e.*, mea-

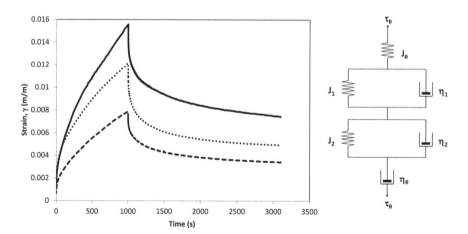

Fig. 9.4 Cheeses are viscoelastic materials. Their viscoelastic properties can be analyzed by creep and relaxation tests (see Chap. 4), which can be characterized by different models. Shown here is a creep test for different mozzarella cheeses that follows a stress step from $\tau = 0$ to τ_0 followed by maintaining τ_0 and measuring the resulting deformation with subsequent relaxation. Such behavior can be modeled by a series of Voigt-Kelvin elements of parallel springs and dashes as shown. (Reprinted with permission of Elsevier from Sharma et al. 2017)

surements of the deformation of a material when a constant stress is applied. The Voigt-Kelvin model has been used to describe the rheology of cheddar and mozzarella cheeses.
- *Peleg model:* This model uses the following expression to describe stress-relaxation and creep curves:

$$\frac{t}{Y(t)} = \frac{1}{ab} + \frac{t}{a} \tag{9.1}$$

Here, $Y(t)$ is the decay parameter, t is the relaxation time, a is the extent of stress decay during relaxation, and b is the rate of stress relaxation (Peleg, 1979). If $a = 0$ and $b = 0$, then the stress does not relax at all (ideal elastic solid). Conversely, if $a = 1$ and $b = 1$, then the stress level reaches zero (ideal liquid). This model has been used to describe the rheology of Edam cheese.

Several parameters influence the final textural attributes of cheese products. Some of the most important ones are the composition of the milk (standardization), milk pretreatments such as homogenization (for some cheeses) and thermal processing (for most cheeses), and operating variables (such as rennet concentration, calcium concentration, sodium concentration, pH, and amount of whey removed). For instance, the fortification of cheese with calcium and phosphorous before rennet addition can result in an increase in hardness, springiness, cohesiveness, resilience, and chewiness (Chevanan et al., 2006).

The ripening phase is crucial for controlling the textural attributes of cheeses (Irudayaraj, 1999). This is because it induces partial proteolysis of casein, which alters the protein network and liberates peptides from the caseins. These processes can increase or decrease the hardness of cheese depending on the pH and the water-to-casein ratio (Lawrence et al., 1987). For instance, a cheddar cheese tends to harden as a result of proteolysis, whereas a camembert cheese tends to soften and develop a creamy texture because of the diffusion of calcium to the outer layers and the increase in pH, which increases protein and decreases calcium solubility. These differences have mainly been related to the water-to-casein ratio, which is low in cheddar and high in Camembert (Irudayaraj, 1999; Schlesser et al., 1992). It should be noted that some cheeses are not ripened (*e.g.,* ricotta, cottage, and fresh pasta filata cheese) and their textural attributes are mainly achieved using other processing steps.

Another important factor that alters the textural attributes of cheese is the pH, which is due to its influence on a number of physicochemical parameters. The electrical charge, solubility, and interactions of milk proteins are influenced by pH, which alters the nature of the 3D protein network formed. Moreover, the water-solubility of calcium depends on pH, which influences its ability to promote interactions between protein molecules. Typically, a less hard structure is obtained for rennet-induced cheeses that have a pH closer to the isoelectric point (pI) of the caseins, because more calcium is solubilized and so the casein network is weakened (Pastorino et al., 2003). For example, when the draining-pH during Camembert production is sufficiently low, more calcium is solubilized within the whey fraction. When the pH increases again during the ripening phase, the cheese softens and can even liquify because there are then fewer calcium ions available to form bridges between the casein molecules. This leads to greater protein solubilization during the ripening phase and therefore a weaker gel strength (Batty et al., 2019).

9.3.4.2 Meltability

The melting behavior is another key quality attribute of many cheese types. Most cheeses tend to soften in the temperature range from 30 to 75 °C (Fig. 9.5) (Karoui et al., 2003; Ray et al., 2016; Schenkel et al., 2013a, b). However, some cheese types do not melt upon heating because they contain strong protein networks that are held together by numerous strong covalent bonds (Lucey et al., 2003). The melting characteristics of cheeses are especially important for those used as pizza toppings. The gel-sol transition responsible for cheese melting is mainly the result of two

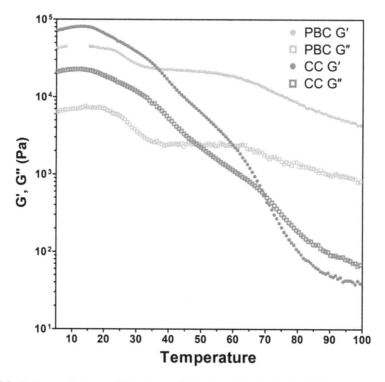

Fig. 9.5 Each type of cheese exhibits its own distinctive melting behavior. Differences in the melting behavior of a cheddar cheese (CC) and a plant-based cheddar cheese (PBC) are shown here by measuring changes in the elastic and shear modulus *versus* temperature. The cheddar cheese transforms from a mainly solid material to a mainly liquid material ($G' = G''$) at around 70 °C. (Reprinted with permission of Elsevier from Mattice and Marangoni 2020)

physicochemical changes that occur during heating (Schenkel et al., 2013b). First, up to 40 °C, the softening is mainly due to the melting of fat crystal networks. If the temperature is raised further, the interactions between the individual casein molecules change because of an increase in the strength of the hydrophobic attractions at elevated temperatures. This causes a contraction and weakening of the 3D casein network and thus favors the softening (Lucey et al., 2003). In general, any changes in the environmental conditions that modify the casein-casein interactions can alter the melting behavior of cheese.

The melting characteristics of cheeses can be controlled by adding emulsifying salts, such as citrates (*e.g.,* sodium citrate) and phosphates (*e.g.,* sodium phosphate). The addition of emulsifying salts increases the meltability of cheeses by reducing the strength of the casein network through several mechanisms: (*i*) reducing calcium bridge formation between casein molecules by binding calcium ions; (*ii*) increasing protein solubility by increasing the pH away from the isoelectric point; (*iii*), increasing protein solubility through a salting-in effect (Guinee, 2017). The type and amount of emulsifying salts used must be carefully controlled to obtain the required physicochemical attributes in the final product.

9.3.4.3 Flavor Profile

The development of highly flavorful cheeses from relatively bland milks involves a series of complex chemical and biochemical transformations, which depend on the ingredients, processes, and ripening conditions used. The origin of the raw milk used for cheese production has an important influence on the final flavor profile. The type (*e.g.,* cow, sheep, goat, or buffalo), feed, age, and health of the animal all play an important role in the development of cheese flavor (Faulkner et al., 2018). As an example, pasture feeding leads to an increase in the concentration of β-carotene in milk, which can degrade into *p*-cresol, giving a "barny"-like flavor (Kilcawley et al., 2018). Moreover, the processing steps used in cheese manufacturing influence the development of the cheese flavor. It is well known that cheeses produced from raw milk, which does not undergo any heat treatment step, develop a different flavor than the ones produced from pasteurized milk. For example, when French cheeses (Brie, Camembert, Saint Nectaire, etc.) are produced from raw milk, the overall aroma intensity and its flavor characteristics are changed. In particular, the sharp, butyric, bitter, goaty, and moldy flavor attributes increased for the non-pasteurized cheeses (Chambers et al., 2010). Another important factor that influences the cheese flavor is the fermentation and transformation of the ingredients during the ripening stage. This step involves the production and release of specific compounds derived from lipids, proteins, and carbohydrates. The enzymes that are involved in this operation can be indigenous enzymes, exogenous enzymes (rennet), and/or produced by the microflora. These enzymes convert the ingredients into various different molecules including carboxylic acids, lactones, ketones, alcohols, and aldehydes, which contribute to the flavor profile (Ianni et al., 2020).

The lipids, proteins, and carbohydrates are all susceptible to enzymatic degradation, leading to different kinds of characteristic flavor compounds being produced during the ripening stage. Free fatty acids are released from lipids by hydrolysis of triacylglycerol molecules. Short-chain fatty acids, which are common in milk fat, directly contribute to the flavor profile of cheese (Collins et al., 2003). Additionally, these free fatty acids are further transformed by reactions such as esterification, intramolecular esterification of hydroxyacids, partial β-oxidation, and thiol reactions. The resulting volatile compounds include butyl esters, γ- and δ-lactones, 2-methyl ketones, and thioesters. However, lipid oxidation reactions are not usually a major driver in the development of volatile compounds in cheese (McSweeney, 2004). Proteins are hydrolyzed into peptides and amino acids by the action of proteases, which influence the taste and aroma (Laska, 2010). Additionally, different enzymes (deaminases, transaminases, lyases, and decarboxylases) catalyze the continuous breakdown of the peptides and amino acids. This leads to the development of volatile compounds such as carbonyls, keto acids, sulfur-containing metabolites, and amines (McSweeney, 2017). Finally, lactose can be degraded by glycolysis, which produces lactate – if the right microorganisms are available. The lactose can subsequently be further transformed into volatile compounds, namely propionate, and acetate (Eugster et al., 2019).

9.3.4.4 Shreddability

Some cheeses are sold in a shredded form because it increases their ease of use. In particular, consumers and food manufacturers use shredded cheeses because they can be evenly applied to the surfaces of foods (*e.g.,* on pizza) and because they increase the melting speed (Apostolopoulos & Marshall, 1994). For a cheese to be shreddable, the viscoelastic behavior needs to be carefully controlled. If a cheese is too soft, has a high moisture content, a low elastic modulus, or a high surface energy, it will exhibit strong blade adhesion during cutting, thereby adversely influencing its shreddability (Childs et al., 2007). On the other hand, if the cheese is too hard, has a low moisture content, and has a high elastic modulus, it is often too gummy and dry for efficient shredding because it undergoes excessive shattering during the shredding process (Kindstedt, 1995). Other studies showed that processing operations that prevented the formation of brittle structures enhanced the shreddability of cheeses (Banville et al., 2013).

For regular commercial cheeses, common shredding problems are the formation of fines and the adhesion of the cheese to the blade (Childs et al., 2007). These challenges should therefore be considered during the product development stage. In the future, it will be important to systematically evaluate the factors that govern the shreddability of plant-based cheeses.

9.4 Plant-Based Cheese Ingredients

Plant-based cheeses are produced from a variety of raw materials. In contrast to plant-based meat alternatives, the primary ingredient is not always protein but can also be starch. In this section, we will provide a brief summary of the main ingredients used to formulate plant-based cheeses and review their properties. For more information, the reader is referred to the chapter on plant-based ingredients (Chap. 2) and recent review articles and books (BeMiller & Whistler, 2009; Day, 2013; Grossmann & Weiss, 2021; Kyriakopoulou et al., 2021; McClements et al., 2019, 2021; Nadathur et al., 2016; Zia-ud-Din et al., 2017).

9.4.1 Polysaccharides

Plant-based cheeses formulated from polysaccharides usually use starches as the main functional ingredient (Table 9.1). Starch is a polysaccharide consisting of chains of glucose molecules linked by α-1–4 (amylose) or α-1–4 and α-1–6 (amylopectin) glycosidic bonds. Amylose is primarily a linear polysaccharide, whereas amylopectin is a highly branched one. The ratio of amylose-to-amylopectin, as well as their molecular weight and degree of branching, vary among different botanical sources, which alters their functional performances in end products. In plant-based cheeses, starches extracted from tapioca, potato, and corn are usually employed:

- Tapioca starch is obtained from the root of *Manihot esculenta*, which is also known as cassava or manioc. The starch granules in the roots have diameters in the range of 4–35 µm. The starch is extracted by washing and cutting the roots, leading to the formation of a slurry. The starch granules are then fractionated from the slurry by a sequence of screens with decreasing pore sizes, hydrocyclones, and decanter centrifuges. The obtained fractions are dried in the last step to yield a starch powder (Breuninger et al., 2009).
- Potato starch is produced using similar operation procedures. The potatoes are ground and the resulting potato juice is fed into a sequence of centrifugal sieves, centrifugal separators, and hydrocyclones to separate the starch granules (1–120 µm) from the fibers (80–500 µm) (Grommers & van der Krogt, 2009).
- Corn starch is generated by immersing corn kernels in an aqueous solution, followed by wet milling, cyclone separation, screening, centrifuging, and filtration steps to liberate the starch from the kernel, mill the endosperm, and separate the granules from the germ, proteins, and fibers (Eckhoff & Watson, 2009).

Starches are primarily used in plant-based cheeses because of their thickening, gelling, and water-holding properties. Upon heating in water, the starch granules swell, the viscosity increases and a gel may be formed. During cooling, water and other ingredients are trapped within the 3D polymer network formed by the hydrogen-bonded starch molecules, which is known as retrogradation or setback (Kasprzak et al., 2018). In more detail, when starch granules are dispersed in water and heated, the starch granules start to absorb water and swell. Swelling causes an increase in the effective volume of the granules, which increases the shear viscosity (Chap. 4). If the temperature is further increased, a critical swelling ratio is obtained and the starch granules start to disintegrate, which results in the release of starch molecules (mainly amylose). The point at which this phenomenon occurs is usually called the 'pasting temperature', which can be determined experimentally as the temperature at which the maximum viscosity is observed during heating. The pasting temperature is influenced by the amylose-to-amylopectin ratio and is therefore not the same for every starch ingredient. The reported pasting temperatures for starches used in plant-based cheeses are 63 °C for tapioca, 64 °C for potato, and 80 °C for corn (Taggart & Mitchell, 2009). When the solution is cooled again, the released amylose molecules associate and form crosslinks with each other through hydrogen bonding. As a result, a 3D network of aggregated starch molecules is formed that provides viscoelastic properties to the end product.

Amylose and amylopectin influence the properties of materials in different ways. Amylose has low thickening properties but is able to form strong irreversible gels due to a high degree of retrogradation. In contrast, amylopectin has high thickening properties but only forms weak and reversible gels (Schirmer et al., 2015). The thickening properties are further influenced by the granule size. Plants that produce larger granules (*e.g.,* potatoes) possess higher thickening properties, especially during heating (Schirmer et al., 2013).

Because starches from different origins have different functional properties, several kinds of starches are often used in the preparation of plant-based cheeses to create an optimum degree of retrogradation, achieve an optimum softening temperature, and final gel strength as well as viscoelastic properties. For plant-based cheese applications, starches should gelatinize into a stretchy and formable mass and build a viscoelastic gel upon cooling *via* partial retrogradation. Recent studies indicate that especially tapioca starch exhibits these desirable attributes (Mattice & Marangoni, 2020). Ultimately, starch functionality depends on granule size, amylose:amylopectin ratio, the molecular characteristics of amylose and amylopectin, and any physical, chemical, or enzymatic modifications of the starch (Breuninger et al., 2009; Schirmer et al., 2015). For instance, some starches naturally have a high tendency for retrogradation (*e.g.,* corn starch), whereas others have a much lower tendency (*e.g.,* potato and tapioca starch), which influences the final texture of plant-based cheeses (Jackson, 2003). Moreover, waxy starches that have high amylopectin contents are commonly employed in plant-based cheese formulations, which is probably because their decreased tendency for retrogradation facilitates the creation of a softer texture. Moreover, such waxy starches have been shown to have excellent melting properties, which is an important property of some cheeses. This kind of behavior has been reported for waxy potato and rice starches but also for tapioca starches, and the melting characteristics can be enhanced by using mixtures of such starches and proteins (Bergsma, 2017). Finally, the employed starches can be modified to various degrees to change their physicochemical and functional properties: the molecular weight of starches can be reduced by acid or enzyme hydrolysis to decrease the paste viscosity during gelatinization and increase the gel strength upon retrogradation; the hydrophobicity of starches can be increased by octenyl succinate derivatization, which reduces oiling-off effects; and the interior of starch granules can be covalently crosslinked to increase their heat and shear stability, which is important for some applications (Klemaszewski et al., 2016; Taggart & Mitchell, 2009).

9.4.2 Proteins

Currently, the main plant proteins used in the formulation of plant-based cheese products are from soy, pea, lupin, potato, nuts, and corn. The most important functional properties that the proteins need to have for plant-based cheese applications are emulsification, gelation, and water holding. However, they may also play an important role in the quality attributes of the end product due to their role as flavor precursors. Consequently, manufacturers should select the most appropriate plant protein, or combination of plant proteins, to obtain the physicochemical, functional, and sensory attributes required in the end product. This often relies on having a good understanding of their solubilities under different pH and ionic strength conditions, their interfacial activity and stabilization behavior, their thermal denaturation temperatures, their gelation properties, their capabilities to retain liquids, and their accessibilities to enzymatic modifications such as proteolysis and crosslinking. The most important functional properties of plant proteins are discussed in more detail in Chaps. 2 and 4, as well as in various review articles (Day, 2013; McClements & Grossmann, 2021; Nadathur et al., 2016). The most important attributes of plant proteins for plant-based cheese applications are summarized in Table 9.3.

Of these proteins, especially zein has recently been proven useful in the formulation of plant-based cheeses. Zein is extracted from the endosperm of corn and is classified as a prolamin that is insoluble in water but soluble in concentrated aqueous ethanol solutions. Industrially, zein is usually obtained from corn by milling followed by solvent extraction, which yields a number of protein subfractions: α-zein, β-zein, γ-zein, and δ-zein. Of these subfractions, α-zein (21–26 kDa) is the most abundant one (Anderson & Lamsal, 2011). Zein is a hydrophobic storage protein that is produced by the plant to store lutein in a triple helix with a hydrophobic core (Anderson & Lamsal, 2011; Momany et al., 2006). In food applications, zein has been proven to be a useful protein for encapsulating hydrophobic compounds such as oil-soluble vitamins and nutraceuticals. Moreover, because zein is insoluble in water, it readily aggregates and forms protein particles that have been utilized to stabilize oil-water interfaces in emulsions through a Pickering effect (Fathi et al., 2018). When present in a dry state, zein has a glass transition temperature of around 139 °C but water plasticizes the protein (Madeka & Kokini, 1996). This enables the creation of a rubbery stretchable mass at temperatures well below the glass transition temperature. This phenomenon has been used to create plant-based cheese analogs with similar textural attributes as cheddar cheese (Mattice & Marangoni, 2020, 2021).

Table 9.3 Overview of some important protein properties that are utilized for the plant-based cheese formation

Source	Solubility	Denaturation	Enzymatic crosslinking	Liquid retention
Legume proteins	Minimum solubility at low ionic strengths: 7S globulins pH 4.5–7.0; 11S globulins pH 4.0–7.5.	Soy glycinin (11S) has a T_d 78–94 °C at pH 7.6; β-conglycinin has a T_d between 67 °C and 87 °C.	Pea, soy, and lupin proteins are susceptible to transglutaminase cross-linking.	Oil binding capacity is commonly highest for soy followed by lupin and pea protein.
	Solubility at high ionic strengths: 7S globulins have a solubility >90% at pH 3–9; solubility of most 11S globulins is low up to a pH of 5–6.	Pea legumin has a T_d 87 °C, total pea T_d 76–82 °C.		
		Lupin globulins denature at 94–114 °C.		
Potato proteins	Patatin has a minimum solubility at a pH of 3.5 and I = 200 mM, higher solubility at low ionic strengths (I = 0.15 mM); 16–25 kDa fraction has a very high solubility across a broad pH-range from 2.5 to 12.	Patatin denatures at T_d 59–60 °C at I = 10 mmol L^{-1} and pH 7; protease inhibitor fraction denatures at T_d 66–68 °C.	Potato proteins are susceptible for cross-linking with enzymes such as transglutaminase, peroxidase, laccase, and tyrosinase.	
Nut proteins	Cashew kernel proteins have a minimum solubility in the pH range from 4.0 to 5.0.	Cashew proteins gel after heat treatment at 100 °C with a least gelling concentration of 6.5 to 13.5%.		Nut proteins have an oil absorption capacity in the region of 3–4 g oil per g of protein, gelling concentration between 4–14%.
Zein proteins		Glass transition temperature of 139 °C.		

Modified from Grossmann and McClements (2021)

9.4.3 Fats

Fats and oils also play an important role in determining the physicochemical, functional, sensorial, and nutritional attributes of plant-based cheeses. In regular animal-milk cheese, the fats are mainly present as triacylglycerols (Brady, 2013). The main fatty acids in dairy fats are saturated (~70%), monounsaturated (~25%), and

9.4 Plant-Based Cheese Ingredients

polyunsaturated (~2.3%). Moreover, milk also contains considerable amounts of short-chain fatty acids, for example, butyric and caprylic acid. These short-chain fatty acids are volatile, which has important consequences for the flavor profile of animal-based cheeses (Macedo & Malcata, 1996; Månsson, 2008). Milk fat also has a high proportion of long-chain saturated fatty acids that tend to be partially crystalline at ambient temperatures, which is critical for the production of semi-solid cheeses. The fats in cheese products are usually present in an emulsified form and so they also contribute to the desirable appearances of these products due to their ability to scatter light waves (Chap. 4). In addition, they can act as solvents for nonpolar flavor molecules, which is important for designing the overall flavor profile of plant-based cheeses. Ideally, the fats in plant-based cheese products should provide similar textural attributes. It is therefore important to understand and control the solid fat content (SFC) of plant-based cheese formulations to obtain the desired characteristics (Fig. 9.6). For creating desirable textural and sensorial attributes in plant-based cheeses, many different plant oils and fats have been utilized, including avocado, canola, cocoa, coconut, corn, palm, safflower, sesame, soybean, and sunflower oils (Sha & Xiong, 2020). Only cocoa, coconut, and palm oil are solid at room temperature and so can be used to generate desirable solid-like textural attributes in cheese analogs when mixed with oils with a lower melting temperature.

In general, there is a great variety of commercially viable plant sources that can be used to obtain edible fats and oils, such as algae, canola, corn, flaxseeds, olives, palm, peanuts, safflowers, sunflowers, and vegetables. Because these plants grow in different regions of the world and store the fats in different parts of the plant, there is also a great diversity in their physicochemical characteristics, such as fatty acid

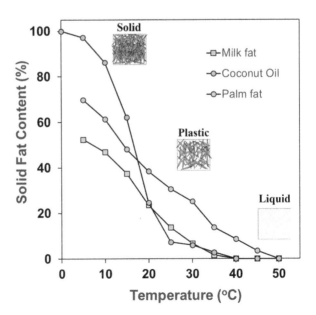

Fig. 9.6 The solid fat content (SFC)-temperature profile of edible fats depends on the fatty acid composition, which is determined by their biological origin. Plant-based fats should mimic the SFC profiles of animal-based ones. (Reprinted with permission of Elsevier from Grossmann and McClements 2021)

composition, melting point, and oxidative stability. The melting points of lipids tend to increase as the fatty acid chain lengths increase and the number of double bonds decreases. Most plant-based oils contain relatively high levels of unsaturated fatty acids, which means they tend to be fluid at room temperature (Chap. 2). As a result, they cannot provide the desirable textural and melting attributes required for certain types of cheese. The melting point of these liquid oils can be increased using hydrogenation to reduce the number of double bonds in the fatty acids. However, this method has become less popular because it results in the formation of unhealthy trans-fatty acids when the process is not carefully controlled (Hu et al., 2001).

Some plant oils naturally contain appreciable levels of saturated fatty acids, such as coconut oil and cocoa butter, which means they can be used to create textural and melting characteristics similar to those normally provided by milk fat. Typically, a plant-based solid fat (such as coconut oil) is mixed with a plant-based liquid oil (such as sunflower oil) to obtain a solid fat content *versus* temperature profile that resembles that of milk fat (Fig. 9.6). However, it is important to recognize that solid fats like coconut oil also contain relatively high amounts of saturated fats that might also have adverse effects on human health (Ludwig et al., 2018b). However, the health impacts of different kinds of saturated fats are still under debate (Chap. 5).

A higher content of unsaturated fatty acids in plant-based cheese formulations influences their nutritional characteristics and shelf life (McClements & Decker, 2017). Unsaturated fats, particularly polyunsaturated ones (like the ones found at high concentrations in flaxseed or algal oils), may have beneficial health effects (Saini & Keum, 2018; Shahidi & Ambigaipalan, 2018). A disadvantage of these fatty acids is that the high unsaturated fatty acid content results in a low melting temperature and a high susceptibility to lipid oxidation, which can adversely affect the texture and flavor (*via* rancidity) of these products (Arab-Tehrany et al., 2012; McClements & Decker, 2017; Nogueira et al., 2019). A number of strategies can be utilized to reduce oxidation of unsaturated lipids: reduce exposure to oxygen, heat, or light by controlling the storage conditions or using suitable packaging materials; reduce pro-oxidant contamination (*e.g.,* transition metal ions or lipoxygenase); incorporate antioxidants and chelating agents (*e.g.,* polyphenols or EDTA); and by utilizing food matrix engineering approaches (Jacobsen, 2015; Jacobsen et al., 2013; McClements & Decker, 2018).

9.5 Production of Plant-Based Cheese

The production of plant-based cheeses typically involves different processing routes compared to the production of animal-based ones. The main goal of these processes is to obtain plant-based products with similar attributes as the conventional cheese products they are designed to replace. This can be achieved by the use of different ingredients and processing operations, which will be reviewed in this section.

9.5.1 Overview of Production Methods

The production of animal-based cheeses begins with the selection of an appropriate feed material for the cows, *e.g.,* grass, soy, or cereals. These materials are broken down in the digestive tract of the cow, thereby releasing the building blocks that are absorbed by the animal and utilized for milk production. The cow is then milked and the fluid milk obtained is transformed into a cheese by using the processing operations discussed earlier: pasteurization, standardization, acidification, coagulation, curd formation, and ripening (depending on the cheese type). The most important process operation involves a sol-gel transition of the fluid milk into a solid cheese, which is initiated by enzymes and/or acidic fermentation. This is in contrast to the production of plant-based cheeses that follow different structuring principles:

- *Fractionation Route*: In this route, extracted and purified plant-based functional ingredients are used. Polysaccharides, proteins, and fats are obtained from different sources and recombined to achieve the desired composition. The proteins used in plant-based cheeses typically come from soybeans, peas, beans, lupines, and potatoes. Starch is the most commonly used polysaccharide, which is usually extracted from corn, pea, tapioca, or potato. Other polysaccharides can also be used as functional ingredients, such as pectin, guar gum, locust bean gum, cellulose, agar, alginate, carrageenan, and xanthan gum. Plant-based liquid oils and solid fats are typically extracted from soybean, sunflower, rapeseed, coconut, oil palm, or cocoa beans. The fractionated protein and/or polysaccharide ingredients are often solubilized in water and then blended with oil to create a plant-based oil-in-water emulsion that contains biopolymers in the aqueous phase. Finally, this mixture may be solidified using a variety of approaches, including heating, cooling, pH adjustment, enzyme addition, or salting-out mechanisms. Some of the most common physicochemical phenomena used to create such structures are: solubilization, denaturation, and aggregation of proteins; gelatinization and retrogradation of starches; coil-helix transitions of gums; and emulsification, melting, or crystallization of oils or fats. Overall, this route follows several kinds of phase transitions from the raw material to the final product when cheese analogs are produced: *solid* (plant raw material) → *liquid* (extraction/fractionation) → *solid* (dried ingredients) → *liquid* (emulsion) → *solid* (plant-based cheese) (Fig. 9.7).
- *Tissue Disruption Route*: In contrast to the fractionation route, the ingredients are not extracted and isolated from the plant material and recombined in the tissue disruption route. Instead, the whole plant material is used for cheese production. For instance, whole nuts have been used as an ingredient for the production of plant-based cheeses that did not undergo a preceding extraction step. In this processing route, the ingredients (commonly seeds) are soaked in aqueous solutions to weaken the shells and the cell wall followed by a homogenization step to produce a concentrated colloidal dispersion. This dispersion contains the compounds found in the raw material, namely oil bodies, plant tissue fragments, dissolved biopolymers, sugars, and salts. An additional separation step can then

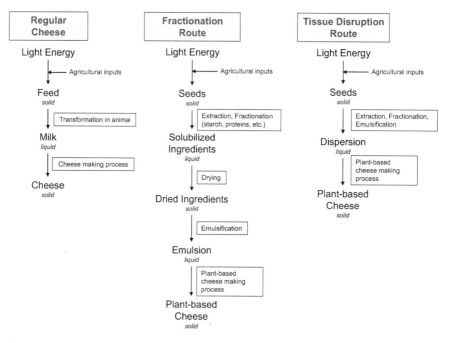

Fig. 9.7 Flow chart comparing the overall processing steps from the raw material to the final regular and plant-based solid viscoelastic cheese based on phase transitions. Notably, the fractionation route employs many phase transitions to obtain the end product. (Reprinted with permission from Elsevier, Grossmann and McClements 2021)

be added to separate any undesirable substances (*e.g.,* fibers) or to fine-tune the composition. To induce a sol-gel transition of this system, the same principles as described for the fractionation route can be employed. An important difference compared to the fractionation route is that only two-phase transitions are involved in this route from the raw material to the plant-cheese: solid (plant raw material) → liquid (dispersion) → solid (plant-based cheese) (Fig. 9.7). This processing route should therefore consume less energy. This was demonstrated in a recent study that investigated the environmental impact (global warming potential) of the various operations available for protein extraction and isolation and concluded that highly refined ingredients have the largest adverse impact on the environment (Lie-Piang et al., 2021).

The main difference between these two processing routes is related to the ingredients principally responsible for the sol-gel transition. In the fractionation route, the sol-gel transition is facilitated by isolated gelling ingredients that are sourced from various plant materials. In the tissue-disruption route, the sol-gel transition is caused by ingredients that originated from the original raw material. Both routes employ various sol-gel transition approaches to obtain a curd by different techniques: self-association; enzyme crosslinking; pH changes; adding salts; or thermal treatment. Some procedures also involve combinations of these different approaches (Table 9.4). In the following sections, we will describe these routes in more detail.

9.5 Production of Plant-Based Cheese

Table 9.4 Overview of studies that investigated the formation of plant-based cheese analogs

Ingredients	Pretreatment	Sol-gel transition	Properties	Reference
Fractionation route: Starch or protein-based emulsion gels (oil/fat crystal dispersed phase)				
Waxy non-modified potato starch, potato protein, sunflower oil		Gelatinization → setback	Semi-hard, melting	Bergsma (2017)
Modified, pregelatinized high-amylose corn starch, shortening, buffering salts		Gelatinization, melting → setback, crystallization	Semi-soft, good shredding, poor meltability	Zwiercan et al. (1987)
Different starches, fat, carrageenan, xanthan, guar gum, buffering salts, and others		Gelatinization, melting → setback, crystallization		Atapattu and Fannon (2014)
Zein, high oleic sunflower oil, coconut oil, starch, xanthan		Self-assembly, plasticizing, melting, gelatinization → plasticized protein network	High stretchability, weaker structure at elevated T	Mattice and Marangoni (2020)
Fractionation route: Protein-based emulsion gel (filled/particulate)				
Zein, potato protein, olive oil		Enzymatic crosslinking (tyrosinase)	Shear-thinning behavior, crumbly texture	Glusac et al. (2018)
Zein, pea protein, corn oil		Enzymatic crosslinking (tyrosinase)	Paste-like structure	Glusac et al. (2019)
Soy tofu, soy protein isolate, maltodextrin, palm oil		Enzymatic crosslinking (transglutaminase)	Cream-cheese style	Lim et al. (2011)
Pea-globulins, sunflower cream fraction, glucose		Enzymatic crosslinking (transglutaminase) + acidification by starter cultures		Holz-Schietinger et al. (2014)

(continued)

Table 9.4 (continued)

Ingredients	Pretreatment	Sol-gel transition	Properties	Reference
Soy globulins, sunflower oil/ moong bean 8S, palm oil, buffering salts/soy globulins, pea globulins, pea prolamins, xanthan	Heat treatment	Heat-aggregation/acidification with starter cultures	Melting, stretching properties	Holz-Schietinger et al. (2014)
Tissue disruption route: Protein-based emulsion gel (filled/particulate)				
Cashew-, soy milk analog	Heat-treatment	Ion gelation with multivalent cations	Semi-hard, protein content in dry matter up to 64%	Oyeyinka et al. (2019)
Lupine paste, oil, emulsifying salts		Heat aggregation	Protein content up to 14.9%, increased oil separation, firmness, and a lower springiness compared to control	Awad et al. (2014)
Soy milk analog	Heat-treatment	Acidification with starter cultures	Further processing to a soft cheese analog	Matias et al. (2014)
Soy milk analog	Heat-treatment	Acidification with starter cultures, ion gelation ($CaSO_4$)	Semi-hard, hardness higher compared to commercial sample	Chumchuere et al. (2000)
Soy milk analog, coconut oil, carrageenan, emulsifying salts	Heat-treatment	Acidification with starter cultures/glucono-delta-lacton	Cream-cheese-type, protein content up to 19.5%	Li et al. (2013)
Soy milk, soy oil, okara	Heat-treatment	Acidification with starter cultures	Cream-cheese-type, protein content up to 17.6%	Giri et al. (2018)
Soy milk	Heat treatment	Acidification with starter cultures	*L. rhamnosus* ferments soy oligosaccharides	Liu et al. (2006)

(continued)

9.5 Production of Plant-Based Cheese

Table 9.4 (continued)

Ingredients	Pretreatment	Sol-gel transition	Properties	Reference
Soy milk, pigeon pea milk	Heat treatment	Acidification with starter cultures, ion gelation with multivalent cations (CaCl$_2$)	Up to 58.7% protein in dry matter	Verma et al. (2005)
Soy milk	Heat treatment	Acidification with starter cultures, ion gelation with multivalent cations (Ca^{2+}-lactate)	Hard-cheese type, fat content of 11.80% after 3 months ripening	El-Ella (1980)
Soy milk	Heat treatment	Ion gelation with multivalent cations (CaSO$_4$), acidification with citric acid	Cream-cheese type after blending with palm oil and polysaccharides	Zulkurnain et al. (2008)
Soy milk	Heat treatment	Ion gelation with multivalent cations (alum)		Chikpah et al. (2015)
Macadamia-almond nut milk analog	Heat treatment	Acidification with starter cultures, enzymatic crosslinking (transglutaminase)	Fresh-, salted-, soft-ripened cheese types	Holz-Schietinger et al. (2014)
Soy milk	Heat treatment	Enzyme-rich *Moringa oleifera* extract	Soft-white cheese	Sánchez-Muñoz et al. (2017)
Soy milk	Heat treatment	Plant and microbial proteases	Only curd production	Murata et al. (1987)
Cashew kernels		Acidification with fermented quinoa dispersion	Brie, red, herb, cheddar, and blue cheese-like types	Chen et al. (2020)

Reprinted with permission of Elsevier from Grossmann and McClements (2021)

It should be noted that some of these methods only describe the formation of a plant-based cheese curd rather than a final cheese-like product (which typically involves an additional ripening stage).

9.5.2 Fractionation Route

The fractionation route involves the formation of an oil-in-water emulsion by blending functional ingredients that have been extracted and refined in previous processes. In plant-based cheeses, these ingredients are typically proteins, fats, and polysaccharides. The emulsion is subsequently treated to induce the sol-gel transition. Most commonly, a mixture of starches and fats, or a combination of proteins and fats is used to produce the desired cheese analog texture (Fig. 9.8).

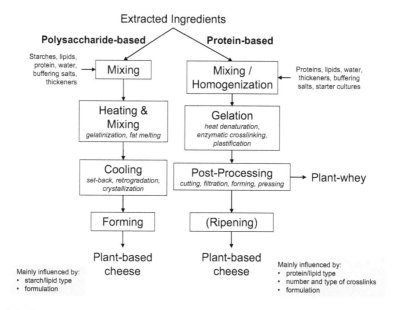

Fig. 9.8 Flow process chart comparing the different processing routes for plant-based cheese production for the fractionation route that is based on extracted polysaccharides (left) or proteins (right) from different raw materials. (Reprinted with permission of Elsevier from Grossmann and McClements 2021)

Polysaccharide-Based Starch is commonly used in polysaccharide-based cheese analogs but other polysaccharides such as alginate, carrageenan, or guar gum might be used as well as a gelling ingredient. The processes that yield such plant-based cheeses have been mainly described in patents and will be reviewed below (Atapattu & Fannon, 2014; Bergsma, 2017; Klemaszewski et al., 2016; Schelle et al., 2020; Zwiercan et al., 1987).

The process of producing plant-based cheeses from starch involves the gelatinization of the starch granules at elevated temperatures followed by a sol-gel transition upon cooling through retrogradation, which enables the setback and subsequent gel formation of the gelatinized starch molecules (Taggart & Mitchell, 2009). Because emulsified oil is incorporated during this process, a viscoelastic emulsion gel is created (network of connected biopolymers with embedded droplets (described as emulsion-filled gels), or a 3-dimensional lipid network formed from flocculated droplets). Often, this network consists of fat droplets and water trapped in a semi-solid 3-dimensional network formed by starch. The fats and oils used play a significant role in determining the final structure. The droplets or crystals can partially coalesce or fuse together, which contributes to the final structure of the product (Fig. 9.9a, b).

9.5 Production of Plant-Based Cheese

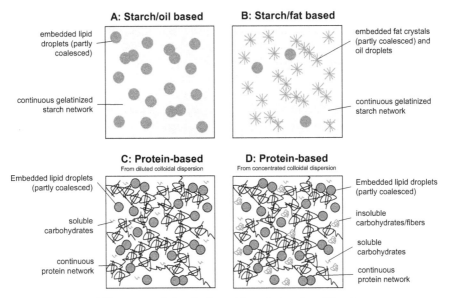

Fig. 9.9 Different viscoelastic emulsion gels are formed during plant-based cheese production. Starch structures are produced from the fractionation route, whereas protein-based gels may be produced from the fractionation or tissue-disruption routes. Depending on the process, protein-based structures can be distinguished as coming from a dilute colloidal dispersion (*e.g.*, a plant-based soymilk) that involves separation of insoluble particles or coming from a concentrated colloidal dispersion (*e.g.*, paste-like ground nuts) using the whole seed material. (Reprinted with permission from Elsevier, Grossmann and McClements 2021)

As mentioned above, the production of plant-based cheeses using starches involves an emulsification and heating step. The type and amount of starch and other ingredients used influence the final attributes of the product. For example, a starch-based cheese analog was obtained by mixing and heating a combination of starch (15–22% waxy potato starch), protein (0.5–8% potato protein), lipids (15–35% shortening), and water (35–75%) to 70–90 °C (Bergsma, 2017). In this patent, the melted and gelatinized emulsion was transferred into molds and the sol-gel transition was induced using a cooling step. Another starch-based cheese analog was patented using a similar approach, which involved mixing pre-gelatinized modified high amylose corn starch (>20%), fat (22.8%), buffering salts, and water above the melting temperature of the fat, and then cooling the dispersion to 4 °C to obtain a viscoelastic material (Zwiercan et al., 1987). Other authors have shown and patented that plant-based cheeses can be obtained using different kinds of starches, including native starch, acid-treated corn, octenyl succinate-modified starch, and hydroxypropyl distarch phosphate (Atapattu & Fannon, 2014). Other hydrocolloids were also added to modify the textural attributes of these products, like carrageenan and guar gum. Moreover, buffering salts were used to ensure the pH of the plant-based cheeses was close to that of real dairy cheeses (around pH 5–6). The authors in this patent utilized a production approach similar to that describe earlier. The hydrocolloids and fats were heated to 83 °C in the presence of water to dissolve and melt them. The ingredients were then blended to form an oil-in-water emulsion. The emulsion was cooled, which caused it to solidify due to the entanglement and crosslinking of the starch molecules and due to the crystallization of the fat phase (which had a melting point between 30–52 °C). The authors stated that the water content could range from 20% to 80% and the fat content from 15% to 30%.

Protein-Based Proteins (mainly caseins) are one of the most important constituents that are responsible for the sol-gel transitions in dairy-based cheeses (Lamichhane et al., 2018). Various plant proteins have been investigated for their potential to mimic the textural attributes provided by caseins in dairy products. However, this has proven challenging because plant proteins are usually large globular molecules, whereas caseins are small flexible molecules that can assemble into complex superstructures (casein micelles). Additionally, the proteins found in casein have posttranslational modifications such as phosphorylation and glycosylation. This enables them to bind calcium, to associate into clusters, and to generate repulsive interactions. The casein micelles have an average diameter of around 200 nm with a range from around 50–600 nm. It is challenging to mimic the properties of the casein molecules and micelles using plant proteins. Individual plant proteins are relatively small, *e.g.*, the 7S and 11S soy globulin protein fractions have a radius of gyration below 6 nm (Glantz et al., 2010; Guo et al., 2012). It is possible to form aggregates with a similar diameter as casein micelles (~200 nm) from these plant proteins by controlled heating at specific pH and ionic strength values. However, the functional attributes of these colloidal particles are usually different from those of casein micelles. These differences can be attributed to differences in their surface chemistries (such as charge or hydrophobicity) that alter their coagulation behavior (Chen et al., 2019).

For this reason, researchers are continuing to investigate the formation and properties of different kinds of plant protein aggregates so as to better mimic the functionality of casein micelles. For example, the textural attributes and the water holding capacity of the gels formed by such aggregates depend on their size, as well as the crosslinking method used to assemble them (Wang et al., 2017; Wu et al., 2019). Different crosslinking methods influence the type and number of crosslinks formed between the protein aggregates, which influences the characteristics of the obtained gels (Ni et al., 2015). This was shown in a study that utilized different gelation methods. Here, the textural attributes of plant-based gels (15% pea protein) depended on the gelation method used (Ben-Harb et al., 2018): (*i*) acid gelation by addition of glucono-delta-lactone; (*ii*) enzymatic gelation by addition of chymosin and transglutaminase; and (*iii*) heat-gelation by thermal treatment. Interestingly, all three gelling methods induced a sol-gel transition but the textural attributes of the gels formed were quite different. For instance, the highest elasticity was observed in the acid-induced gels, while the highest resistance to strain occurred in the

9.5 Production of Plant-Based Cheese

enzymatic-induced gels. Individual or combined approaches can therefore be used to create plant-based gels with different textural and sensorial characteristics, which has important consequences for designing plant-based cheeses that range from soft to hard. The same study also showed that those plant proteins could be used to obtain semi-solid cheese-like materials (Ben-Harb et al., 2018). It is therefore important to further investigate the mechanisms used to promote plant protein aggregation so as to obtain plant-based ingredients that can more closely mimic the properties of casein micelles.

Several processing techniques have been described to obtain protein-based cheese analogs using the fractionation route (Fig. 9.8). These processing techniques result in semisolid emulsion gels that are comprised of a three-dimensional continuous protein network with embedded fat droplets, water, and other ingredients (Fig. 9.9). The techniques employed can be categorized as self-association, enzymatic crosslinking, and acidification, which will be described in more detail below:

- Self-association techniques rely on the tendency for certain types of proteins to spontaneously associate with their neighbors when they are dispersed in water due to hydrophobic attraction, which is similar to casein when the glycomacropeptide is released. The most common food protein that aggregates based on this mechanism is zein (Glusac et al., 2018, 2019; Mattice & Marangoni, 2020). For example, Mattice and Marangoni (2020) mixed 30% zein with 1.5% fat, 2.8% starch, and 0.7% xanthan. This dispersion was then plasticized at 80 °C for 5 min, which led to the formation of a stretchable protein-rich material. Interestingly, the plant-based cheese obtained from this process had similar textural attributes as cheddar cheese when characterized by texture profile analysis and stretchability tests. However, one drawback of using zein for plant-based cheese formulations is its rather low level of essential amino acids and thus low protein quality (PDCAAS or DIAAS scores) (Boye et al., 2012).
- Enzymatic crosslinking utilizes specific enzymes (especially transglutaminase and tyrosinase) to promote the formation of protein-protein crosslinks and gelation. In a series of studies, 40% oil-in-water emulsions were produced by homogenizing an aqueous solution containing pea proteins with an oil phase containing zein (0.8%) (Glusac et al., 2018, 2019). The emulsions were then incubated with tyrosinase, because transglutaminase does not crosslink zein (Mattice & Marangoni, 2021). The presence of the zein and the addition of the tyrosinase considerably increased the elasticity of the emulsion gels. In a similar study, a mixture of potato protein and zein resulted in a non-crumbly paste that was suitable to create cheese analogs (Glusac et al., 2018). Soy proteins (10–20%) crosslinked with transglutaminase have also been used as a base material to form cheese analogs (Lim et al., 2011). Other researchers used a mixture of pea globulin (4%), sunflower cream fraction (20%), and starter cultures to create cheese analogs (Holz-Schietinger et al., 2014). In this patent, the transglutaminase was added after 1 h incubation with starter cultures. The system was then incubated until it reached a final pH of 4.2, the plant whey fraction was removed, and it was stored at 4 °C.

- Heat treatments can also be used to obtain semi-solid plant-based cheese analogs. For instance, a cheese analog was produced in a patent by blending 6% soy protein globulins and 20% of sunflower oil with water. The resulting oil-in-water emulsion was then heated and cooled down to obtain a plant-based cheese analog that was able to melt (Holz-Schietinger et al., 2014). The same authors combined acidification with a thermal treatment by blending 4% moong 8S proteins at pH 7.4 and 50 mM NaCl, mixed with 20% palm oil. The emulsion was inoculated with starter cultures and 3% sodium citrate after heating to 95 °C. Surprisingly, this procedure resulted in a reversibly melting plant-based cheese. In contrast, a stretchy cheese was produced by mixing soy globulins (4%), pea globulins (2%), and pea prolamins (2%), heating the mixture to 95 °C, and then adding starter cultures after the cooling phase. The high stretchability and melting behavior were linked to the addition of the prolamins.

In conclusion, a number of different fractionated ingredients and processing operations can be used to produce plant-based cheeses using the fractionation route. The processes discussed in this section lead to the formation of semi-solid emulsion gels consisting of starch or protein networks that trap fat droplets and fluids. It is likely that plant-based cheese analogs can be assembled from various other types and combinations of plant-derived ingredients, and further work is clearly needed in this area to create products that more closely mimic real cheeses.

9.5.3 Tissue Disruption Route

In the tissue disruption route, whole plant materials (such as seeds) are used to create plant-based cheese analogs (Fig. 9.10). Typically, the seeds are first soaked and then broken down to form a diluted (plant-based milk, Fig. 9.9c) or concentrated (plant-based cream, Fig. 9.9d) colloidal dispersion, which is subsequently transformed into a solid viscoelastic emulsion gel. The final dispersion may contain all of the components in the initial seeds or it may have undergone minimal processing to remove some undesirable components, such as insoluble particles or fibers (McClements et al., 2019). The colloidal dispersions can then be transformed into a cheese analog by inducing a sol-gel transition. Most commonly, a plant-based milk is used as a starting material (Table 9.4) and the sol-gel transition is triggered by heating, acidification, enzymatic crosslinking, salting-out, or some combination of these treatments. Common starting materials for forming cheese analogs using this approach are soy, pea, lupin, oat, and nut milks.

The majority of studies in this area have utilized a sol-gel transition that involves a combination of a heat treatment and one other technique, for example, a heat treatment combined with an acidification step (Table 9.4). In this procedure, a protein dispersion is first heated followed by a cold gelation step. The heat treatment causes the globular plant proteins to unfold, exposing functional groups to the surface, which causes them to aggregate with each other through hydrophobic and other

9.5 Production of Plant-Based Cheese

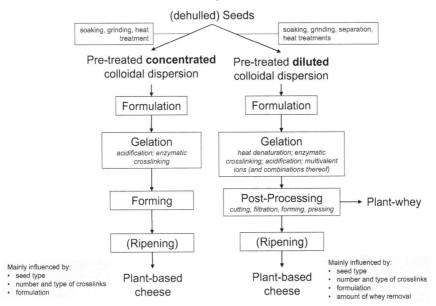

Fig. 9.10 Flow process chart comparing the different processing routes for plant-based cheese production for the tissue disruption route based on concentrated (i.e., a seed paste often based on nuts; left) or diluted (i.e., a plant-based milk; right) colloidal dispersions. (Reprinted with permission of Elsevier from Grossmann and McClements 2021)

attractive interactions (e.g., salt bridges) (Fig. 9.11) (Ni et al., 2015). The heat treatment also inactivates undesirable enzymes, like lipoxygenases that can produce off-flavors. It is important that the heating conditions (time and temperature) are carefully controlled to ensure that excessive protein aggregation and gelation do not occur during the heating step. The final sol-gel transition is typically promoted using another process that increases the attraction between the denatured or partially denatured protein molecules, such as adding mineral ions or altering the pH closer to the isoelectric point. These processes promote aggregation by reducing the electrostatic repulsion between the protein molecules, which have been 'activated' by the heat treatment (exposure of functional groups) (Zhang et al., 2018; Zheng et al., 2020).

It is possible to promote a sol-gel transition by only carrying out a heat treatment, without any subsequent steps as shown in the study by Awad et al. (2014). The authors employed lupin seeds as protein-rich raw materials for producing a plant-based cheese analog. The lupin seeds were soaked and ground to form a colloidal dispersion. Subsequently, oils (17.5%), water (11.8%), and emulsifying salts were incorporated and the dispersion was heated to 85–90 °C and held for 10 min. The hot liquid dispersion formed was then poured into molds and cooled down. However, the lupin cheese formed using this process was reported to have increased oil

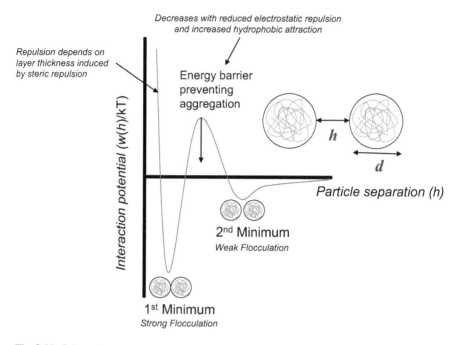

Fig. 9.11 Schematic representation of the interaction potential between two proteins modeled as colloidal spherical particles showing the primary minimum, secondary minimum, and energy barrier. For protein-based plant-based cheese production, the energy barrier has to be lowered by reducing electrostatic repulsion (salt addition or pH adjustment) and/or increasing hydrophobic interactions (thermal treatment) to induce protein aggregation by allowing them to come into close proximity. (Reprinted with permission of Elsevier from Grossmann and McClements 2021)

separation, higher firmness, lower springiness, and worse sensorial attributes than regular cheese.

Many researchers have utilized a two-step process to more closely mimic the properties of regular cheese. Fermented tofu is a typical product that has similar properties as cheese and similar processing operations are utilized during its production (Zhang et al., 2018; Zheng et al., 2020). Here, the soymilk is first heat-treated followed by an ion-gelation or an acidification approach. The soymilk is first heated to 65–95 °C to induce partial denaturation of the globular proteins and to expose the hydrophobic and other functional groups. Researchers have reported that disulfide bridges are not a major contributor to the tofu structure formed by this procedure (Kohyama et al., 1995). However, this process promotes protein unfolding and aggregation, but does not lead to the formation of a gel. The sol-gel transformation is then achieved by adding calcium ions and/or by adjusting the pH of the solution containing the heat-denatured soy proteins (Zhang et al., 2018; Zheng et al., 2020). After gel formation, the soy curd is pressed to release the soy whey, which is removed to yield a soft or hard tofu, depending on the amount of whey that is removed (Kao et al., 2003). The curd is further pressed and inoculated with strains of *Actinomucor* spp., *Rhizopus* spp., or mold *Mucor* spp. Similar to regular cheese, the starter cultures induce proteolysis of the soy protein, as well as promoting other chemical reactions, which leads to the generation of peptides, amino acids, and various flavor compounds (Liu et al., 2018). The utilization of different starting materials and microorganisms means that the sensorial properties of plant-based cheeses are different from those of animal-based ones. For this reason, researchers are examining new fermentation strategies to create cheese analogs that more closely match the flavor profiles of real cheeses.

Similar approaches that involve heating followed by ion gelation have been used to convert other kinds of plant-based milks into cheese analogs. In one study, soy and cashew nut milks were combined to improve the nutritional value of the final product (Oyeyinka et al., 2019). The processing scheme used to produce the plant milks involved soaking, milling, and then filtering the seeds to remove any insoluble compounds. The authors found a maximum in sensorial acceptance when 60% of soymilk was mixed with 40% of cashew nut milk. Similar to tofu production, the plant-based milks were boiled first and then ammonium aluminum sulfate was added to induce coagulation (Table 9.4). The proximate composition of the plant-based cheese produced was 64% protein and 6% fat (dry matter basis). This study suggests that cheese analogs with improved properties may be created by combining two or more plant-based raw materials.

Lactic acid fermentation (*i.e.*, acid curd preparation) is another approach for forming viscoelastic gels from colloidal dispersions of plant materials. Matias et al. (2014) have shown that gelation can be achieved by inoculating soymilk with appropriate starter cultures. For example, a plant-based cheese was obtained by heating soymilk to 95 °C for 10 min, cooling, and then adding *Bifidobacterium animalis* subsp. *lactis* Bb-12, *L. acidophilus* La-5, and *S. thermophilus*. Different structures and textures could be obtained by allowing the cultures to reach different final pH values. For example, a soft quark-like cheese was obtained at pH 5.7, while a harder cheese was obtained at pH 4.8 when the plant whey fraction was removed (Matias et al., 2014). The final pH is important because it influences the magnitude of the electrostatic interactions between the protein molecules, which influences their structural organization and resistance to deformation. This study showed that the textural properties of cheese analogs can be tuned by controlling the fermentation conditions and by the final pH used. The researchers also showed that the harder cheese could be mixed with soy cream and hydrocolloids to obtain a cream cheese analog. In another study, *Streptococcus thermophilus* has been combined with *Lactobacillus fermentum* to ferment soymilk, which was then further processed into a semi-hard cheese analog. The soymilk was held at 63 °C for 30 min, fermented to pH 4.5, and then the resulting soy curd was pressed to remove the soy whey. The plant-based cheese obtained had similar textural attributes as its dairy counterpart but the sensorial scores were lower (Chumchuere et al., 2000). Using a similar approach, soymilk was fermented with *Lactobacillus casei* ssp. *casei* to a final pH of 6.3 or with a mixture of *Lactobacillus acidophilus* and *Bifidobacterium animalis* subsp. *lactis* in combination with glucono-delta-lactone to obtain a final pH between 4.7 and 7.2, which led to the formation of either spreadable or firm cheese types depending on the conditions used (soybean oil was added to achieve the desired spreadability) (Giri et al., 2018; Li et al., 2013). Li et al. (2013) further showed that the protease papain is useful for improving the mouthfeel of soy-based cheese spreads, which was attributed to its ability to decrease the particle size of protein aggregates, thereby increasing the creamy mouthfeel. Moreover, cheese analogs have been produced by fermenting soymilk with lactic acid bacteria (*Lactobacillus rhamnosus*) to a final pH between 4.7 and 5.2, and then the resulting curd was cut, pressed, and salted (Liu et al., 2006).

The structural and physicochemical properties of cheese analogs produced by a heat/pH-precipitation treatment are different from those of real cheeses produced by a rennet treatment, especially in terms of the molecular and colloidal interactions involved, as well as the structures formed. For example, the three-dimensional structure in rennet cheese is stabilized by a combination of calcium bridges, hydrogen bonds, and hydrophobic interactions (Paula Vilela et al., 2020). This is in contrast to the interactions that are present in cheese analogs, with the hydrophobic attraction being the main driver for protein-protein interactions because the proteins are unfolded by the heat-treatment and the net charge is minimized because fermentation reduces the pH to around the isoelectric point of the proteins (Fig. 9.11).

To more closely mimic the molecular interactions in dairy-based cheeses, multivalent cations can be added to heat-treated plant-based milks to play the role that calcium plays in real milk (Chikpah et al., 2015). In addition, ion-gelation can be combined with acid-gelation to create plant-based cheeses with different properties (Table 9.4) (El-Ella, 1980; Verma et al., 2005; Zulkurnain et al., 2008). The combination of ion- and acid-gelation affects the textural attributes of plant-based cheese analogs because the number of deprotonated carboxyl groups ($-COO^-$) decreases when the pH is decreased, which reduces the number of calcium bridges formed between the protein molecules (Canabady-Rochelle et al., 2009). This approach was used to create cheese analogs from fermented heat-treated soymilk using *Lactococcus lactis* (El-Ella, 1980). The authors used calcium lactate to introduce calcium ions into the fermented soymilk, which induced a sol-gel transition. Subsequently, the curd was pressed, salted, and stored under controlled conditions for up to 3 months. A sensorial analysis revealed that the flavor intensity of the cheese increased during storage. The cheese analog had a relatively bland flavor profile after 4 weeks of ripening, which developed into a slightly acid and cheesy flavor after 2 to 3 months. A similar approach was utilized by Verma et al. (Verma et al., 2005) who combined acidification, calcium addition, and heat treatment to produce a plant-based cheese. A mixture of pigeon pea milk and soy milk was inoculated with *Streptococcus thermophilus* and *Lactobacillus delbrueckii* subsp. *bulgaricus*. After fermentation, curd formation was induced by the addition of calcium (0.02% $CaCl_2$) and a heat treatment (up to 95 °C, 15 min). The plant whey was subsequently removed by filtration and pressing. Overall, these studies give important insights into combining different techniques to obtain plant-based cheeses but more research is necessary in this area. In particular, research is needed to identify the nature of the molecular interactions involved, as well as the types of structures formed, and then to relate these to the physicochemical and sensory attributes of the end products, such as their appearances, textures, functionalities, and flavor profiles.

Enzyme crosslinking can also be used to coagulate plant-based milks and transform them into a curd. Several research articles and patents have described different kinds of enzymes that can be used to induce the sol-gel transition (Brown et al., 2013; Holz-Schietinger et al., 2014; Murata et al., 1987; Sánchez-Muñoz et al., 2017). For example, a mixture of macadamia and almond nuts was first blanched in boiling water for 30 s, softened for 16 h, and then disintegrated by milling (Brown et al., 2013; Holz-Schietinger et al., 2014). After separation of insoluble particles by centrifugation, the skim- and cream-phases were collected and recombined at defined ratios similar to the standardization carried out during the production of dairy cheese to obtain the desired fat and dry matter content. Additionally, a heat-treatment step was used to pasteurize the standardized plant milk before the cheese analog production. Based on this standardized milk, different cheese types were produced following the enzymatic cross-linking approach, for example fresh, soft-ripened, and salted (cheddar-like) plant cheeses. To prepare these, the initial processing was similar. The standardized milk was inoculated with mesophilic starter cultures, namely *Geotrichum candidum*, *Penicillium candidum*, and *Debaromyces hansenii* for soft-ripened cheese. The starter cultures initiated a drop in pH, which reached 5.6 after 1 h and 4.4 after 12 h. Subsequently, microbial transglutaminase was blended into the fermented milk. The enzyme induced the sol-gel transition by facilitating the formation of an isopeptide bond between lysine and glutamine residues on different protein molecules. This reaction was carried out for 12 h at room temperature, which led to the successful development of a plant-based curd. The curd was then cut into pieces of around 1.3 cm. The plant-whey released (40–50%) was separated from the curd and then the remaining curd was mixed again using a simple whisking operation. This material was then used to form different kinds of plant-based cheeses. For fresh and salted cheese, the curd was molded and then pressed. For soft-ripened cheese, the curd was molded without the application of pressure. After molding, the cheeses were transferred into brine solutions for 30 min at 10 °C. Subsequently, the fresh cheese types were directly transferred into packaging materials. The soft-ripened cheese was ripened at temperatures from 10 to 16 °C and at relative humidities from 75 to 90% for 17 days. The salted cheese was ripened at 13 °C and 55% relative humidity for 3 weeks followed by smoking or waxing. Moreover, the addition of exogenous proteases and lipases enhanced the textural and sensorial characteristics of the plant-based cheese, which was probably by similar chemical pathways as found in dairy-based cheeses. Finally, fermentation with *Staphylococcus xylosus* and *Brevibacterium* was shown in this patent to yield volatile compounds typically found in real cheese, such as 3-methyl butanoic acid and dimethyl trisulfide (Holz-Schietinger et al., 2014).

A different enzymatic technique was employed by Murata et al. (1987). In this study, the authors partially hydrolyzed the plant proteins with different proteases instead of generating crosslinks between them, which is more similar to the production of dairy cheeses by using rennet to partially hydrolyze the κ-casein. The authors showed that it was possible to obtain a curd by proteolysis with most of the proteases investigated when the proteolysis was carried out at pH 6.1 in soymilk, except for *Aspergillus saitoi* proteases, rennin, and pepsin. Thus, it is interesting to note that two entirely different approaches (crosslinking proteins *vs.* hydrolyzing proteins) can both yield plant-based cheese curds. There is certainly a need for more research to understand how these different approaches affect the textural and sensorial attributes of plant-based cheeses, as well as to establish their potential for scaling the production to commercial levels.

So far, we have discussed different methods of producing plant-based cheese curds and cheeses from dilute plant-based milk dispersions but it is also possible to produce them from concentrated ones. In this case, the whole plant material is utilized to prepare the cheese analogs (Fig. 9.10). This approach should be more sustainable because less processing is required. Moreover, beneficial functional or nutritional components can be retained within the end product, such as phospholipids and fibers that are known to improve emulsifying and oil-holding properties, respectively (Nikiforidis, 2019; Sánchez-Zapata et al., 2009). Fermented nut pastes are widely used for this purpose. The cheese types produced using this approach are often camembert-style cheeses. To obtain plant-based cheese from fermented nut paste, the nuts are hydrated in hot water to soften the cell walls and to inactivate the microorganisms. The softened nuts are then milled into a concentrated colloidal dispersion and starter mesophilic bacteria and fungi are added, such as *Lactococcus lactis* subsp. *lactis*, *Lactococcus lactis* subsp. *Cremoris*, and *Penicillium camemberti*. The paste is then fermented for around 24 h, formed, and further fermented for several weeks at a temperature from around 12 to 18 °C and a relative humidity >80%. During this time, the plant-based cheese develops its characteristic flavor profile, as well as its characteristic appearance due to the formation of a surface mold. This process has been used to produce a variety of cheese analogs from milled cashew kernels that were fermented with different starter cultures (Chen et al., 2020). The authors showed that *Lactococcus*, *Pediococcus*, and *Weissella* genera were the main forms of bacteria growing in the 'brie' and 'blue' cashew cheese types. The authors also reported that the allergenicity of the proteins decreased as a result of fermentation. However, further research is still required to better understand the structural composition and optimize this process. Probably, a viscoelastic filled-emulsion gel is created by using this approach in which the fat droplets are embedded in a coagulated protein matrix that holds fibers within the structure. The presence of high levels of fiber would be a nutritional advantage for this kind of cheese analog. As mentioned earlier, these plant-based cheese analogs contain similar constituents as the raw materials because no separation or extraction processes are used during their manufacture. As a result, no plant whey is generated and the final nut-based cheese is rich in proteins, fats, and fibers (Table 9.1).

In conclusion, the tissue disruption route is also suitable for producing certain kinds of plant-based cheese analogs. It involves less processing than the fractionation route because the constituents do not need to be isolated from the raw materials. However, the disadvantage is that the functionality of the ingredients in many plant-based materials is not sufficient to achieve the desired textural and sensorial attributes and therefore other ingredients often need to be added to achieve a consistent high quality that is acceptable to consumers (Table 9.1). In the future, the textural attributes of cheese analogs could be further optimized by selecting combinations of raw materials with the desired properties *e.g.*, by mixing different plant milks. Moreover, the sol-gel and ripening processes need to be further optimized by gaining a better understanding of the colloidal interactions and chemical reactions involved.

9.6 Sustainability and Health Considerations

This section provides a brief overview of the potential impact of replacing dairy cheeses with plant-based analogs on environmental sustainability and human health.

9.6.1 Greenhouse Gas Emissions

Cow's cheese is the most commonly consumed cheese type and it will therefore be used as a benchmark. Several studies indicate that most of the greenhouse gas emissions in the manufacture of cheeses are emitted during the production of the milk by the cow. Depending on the assumptions used in the study (such as cheese type, boundary conditions, *etc.*), the milk production step has been estimated to be responsible for 65 to 98% (average 82%) of the total CO_2-eq emissions (Bava et al., 2018; Finnegan et al., 2018; González-García et al., 2013; Üçtuğ, 2019; van Middelaar et al., 2011). In contrast, the influence of the conversion of milk into cheese, as well as the retail part, is much smaller. These parts of the supply chain are impacted by factors such as the use of renewable energies, sewage treatment, packaging, and logistics (Bava et al., 2018; Dalla Riva et al., 2017; González-García et al., 2013; Tarighaleslami et al., 2020). Thus, cow's milk itself makes a substantial contribution to the overall greenhouse gas emissions of cheeses. This is also because around 4 to 10 L of milk is required to produce 1 kg of cheese. A kilogram of cow's milk has a reported global warming potential of around 1.39 kg CO_2-eq per L (min: 0.54; max: 7.50 kg CO_2-eq per L) (Clune et al., 2017). Thus, the impact of the greenhouse gas emissions associated with the production of raw milk for cheese ranges from around 1.62 (for fresh cheese) to 8.3 (for semi-hard cheese) kg CO_2-eq per kg of cheese, resulting in total product average emissions of 8.86 (min: 5.33; max: 16.4) kg CO_2-eq per kg of cheese (Clune et al., 2017; Finnegan et al., 2018).

In comparison, the ingredients used for plant-based cheeses typically have much lower greenhouse gas emissions (Table 9.5). However, these values have to be interpreted with caution until comprehensive life cycle assessments have been carried out for plant-based cheeses. Moreover, the ingredients used to formulate plant-based cheeses do not have the same nutritional value as those in dairy cheeses, and so it is often difficult to accurately calculate their environmental impact on an equivalent nutritional basis. Moreover, the conversion ratios of plant-based milks into plant-based cheeses have not been reported yet, which can also have a considerable impact on the assessment of the environmental sustainability.

Table 9.5 Reported greenhouse gas emissions of selected ingredients used for the production of plant-based cheese alternatives and cow milk

Ingredient	Greenhouse gas emissions in kg CO_2-eq per kg or L
Almond and coconut milk	0.42 (0.39–0.44)
Soy milk	0.88 (0.66–1.40)
Soy protein isolate	2.4 per kg protein
Pea milk	0.39
Cassava starch	0.59
Tree nuts	1.42 (0.43–3.77)
Palm oil	1.4–2.0
Sunflower oil	0.8
Tofu	0.98 (0.87–1.1)
Cow milk	1.39 (0.54–7.50)
Cheese	8.86 (5.33–16.35)

Typically, 4–10 L of milk is needed to obtain 1 kg of cheese. The data in brackets show the calculated minimum to maximum range of the values. (Taken from Braun et al. 2016; Clune et al., 2017; Henderson & Unnasch, 2017; Schmidt, 2015; Usubharatana & Phungrassami, 2015)

Having said that, the greenhouse gas emissions of plant-based milks are commonly reported to be lower than that of cow's milk. This would result in lower greenhouse gas emissions for plant-based cheeses if similar milk-to-cheese conversion ratios are assumed. However, further research is still needed to generate the data needed to reliably assess this claim. Nonetheless, some preliminary assumptions can already be drawn from existing products and value chains. For example, cheeses based on nuts are produced following the tissue disruption route without the separation of whey. This means that there is no conversion ratio and 1 kg of plant-based milk will result in around 1 kg of plant-based cheese (on a dry weight basis). Thus, a plant-based cheese is expected to emit 5.1 to 6.9 kg less of CO_2-eq per kg of soft to semi-hard cheese, respectively (assuming 1.42 kg CO_2-eq for 1 kg of nuts vs. 6.5–8.3 kg CO_2-eq for the cheese (Clune et al., 2017; Finnegan et al., 2018). This preliminary data is further supported by the fact that soy tofu has a much lower global warming potential than cheese but involves similar manufacturing operations. To produce 1 kg of tofu 0.98 kg CO_2-eq are emitted including the emissions related to agricultural production and tofu factory processing (Mejia et al., 2018). This is in contrast to the emissions produced by cheese, which were reported to be around 8.86 (min: 5.33; max: 16.35) kg CO_2-eq per kg cheese (Clune et al., 2017). As already mentioned, the actual emissions might be somewhat different when the nutritional value (*e.g.*, content of essential amino acids) is considered and similar product categories are compared (Tessari et al., 2016).

In summary, plant-based cheeses have the potential to produce less greenhouse gas emissions than animal milk-based cheeses but further life cycle assessments are needed to establish this. Moreover, based on the life cycle analyses for plant-based milks discussed in Chap. 8, one would expect cheese analogs to lead to less pollution, land use, water use, and biodiversity loss than real cheeses.

9.6.2 Health Aspects

Cheese is a nutrient-rich food containing high proportions of macronutrients and micronutrients that may be beneficial to human health, including proteins, lipids, vitamins, and minerals. There may therefore be nutritional implications when people switch from animal- to plant-based cheeses. For this reason, we discuss the possible nutritional and health consequences associated with consuming plant-based cheeses in this section.

Incorporating 'healthy' plant-based foods into the human diet is associated with positive health outcomes in observational studies (Kim et al., 2018, 2019; Satija et al., 2017). For example, people that consume more healthy plant-based foods have been found to have a decreased risk of all-cause and cardiovascular disease mortality (Kim et al., 2018). In these studies, healthy plant-based foods are considered to be whole grains, fruits, vegetables, nuts, legumes, tea, and coffee. In contrast, unhealthy plant-based foods are considered to be refined grains, sweets, snacks, baked goods, fruit juices, and sugar-sweetened beverages. Other studies have reported that people who follow a healthy vegetarian diet are less likely to die from ischemic heart disease but there was no correlation between all-cause mortality and cancer mortality (Dinu et al., 2017). This contrasts with some population-based studies that could find no such effects and even associated vegetarian-based diets with increased risk for certain diseases, such as stroke (Appleby et al., 2016; Mihrshahi et al., 2017; Orlich et al., 2013; Tong et al., 2019). This effect is most likely related to the fact that a vegetarian diet does not necessarily contain only healthy foods (Magkos et al., 2020; Mihrshahi et al., 2017). A diet rich in sugary and starchy foods is still considered to be vegetarian but might have negative health outcomes, which may not be reflected in large uncontrolled studies. Such methodological gaps have been closed by using plant-based diet indices and grouping foods in different categories according to their possible effects on human health, which showed that an increase in consumption of healthy plant-based foods is related to positive health outcomes, as mentioned above (Kim et al., 2018, 2019; Satija et al., 2017).

In this framework, it is important to address the potential impacts of plant-based cheeses on human health and wellbeing. Plant-based cheeses often contain relatively high amounts of processed ingredients, such as starches and fats (Table 9.1). For conventional dairy cheese, most observational studies have found no association between cheese consumption and higher mortality or higher risk of cardiovascular diseases (de Goede et al., 2016; Farvid et al., 2017; Guo et al., 2017; Hjerpsted & Tholstrup, 2016; Mazidi et al., 2019; Pala et al., 2019). However, these results should be interpreted with caution because rigorous randomized controlled studies are still missing (Sacks et al., 2017). Nonetheless, a recent meta-analysis of observational studies reported that individuals who had increased blood biomarkers of dairy product consumption had less cardiovascular disease (Trieu et al., 2021). In addition, most of the nutritional studies seem to suggest that dairy cheese consumption has positive or neutral effects on human health (de Goede et al., 2016; Farvid et al., 2017; Guo et al., 2017; Hjerpsted & Tholstrup, 2016; Mazidi et al., 2019; Pala et al., 2019). This is possibly because cheese contains a range of ingredients that might promote health, including high levels of calcium, protein, beneficial fatty acids (*e.g.*, conjugated linoleic acid), vitamins, and other compounds that may have beneficial effects on the body (Hjerpsted & Tholstrup, 2016; Magkos et al., 2020). These findings are somewhat surprising because dairy products like cheese have a high saturated fatty acid content. Saturated fats have been associated with higher all-cause mortality and an increased risk for cardiovascular diseases because they increase the LDL cholesterol concentration in the blood, especially when saturated fatty acids replace monounsaturated or polyunsaturated fatty acids in the diet (but not when they replace carbohydrates) (Sacks et al., 2017). However, there is still much debate about the impact of different types of saturated fatty acids on human health but a recent meta-analysis concluded that diets high in saturated fat were associated with higher mortality from all-causes, CVD, and cancer, whereas diets high in polyunsaturated fat were associated with lower mortality from all-causes, CVD, and cancer (Heileson, 2020; Kim et al., 2021; Lawrence, 2021).

There is much fewer data available on the health effects of consuming plant-based cheeses and so the discussion will mainly focus on the main ingredients used to formulate these products (Table 9.1). Bovine cheese is rich in protein, fats, calcium (8–79%DV per 100g), phosphorus (16–57%DV per 100g), vitamin A (3–20%DV per 100g), vitamin B_{12} (7–56%DV per 100g) and others (Górska-Warsewicz et al., 2019). Moreover, these nutrients are also of high quality and bioavailability. Caseins have a high protein quality with a digestible indispensable amino acid score (DIAAS) of 1.29, which is higher than that reported for plant proteins (Guillin et al., 2021; Mathai et al., 2017). Moreover, the calcium found in milk is readily absorbed by the human body with around 30% being taken up into the bloodstream, which is higher than the calcium found in most plant sources (Yang et al., 2012). One reason for the lower bioavailability of the nutrients in plants is the presence of antinutrients that inhibit digestion and absorption, such as trypsin inhibitors and phytates (Chap. 5).

In contrast, many plant-based cheese formulations do not contain the same nutrients as found in dairy cheese (Table 9.1). This is especially true for the formulations that follow the fractionation route and are based on starch and oils. These plant-based cheeses do not contribute considerable amounts of calcium, protein, vitamin A, and vitamin B_{12} to the human diet (unless fortified). Moreover, these formulations may contain relatively high amounts of saturated fatty acids from the plant-derived fats used to mimic milk fat in these products, such as coconut oil (>90% saturated fatty acids). In addition, they may contain high levels of refined carbohydrates (rapidly digestible starch), which may be detrimental to human health if eaten on a regular basis due to the dysregulation of the insulin response (Ludwig et al., 2018a). However, there is some controversy about whether coconut oils (which contain high levels of medium-chain saturated fatty acids) have similar negative effects on LDL levels as animal fats (which contain high levels of long-chain saturated fatty acids) (Eyres et al., 2016; Hewlings, 2020). Moreover, there is even a debate about whether saturated fats in general are responsible for adverse effects on heart disease as mentioned before.

It is possible to create plant-based cheese products with more healthy nutritional profiles, which should be an important focus of future research and development. Cheese analogs based on nuts are a good example of a nutritious formulation. Nuts are rich in several important nutrients, low in saturated fatty acids, and have been positively linked to improved human health (Chen et al., 2017; Fardet & Boirie, 2014; Schwingshackl et al., 2017). For example, cashew nuts contain high levels of oleic acid, protein, and dietary fibers. They also contain various vitamins per 100g (such as B_1 (37%DV), B_5 (17%DV), B_6 (32%DV), and K (32%DV)) and minerals (such as zinc (61%DV), iron (51%DV), magnesium (82%DV), copper (110%DV), selenium (28%DV), and phosphorus (85%DV)) at relatively high levels (USDA, 2021). Despite being nutrient-dense foods, nuts do lack some of the important nutrients normally found in cheese, such as calcium, vitamin A, and vitamin B_{12}. To overcome this deficiency, plant-based cheeses could be fortified with these nutrients, or consumers could be educated about what kind of foods they should eat to ensure a sufficient uptake of all the required nutrients.

Another important nutrient that needs to be considered is salt. Many kinds of cheese contain high levels of salt (USDA, 2021). For example, cheddar, feta, gouda, Camembert, and Parmigiano-Reggiano contain relatively high sodium concentrations, *i.e.,* around 600–1500 mg of sodium per 100 g of cheese. Some others are lower in salt, for example, Swiss cheese and ricotta typically contain 100–200 mg of sodium per 100 g of cheese. The intake of high levels of sodium is associated with increased blood pressure, which can lead to higher rates of coronary heart disease, stroke, and kidney failure in some individuals (Chobanian & Hill, 2000; Farquhar et al., 2015; Wang et al., 2020). The blood pressure raising effect is especially strong in individuals that fall into the group of "salt-sensitive", which affects 30–50% of the hypertensive and 25% of the normotensive subjects (Balafa & Kalaitzidis, 2021; Gholami et al., 2020). Very low sodium intakes are also detrimental to human health and should therefore be avoided (Adedinsewo et al., 2021; Messerli et al., 2020). However, the daily recommendation of salt intake is around 2.3 g for adults and so cheese consumption may significantly contribute to the total salt intake (Santos et al., 2021). The same concerns are relevant to the plant-based cheeses discussed in this chapter (Table 9.1), which also contain considerable amounts of salt and sodium. Consequently, future research and development should focus on the production of plant-based cheese analogs that contain healthy levels of salts and other nutrients.

9.7 Other Dairy Alternatives

9.7.1 Yogurt

Dairy yogurt is an acidified food that is produced by fermenting milk with (typically) *Streptococcus thermophilus* and *Lactobacillus delbrueckii* subsp. *bulgaricus* until the pH falls below pH 4.5 and the final lactic acid bacteria (LAB) concentration exceeds 10^8 colony-forming units/g (Montemurro et al., 2021). Yogurt is typically produced using a series of steps. The milk is standardized to reach the required initial composition (fat and dry matter content), homogenized, heat-treated, and then inoculated with LAB and fermented at 40–44 °C to induce gelation. The sol-gel transition is caused by a decrease in electrostatic and steric repulsion (release of κ-casein from the micelle) between the protein molecules (Sinaga et al., 2017). During fermentation, the LAB convert some of the lactose to lactic acid, which causes the pH of the aqueous phase to decrease towards the isoelectric point of the casein (around 4.6). As a result, the negative charge on the casein micelles decreases and there is a collapse of the κ-casein 'hairs' on the micelle's surfaces. In addition, calcium ions are released from the micelles because of the reduction in the number of anionic carboxyl groups available to bind them. As a result, the micelles dissociate and the proteins aggregate with their neighbors, thereby forming a delicate 3D protein network (Fig. 9.12) that provides some mechanical strength (Lucey, 2020). Therefore, yogurts have a yield stress upon the application of stresses. Once deformation and/or flow is initiated, yogurts show shear-thinning properties because the aggregated biopolymers align and orientate in the shear field (Fig. 9.14), which is best described by the Herschel-Bulkley model (Hassan et al., 2003):

$$\tau = \tau_y + K\dot{\gamma}^n \tag{9.2}$$

Here, τ being the shear stress (Pa), τ_y is the yield stress, K is the consistency index that is also called "Herschel/Bulkley viscosity" (Pa s^n), $\dot{\gamma}$ is the shear rate (s^{-1}), and n is the flow index. For $n < 1$ the fluid is shear-thinning, whereas for $n > 1$ the fluid is shear-thickening. For example, this equation has been used to describe the textural attributes of yogurts that have been fermented with nonropy, non-capsule-forming *S. thermophilus* and nonropy, non-capsule-forming *L. delbrueckii* ssp. *Bulgaricus*, which resulted in values of τ_y = 15.4 Pa, K = 0.9 Pa s^n, and a flow index

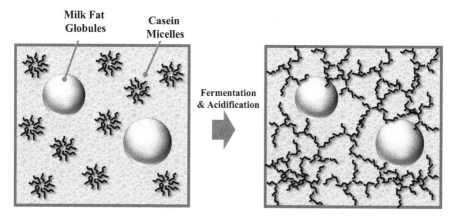

Fig. 9.12 Yogurt can be produced by adding starter cultures to milk to acidify the system, which promotes dissociation of the casein micelles and the formation of a delicate 3D protein network that traps fat droplets inside

of 0.65 (Hassan et al., 2003). Lastly, it has to be mentioned yogurt gels can be broken by stirring, which results in a 'stirred yogurt' that has a lower yield stress.

Fermentation approaches are also used for the production of plant-based yogurts. Plant-based yogurt products currently on the market are typically based on soy, oat, pea, almond, cashew, coconut, and lupin seeds (Boeck et al., 2021; Montemurro et al., 2021). Obtaining a stable gel by fermentation from these raw materials has been proven difficult because the gels do not exhibit the same textural and fluid retention properties as dairy yogurt. This is because the molecular structure and interactions of plant proteins are different from those of casein, and the total protein content of some plant-based milks is lower than that of cow's milk, especially for cereal-based yogurt analogs (Bernat et al., 2014). As discussed earlier, casein molecules and micelles have different structural properties compared to globular plant proteins and their aggregates. Thus, acidification leads to different interactions and structural organizations for plant proteins than for casein, which leads to different gel properties. To overcome this, hydrocolloids such as starch, gellan gum, guar gum, locust bean gum, xanthan gum, pectins, agar, or carrageenan are often incorporated into plant-based yogurts to enhance their structural and mechanical properties (Boeck et al., 2021). Moreover, a heat treatment that denatures or partially denatures the plant proteins has been shown to enhance the structure forming abilities of acid fermented products (Montemurro et al., 2021). For example, partial denaturation was shown to be beneficial for the structure formation in pea protein-based yogurt analogs. This is because a slight denaturation facilitates the release of functional groups and if the right temperature-time combinations are employed an excessive aggregation can be prevented during the heat treatment, allowing for more efficient gelation during the following acidification (Klost & Drusch, 2019).

A representative processing scheme for creating a plant-based yogurt is shown in Fig. 9.13. Initially, a plant-based 'milk' is prepared from one or more starting materials, such as flours, flakes, and isolates/concentrates derived from plants. This process usually involves mechanical size reduction, mixing, and water dispersion steps (Chap. 8). After standardization and formulation, this milk is typically homogenized and pasteurized, which may cause starch gelatinization (if present). A filtration step can be used to concentrate the plant-based milk or dispersion to obtain the desired dry matter content, which is important to achieve the optimum structural properties. The plant-based dispersion is then inoculated with starter cultures to induce acidification. In principle, the same starter cultures used in dairy yogurt (such as lactic acid bacteria) can be employed for the production of plant-based yogurts, namely *Streptococcus thermophilus* and *Lactobacillus delbrueckii* subsp. *bulgaricus* (Table 9.6). However, these species are often less suited to achieve optimum growth rates, acidity levels, and volatile compound production under the different conditions present in plant milks compared to dairy milks. For example, a drop in pH from 6.6 to 4.7 (after 18 h) was reported for fermenting a pea protein dispersion with *Lactobacillus delbrueckii* ssp. *bulgaricus* and *Streptococcus thermophilus* (Klost & Drusch, 2019). This pH is considerably higher than for dairy yogurt, which typically has a pH below 4.5 (Dahlan & Sani, 2017). This is partly because cow's milk contains lactose that can be used for microbial fermentation by the LAB,

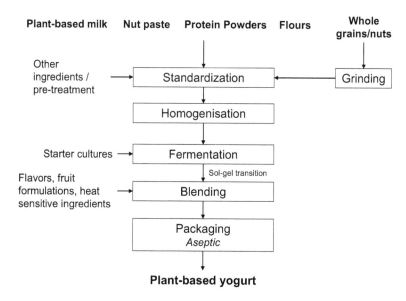

Fig. 9.13 Possible processing operations and raw materials involved in the production of plant-based yogurt alternatives. Different raw materials can be used to obtain a plant-based yogurt. A heat treatment can be added if adverse microbials are present or the product is stored at room temperature. (Modified from Montemurro et al. 2021)

whereas the same bacteria had to use sucrose added to the pea protein dispersion. Therefore, organic acids such as citric acid or malic acid are added to some formulations to reach the desired final pH (Boeck et al., 2021).

Therefore, studies have been carried out to assess the efficacy of other bacteria that may be more suitable to ferment plant-based raw materials. Potential bacteria for this purpose can be identified by letting the desired plant-based material spontaneously ferment and then selecting specific bacteria that were able to grow during this process for further studies. For example, *Lactiplantibacillus plantarum* was obtained from a spontaneously fermented quinoa broth and showed enhanced acidification properties and release of total phenols compared to *Weissella confuse* DSM 20194 (Lorusso et al., 2018). Other starter cultures have been utilized to produce several kinds of plant-based yogurts, as shown in Table 9.6. These bacteria can also be selected to produce exopolysaccharides to enhance the textural attributes of the plant-based yogurts (Fig. 9.14), which is beneficial if the initial plant materials do not provide the required mechanical properties (Montemurro et al., 2021).

One major advantage of fermenting plant-based raw materials is the reduction of antinutritional compounds. These compounds include but are not limited to, raffinose, phytic acid, condensed tannins, alkaloids, lectins, pyrimidine glycosides, and protease inhibitors (Montemurro et al., 2021). It is known that fermentation can reduce such antinutritional compounds (Tangyu et al., 2019). For example, it was shown that fermenting rice, chickpea, and lentil flour with *Lactoplantibacillus*

9.7 Other Dairy Alternatives

Table 9.6 Overview of some selected ingredients, processing parameters, and starter cultures used in plant-based yogurt production

Main ingredient	Starter culture used	Pre-treatment	Status	Reference
Oat protein concentrate (15% w/w)	*Streptococcus thermophilus* and *Lactobacillus delbrueckii* subsp. *bulgaricus* (commercial strains for yogurt production)	Heat treatment at 90 °C for 30 min	Experimental	Brückner-Gühmann et al. (2019)
Potato protein isolate (5% w/v)	*Streptococcus thermophilus* and *Lactobacillus delbrueckii* subsp. *bulgaricus* (commercial strains for yogurt production)	High-pressure homogenization (200 MPa)	Experimental	Levy et al. (2021)
Pea protein isolate (10% w/w)	*Streptococcus thermophilus* and *Lactobacillus delbrueckii* subsp. *bulgaricus* (commercial strains for yogurt production)	Heat treatment 60 °C for 60 min and high-pressure homogenisation (3 MPa)	Experimental	Klost and Drusch (2019)
Soymilk (6.8% solids)	*Streptococcus thermophilus* and *Lactobacillus delbrueckii* subsp. *bulgaricus* (commercial strains for yogurt production)	Concentration (heat treatment at 90 °C for 15 min), addition of strawberry or orange jam (30% w/w)	Experimental	Al-Nabulsi et al. (2014)
Brown rice, soaked rice, or germinated rice (22% w/v)	Commercial thermophilic starters	Gelatin supplementation, heat treatment at 95 °C for 30 min, filtration	Experimental	Cáceres et al. (2019)
Soymilk	*Streptococcus thermophilus* St1342, *Lactobacillus delbrueckii* subsp. *bulgaricus* Lb1466 and a probiotic strain (*Lactobacillus acidophilus* L10, *Lacticaseibacillus paracasei* L26, *Bifidobacterium lactis* B94)	Heat treatment at 90 °C for 30 min	Experimental	Donkor et al. (2005)
Defatted soy flour (11.6% w/w)	*Streptococcus thermophilus* ATCC 19987 and *Lacticaseibacillus casei* ATCC 393	Heat treatment at 121 °C for 15 min and supplementation with gelatin	Experimental	Cheng et al. (1990)
Millet flour (8% w/v)	*Lacticaseibacillus rhamnosus* GR-1 and *Streptococcus thermophilus* C106	Heat treatment at 90–95 °C for 60 min	Experimental	Stefano et al. (2017)

(continued)

Table 9.6 (continued)

Main ingredient	Starter culture used	Pre-treatment	Status	Reference
Almond (8% w/w)	*Limosilactobacillus reuteri* ATCC 55730 (probiotic) and *Streptococcus thermophilus* CECT 986	High pressure homogenisation (172 MPa for 2–4 s) and heat treatment at 85 °C for 30 min	Experimental	Bernat et al. (2015)
Emmer flour (30% w/v)	*Lactiplantibacillus plantarum* 6E, *Lacticaseibacillus rhamnosus* SP1, *Weissella cibaria* WC4 (EPS-producer)	Starch gelatinization at 60° for 30 min, use of EPS-producer LAB strain	Experimental	Coda et al. (2011)
Quinoa (35% w/v)	*Lactiplantibacillus plantarum* T6B10, *Lacticaseibacillus rhamnosus* SP1 (probiotic), *Weissella confusa* DSM 20194, (EPS-producer)	Starch gelatinization at 63 °C for ca. 19 min	Experimental	Lorusso et al. (2018)
Lupin protein isolate (2% w/v)	*Lactiplantibacillus plantarum* TMW 1.460 and TMW 1.1468, or *Pediococcus pentosaceus* BGT B34 and *Levilactobacillus brevis* BGT L150	Heat treatment (140 °C for 10 s or 80 °C for 60 s) and EPS-producer LAB strain	Experimental	Hickisch et al. (2016)
Oat flakes (25% w/w)	*Lactiplantibacillus plantarum* LP09	Enzymatic treatments (xylanase and α-amylase)	Experimental	Luana et al. (2014)
Rice (10% w/w), lentil (5% w/w), and chickpea (5% w/w) flours	*Lactiplantibacillus plantarum* DSM33326, *Levilactobacillus brevis* DSM33325, *Lacticaseibacillus rhamnosus* SP1 (probiotic)	Heat treatment at 80 °C for 15 min	Experimental	Pontonio et al. (2020)
Quinoa flour (14.3% w/w)	*Weissella cibaria* MG1 (EPS producer)	Heat treatment at 121 °C for 15 min, α-amylase and protease treatments, high-pressure homogenisation (180 MPa)	Experimental	Zannini et al. (2018)
Soy (10% w/v)	*Lactiplantibacillus plantarum* B1–6	Heat treatment at 108 °C for 15 min	Experimental	Rui et al. (2019)
Soy, soaked soy, or germinated soy (10% w/v)	*Levilactobacillus brevis* KCTC 3320	Heat treatment at 121 °C for 15 min	Experimental	Hwang et al. (2018)

(continued)

9.7 Other Dairy Alternatives

Table 9.6 (continued)

Main ingredient	Starter culture used	Pre-treatment	Status	Reference
Peanut (16.7% w/w)	*Enterococcus faecalis* T110 (probiotic)	Heat treatment in autoclave at 121 °C and 15 psi for 3–5 min	Experimental	Bansal et al. (2016)
Soymilk (12.5% w/w)	*Bifidobacterium longum* SPM1205	Heat treatment at 95 °C for 5 min, supplementation with agar, strawberry syrup (20% w/w) and 0.05% (w/w) of freeze-dried diced strawberry	Experimental	Park et al. (2012)
Soy and a pigment rich extract (red beetroot, hibiscus, opuntia, red radish)	–	–	Experimental	Dias et al. (2020)
Hulled soy beans (7.9% w/v)	*Streptococcus thermophilus* and *Lactobacillus delbrueckii* subsp. *bulgaricus* (commercial strains for yogurt production)	Supplementation with pectin	Commercial	Grasso et al. (2020)
Hulled soy beans (9% w/v)	*Streptococcus thermophilus* and *Lactobacillus delbrueckii* subsp. *bulgaricus* (commercial strains for yogurt production)	–	Commercial	Grasso et al. (2020)
Coconut cream (20% w/v) and modified maize starch	*Streptococcus thermophilus* and *Lactobacillus delbrueckii* subsp. *bulgaricus* (commercial strains for yogurt production)	Supplementation with pectin	Commercial	Grasso et al. (2020)
Cashew "milk" (97% v/v) and tapioca starch	*Streptococcus thermophilus* and *Lactobacillus delbrueckii* subsp. *bulgaricus* (commercial strains for yogurt production)	Supplementation with carob gum	Commercial	Grasso et al. (2020)

(continued)

Table 9.6 (continued)

Main ingredient	Starter culture used	Pre-treatment	Status	Reference
Almond "milk" (95% v/v) and tapioca starch	*Streptococcus thermophilus* and *Lactobacillus delbrueckii* subsp. *bulgaricus* (commercial strains for yogurt production)	Supplementation with carob gum	Commercial	Grasso et al. (2020)
Hemp juice 96% (water, hemp seed 3% w/v) and rice starch	Selected strain of *Bifidobacterium* and *Lactobacillus acidophilus*	Supplementation with agar	Commercial	Grasso et al. (2020)
Oat 12% (w/v)	–	Supplementation with potato starch and potato protein	Commercial	Greis et al. (2020)
Oat 8.5% (w/v)	–	Supplementation with modified starch, pectin	Commercial	Greis et al. (2020)
Oat 8% (w/v)	–	Supplementation with potato protein, starch (corn, potato), pectin	Commercial	Greis et al. (2020)
Oat 12% (w/v)	–	Supplementation with potato protein, tapioca starch, potato starch, xanthan, locust bean gum	Commercial	Greis et al. (2020)
Oat	–	Supplementation with pea protein, modified potato starch	Commercial	Greis et al. (2020)
Oat 12% (w/v) (OATLY®)	Commercial strains for yogurt production	Supplementation with potato starch	Commercial	Oatly AB, Sweden
Soy 10.7% (w/v) (ALPRO®)	*Streptococcus thermophilus* and *Lactobacillus delbrueckii* subsp. *bulgaricus* (commercial strains for yogurt production)	Supplementation with pectin	Commercial	Alpro Comm. VA, Belgium
Oat 8% (w/v) (YOSA®)	*Bifidobacterium* BB12 and *Lacticaseibacillus rhamnosus* GG	Supplementation with pectin	Commercial	Fazer Oy, Finland

Adopted from Montemurro et al. (2021), under CC BY 4.0 (http://creativecommons.org/licenses/by/4.0/)

plantarum DSM33326 and *Levilactobacillus brevis* DSM33325 reduced the phytic acid, condensed tannins, saponins, and raffinose content (Pontonio et al., 2020). Moreover, the activity of trypsin inhibitors can also be reduced by fermentation (Montemurro et al., 2021). Thus, the production of yogurt from plant-based raw

9.7 Other Dairy Alternatives

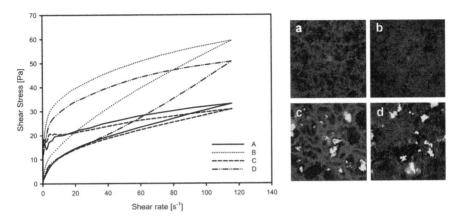

Fig. 9.14 The textural attributes of plant-based yogurts should mimic that of yogurt prepared from animal milk. Yogurt is a non-Newtonian material with shear-thinning properties. If specific starter cultures are used, the textural properties of the yogurt can be modified. In this case, a cow milk-based yogurt was fermented with different *(a–d)* strains of exopolysaccharide-forming starter cultures (*S. thermophilus* and *L. delbrueckii* ssp. *bulgaricus*). Red indicates protein, green exopolysaccharides. (Reprinted with permission of Elsevier from Hassan et al. 2003)

materials seems to be beneficial to obtain a food product that is low in antinutritional compounds.

One of the major challenges associated with the production of plant-based yogurts is obtaining a sensory profile that is desirable and comparable to that found in dairy yogurts. The main aromatic compounds that influence the flavor of dairy yogurt are acetaldehyde, acetone, acetoin, and diacetyl, as well as acetic, formic, butanoic, and propanoic acids (Routray & Mishra, 2011). In contrast, plant-based yogurts are associated with 'beany' and 'raisin' aromas, as well as 'bitter taste' and 'astringency', especially soy-based yogurts (Montemurro et al., 2021). Similarly, 'bitterness', 'astringency', and 'greeny' notes were also reported for cereal-based yogurt formulations, which is a result of the presence of phenolic compounds and lipid oxidation accelerated by lipoxygenase (Doehlert et al., 2010; Montemurro et al., 2021). To overcome this challenge, several approaches can be employed to reduce or remove off-flavors such as plant-breeding, separation of pre-cursors, thermal processing, vacuum distillation, fermentation, and flavor masking (Tangyu et al., 2019).

Lastly, the nutritional value of plant-based yogurts differs from that of dairy yogurt, which may have health consequences. The nutritional profile of yogurt analogs can vary considerably depending on the raw materials, processing operations, and additives used. As a result, they may contain more or fewer sugars, proteins, and saturated fats than their dairy counterparts. Moreover, the plant materials used to form yogurt analogs typically contain less calcium and vitamin B_{12} than milk, and so plant-based products are often enriched with these micronutrients (Boeck et al., 2021).

9.7.2 Ice Cream

Ice cream is a structurally complex colloidal food, which contains different kinds of particles (crystalline fat droplets, air bubbles, and ice crystals) dispersed in a freeze-concentrated aqueous solution (Fig. 9.15) (Goff, 1997). It is usually produced from a mixture of milk, additional fat (cream or other fats), non-fat milk solids, emulsifiers, hydrocolloids, flavoring agents, and other ingredients. An ice cream mix is produced by blending all the ingredients in water. The ice cream mix is then pasteurized, homogenized, cooled to 4 °C, aged, frozen to −16 to −20 °C (often with mechanical agitation), and then stored at −25 to −40 °C.

Each processing step and ingredient is used to fulfill a specific task in this sequence (Goff, 1997):

- *Pasteurization*: food safety
- *Homogenization*: emulsion formation, improved whippability, smoother product
- *Cooling*: fat droplet crystallization, ingredient hydration
- *Aging*: fat droplet interface modification
- *Freezing*: ice crystal formation, air incorporation, solid structure formation

During homogenization, small fat globules are produced that are coated by a layer of proteins (usually caseins). During cooling, the lipid phase inside the fat droplets partly crystallizes. During aging, low molecular weight surfactants (such as lecithin, mono-/di-acylglycerides, or polysorbates) are added, which partly displace the proteins from the fat droplet interfaces. This is important because it promotes partial

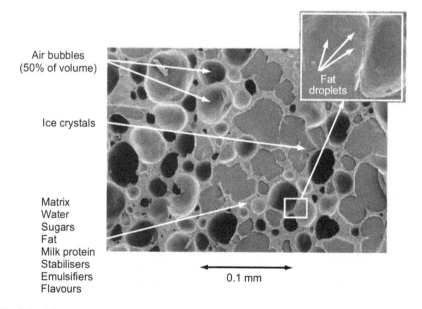

Fig. 9.15 Microstructure illustrated by scanning electron microscope of ice cream that consists of particles (crystalline fat droplets, air bubbles, and ice crystals) dispersed in a freeze-concentrated aqueous solution. (Reprinted with permission of Elsevier from Cebula and Hoddle 2009)

coalescence during the subsequent cooling/freezing and shearing stages. The thick protein layer that was originally around the fat globules inhibits this process. During partial coalescence, the partly crystalline fat droplets aggregate with each other because solid fat crystals from one droplet penetrate into a liquid oil region in another partially crystalline droplet. This is important in ice cream production since it leads to a network of crystalline fat droplets around the air bubbles and within the aqueous phase, which contributes to the mechanical strength of the final product. The rate of partial coalescence depends on the solid fat content of the lipid phase, which depends on the temperature. Partial coalescence typically occurs at a temperature where the fat droplets are partially crystalline. Thus, it is critical to identify plant fats that exhibit similar crystallization behavior as milk fat when developing plant-based ice creams, as this will lead to similar mechanical properties, melting behavior, and mouthfeel in the end product.

Another important aspect is the addition of low molecular weight compounds (*e.g.*, sugars and minerals), which contribute to the taste of ice cream and lower the melting point of the aqueous phase, which is important for controlling the hardness and melting behavior of ice cream. This is often controlled by the addition of sugar and is further enhanced by the addition of non-fat milk solids, which contain lactose and minerals. The presence of these solutes leads to the formation of a freeze-concentrated aqueous phase during freezing, which contributes to the desirable soft texture of ice creams. This non-frozen aqueous phase is typically a highly viscous concentrated solution in which the ice crystals and other particles are embedded. Hydrocolloids (*e.g.*, carrageenan, locust bean gum, or xanthan gum) may also be added to slow down ice and lactose crystal growth, so as to avoid the formation of large crystals that would lead to a gritty mouthfeel (Cook & Hartel, 2010) and to bind water to slow down dripping. Ideally, the ice and lactose crystals should have diameters of around 10–20 µm to produce a smooth mouthfeel in the end product.

Finally, the whipping of air into the ice cream mix during freezing is important to obtain a smooth and soft texture. The air bubbles have a mean diameter of around 50 µm (Clarke, 2007; Goff & Hartel, 2013). These air bubbles are stabilized by partially coalesced fat globules and proteins. The amount of air incorporated into the product is usually described as the 'overrun', which is the increase in volume compared to the initial volume of the mix:

$$\% \text{Overrun} = 100 \times \frac{V_\text{P} - V_\text{M}}{V_\text{M}} \qquad (9.3)$$

Here, V_M is the volume of the initial ice cream mix and V_P is the volume of the final ice cream product. Commonly, overruns are somewhere between 25–120% with higher overruns being used for cheaper products (Clarke, 2007; Goff & Hartel, 2013).

Plant-based ice creams can be produced using similar processing steps as dairy ice cream but the ingredients used are different. Ice cream analogs can be produced from plant-based milks (such as cashew, pea, or oat milk) or from refined ingredients that are blended into an ice cream mix (*e.g.*, flours, concentrates, or isolates). Typically, a mixture of proteins, lipids, sugars, salts, hydrocolloids, emulsifiers, colors, and flavors are blended together. The plant-based ingredients used should fulfill the same functional properties as their dairy equivalents, but they often behave differently because of their different molecular characteristics.

The main difference between the production of dairy ice creams and their plant-based counterparts is that plant proteins and fats are used (Fig. 9.16). Ideally, it is important to utilize ingredients that exhibit similar functionalities as their dairy equivalents so as to achieve similar desirable quality attributes in the final ice cream. As discussed earlier, dairy proteins are utilized to facilitate emulsification, water holding, and whippability of ice creams. Therefore, plant proteins should be selected that show similar emulsifying and water-holding properties as dairy ones to ensure similar fat globule diameters and viscosities of the ice cream mix. A lack of these properties can be overcome by adding other plant-based emulsifiers or by incorporating hydrocolloids such as gums, starches, xanthan, or carrageenan. Dairy fat is essential for air bubble stabilization, texture, mouthfeel, and melting behavior. Therefore, plant-based fats should mimic these characteristics. Typically, this means that the solid fat content *versus* temperature profile of the plant-based fats should match that of the dairy fat. This could be achieved by using plant-based oils that are

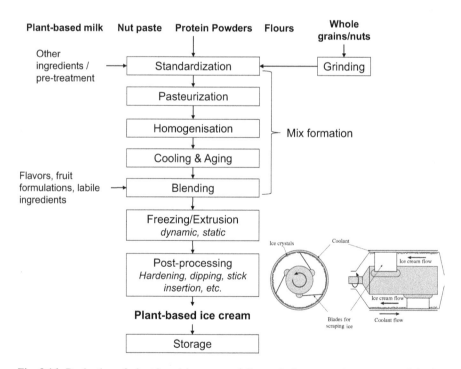

Fig. 9.16 Production of plant-based ice-creams follows similar processing routes as dairy ice creams but the process needs to be adjusted for the different raw materials. Other ingredients added may include sugar, emulsifiers, texturing agents, fat/oils and pre-processing may include blending, pasteurization, and primary emulsion formation. Image shows the working principle of a continuous ice cream freezer, which freezes ice cream and incorporates air by agitation and wall scraping. (Reprinted with permission of Elsevier from Aldazabal et al., 2006)

9.7 Other Dairy Alternatives

partially solid at room temperature, such as palm or coconut oil (Fig. 9.6). Palm oil has a solid fat content of around 60% at 5 °C, which is fairly similar to that of milk fat, which has a solid fat content of around 56% at 4 °C (Lopez et al., 2006; Noor Lida et al., 2002). Coconut oil can also be used but it has a considerably higher solid fat content than dairy fat (~80% at 5 °C) (Smith, 2015). For this reason, plant-based solid fats (such as palm or coconut oil) can be blended with plant-based liquid oils (such as canola, sunflower, or soybean oil) to obtain an appropriate solid fat content *versus* temperature profile. Thus, by choosing and blending the right plant-based ingredients, optimized formulations for designing plant-based ice creams can be achieved. Two examples of the ingredients used to produce commercial products (ingredient list taken in October 2021) are highlighted below:

- *Ben & Jerry's Chocolate Fudge Brownie* (Ben & Jerry's Homemade Holdings Inc., USA): Almond milk (water, almonds), liquid sugar (sugar, water), coconut oil, sugar, cocoa (processed with alkali), wheat flour, corn syrup solids, soybean oil, cocoa powder, corn syrup, pea protein, sunflower lecithin, corn starch, guar gum, vanilla extract, locust bean gum, salt, baking soda, natural flavor (coconut), soy lecithin, barley malt.
- *Oatly Chocolate* (Oatly AB, Sweden): Water, oats, sugar, dextrose, rapeseed oil, glucose syrup, fully hydrogenated vegetable oils (coconut, rapeseed), cocoa 2.5%, coconut oil, emulsifier (mono- and diglycerides of fatty acids), stabilizer (locust bean gum, guar gum), salt, natural flavor.

As with other plant-based products, the nutritional profile of ice cream analogs is different from that of dairy ice cream and depends on the types and amounts of different ingredients used to formulate them. In the future, it will be important to compare the nutritional profiles of dairy and plant-based ice creams and their potential impacts on human health. However, as ice cream is mainly consumed for pleasure, the nutritional profile might be less important than for other products. There is also a need for more research on comparing the environmental impacts of dairy and plant-based ice creams.

9.7.3 Whipping Cream

Whipped cream is also a complex colloidal dispersion (Fig. 9.17), which consists of air bubbles suspended in an aqueous medium. The air bubbles are stabilized by a mixture of protein molecules and partially crystalline fat droplets. The fat droplets aggregate with each other and form a shell around the air bubbles that provides them with some mechanical rigidity and stability.

Whipped cream is typically created by beating cream with air at relatively low temperatures (0–7 °C) where the fat phase is partially crystalline, as this leads to partial coalescence of the fat droplets and network formation. The aggregated fat droplets form a 3D network in the continuous phase, which contributes to the texture and stability of the end product. As mentioned, the mechanical forces generated

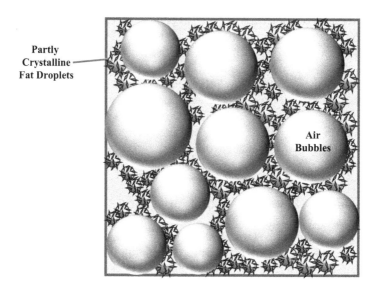

Fig. 9.17 Whipping cream is a complex colloidal dispersion consisting of air bubbles that are coated by proteins and aggregated fat droplets that are dispersed in an aqueous phase

during whipping introduce air bubbles into the aqueous phase, which are then stabilized by a mixture of adsorbed proteins and aggregated fat globules. Globular proteins may partially unfold after adsorption at the interface and then interact with each other, which increases the mechanical strength of the interfacial layer, thereby enhancing the resistance of the air bubbles to coalescence or breakdown. Hydrocolloids, such as gelatin or gums, are often added to dairy versions of whipping cream to thicken the aqueous phase, which improves their stability by retarding the movement of the air bubbles. The inclusion of sugars can increase or decrease foam quality depending on when they are introduced into the whipping cream. When the sugars are included before whipping, the foam volume and stability are often reduced since the sugars increase the stability of the globular whey proteins, which inhibits their unfolding and aggregation at the air-water interface, thereby reducing their tendency to form a protective coating around the air bubbles. Conversely, when the sugars are included after whipping, the foam volume and stability are often improved since they increase the strength of the attractive interactions between the adsorbed protein molecules through an osmotic effect, as well as increasing the viscosity of the aqueous phase, thereby inhibiting air bubble movement.

The creation of a high-quality plant-based whipping cream depends on mimicking the behavior of the milk fat globules in diary cream during the whipping process. As with ice cream, this largely depends on creating an interfacial layer around the fat droplets that is similar to that found around milk fat globules, as well as using a lipid phase that has a similar solid fat content *versus* temperature profile and crystallization behavior as milk fat. This can be achieved by using appropriate plant-based emulsifiers (such as proteins and phospholipids), as well as by using a blend of plant-based oils (such as coconut oil and sunflower oil) that provides the correct solid fat content profile (Fig. 9.6). Studies have shown that plant-based whipping creams can be produced from soybean ingredients, however, partial hydrolysis of the soy proteins was required to obtain a suitable foam volume and stability (Fu et al., 2020).

9.7.4 Butter

The butter made from cow's milk is another complex colloidal system. Compositionally, it consists of about 80% milkfat, 18% water, and 2% milk solids. Structurally, it consists of water droplets dispersed in a partially crystalline lipid phase, which consists of a 3D network of aggregated fat crystals suspended in liquid oil (Fig. 9.18).

Typically, butter is formed from pasteurized dairy cream using a controlled phase inversion process known as churning. This involves a specific cooling and heating process to obtain an optimum ratio of crystallized fat (for example 8 °C → 21 °C → 20 °C for very solid fat as starting material (Kessler, 2002)), then applying mechanical agitation using a votator device. During this operation, cream, which is an oil-in-water emulsion, is converted into butter, which is a water-in-oil emulsion, through a phase inversion process (Fig. 9.18). The partially crystalline fat globules clump together during churning due to a phenomenon known as partial coalescence. These clumps grow larger and larger as the process continues and eventually phase inversion occurs, leading to a system containing water droplets and some air bubbles trapped in a 3D fat crystal network. The textural characteristics of butter depend on the solid fat content and fat crystal morphology, which are governed by the initial fatty acid composition of the milk fat and the thermal-mechanical history of the product. Butters are usually designed to exhibit

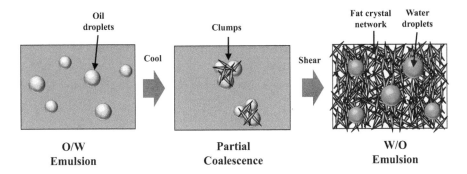

Fig. 9.18 The production of butter or plant-based butter analogs involves the conversion of an oil-in-water emulsion into a water-in-oil emulsion, which is achieved by cooling and shearing to promote partial coalescence

plastic-like rheological characteristics over a specific range of temperatures. Thus, their textural properties are characterized by a yield stress, which should be high enough to prevent the butter from collapsing under its own weight but low enough to ensure it is spreadable. It is therefore important to carefully control the solid fat content *versus* temperature profile, as well as the fat crystal morphology and interactions, so as to obtain the desirable textural and sensory attributes in the final product (Fig. 9.19). For instance, it should spread when it comes out of the refrigerator, not collapse when held at room temperature, and melt in the mouth.

Plant-based alternatives to butter have been available for many years in the form of margarine products. These were originally developed as a cheaper alternative to butter but are now being developed to meet the growing demand for plant-based foods. These products can be manufactured using a similar process to butter production, such as cooling and mechanical agitation of an oil-in-water emulsion to induce phase inversion and to create a water-in-oil emulsion with a partially solidified fat phase. However, it is important to select an appropriate plant-based emulsifier (e.g., mono- and diacylglycerols, lecithins) to form the original oil-in-water emulsion, as well as to select an appropriate plant-based lipid (e.g., palm, coconut, shea, sunflower), that will give the required solid fat content and fat crystal characteristics.

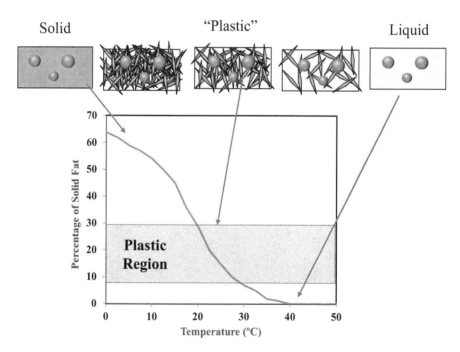

Fig. 9.19 The desirable functional attributes of butter or its plant-based analogs depends on the solid fat content versus temperature profile, as well as the morphology and interactions of the fat crystals

9.8 Future Considerations

Considerable progress has already been made in the commercial production of plant-based dairy analogs, with a number of successful products that already have been released to the market. However, research and development are still required to create high-quality plant-based dairy products that consumers want to incorporate into their diets, especially cheese analogs. Some of these products should be designed to accurately simulate the desirable physicochemical and sensory attributes of existing dairy products (such as a cheddar or ricotta cheese). However, other products may be designed to have their own unique characteristics. For instance, they could be designed to exhibit functional characteristics somewhat similar to existing dairy products (such as tastiness, hardness, yield stress, meltability, and stretchability), rather than faithfully mimicking their sensory attributes. Designing plant-based dairy products on a functionality-first approach would allow new product categories to emerge.

Based on the information presented in this chapter, we highlight a number of important areas where we believe future developments are still required:

- *Sensorial attributes*: There is still a pressing need to better understand and control the sensory attributes of plant-based dairy alternatives so as to increase their consumer acceptance. Undesirable flavors and mouthfeels are often associated with this product category. For instance, plant-based cheeses have been reported to have beany and gritty flavors (especially those formulated from soybeans), whereas plant-based yogurts have been reported to have beany, raisiny, bitter, and astringent flavors (Montemurro et al., 2021; Short et al., 2021). Some strategies for enhancing the sensorial attributes of dairy products have been discussed in this chapter, including fermentation, thermal treatments, and plant breeding but further developments are still required to achieve more desirable flavor profiles. Having said this, as plant-based foods become more familiar to consumers, they might become more accepting of the unique flavor profiles associated with them.
- *Protein gels*: Dairy products like cheese and yogurt contain 3D networks of aggregated milk proteins, which give them their unique textural, fluid retention, and other characteristic functional attributes. As discussed in this chapter, it is difficult to mimic the structural and physicochemical properties of dairy gels using plant proteins because of their different molecular features. Instead, plant proteins typically have to be used in combination with hydrocolloids to achieve functional properties somewhat similar to those in dairy gels. Consequently, plant-based products are often lower in protein content than their dairy counterparts. To overcome this problem, there is a need to better understand the gelling properties of plant proteins, either alone or in combination with other plant-derived ingredients. In addition, more collaborative work between agricultural and food scientists is required to produce ingredients with the desired functional attributes, such as proteins with a higher degree of hydrophobic amino acids at their surfaces, as this may enhance their gelation properties.

- *Whole ingredients*: Many of the processing routes presented in this chapter are based on extracted and isolated ingredients. Valuable health-promoting components, such as fibers and micronutrients, are often removed from these ingredients during their isolation. Moreover, processing steps that utilize energy, water, and chemicals are often required to isolate these ingredients, which results in a process with lower sustainability. In the future, it would be beneficial to obtain plant-based dairy products from whole ingredients that are only minimally processed. Consequently, health-promoting nutrients are retained and less energy and resources are expended. It would therefore be advantageous to design new processing operations that allow for the transformation of whole plant-derived ingredients into plant-based dairy alternatives. An example of this approach is the cheese analogs formed from nuts. Of course, it should be kept in mind that plants often contain antinutritional and potentially toxic compounds that need to be removed or inactivated during processing.
- *Health and Sustainability*: There is a need for rigorous assessments of the nutritional and sustainability profiles of dairy analogs so as to assess any health and environmental advantages or disadvantages they may have when compared to real dairy products. These comparisons should be carried out using standardized validated methods by independent researchers who do not have a vested interest in the outcome of the results.

References

Adedinsewo, D. A., Pollak, A. W., & Carter, R. E. (2021). Dietary sodium and mortality: How much do we really know? *European Heart Journal, 00*, 1–3. https://doi.org/10.1093/eurheartj/ehaa1086

Aldazabal, J., Martín-Meizoso, A., Martínez-Esnaola, J. M., & Farr, R. (2006). Deterministic model for ice cream solidification. *Computational Materials Science, 38*(1), 9–21. https://doi.org/10.1016/j.commatsci.2005.12.033

Alinovi, M., Mucchetti, G., Andersen, U., Rovers, T. A. M., Mikkelsen, B., Wiking, L., & Corredig, M. (2020). Applicability of confocal Raman microscopy to observe microstructural modifications of cream cheeses as influenced by freezing. *Food, 9*(5), 679. https://doi.org/10.3390/foods9050679

Al-Nabulsi, A., Shaker, R., Osaili, T., Al-Taani, M., Olaimat, A., Awaisheh, S., … Holley, R. (2014). Sensory evaluation of flavored soy milk-based yogurt: A comparison between Jordanian and Malaysian consumers. *Journal of Food Science and Engineering, 4*, 27–35.

Anderson, T. J., & Lamsal, B. P. (2011). Zein extraction from corn, corn products, and coproducts and modifications for various applications: A review. *Cereal Chemistry, 88*(2), 159–173. https://doi.org/10.1094/cchem-06-10-0091

Apostolopoulos, C., & Marshall, R. J. (1994). A quantitative method for the determination of shreddability of cheese. *Journal of Food Quality, 17*(2), 115–158. https://doi.org/10.1111/j.1745-4557.1994.tb00137.x

Appleby, P. N., Crowe, F. L., Bradbury, K. E., Travis, R. C., & Key, T. J. (2016). Mortality in vegetarians and comparable nonvegetarians in the United Kingdom. *The American Journal of Clinical Nutrition, 103*(1), 218–230. https://doi.org/10.3945/ajcn.115.119461

Arab-Tehrany, E., Jacquot, M., Gaiani, C., Imran, M., Desobry, S., & Linder, M. (2012). Beneficial effects and oxidative stability of omega-3 long-chain polyunsaturated fatty acids. *Trends in Food Science & Technology, 25*(1), 24–33. https://doi.org/10.1016/j.tifs.2011.12.002

References

Atapattu, C., & Fannon, J. (2014). *Improved dry blend for making cheese analogue*. Retrieved from: https://patentscope.wipo.int/search/en/detail.jsf?docId=WO2014085250. Accessed Mar 2021.

Awad, R. A., Salama, W. M., & Farahat, A. M. (2014). Effect of lupine as cheese base substitution on technological and nutritional properties of processed cheese analogue. *Acta Scientiarum Polonorum Technologia Alimentaria, 13*(1), 55–64. https://doi.org/10.17306/J.AFS.2014.1.5

Balafa, O., & Kalaitzidis, R. G. (2021). Salt sensitivity and hypertension. *Journal of Human Hypertension, 35*(3), 184–192. https://doi.org/10.1038/s41371-020-00407-1

Bansal, S., Mangal, M., Sharma, S. K., Yadav, D. N., & Gupta, R. K. (2016). Optimization of process conditions for developing yoghurt like probiotic product from peanut. *LWT, 73*, 6–12. https://doi.org/10.1016/j.lwt.2016.04.059

Banville, V., Morin, P., Pouliot, Y., & Britten, M. (2013). Physical properties of pizza Mozzarella cheese manufactured under different cheese-making conditions. *Journal of Dairy Science, 96*(8), 4804–4815. https://doi.org/10.3168/jds.2012-6314

Batty, D., Waite-Cusic, J. G., & Meunier-Goddik, L. (2019). Influence of cheese-making recipes on the composition and characteristics of Camembert-type cheese. *Journal of Dairy Science, 102*(1), 164–176. https://doi.org/10.3168/jds.2018-14964

Bava, L., Bacenetti, J., Gislon, G., Pellegrino, L., D'Incecco, P., Sandrucci, A., … Zucali, M. (2018). Impact assessment of traditional food manufacturing: The case of Grana Padano cheese. *Science of the Total Environment, 626*, 1200–1209. https://doi.org/10.1016/j.scitotenv.2018.01.143

BeMiller, J. N., & Whistler, R. L. (Eds.). (2009). *Starch: Chemistry and technology* (3rd ed.). Academic.

Ben-Harb, S., Panouillé, M., Huc-Mathis, D., Moulin, G., Saint-Eve, A., Irlinger, F., … Souchon, I. (2018). The rheological and microstructural properties of pea, milk, mixed pea/milk gels and gelled emulsions designed by thermal, acid, and enzyme treatments. *Food Hydrocolloids, 77*, 75–84. https://doi.org/10.1016/j.foodhyd.2017.09.022

Bergsma, J. (2017). Vegan cheese analogue. Patent number: WO2017150973A1. Retrieved from: https://patents.google.com/patent/WO2017150973A1/en?oq=plant+based+cheese+starch. Accessed Mar 2021.

Bernat, N., Cháfer, M., Chiralt, A., & González-Martínez, C. (2014). Vegetable milks and their fermented derivative products. *International Journal of Food Studies, 3*(1). https://doi.org/10.7455/ijfs/3.1.2014.a9

Bernat, N., Cháfer, M., Chiralt, A., & González-Martínez, C. (2015). Probiotic fermented almond "milk" as an alternative to cow-milk yoghurt. *International Journal of Food Studies, 4*(2). https://doi.org/10.7455/ijfs/4.2.2015.a8

Boeck, T., Sahin, A. W., Zannini, E., & Arendt, E. K. (2021). Nutritional properties and health aspects of pulses and their use in plant-based yogurt alternatives. *Comprehensive Reviews in Food Science and Food Safety, 20*(4), 3858–3880. https://doi.org/10.1111/1541-4337.12778

Boye, J., Wijesinha-Bettoni, R., & Burlingame, B. (2012). Protein quality evaluation twenty years after the introduction of the protein digestibility corrected amino acid score method. *British Journal of Nutrition, 108*(S2), S183–S211. https://doi.org/10.1017/S0007114512002309

Brady, J. (2013). *Introductory food chemistry* (Illustrated ed.). Comstock Publishing Associates.

Braun, M., Muñoz, I., Schmidt, J. H., & Thrane, M. (2016). Sustainability of soy protein from life cycle assessment. *The FASEB Journal, 30*(S1), 894.5–894.5. https://doi.org/10.1096/fasebj.30.1_supplement.894.5

Breuninger, W. F., Piyachomkwan, K., & Sriroth, K. (2009). Chapter 12 – Tapioca/cassava starch: Production and use. In J. BeMiller & R. Whistler (Eds.), *Starch* (3rd ed., pp. 541–568). Academic. https://doi.org/10.1016/B978-0-12-746275-2.00012-4

Brown, P. O., Casino, M., Voccola, L. S., & Varadan, R. (2013). Methods and compositions for consumables. Patent number: CA2841470A1. Retrieved from: https://patents.google.com/patent/CA2841470A1/pt. Accessed Mar 2021.

Brückner-Gühmann, M., Vasil'eva, E., Culetu, A., Duta, D., Sozer, N., & Drusch, S. (2019). Oat protein concentrate as alternative ingredient for non-dairy yoghurt-type product. *Journal of the Science of Food and Agriculture, 99*(13), 5852–5857. https://doi.org/10.1002/jsfa.9858

Cáceres, P. J., Peñas, E., Martínez-Villaluenga, C., García-Mora, P., & Frías, J. (2019). Development of a multifunctional yogurt-like product from germinated brown rice. *LWT, 99*, 306–312. https://doi.org/10.1016/j.lwt.2018.10.008

Canabady-Rochelle, L.-S., Sanchez, C., Mellema, M., & Banon, S. (2009). Study of calcium–soy protein interactions by isothermal titration calorimetry and pH cycle. *Journal of Agricultural and Food Chemistry, 57*(13), 5939–5947. https://doi.org/10.1021/jf900424b

Cebula, D. J., & Hoddle, A. (2009). Chapter 9 – Chocolate and couvertures: Applications in ice cream. In G. Talbot (Ed.), *Science and technology of enrobed and filled chocolate, confectionery and bakery products* (pp. 163–182). Woodhead Publishing. https://doi.org/10.1533/9781845696436.1.163

Chambers, D. H., Esteve, E., & Retiveau, A. (2010). Effect of milk pasteurization on flavor properties of seven commercially available French cheese types. *Journal of Sensory Studies, 25*(4), 494–511. https://doi.org/10.1111/j.1745-459x.2010.00282.x

Chen, G.-C., Zhang, R., Martínez-González, M. A., Zhang, Z.-L., Bonaccio, M., van Dam, R. M., & Qin, L.-Q. (2017). Nut consumption in relation to all-cause and cause-specific mortality: A meta-analysis 18 prospective studies. *Food & Function, 8*(11), 3893–3905. https://doi.org/10.1039/C7FO00915A

Chen, J. M., Al, K. F., Craven, L. J., Seney, S., Coons, M., McCormick, H., … Burton, J. P. (2020). Nutritional, microbial, and allergenic changes during the fermentation of cashew 'cheese' product using a quinoa-based rejuvelac starter culture. *Nutrients, 12*(3), 648. https://doi.org/10.3390/nu12030648

Chen, M., Lu, J., Liu, F., Nsor-Atindana, J., Xu, F., Goff, H. D., … Zhong, F. (2019). Study on the emulsifying stability and interfacial adsorption of pea proteins. *Food Hydrocolloids, 88*, 247–255. https://doi.org/10.1016/j.foodhyd.2018.09.003

Cheng, Y. J., Thompson, L. D., & Brittin, H. C. (1990). Sogurt, a yogurt-like soybean product: Development and properties. *Journal of Food Science, 55*(4), 1178–1179. https://doi.org/10.1111/j.1365-2621.1990.tb01631.x

Chevanan, N., Muthukumarappan, K., Upreti, P., & Metzger, L. E. (2006). Effect of calcium and phosphorus, residual lactose and salt-to-moisture ratio on textural properties of Cheddar cheese during ripening. *Journal of Texture Studies, 37*(6), 711–730. https://doi.org/10.1111/j.1745-4603.2006.00080.x

Chikpah, S. K., Teye, M., Annor, J. A. F., & Teye, G. A. (2015). Potentials of sodom apple (Calotropis procera) extract as a coagulant to substitute alum in soy cheese production in Ghana. *Elixir Food Science, 79*, 30166–30170.

Childs, J. L., Daubert, C. R., Stefanski, L., & Foegeding, E. A. (2007). Factors regulating cheese shreddability. *Journal of Dairy Science, 90*(5), 2163–2174. https://doi.org/10.3168/jds.2006-618

Chobanian, A. V., & Hill, M. (2000). National Heart, Lung, and Blood Institute workshop on sodium and blood pressure. *Hypertension, 35*(4), 858–863. https://doi.org/10.1161/01.HYP.35.4.858

Chumchuere, S., MacDougall, D. B., & Robinson, R. K. (2000). Production and properties of a semi-hard cheese made from soya milk. *International Journal of Food Science & Technology, 35*(6), 577–581. https://doi.org/10.1111/j.1365-2621.2000.00414.x

Clarke, C. (2007). *The science of ice cream*. Royal Society of Chemistry.

Clune, S., Crossin, E., & Verghese, K. (2017). Systematic review of greenhouse gas emissions for different fresh food categories. *Journal of Cleaner Production, 140*, 766–783. https://doi.org/10.1016/j.jclepro.2016.04.082

Coda, R., Rizzello, C. G., Trani, A., & Gobbetti, M. (2011). Manufacture and characterization of functional emmer beverages fermented by selected lactic acid bacteria. *Food Microbiology, 28*(3), 526–536. https://doi.org/10.1016/j.fm.2010.11.001

Coker, C. J., Crawford, R. A., Johnston, K. A., Singh, H., & Creamer, L. K. (2005). Towards the classification of cheese variety and maturity on the basis of statistical analysis of proteolysis data – A review. *International Dairy Journal, 15*(6), 631–643. https://doi.org/10.1016/j.idairyj.2004.10.011

Collins, Y. F., McSweeney, P. L. H., Wilkinson, M. G. (2003). Lipolysis and free fatty acid catabolism in cheese: a review of current knowledge. *International Dairy Journal. 13*(11): 841–866. https://doi.org/10.1016/S0958-6946(03)00109-2

Cook, K. L. K., & Hartel, R. W. (2010). Mechanisms of ice crystallization in ice cream production. *Comprehensive Reviews in Food Science and Food Safety, 9*(2), 213–222. https://doi.org/10.1111/j.1541-4337.2009.00101.x

D'Incecco, P., Limbo, S., Hogenboom, J., Rosi, V., Gobbi, S., & Pellegrino, L. (2020). Impact of extending hard-cheese ripening: A multiparameter characterization of Parmigiano Reggiano cheese ripened up to 50 months. *Food, 9*(3), 268. https://doi.org/10.3390/foods9030268

Dahlan, H. A., & Sani, N. A. (2017). The interaction effect of mixing starter cultures on homemade natural yogurt's pH and viscosity. *International Journal of Food Studies, 6*(2). https://doi.org/10.7455/ijfs/6.2.2017.a3

Dalgleish, D. (2011). On the structural models of bovine casein micelles – Review and possible improvements. *Soft Matter, 7*(6), 2265–2272. https://doi.org/10.1039/C0SM00806K

Dalla Riva, A., Burek, J., Kim, D., Thoma, G., Cassandro, M., & De Marchi, M. (2017). Environmental life cycle assessment of Italian mozzarella cheese: Hotspots and improvement opportunities. *Journal of Dairy Science, 100*(10), 7933–7952. https://doi.org/10.3168/jds.2016-12396

Day, L. (2013). Proteins from land plants – Potential resources for human nutrition and food security. *Trends in Food Science & Technology, 32*(1), 25–42. https://doi.org/10.1016/j.tifs.2013.05.005

de Goede, J., Soedamah-Muthu, S. S., Pan, A., Gijsbers, L., & Geleijnse, J. M. (2016). Dairy consumption and risk of stroke: A systematic review and updated dose–response meta-analysis of prospective cohort studies. *Journal of the American Heart Association: Cardiovascular and Cerebrovascular Disease, 5*(5). https://doi.org/10.1161/JAHA.115.002787

Dias, S., Castanheira, E. M. S., Fortes, A. G., Pereira, D. M., & Gonçalves, M. S. T. (2020). Natural pigments of anthocyanin and betalain for coloring soy-based yogurt alternative. *Food, 9*(6), 771. https://doi.org/10.3390/foods9060771

Dinu, M., Abbate, R., Gensini, G. F., Casini, A., & Sofi, F. (2017). Vegetarian, vegan diets and multiple health outcomes: A systematic review with meta-analysis of observational studies. *Critical Reviews in Food Science and Nutrition, 57*(17), 3640–3649. https://doi.org/10.1080/10408398.2016.1138447

Doehlert, D. C., Angelikousis, S., & Vick, B. (2010). Accumulation of oxygenated fatty acids in oat lipids during storage. *Cereal Chemistry, 87*(6), 532–537. https://doi.org/10.1094/CCHEM-05-10-0074

Donkor, O. N., Henriksson, A., Vasiljevic, T., & Shah, N. P. (2005). Probiotic strains as starter cultures improve angiotensin-converting enzyme inhibitory activity in soy yogurt. *Journal of Food Science, 70*(8), m375–m381. https://doi.org/10.1111/j.1365-2621.2005.tb11522.x

Dugat-Bony, E., Straub, C., Teissandier, A., Onésime, D., Loux, V., Monnet, C., ... Bonnarme, P. (2015). Overview of a surface-ripened cheese community functioning by meta-omics analyses. *PLoS One, 10*(4), e0124360. https://doi.org/10.1371/journal.pone.0124360

Eckhoff, S. R., & Watson, S. A. (2009). Chapter 9 – Corn and sorghum starches: Production. In J. BeMiller & R. Whistler (Eds.), *Starch* (3rd ed., pp. 373–439). Academic. https://doi.org/10.1016/B978-0-12-746275-2.00009-4

El-Ella, W. M. A. (1980). Hard cheese substitute from soy milk. *Journal of Food Science, 45*(6), 1777–1778. https://doi.org/10.1111/j.1365-2621.1980.tb07610.x

Eugster, E., Fuchsmann, P., Schlichtherle-Cerny, H., Bütikofer, U., & Irmler, S. (2019). Formation of alanine, α-aminobutyrate, acetate, and 2-butanol during cheese ripening by Pediococcus acidilactici FAM18098. *International Dairy Journal, 96*, 21–28. https://doi.org/10.1016/j.idairyj.2019.04.001

Eyres, L., Eyres, M. F., Chisholm, A., & Brown, R. C. (2016). Coconut oil consumption and cardiovascular risk factors in humans. *Nutrition Reviews, 74*(4), 267–280. https://doi.org/10.1093/nutrit/nuw002

Faccia, M., Gambacorta, G., Martemucci, G., Difonzo, G., & D'Alessandro, A. G. (2019). Chemical-sensory traits of fresh cheese made by enzymatic coagulation of donkey milk. *Foods (Basel, Switzerland), 9*(1), 1–13. https://doi.org/10.3390/foods9010016

Fardet, A., & Boirie, Y. (2014). Associations between food and beverage groups and major diet-related chronic diseases: An exhaustive review of pooled/meta-analyses and systematic reviews. *Nutrition Reviews, 72*(12), 741–762. https://doi.org/10.1111/nure.12153

Farquhar, W. B., Edwards, D. G., Jurkovitz, C. T., & Weintraub, W. S. (2015). Dietary sodium and health. *Journal of the American College of Cardiology, 65*(10), 1042–1050. https://doi.org/10.1016/j.jacc.2014.12.039

Farrell, H. M., Jimenez-Flores, R., Bleck, G. T., Brown, E. M., Butler, J. E., Creamer, L. K., … Swaisgood, H. E. (2004). Nomenclature of the proteins of cows' milk – Sixth revision. *Journal of Dairy Science, 87*(6), 1641–1674. https://doi.org/10.3168/jds.S0022-0302(04)73319-6

Farvid, M. S., Malekshah, A. F., Pourshams, A., Poustchi, H., Sepanlou, S. G., Sharafkhah, M., … Malekzadeh, R. (2017). Dairy food intake and all-cause, cardiovascular disease, and cancer mortality. *American Journal of Epidemiology, 185*(8), 697–711. https://doi.org/10.1093/aje/kww139

Fathi, M., Donsi, F., & McClements, D. J. (2018). Protein-based delivery systems for the nano-encapsulation of food ingredients. *Comprehensive Reviews in Food Science and Food Safety, 17*(4), 920–936. https://doi.org/10.1111/1541-4337.12360

Faulkner, H., O'Callaghan, T. F., McAuliffe, S., Hennessy, D., Stanton, C., O'Sullivan, M. G., … Kilcawley, K. N. (2018). Effect of different forage types on the volatile and sensory properties of bovine milk. *Journal of Dairy Science, 101*(2), 1034–1047. https://doi.org/10.3168/jds.2017-13141

Finnegan, W., Yan, M., Holden, N. M., & Goggins, J. (2018). A review of environmental life cycle assessment studies examining cheese production. *The International Journal of Life Cycle Assessment, 23*(9), 1773–1787. https://doi.org/10.1007/s11367-017-1407-7

Fox, P. F., Guinee, T. P., Cogan, T. M., & McSweeney, P. L. H. (2017). Cheese: Structure, rheology and texture. In P. F. Fox, T. P. Guinee, T. M. Cogan, & P. L. H. McSweeney (Eds.), *Fundamentals of cheese science* (pp. 475–532). Springer. https://doi.org/10.1007/978-1-4899-7681-9_14

Fox, P. F., & McSweeney, P. L. H. (2017). Chapter 1 – Cheese: An overview. In P. L. H. McSweeney, P. F. Fox, P. D. Cotter, & D. W. Everett (Eds.), *Cheese* (4th ed., pp. 5–21). Academic. https://doi.org/10.1016/B978-0-12-417012-4.00001-6

Franzoi, M., Niero, G., Visentin, G., Penasa, M., Cassandro, M., & De Marchi, M. (2019). Variation of detailed protein composition of cow milk predicted from a large database of mid-infrared spectra. *Animals, 9*(4), 176. https://doi.org/10.3390/ani9040176

Fu, L., He, Z., Zeng, M., Qin, F., & Chen, J. (2020). Effects of soy protein composition in recombined soy-based cream on the stability and physical properties of whipping cream. *Journal of the Science of Food and Agriculture, 100*(6), 2732–2741. https://doi.org/10.1002/jsfa.10305

GFI. (2020). *2020 state of the industry report*. The Good Food Institute. Retrieved from: https://gfi.org/resource/plant-based-retail-report/. Accessed Aug 2021

Gholami, A., Rezaei, S., Jahromi, L. M., Baradaran, H. R., Ghanbari, A., Djalalinia, S., … Farzadfar, F. (2020). Is salt intake reduction a universal intervention for both normotensive and hypertensive people: A case from Iran STEPS survey 2016. *European Journal of Nutrition, 59*(7), 3149–3161. https://doi.org/10.1007/s00394-019-02153-8

Giri, S. K., Tripathi, M. K., & Kotwaliwale, N. (2018). Effect of composition and storage time on some physico-chemical and rheological properties of probiotic soy-cheese spread. *Journal of Food Science and Technology, 55*(5), 1667–1674. https://doi.org/10.1007/s13197-018-3078-1

Glantz, M., Devold, T. G., Vegarud, G. E., Lindmark Månsson, H., Stålhammar, H., & Paulsson, M. (2010). Importance of casein micelle size and milk composition for milk gelation. *Journal of Dairy Science, 93*(4), 1444–1451. https://doi.org/10.3168/jds.2009-2856

Glusac, J., Davidesko-Vardi, I., Isaschar-Ovdat, S., Kukavica, B., & Fishman, A. (2018). Gel-like emulsions stabilized by tyrosinase-crosslinked potato and zein proteins. *Food Hydrocolloids, 82*, 53–63. https://doi.org/10.1016/j.foodhyd.2018.03.046

Glusac, J., Davidesko-Vardi, I., Isaschar-Ovdat, S., Kukavica, B., & Fishman, A. (2019). Tyrosinase-crosslinked pea protein emulsions: Impact of zein incorporation. *Food Research International, 116*, 370–378. https://doi.org/10.1016/j.foodres.2018.08.050

Goff, H. D. (1997). Colloidal aspects of ice cream – A review. *International Dairy Journal, 7*(6), 363–373. https://doi.org/10.1016/S0958-6946(97)00040-X

Goff, H. D., & Hartel, R. W. (2013). *Ice cream* (7th ed.). Springer. https://doi.org/10.1007/978-1-4614-6096-1

González-García, S., Castanheira, É. G., Dias, A. C., & Arroja, L. (2013). Environmental performance of a Portuguese mature cheese-making dairy mill. *Journal of Cleaner Production, 41*, 65–73. https://doi.org/10.1016/j.jclepro.2012.10.010

Górska-Warsewicz, H., Rejman, K., Laskowski, W., & Czeczotko, M. (2019). Milk and dairy products and their nutritional contribution to the average polish diet. *Nutrients, 11*(8), 1–19. https://doi.org/10.3390/nu11081771

Grasso, N., Alonso-Miravalles, L., & O'Mahony, J. A. (2020). Composition, physicochemical and sensorial properties of commercial plant-based yogurts. *Food, 9*(3), 252. https://doi.org/10.3390/foods9030252

Greis, M., Sainio, T., Katina, K., Kinchla, A. J., Nolden, A., Partanen, R., & Seppä, L. (2020). Dynamic texture perception in plant-based yogurt alternatives: Identifying temporal drivers of liking by TDS. *Food Quality and Preference, 86*, 104019. https://doi.org/10.1016/j.foodqual.2020.104019

Grommers, H. E., & van der Krogt, D. A. (2009). Chapter 11 – Potato starch: Production, modifications and uses. In J. BeMiller & R. Whistler (Eds.), *Starch* (3rd ed., pp. 511–539). Academic. https://doi.org/10.1016/B978-0-12-746275-2.00011-2

Grossmann, L., & McClements, D. J. (2021). The science of plant-based foods: Approaches to create nutritious and sustainable plant-based cheese analogs. *Trends in Food Science & Technology, 118*, 207–229. https://doi.org/10.1016/j.tifs.2021.10.004

Grossmann, L., & Weiss, J. (2021). Alternative protein sources as technofunctional food ingredients. *Annual Review of Food Science and Technology, 12*(1), 93–117. https://doi.org/10.1146/annurev-food-062520-093642

Guillin, F. M., Gaudichon, C., Guérin-Deremaux, L., Lefranc-Millot, C., Azzout-Marniche, D., Khodorova, N., & Calvez, J. (2021). Multi-criteria assessment of pea protein quality in rats: A comparison between casein, gluten and pea protein alone or supplemented with methionine. *British Journal of Nutrition, 125*(4), 389–397. https://doi.org/10.1017/S0007114520002883

Guinee, T. P. (2017). Chapter 46 – Pasteurized processed and imitation cheese products. In P. L. H. McSweeney, P. F. Fox, P. D. Cotter, & D. W. Everett (Eds.), *Cheese* (4th ed., pp. 1133–1184). Academic. https://doi.org/10.1016/B978-0-12-417012-4.00046-6

Guo, J., Astrup, A., Lovegrove, J. A., Gijsbers, L., Givens, D. I., & Soedamah-Muthu, S. S. (2017). Milk and dairy consumption and risk of cardiovascular diseases and all-cause mortality: Dose–response meta-analysis of prospective cohort studies. *European Journal of Epidemiology, 32*(4), 269–287. https://doi.org/10.1007/s10654-017-0243-1

Guo, J., Yang, X.-Q., He, X.-T., Wu, N.-N., Wang, J.-M., Gu, W., & Zhang, Y.-Y. (2012). Limited aggregation behavior of β-conglycinin and its terminating effect on glycinin aggregation during heating at pH 7.0. *Journal of Agricultural and Food Chemistry, 60*(14), 3782–3791. https://doi.org/10.1021/jf300409y

Hallén, E., Lundén, A., Tyrisevä, A.-M., Westerlind, M., & Andrén, A. (2010). Composition of poorly and non-coagulating bovine milk and effect of calcium addition. *Journal of Dairy Research, 77*(4), 398–403. https://doi.org/10.1017/S0022029910000671

Hassan, A. N., Ipsen, R., Janzen, T., & Qvist, K. B. (2003). Microstructure and rheology of yogurt made with cultures differing only in their ability to produce exopolysaccharides. *Journal of Dairy Science, 86*(5), 1632–1638. https://doi.org/10.3168/jds.S0022-0302(03)73748-5

Heileson, J. L. (2020). Dietary saturated fat and heart disease: A narrative review. *Nutrition Reviews, 78*(6), 474–485. https://doi.org/10.1093/nutrit/nuz091

Henderson, P. A., & Unnasch, S. (2017). *Life cycle assessment of ripple non-dairy milk*. Retrieved from: https://www.ripplefoods.com/pdf/Ripple_LCA_Report.pdf

Hewlings, S. (2020). Coconuts and health: Different chain lengths of saturated fats require different consideration. *Journal of Cardiovascular Development and Disease, 7*(4), 1–15. https://doi.org/10.3390/jcdd7040059

Hickisch, A., Beer, R., Vogel, R. F., & Toelstede, S. (2016). Influence of lupin-based milk alternative heat treatment and exopolysaccharide-producing lactic acid bacteria on the physical characteristics of lupin-based yogurt alternatives. *Food Research International, 84*, 180–188. https://doi.org/10.1016/j.foodres.2016.03.037

Hjerpsted, J., & Tholstrup, T. (2016). Cheese and cardiovascular disease risk: A review of the evidence and discussion of possible mechanisms. *Critical Reviews in Food Science and Nutrition, 56*(8), 1389–1403. https://doi.org/10.1080/10408398.2013.769197

Holsinger, V. H., Smith, P. W., & Tunick, M. H. (1995). Overview: Cheese chemistry and rheology. In E. L. Malin & M. H. Tunick (Eds.), *Chemistry of structure-function relationships in cheese* (pp. 1–6). Springer. https://doi.org/10.1007/978-1-4615-1913-3_1

Holz-Schietinger, C., Klapholz, S., Wardan, R., Casino, M., Brown, P. O., & Eisen, M. (2014). Non-dairy cheese replica comprising a coacervate. Patent number: WO2014110540A1. Retrieved from: https://patents.google.com/patent/WO2014110540A1/en?oq=vegan+cheese+pea. Accessed Mar 2021.

Hougaard, A. B., Ardö, Y., & Ipsen, R. H. (2010). Cheese made from instant infusion pasteurized milk: Rennet coagulation, cheese composition, texture and ripening. *International Dairy Journal, 20*(7), 449–458. https://doi.org/10.1016/j.idairyj.2010.01.005

Hu, F. B., Manson, J. E., & Willett, W. C. (2001). Types of dietary fat and risk of coronary heart disease: A critical review. *Journal of the American College of Nutrition, 20*(1), 5–19. https://doi.org/10.1080/07315724.2001.10719008

Hwang, C. E., Haque, M. A., Lee, J. H., Song, Y. H., Lee, H. Y., Kim, S. C., & Cho, K. M. (2018). Bioconversion of γ-aminobutyric acid and isoflavone contents during the fermentation of high-protein soy powder yogurt with Lactobacillus brevis. *Applied Biological Chemistry, 61*(4), 409–421. https://doi.org/10.1007/s13765-018-0366-4

Ianni, A., Bennato, F., Martino, C., Grotta, L., & Martino, G. (2020). Volatile flavor compounds in cheese as affected by ruminant diet. *Molecules, 25*(3), 1–16. https://doi.org/10.3390/molecules25030461

Irudayaraj, J. (1999). Texture development in cheddar cheese during ripening. *Canadian Biosystems Engineering/Le Genie des biosystems au Canada, 41*(4), 253–258.

Jackson, D. S. (2003). Starch | functional properties. In B. Caballero (Ed.), *Encyclopedia of food sciences and nutrition* (2nd ed., pp. 5572–5575). Academic. https://doi.org/10.1016/B0-12-227055-X/01143-3

Jacobsen, C. (2015). Some strategies for the stabilization of long chain n-3 PUFA-enriched foods: A review. *European Journal of Lipid Science and Technology, 117*(11), 1853–1866. https://doi.org/10.1002/ejlt.201500137

Jacobsen, C., Sørensen, A.-D. M., & Nielsen, N. S. (2013). Chapter 4 – Stabilization of omega-3 oils and enriched foods using antioxidants. In C. Jacobsen, N. S. Nielsen, A. F. Horn, & A.-D. M. Sørensen (Eds.), *Food enrichment with omega-3 fatty acids* (pp. 130–149). Woodhead Publishing. https://doi.org/10.1533/9780857098863.2.130

Johnson, M. E. (2017). A 100-year review: Cheese production and quality. *Journal of Dairy Science, 100*(12), 9952–9965. https://doi.org/10.3168/jds.2017-12979

Kammerlehner, J. (2009). *Cheese technology*. B&L MedienGesellschaft mbH & Co. KG.

Kao, F.-J., Su, N.-W., & Lee, M.-H. (2003). Effect of calcium sulfate concentration in soymilk on the microstructure of firm tofu and the protein constitutions in Tofu whey. *Journal of Agricultural and Food Chemistry, 51*(21), 6211–6216. https://doi.org/10.1021/jf0342021

Karoui, R., Laguet, A., & Dufour, É. (2003). Fluorescence spectroscopy: A tool for the investigation of cheese melting – Correlation with rheological characteristics. *Le Lait, 83*(3), 251–264. https://doi.org/10.1051/lait:2003014

References

Kasprzak, M. M., Macnaughtan, W., Harding, S., Wilde, P., & Wolf, B. (2018). Stabilisation of oil-in-water emulsions with non-chemical modified gelatinised starch. *Food Hydrocolloids, 81*, 409–418. https://doi.org/10.1016/j.foodhyd.2018.03.002

Kessler, H. G. (2002). *Food and bio process engineering – Dairy technology.* Kessler, N. Verlag A. Kessler.

Kilcawley, K. N., Faulkner, H., Clarke, H. J., O'Sullivan, M. G., & Kerry, J. P. (2018). Factors influencing the flavour of bovine milk and cheese from grass based versus non-grass based milk production systems. *Foods (Basel, Switzerland), 7*(3), 1–43. https://doi.org/10.3390/foods7030037

Kim, H., Caulfield, L. E., Garcia-Larsen, V., Steffen, L. M., Coresh, J., & Rebholz, C. M. (2019). Plant-based diets are associated with a lower risk of incident cardiovascular disease, cardiovascular disease mortality, and all-cause mortality in a general population of middle-aged adults. *Journal of the American Heart Association, 8*(16), e012865. https://doi.org/10.1161/JAHA.119.012865

Kim, H., Caulfield, L. E., & Rebholz, C. M. (2018). Healthy plant-based diets are associated with lower risk of all-cause mortality in US adults. *The Journal of Nutrition, 148*(4), 624–631. https://doi.org/10.1093/jn/nxy019

Kim, Y., Je, Y., & Giovannucci, E. L. (2021). Association between dietary fat intake and mortality from all-causes, cardiovascular disease, and cancer: A systematic review and meta-analysis of prospective cohort studies. *Clinical Nutrition, 40*(3), 1060–1070. https://doi.org/10.1016/j.clnu.2020.07.007

Kindstedt, P. S. (1995). Factors affecting the functional characteristics of unmelted and melted Mozzarella cheese. In E. L. Malin & M. H. Tunick (Eds.), *Chemistry of structure-function relationships in cheese* (pp. 27–41). Springer. https://doi.org/10.1007/978-1-4615-1913-3_4

Klemaszewski, J. L., Fonteyn, D., Bouron, F., & Lemonnier, L. (2016). Cheese product with modified starches. Patent number: WO2016195814A1. Retrieved from: https://patents.google.com/patent/WO2016195814A1/en. Accessed Mar 2021.

Klost, M., & Drusch, S. (2019). Structure formation and rheological properties of pea protein-based gels. *Food Hydrocolloids, 94*, 622–630. https://doi.org/10.1016/j.foodhyd.2019.03.030

Kohyama, K., Sano, Y., & Doi, E. (1995). Rheological characteristics and gelation mechanism of tofu (soybean curd). *Journal of Agricultural and Food Chemistry, 43*(7), 1808–1812. https://doi.org/10.1021/jf00055a011

Konuspayeva, G., Camier, B., Aleilawi, N., Al-Shumeimyri, M., Al-Hammad, K., Algruin, K., … Faye, B. (2017). Manufacture of dry- and brine-salted soft camel cheeses for the camel dairy industry. *International Journal of Dairy Technology, 70*(1), 92–101. https://doi.org/10.1111/1471-0307.12319

Kyriakopoulou, K., Keppler, J. K., & van der Goot, A. J. (2021). Functionality of ingredients and additives in plant-based meat analogues. *Food, 10*(3), 600. https://doi.org/10.3390/foods10030600

Lamichhane, P., Kelly, A. L., & Sheehan, J. J. (2018). Symposium review: Structure-function relationships in cheese. *Journal of Dairy Science, 101*(3), 2692–2709. https://doi.org/10.3168/jds.2017-13386

Laska, M. (2010). Olfactory perception of 6 amino acids by human subjects. *Chemical Senses, 35*(4), 279–287. https://doi.org/10.1093/chemse/bjq017

Lawrence, G. D. (2021). Perspective: The saturated fat–unsaturated oil dilemma: Relations of dietary fatty acids and serum cholesterol, atherosclerosis, inflammation, cancer, and all-cause mortality. *Advances in Nutrition, 0*, 1–10. https://doi.org/10.1093/advances/nmab013

Lawrence, R. C., Creamer, L. K., & Gilles, J. (1987). Texture development during cheese ripening. *Journal of Dairy Science, 70*(8), 1748–1760. https://doi.org/10.3168/jds.S0022-0302(87)80207-2

Levy, R., Okun, Z., Davidovich-Pinhas, M., & Shpigelman, A. (2021). Utilization of high-pressure homogenization of potato protein isolate for the production of dairy-free yogurt-like fermented product. *Food Hydrocolloids, 113*, 106442. https://doi.org/10.1016/j.foodhyd.2020.106442

Li, Q., Xia, Y., Zhou, L., & Xie, J. (2013). Evaluation of the rheological, textural, microstructural and sensory properties of soy cheese spreads. *Food and Bioproducts Processing, 91*(4), 429–439. https://doi.org/10.1016/j.fbp.2013.03.001

Lie-Piang, A., Braconi, N., Boom, R. M., & van der Padt, A. (2021). Less refined ingredients have lower environmental impact – A life cycle assessment of protein-rich ingredients from oil- and starch-bearing crops. *Journal of Cleaner Production, 292*, 126046. https://doi.org/10.1016/j.jclepro.2021.126046

Lim, T., Easa, A., Karim, A., Bhat, R., & Liong, M. (2011). Development of soy-based cream cheese via the addition of microbial transglutaminase, soy protein isolate and maltodextrin. *British Food Journal, 113*(9), 1147–1172. https://doi.org/10.1108/00070701111174587

Liu, D.-M., Li, L., Yang, X.-Q., Liang, S.-Z., & Wang, J.-S. (2006). Survivability of Lactobacillus rhamnosus during the preparation of soy cheese. *Food Technology and Biotechnology, 44*(3), 417–422.

Liu, J., Han, B., Deng, S., Sun, S., & Chen, J. (2018). Changes in proteases and chemical compounds in the exterior and interior of sufu, a Chinese fermented soybean food, during manufacture. *LWT, 87*, 210–216. https://doi.org/10.1016/j.lwt.2017.08.047

Lopez, C., Briard-Bion, V., Camier, B., & Gassi, J.-Y. (2006). Milk fat thermal properties and solid fat content in emmental cheese: A differential scanning calorimetry study. *Journal of Dairy Science, 89*(8), 2894–2910. https://doi.org/10.3168/jds.S0022-0302(06)72562-0

Lorusso, A., Coda, R., Montemurro, M., & Rizzello, C. G. (2018). Use of selected lactic acid bacteria and quinoa flour for manufacturing novel yogurt-like beverages. *Food, 7*(4), 51. https://doi.org/10.3390/foods7040051

Luana, N., Rossana, C., Curiel, J. A., Kaisa, P., Marco, G., & Rizzello, C. G. (2014). Manufacture and characterization of a yogurt-like beverage made with oat flakes fermented by selected lactic acid bacteria. *International Journal of Food Microbiology, 185*, 17–26. https://doi.org/10.1016/j.ijfoodmicro.2014.05.004

Lucey, J. A. (2020). Chapter 16 – Milk protein gels. In M. Boland & H. Singh (Eds.), *Milk proteins* (3rd ed., pp. 599–632). Academic. https://doi.org/10.1016/B978-0-12-815251-5.00016-5

Lucey, J. A., & Horne, D. S. (2018). Perspectives on casein interactions. *International Dairy Journal, 85*, 56–65. https://doi.org/10.1016/j.idairyj.2018.04.010

Lucey, J. A., Johnson, M. E., & Horne, D. S. (2003). Invited review: Perspectives on the basis of the rheology and texture properties of cheese. *Journal of Dairy Science, 86*(9), 2725–2743. https://doi.org/10.3168/jds.S0022-0302(03)73869-7

Ludwig, D. S., Hu, F. B., Tappy, L., & Brand-Miller, J. (2018a). Dietary carbohydrates: Role of quality and quantity in chronic disease. *The BMJ, 361*, k2340. https://doi.org/10.1136/bmj.k2340

Ludwig, D. S., Willett, W. C., Volek, J. S., & Neuhouser, M. L. (2018b). Dietary fat: From foe to friend? *Science, 362*(6416), 764–770. https://doi.org/10.1126/science.aau2096

Macedo, A. C., & Malcata, F. X. (1996). Changes in the major free fatty acids in Serra cheese throughout ripening. *International Dairy Journal, 6*(11), 1087–1097. https://doi.org/10.1016/S0958-6946(96)00032-5

Madeka, H., & Kokini, J. L. (1996). *Effect of glass transition and cross-linking on rheological properties of zein: Development of a preliminary state diagram*. Cereal Chemistry (USA).

Magkos, F., Tetens, I., Bügel, S. G., Felby, C., Schacht, S. R., Hill, J. O., … Astrup, A. (2020). A perspective on the transition to plant-based diets: a diet change may attenuate climate change, but can it also attenuate obesity and chronic disease risk? *Advances in Nutrition, 11*(1), 1–9. https://doi.org/10.1093/advances/nmz090

Månsson, H. L. (2008). Fatty acids in bovine milk fat. *Food & Nutrition Research, 52*. https://doi.org/10.3402/fnr.v52i0.1821

Mathai, J. K., Liu, Y., & Stein, H. H. (2017). Values for digestible indispensable amino acid scores (DIAAS) for some dairy and plant proteins may better describe protein quality than values calculated using the concept for protein digestibility-corrected amino acid scores (PDCAAS). *British Journal of Nutrition, 117*(4), 490–499. https://doi.org/10.1017/S0007114517000125

Matias, N. S., Bedani, R., Castro, I. A., & Saad, S. M. I. (2014). A probiotic soy-based innovative product as an alternative to petit-suisse cheese. *LWT – Food Science and Technology, 59*(1), 411–417. https://doi.org/10.1016/j.lwt.2014.05.054

Mattice, K. D., & Marangoni, A. G. (2020). Physical properties of plant-based cheese products produced with zein. *Food Hydrocolloids, 105*, 105746. https://doi.org/10.1016/j.foodhyd.2020.105746

Mattice, K. D., & Marangoni, A. G. (2021). Physical properties of zein networks treated with microbial transglutaminase. *Food Chemistry, 338*, 128010. https://doi.org/10.1016/j.foodchem.2020.128010

Mazidi, M., Mikhailidis, D. P., Sattar, N., Howard, G., Graham, I., & Banach, M. (2019). Consumption of dairy product and its association with total and cause specific mortality – A population-based cohort study and meta-analysis. *Clinical Nutrition, 38*(6), 2833–2845. https://doi.org/10.1016/j.clnu.2018.12.015

McClements, D. J., & Decker, E. (2018). Interfacial antioxidants: A review of natural and synthetic emulsifiers and coemulsifiers that can inhibit lipid oxidation. *Journal of Agricultural and Food Chemistry, 66*(1), 20–35. https://doi.org/10.1021/acs.jafc.7b05066

McClements, D. J., & Decker, E. A. (2017). Chapter 5 – Lipids. In *Fennema's food chemistry* (5th ed.). CRC Press. https://doi.org/10.1201/9781315372914-5

McClements, D. J., & Grossmann, L. (2021). The science of plant-based foods: Constructing next-generation meat, fish, milk, and egg analogs. *Comprehensive Reviews in Food Science and Food Safety, 20*(4), 1–52. https://doi.org/10.1111/1541-4337.12771

McClements, D. J., Newman, E., & McClements, I. F. (2019). Plant-based milks: A review of the science underpinning their design, fabrication, and performance. *Comprehensive Reviews in Food Science and Food Safety, 18*(6), 2047–2067. https://doi.org/10.1111/1541-4337.12505

McClements, D. J., Weiss, J., Kinchla, A. J., Nolden, A. A., & Grossmann, L. (2021). Methods for testing the quality attributes of plant-based foods: Meat- and processed-meat analogs. *Food, 10*(2), 260. https://doi.org/10.3390/foods10020260

McSweeney, P. L. H. (2004). Biochemistry of cheese ripening. *International Journal of Dairy Technology, 57*(2), 19.

McSweeney, P. L. H. (2017). Chapter 14 – Biochemistry of cheese ripening: Introduction and overview. In P. L. H. McSweeney, P. F. Fox, P. D. Cotter, & D. W. Everett (Eds.), *Cheese* (4th ed., pp. 379–387). Academic. https://doi.org/10.1016/B978-0-12-417012-4.00014-4

McSweeney, P. L. H., Fox, P. F., Cotter, P. D., & Everett, D. W. (2017). *Cheese: Chemistry, physics and microbiology* (4th ed.). Academic.

Mejia, A., Harwatt, H., Jaceldo-Siegl, K., Sranacharoenpong, K., Soret, S., & Sabaté, J. (2018). Greenhouse gas emissions generated by tofu production: A case study. *Journal of Hunger & Environmental Nutrition, 13*(1), 131–142. https://doi.org/10.1080/19320248.2017.1315323

Messerli, F. H., Hofstetter, L., Syrogiannouli, L., Rexhaj, E., Siontis, G. C. M., Seiler, C., & Bangalore, S. (2020). Sodium intake, life expectancy, and all-cause mortality. *European Heart Journal, 00*, 1–10. https://doi.org/10.1093/eurheartj/ehaa947

Metz, M., Sheehan, J., & Feng, P. C. H. (2020). Use of indicator bacteria for monitoring sanitary quality of raw milk cheeses – A literature review. *Food Microbiology, 85*, 103283. https://doi.org/10.1016/j.fm.2019.103283

Mihrshahi, S., Ding, D., Gale, J., Allman-Farinelli, M., Banks, E., & Bauman, A. E. (2017). Vegetarian diet and all-cause mortality: Evidence from a large population-based Australian cohort – the 45 and Up Study. *Preventive Medicine, 97*, 1–7. https://doi.org/10.1016/j.ypmed.2016.12.044

Momany, F. A., Sessa, D. J., Lawton, J. W., Selling, G. W., Hamaker, S. A. H., & Willett, J. L. (2006). Structural characterization of α-Zein. *Journal of Agricultural and Food Chemistry, 54*(2), 543–547. https://doi.org/10.1021/jf058135h

Montemurro, M., Pontonio, E., Coda, R., & Rizzello, C. G. (2021). Plant-based alternatives to yogurt: State-of-the-art and perspectives of new biotechnological challenges. *Food, 10*(2), 316. https://doi.org/10.3390/foods10020316

Murata, K., Kusakabe, I., Kobayashi, H., Akaike, M., Park, Y. W., & Murakami, K. (1987). Studies on the coagulation of soymilk-protein by commercial proteinases. *Agricultural and Biological Chemistry, 51*(2), 385–389. https://doi.org/10.1080/00021369.1987.10868052

Muthukumarappan, K., & Swamy, G. J. (2017). Chapter 10 – Rheology, microstructure, and functionality of cheese. In J. Ahmed, P. Ptaszek, & S. Basu (Eds.), *Advances in food rheology and its applications* (pp. 245–276). Woodhead Publishing. https://doi.org/10.1016/B978-0-08-100431-9.00010-3

Nadathur, S., Wanasundara, D. J. P. D., & Scanlin, L. (2016). *Sustainable protein sources.* Academic.

Ng-Kwai-Hang, K. F., Hayes, J. F., Moxley, J. E., & Monardes, H. G. (1982). Environmental influences on protein content and composition of bovine milk. *Journal of Dairy Science, 65*(10), 1993–1998. https://doi.org/10.3168/jds.S0022-0302(82)82449-1

Ni, Y., Wen, L., Wang, L., Dang, Y., Zhou, P., & Liang, L. (2015). Effect of temperature, calcium and protein concentration on aggregation of whey protein isolate: Formation of gel-like microparticles. *International Dairy Journal, 51*, 8–15. https://doi.org/10.1016/j.idairyj.2015.07.003

Nikiforidis, C. V. (2019). Structure and functions of oleosomes (oil bodies). *Advances in Colloid and Interface Science, 274*, 102039. https://doi.org/10.1016/j.cis.2019.102039

Nogueira, M. S., Scolaro, B., Milne, G. L., & Castro, I. A. (2019). Oxidation products from omega-3 and omega-6 fatty acids during a simulated shelf life of edible oils. *LWT, 101*, 113–122. https://doi.org/10.1016/j.lwt.2018.11.044

Noor Lida, H. M. D., Sundram, K., Siew, W. L., Aminah, A., & Mamot, S. (2002). TAG composition and solid fat content of palm oil, sunflower oil, and palm kernel olein belends before and after chemical interesterification. *Journal of the American Oil Chemists' Society, 79*(11), 1137–1144. https://doi.org/10.1007/s11746-002-0617-0

Orlich, M. J., Singh, P. N., Sabate, J., Jaceldo-Siegl, K., Fan, J., Knutsen, S., Beeson, W. L., & Fraser, G. E. (2013). Vegetarian dietary patterns and mortality in Adventist Health Study 2. *JAMA Internal Medicine, 173*(13). https://doi.org/10.1001/jamainternmed.2013.6473

Oyeyinka, A. T., Odukoya, J. O., & Adebayo, Y. S. (2019). Nutritional composition and consumer acceptability of cheese analog from soy and cashew nut milk. *Journal of Food Processing and Preservation, 43*(12), e14285. https://doi.org/10.1111/jfpp.14285

Pala, V., Sieri, S., Chiodini, P., Masala, G., Palli, D., Mattiello, A., … Krogh, V. (2019). Associations of dairy product consumption with mortality in the European Prospective Investigation into Cancer and Nutrition (EPIC)-Italy cohort. *The American Journal of Clinical Nutrition, 110*(5), 1220–1230. https://doi.org/10.1093/ajcn/nqz183

Park, S. Y., Lee, D. K., An, H. M., Kim, J. R., Kim, M. J., Cha, M. K., … Ha, N. J. (2012). Producing functional soy-based yogurt incubated with Bifidobacterium Longum SPM1205 isolated from healthy adult Koreans. *Biotechnology & Biotechnological Equipment, 26*(1), 2759–2764. https://doi.org/10.5504/BBEQ.2011.0152

Pastorino, A. J., Hansen, C. L., & McMahon, D. J. (2003). Effect of pH on the chemical composition and structure-function relationships of Cheddar cheese. *Journal of Dairy Science, 86*(9), 2751–2760. https://doi.org/10.3168/jds.S0022-0302(03)73871-5

Paula Vilela, T., Gomes, A. M., & Ferreira, J. P. (2020). Probing the structure-holding interactions in cheeses by dissociating agents – A review and an experimental evaluation with emmental cheese. *Current Research in Food Science, 3*, 201–206. https://doi.org/10.1016/j.crfs.2020.07.001

Peleg, M. (1979). Characterization of the stress relaxation curves of solid foods. *Journal of Food Science, 44*(1), 277–281. https://doi.org/10.1111/j.1365-2621.1979.tb10062.x

Pontonio, E., Raho, S., Dingeo, C., Centrone, D., Carofiglio, V. E., & Rizzello, C. G. (2020). Nutritional, functional, and technological characterization of a novel gluten- and lactose-free yogurt-style snack produced with selected lactic acid bacteria and leguminosae flours. *Frontiers in Microbiology, 11*, 1664. https://doi.org/10.3389/fmicb.2020.01664

Ray, C. A., Gholamhosseinpour, A., Ipsen, R., & Hougaard, A. B. (2016). The effect of age on Cheddar cheese melting, rheology and structure, and on the stability of feed for cheese powder manufacture. *International Dairy Journal, 55*, 38–43. https://doi.org/10.1016/j.idairyj.2015.11.009

Routray, W., & Mishra, H. N. (2011). Scientific and technical aspects of yogurt aroma and taste: A review. *Comprehensive Reviews in Food Science and Food Safety, 10*(4), 208–220. https://doi.org/10.1111/j.1541-4337.2011.00151.x

Rui, X., Zhang, Q., Huang, J., Li, W., Chen, X., Jiang, M., & Dong, M. (2019). Does lactic fermentation influence soy yogurt protein digestibility: A comparative study between soymilk and soy yogurt at different pH. *Journal of the Science of Food and Agriculture, 99*(2), 861–867. https://doi.org/10.1002/jsfa.9256

Sacks, F. M., Lichtenstein, A. H., Wu, J. W. Y., Appel, L. J., Creager, M. A., Kris-Etherton, P. M., ... van Horn, L. V. (2017). Dietary fats and cardiovascular disease: A presidential advisory from the American Heart Association. *Circulation, 136*(3), e1–e23. https://doi.org/10.1161/CIR.0000000000000510

Saini, R. K., & Keum, Y.-S. (2018). Omega-3 and omega-6 polyunsaturated fatty acids: Dietary sources, metabolism, and significance – A review. *Life Sciences, 203*, 255–267. https://doi.org/10.1016/j.lfs.2018.04.049

Sánchez-Muñoz, M. A., Valdez-Solana, M. A., Avitia-Domínguez, C., Ramírez-Baca, P., Candelas-Cadillo, M. G., Aguilera-Ortíz, M., ... Sierra-Campos, E. (2017). Utility of Milk coagulant enzyme of Moringa oleifera seed in cheese production from soy and skim milks. *Food, 6*(8), 62. https://doi.org/10.3390/foods6080062

Sánchez-Zapata, E., Fuentes-Zaragoza, E., Fernández-López, J., Sendra, E., Sayas, E., Navarro, C., & Pérez-Alvarez, J. A. (2009). Preparation of dietary fiber powder from tiger nut (Cyperus esculentus) milk ("Horchata") byproducts and its physicochemical properties. *Journal of Agricultural and Food Chemistry, 57*(17), 7719–7725. https://doi.org/10.1021/jf901687r

Santos, J. A., Tekle, D., Rosewarne, E., Flexner, N., Cobb, L., Al-Jawaldeh, A., ... Trieu, K. (2021). A systematic review of salt reduction initiatives around the world: A midterm evaluation of progress towards the 2025 global non-communicable diseases salt reduction target. *Advances in Nutrition, 00*, 1–13. https://doi.org/10.1093/advances/nmab008

Satija, A., Bhupathiraju, S. N., Spiegelman, D., Chiuve, S. E., Manson, J. E., Willett, W., ... Hu, F. B. (2017). Healthful and unhealthful plant-based diets and the risk of coronary heart disease in US adults. *Journal of the American College of Cardiology, 70*(4), 411–422. https://doi.org/10.1016/j.jacc.2017.05.047

Schelle, M., Cowperthwaite, S., Kizer, L., & Renninger, N. (2020). Compressible non-dairy cheese analogs, formulations and processes for making same. Patent number: US20200323231A1. Retrieved from: https://patents.google.com/patent/US20200323231A1/en?q=ingredion+cheese+starch&oq=ingredion+cheese+starch. Accessed Mar 2021.

Schenkel, P., Samudrala, R., & Hinrichs, J. (2013a). The effect of adding whey protein particles as inert filler on thermophysical properties of fat-reduced semihard cheese type Gouda. *International Journal of Dairy Technology, 66*(2), 220–230. https://doi.org/10.1111/1471-0307.12036

Schenkel, P., Samudrala, R., & Hinrichs, J. (2013b). Thermo-physical properties of semi-hard cheese made with different fat fractions: Influence of melting point and fat globule size. *International Dairy Journal, 30*(2), 79–87. https://doi.org/10.1016/j.idairyj.2012.11.014

Schirmer, M., Höchstötter, A., Jekle, M., Arendt, E., & Becker, T. (2013). Physicochemical and morphological characterization of different starches with variable amylose/amylopectin ratio. *Food Hydrocolloids, 32*(1), 52–63. https://doi.org/10.1016/j.foodhyd.2012.11.032

Schirmer, M., Jekle, M., & Becker, T. (2015). Starch gelatinization and its complexity for analysis. *Starch – Stärke, 67*(1–2), 30–41. https://doi.org/10.1002/star.201400071

Schlesser, J. E., Schmidt, S. J., & Speckman, R. (1992). Characterization of chemical and physical changes in camembert cheese during ripening. *Journal of Dairy Science, 75*(7), 1753–1760. https://doi.org/10.3168/jds.S0022-0302(92)77934-X

Schmidt, J. H. (2015). Life cycle assessment of five vegetable oils. *Journal of Cleaner Production, 87*, 130–138. https://doi.org/10.1016/j.jclepro.2014.10.011

Schwingshackl, L., Schwedhelm, C., Hoffmann, G., Lampousi, A.-M., Knüppel, S., Iqbal, K., … Boeing, H. (2017). Food groups and risk of all-cause mortality: A systematic review and meta-analysis of prospective studies. *The American Journal of Clinical Nutrition, 105*(6), 1462–1473. https://doi.org/10.3945/ajcn.117.153148

Sha, L., & Xiong, Y. L. (2020). Plant protein-based alternatives of reconstructed meat: Science, technology, and challenges. *Trends in Food Science & Technology, 102*, 51–61. https://doi.org/10.1016/j.tifs.2020.05.022

Shahidi, F., & Ambigaipalan, P. (2018). Omega-3 polyunsaturated fatty acids and their health benefits. *Annual Review of Food Science and Technology, 9*(1), 345–381. https://doi.org/10.1146/annurev-food-111317-095850

Sharma, P., Munro, P. A., Gillies, G., Wiles, P. G., & Dessev, T. T. (2017). Changes in creep behavior and microstructure of model Mozzarella cheese during working. *LWT – Food Science and Technology, 83*, 184–192. https://doi.org/10.1016/j.lwt.2017.05.003

Short, E. C., Kinchla, A. J., & Nolden, A. A. (2021). Plant-based cheeses: A systematic review of sensory evaluation studies and strategies to increase consumer acceptance. *Food, 10*(4), 725. https://doi.org/10.3390/foods10040725

Shurtleff, W., & Aoyagi, A. (2011). *History of fermented tofu: A healthy nondairy/vegan cheese (1610–2011): Extensively annotated bibliography and sourcebook*. Soyinfo Center.

Sinaga, H., Bansal, N., & Bhandari, B. (2017). Effects of milk pH alteration on casein micelle size and gelation properties of milk. *International Journal of Food Properties, 20*(1), 179–197. https://doi.org/10.1080/10942912.2016.1152480

Smith, K. W. (2015). Specialty oils and fats in ice cream. In G. Talbot (Ed.), *Specialty oils and fats in food and nutrition* (Vol. 11, pp. 271–284). Woodhead Publishing. https://doi.org/10.1016/B978-1-78242-376-8.00011-9

Stefano, D. E., White, J., Seney, S., Hekmat, S., McDowell, T., Sumarah, M., & Reid, G. (2017). A novel millet-based probiotic fermented food for the developing world. *Nutrients, 9*(5), 529. https://doi.org/10.3390/nu9050529

Taggart, P., & Mitchell, J. R. (2009). 5 – Starch. In G. O. Phillips & P. A. Williams (Eds.), *Handbook of hydrocolloids* (2nd ed., pp. 108–141). Woodhead Publishing. https://doi.org/10.1533/9781845695873.108

Tangyu, M., Muller, J., Bolten, C. J., & Wittmann, C. (2019). Fermentation of plant-based milk alternatives for improved flavour and nutritional value. *Applied Microbiology and Biotechnology, 103*(23), 9263–9275. https://doi.org/10.1007/s00253-019-10175-9

Tarighaleslami, A. H., Ghannadzadeh, A., Atkins, M. J., & Walmsley, M. R. W. (2020). Environmental life cycle assessment for a cheese production plant towards sustainable energy transition: Natural gas to biomass vs. natural gas to geothermal. *Journal of Cleaner Production, 275*, 122999. https://doi.org/10.1016/j.jclepro.2020.122999

Tessari, P., Lante, A., & Mosca, G. (2016). Essential amino acids: Master regulators of nutrition and environmental footprint? *Scientific Reports, 6*(1), 26074. https://doi.org/10.1038/srep26074

Tong, T. Y. N., Appleby, P. N., Bradbury, K. E., Perez-Cornago, A., Travis, R. C., Clarke, R., & Key, T. J. (2019). Risks of ischaemic heart disease and stroke in meat eaters, fish eaters, and vegetarians over 18 years of follow-up: Results from the prospective EPIC-Oxford study. *BMJ, 366*, l4897. https://doi.org/10.1136/bmj.l4897

Trieu, K., Bhat, S., Dai, Z., Leander, K., Gigante, B., Qian, F., … Marklund, M. (2021). Biomarkers of dairy fat intake, incident cardiovascular disease, and all-cause mortality: A cohort study, systematic review, and meta-analysis. *PLoS Medicine, 18*(9), e1003763. https://doi.org/10.1371/journal.pmed.1003763

Üçtuğ, F. G. (2019). The environmental life cycle assessment of dairy products. *Food Engineering Reviews, 11*(2), 104–121. https://doi.org/10.1007/s12393-019-9187-4

USDA. (2021). *FoodData Central*. Retrieved from: https://fdc.nal.usda.gov/. Accessed Mar 2021.

References

Usubharatana, P., & Phungrassami, H. (2015). Carbon footprint of cassava starch production in North-Eastern Thailand. *Procedia CIRP, 29*, 462–467. https://doi.org/10.1016/j.procir.2015.02.031

van Middelaar, C. E., Berentsen, P. B. M., Dolman, M. A., & de Boer, I. J. M. (2011). Eco-efficiency in the production chain of Dutch semi-hard cheese. *Livestock Science, 139*(1), 91–99. https://doi.org/10.1016/j.livsci.2011.03.013

Verma, P., Agrawal, U. S., Sharma, A. K., Sarkar, B. C., & Sharma, H. K. (2005). Optimization of process parameters for the development of a cheese analogue from pigeon pea (Cajanus cajan) and soy milk using response surface methodology. *International Journal of Dairy Technology, 58*(1), 51–58. https://doi.org/10.1111/j.1471-0307.2005.00182.x

Wang, X., He, Z., Zeng, M., Qin, F., Adhikari, B., & Chen, J. (2017). Effects of the size and content of protein aggregates on the rheological and structural properties of soy protein isolate emulsion gels induced by CaSO4. *Food Chemistry, 221*, 130–138. https://doi.org/10.1016/j.foodchem.2016.10.019

Wang, Y.-J., Yeh, T.-L., Shih, M.-C., Tu, Y.-K., & Chien, K.-L. (2020). Dietary sodium intake and risk of cardiovascular disease: A systematic review and dose-response meta-analysis. *Nutrients, 12*(10), 2934. https://doi.org/10.3390/nu12102934

Wolfschoon Pombo, A. F. (2021). Cream cheese: Historical, manufacturing, and physicochemical aspects. *International Dairy Journal, 117*, 104948. https://doi.org/10.1016/j.idairyj.2020.104948

Wu, C., Ma, W., Chen, Y., Navicha, W. B., Wu, D., & Du, M. (2019). The water holding capacity and storage modulus of chemical cross-linked soy protein gels directly related to aggregates size. *LWT, 103*, 125–130. https://doi.org/10.1016/j.lwt.2018.12.064

Yang, J., Punshon, T., Guerinot, M. L., & Hirschi, K. D. (2012). Plant calcium content: Ready to remodel. *Nutrients, 4*(8), 1120–1136. https://doi.org/10.3390/nu4081120

Zannini, E., Jeske, S., Lynch, K. M., & Arendt, E. K. (2018). Development of novel quinoa-based yoghurt fermented with dextran producer Weissella cibaria MG1. *International Journal of Food Microbiology, 268*, 19–26. https://doi.org/10.1016/j.ijfoodmicro.2018.01.001

Zhang, Q., Wang, C., Li, B., Li, L., Lin, D., Chen, H., … Yang, W. (2018). Research progress in tofu processing: From raw materials to processing conditions. *Critical Reviews in Food Science and Nutrition, 58*(9), 1448–1467. https://doi.org/10.1080/10408398.2016.1263823

Zheng, L., Regenstein, J. M., Teng, F., & Li, Y. (2020). Tofu products: A review of their raw materials, processing conditions, and packaging. *Comprehensive Reviews in Food Science and Food Safety, 19*(6), 3683–3714. https://doi.org/10.1111/1541-4337.12640

Zia-ud-Din, Xiong, H., & Fei, P. (2017). Physical and chemical modification of starches: A review. *Critical Reviews in Food Science and Nutrition, 57*(12), 2691–2705. https://doi.org/10.1080/10408398.2015.1087379

Zulkurnain, M., Goh, M.-H., Karim, A. A., & Liong, M.-T. (2008). Development of a soy-based cream cheese. *Journal of Texture Studies, 39*(6), 635–654. https://doi.org/10.1111/j.1745-4603.2008.00163.x

Zwiercan, G. A., Lacourse, N. L., & Lenchin, J. M. (1987). Imitation cheese products containing high amylose starch as total caseinate replacement. Patent number: US4695475A. Retrieved from: https://patents.google.com/patent/US4695475A/en. Accessed Mar 2021.

Yvonne F., Collins Paul L.H., McSweeney Martin G., Wilkinson (2003) Lipolysis and free fatty acid catabolism in cheese: a review of current knowledge. International Dairy Journal 13(11) 841-866 10.1016/S0958-6946(03)00109-2

Chapter 10
Facilitating the Transition to a Plant-Based Diet

10.1 Introduction

As highlighted throughout this book, there is growing interest from many consumers in adopting a more plant-based diet, which has led many food manufacturers to create new plant-based products. Indeed, there have already been several highly successful commercial products in this market segment, including those by Beyond Meat and Impossible Foods (Figs. 10.1 and 10.2). Nevertheless, there is still considerable room for growth in this field. Currently, only a small fraction of consumers has adopted a predominantly plant-based diet, and plant-based foods only have a small market share in all the categories they are competing in (such as meat, seafood, eggs, milk, yogurt, and cheese). There are still numerous challenges that need to be addressed to facilitate the transition to a more plant-based diet, which is opening new opportunities to the food industry, entrepreneurs, and investors. In this final chapter, we highlight some of the most important issues that we believe should be addressed to promote this transition. These include supply-side issues, such as increasing the availability of high-quality plant-based foods, as well as demand-side issues, such as increasing consumer awareness and desire for plant-based foods.

10.2 Research

10.2.1 Ingredient Innovation

The design and production of the next-generation of high-quality plant-based foods would be greatly facilitated by increasing the quantity, quality, and diversity of plant-derived ingredients with well-defined and consistent functional performances. At present, there is only a relatively limited supply of plant-derived ingredients

Fig. 10.1 Images of plant-based burgers before and after cooking. (Images kindly supplied by Impossible Foods and used with permission)

Fig. 10.2 Images of plant-based burgers, sausages, and chicken nuggets. (Images kindly supplied by Beyond Meat and used with permission)

(particularly proteins) available to formulate plant-based foods, and their functional properties are often unsuitable for the required application, or their functional performance varies considerably from batch-to-batch. There are several areas where innovation is required in this area:

- The identification of new sustainable sources of plant-based ingredients that can be produced economically at a sufficiently large scale for application in the food industry, such as land crops, aquatic plants, industrial side streams, or agricultural residues.

- The development of new strains of agricultural crops using breeding or genetic engineering approaches that contain higher levels of high-quality functional ingredients.
- The creation of innovative processing methods, or the optimization of existing ones, to sustainably isolate plant-based ingredients with a sufficiently high purity and functional performance.
- The identification of functional ingredients that require less processing to produce them, which should reduce waste and increase sustainability.
- The development of standardized analytical methods to characterize and compare the properties of plant-based ingredients so that their suitability for specific applications can be critically assessed. Ideally, these methods should be affordable, rapid, and simple so that they can easily be adopted by industry.

10.2.2 Food Quality Design

Manufacturers of the first generation of plant-based foods focused on simulating the look, feel, and taste of restructured products, like burgers, sausages, nuggets, and meatballs because their sensory attributes are critical to consumer acceptance (Figs. 10.1 and 10.2). Moreover, these products are widely consumed foods whose structures, physicochemical properties, and quality attributes are easier to mimic than those of whole muscle foods like beef steaks, pork chops, chicken breasts, and fish fillets. Academic and industry scientists are already working to develop the science and technology required to produce high-quality whole muscle foods, but further work is still required. This will depend on developing a better understanding of plant-based ingredients, processing and manufacturing operations, and structural design principles. It will also depend on the identification of new functional ingredients, as well as the development of innovative processing technologies, soft matter physics principles, and computational design models. Ideally, open access databases are required that provide information about the functional performance of different plant-based ingredients in a standardized way so that researchers and manufacturers can select the most appropriate ones for particular applications (Clayton and Specht 2021). These databases would provide information about solubility, thickening, binding, foaming, and gelling properties of different proteins, as well as how environmental conditions (such as pH, ionic strength, and temperature) affect these properties. Moreover, structure-function relationships should be developed that link the molecular characteristics of proteins to their functional performance in foods. Advanced computational techniques may play an important role in establishing these relationships and then using them to design plant-based foods in a more rational and systematic fashion. In particular, they may facilitate the simulation of the complex hierarchical structures found in whole muscle foods using plant-derived ingredients.

10.2.3 Nutritional Implications

If the plant-based food movement is successful, a large fraction of the global population will replace animal-based foods with plant-based alternatives. As discussed earlier in this book, the nutritional profile and health effects of plant-based foods are different from those of animal-based ones (Chap. 5). Consequently, the transition to a more plant-based diet could have important health implications for the general population. The first generation of plant-based foods largely focused on creating products that look, feel, and taste delicious, as well as closely mimicking the properties of real animal-based foods. This was critical, because taste is the most important factor consumers consider when purchasing these foods (Parry and Szejda 2019). However, it will be critical to consider nutrition and health implications when designing the next generation of plant-based foods. The modern food industry has been highly successful at producing affordable, convenient, abundant, and delicious foods but has been criticized for producing unhealthy foods (high in fat, salt, and sugar) that are linked to increases in chronic diseases such as obesity, diabetes, hypertension, and heart disease. The transition from animal- to plant-based foods is an ideal opportunity for the food industry to increase the healthiness of its products and promote widespread improvements in human health and wellbeing. Consequently, there is a need for more research to understand the relationship between the nutritional profiles, gastrointestinal fate, and health effects of plant-based foods relative to the animal-based ones they are designed to replace. Studies of the digestibility of plant-based foods, such as the rate and extent of macronutrient digestion, as well as the pharmacokinetics of nutrient and nutraceutical absorption need to be carried out, as well as the impact of plant-based foods on the gut microbiome and human health. This knowledge can then be used to design healthier versions of plant-based foods that will ensure that the next generation of these products can contribute to a healthier society.

10.2.4 Environmental Impacts

Numerous life cycle analyses have shown that plant-based foods have a much lower impact on the environment than equivalent animal-based ones. However, their environmental impact and sustainability can still be improved. For instance, an improved understanding of ingredient functionality may allow food manufacturers to create plant-based foods from ingredients that are sourced from local regions, rather than being shipped across the world. This would reduce the number of miles that ingredients and foods need to be transported, thereby reducing fossil fuel use, as well as improving economic viability. Moreover, the ability to process foods close to the place where the ingredients are grown could stimulate local economies, especially rural ones, by creating jobs in agriculture and industry. To achieve this goal, it will be important to improve our understanding of how ingredients from different agricultural crops behave and how they can be extracted and processed to produce the functional properties required.

Various other approaches can also be used to increase the sustainability of plant-based food production (Clayton and Specht 2021). Currently, the agricultural crops used as a source of plant-based ingredients are not optimized for this purpose. Instead, they have been optimized to produce oils or starches. Traditional selective breeding or modern genetic engineering approaches can be used to redesign these crops to increase the amounts of useful ingredients they contain, as well as to improve their extractability and functionality, which would reduce costs and environmental impacts. Moreover, there is a need to ensure that as much of possible of the crops produced are converted into valuable ingredients, rather than wasted. Consequently, sustainable and economic uses for any waste streams generated should be identified. Furthermore, optimization of existing manufacturing operations, as well as the development of new more energy and resource efficient ones, is required to improve the sustainability and reduce the environmental impact of plant-based food and ingredient production.

Increased information about the environmental benefits of switching from an animal- to a plant-based diet would help inform consumer choices and government decisions. There is therefore a need for more comprehensive life cycle analysis (LCA) studies of specific kinds of plant-based food products, such as meat, seafood, eggs, and milk. These studies should be carried out by independent scientists who do not have a vested interest in the outcome. Improved information about the relative impact of animal- and plant-based products on greenhouse gas emissions, water use, land use, pollution, and biodiversity loss would be important for government, industry, and consumers. Ideally, a standardized LCA protocol should be developed that is easy to adopt by industry, regulate by government, and understand by consumers. Moreover, the principles and procedures behind this protocol should be developed by experts in the field, assessed by independent experts, transparent, and acceptable to the various stakeholders within the food industry. The availability of these LCAs would facilitate the identification of ingredients and processes leading to a more sustainable food production system. Moreover, if LCA information was clearly displayed on food labels in a consistent manner, consumers could make informed choices about which items to select from specific food categories based on their environmental impacts.

The environmental impact of producing plant-based foods may also be reduced by using modern technological innovations throughout the supply chain, such as automation, artificial intelligence, big data, block chain, biotechnology, nanotechnology, and structural design approaches (McClements 2019).

10.2.5 Socioeconomic Impacts

It has been predicted that switching to a modern plant-based diet will have huge socioeconomic impacts (Tubb and Seba 2020). The sales of the meat, fish, egg, and dairy industries may greatly decline, as well as those of other industrial sectors that depend on them. For instance, crop farmers, whose produce is mainly used at present to feed livestock animals, would see their sales plummet. Moreover, the producers of the seeds, fertilizers, pesticides, tractors, and other farm equipment used by these crop farmers will also see their sales decline. There will be much less demand for slaughterhouses, meat processing facilities, and renderers. As a result, many people will lose their jobs and livelihoods, which will have a major impact on the communities they live in. Consequently, research is required to better understand the socioeconomic consequences of transforming the modern food system and to identify effective strategies to tackle the inevitable disruptions caused. This is likely to be a relatively slow transition due to the increasing demands for protein-rich foods as the global population continues to grow but action is still urgently required so these issues can be addressed as soon as possible. In addition, it will be important to bring down the price of plant-based foods so that a broader range of the public can afford to incorporate them into their diet.

10.3 Education

The growth of the plant-based food industry will depend on a workforce with the knowledge and skills required to design, develop, produce, and test its products. Much of the knowledge required is part of traditional food science and engineering courses, but there are also aspects that are not currently covered in a comprehensive and integrated manner. Moreover, many of the people working in this area have been trained in other disciplines and do not have the fundamental knowledge of the physical, chemical, biological, and engineering principles relevant to plant-based foods. There is therefore a need for innovation in education. At the university level, this could involve revising existing courses, such as adding plant-based food sections to food chemistry, food analysis, food microbiology, and food engineering courses. Alternatively, it could involve the creation of new courses at the undergraduate and/or graduate levels that specifically focus on the science and technology of plant-based foods. Indeed, we hope that this book will serve as a resource for this type of course. Some universities, such as UC Berkeley in California, have developed innovative courses that integrate the business and science of plant-based foods, bringing in external speakers from academia and industry working in this area, as well as involving students in product development projects. Moreover, the Good Food Institute, a non-profit organization, has free on-line materials that can be used to develop courses in this area (gfi.org/resource/alt-protein-curriculum-development-and-support/). There is also a need for short courses on plant-based foods for people already working in the food industry who need to learn more about the science and technology behind the ingredients, processing operations, physicochemical properties, sensory attributes, and nutritional properties of plant-based foods. In addition to food and nutrition scientists, there is also a need for education and outreach in other disciplines relevant to the plant-based food industry, including chemical engineering, sociology, economics, consumer science, and marketing. To reach a broader audience, academic and workforce training courses should be taught in-person and on-line, with the option for on-demand viewing to increase flexibility for people with busy schedules.

10.4 Consumer Awareness

Although there has been a rapid increase in the number of people including plant-based foods in their diets, most consumers still do not regularly consume these products. There is therefore a need for more research to understand the factors that impact consumer preferences and choices. This knowledge can then be used to design products that better meet consumer needs, as well as to create messaging that will encourage more consumers to try and adopt plant-based foods, thereby changing their long-term consumption behaviors. At present, the early adopters of plant-based foods are younger, more urban, and more ethnically diverse than the general population. In future, it will be important to encourage other demographic groups to adopt plant-based foods. As an example of the kind of approach required, the Good Food Institute commissioned a report that examined the factors effecting consumer perceptions of plant-based foods in the general population (Parry and Szejda 2019). The authors of this report found that taste was the most important factor in consumer decisions, but other factors also contributed, including product familiarity, food traditions, health, nutrition, animal welfare, and environmental concerns. They also identified several factors that contributed to people's potential selection and liking of plant-based foods, such as packaging and marketing materials. An improved understanding of consumer perceptions of plant-based foods, as well as the factors impacting their eating decisions, could lead to marketing strategies that would increase the number of people adopting more sustainable and healthy plant-based diets.

10.5 Government Support

The adverse effects of the modern food supply on human health and the environment make it an urgent matter for public policy. There are several things that governments can do to support the transition to a healthier and more sustainable plant-based food supply. More funds are needed to support basic research on the development of plant-based foods. As highlighted throughout this book, these foods are highly complex materials whose properties are still poorly understood, which is holding back the design and production of the next generation of high quality, healthy, and sustainable products. Support is needed for research across the entire spectrum: the creation of new agricultural crops and practices that will provide new sources of plant-based ingredients; the development of new isolation and purification methods for extracting plant-based ingredients from these crops; the establishment of structural design rules to assemble plant-based ingredients into high quality foods with the required physicochemical, functional, and sensory attributes; the development and optimization of processing technologies to economically produce plant-based foods on a large scale; improved knowledge of the interaction of plant-based foods with the human senses (smell, taste, and mouthfeel) to improve consumer acceptance; and increased knowledge of the gastrointestinal fate and health impacts of plant-based foods compared to animal-based ones.

Governments could also facilitate the transition to a more plant-based diet by providing greater subsidies and tax incentives to producers of plant-based ingredients and foods. In addition, they could use educational campaigns highlighting the potential health, sustainability, and ethical benefits of consuming plant-based foods. Local and national governments could encourage the adoption of plant-based foods in their food programs and in large institutions under their control, such as schools, hospitals, and military bases. These activities could be part of healthy eating campaigns, which could have large benefits in terms of health, sustainability, and economics. Improved eating habits would reduce health care costs associated with many diet-related chronic diseases, as well as reducing workdays lost through illness.

Government is also needed to establish a clear regulatory framework that will enable the emerging plant-based food companies to bring new products to market. For instance, there has been push back from some segments of the traditional animal-based food industry against calling plant-based products by names such as "meat", "fish", "milk", "eggs" or "cheese". They claim these terms should be reserved only for foods that are derived from animals. In some countries, this has caused challenges to the development and marketing of alternatives to animal-based foods. Finally, governments need to facilitate the establishment of the national and international infrastructure needed to transition from animal- to plant-based foods, such as new production, distribution, and storage systems, as well as new standards and testing regimes.

In general, governments need to develop national strategies that will promote a transition to a food system that prioritizes sustainability, human health and nutrition, the environment, animal welfare, and equity. Such a strategy would ensure that government policies (subsidies, incentives, taxes, and other policy tools) are achieving these goals.

10.6 Food Systems Approaches

The transition to a more plant-based diet would benefit from adopting an integrated food systems approach that linked breeders, farmers, producers, retailers, consumers, the medical profession, and the government. All stakeholders should be encouraged to actively participate in the development of an effective strategy to facilitate this transition because there will be important repercussions at many levels of society. As mentioned earlier, people currently working in the meat, fish, egg, or dairy industries may lose their jobs, which would have a major impact on their lives, as well as the communities where they live. In some cases, it may be possible for the existing workforce, production facilities, and distribution network to be repurposed to produce plant-based foods, but this will not always be the case. It will therefore be important for government policies to be crafted to address these important issues. For instance, the government could use education programs, taxes, or subsidies to support farmers make the transition from raising livestock animals to growing agricultural crops.

It should be noted that most of the plant-based crops (like soybean) are currently grown to feed animals. If there was a switch to more plant-based diets then a lot fewer agricultural crops would need to be grown, which would also affect the profits and livelihoods of these farmers. In the future, it may be possible for farmers to grow fewer crops but increase their value by selecting breeds that provide physicochemical and nutritional attributes that are more suitable for formulating plant-based foods for humans rather than animals. Hence, the total amount of agricultural crops grown may be reduced but their value may be increased. Again, this may require taxes, subsidies, or incentives from government to promote this transition. In the future, there may be a premium on artisanal farming, which produces relatively small quantities of high-quality animal-based foods, which are better for human health and the environment, as well as less harmful to animal welfare.

10.7 Final Thoughts

The critical role that the food supply plays on our health and environment is being increasingly recognized by consumers, industry, and governments. It is now widely accepted that the modern food production system needs to be transformed to address the adverse impacts it is having on global warming, biodiversity loss, pollution, animal welfare, and human health. Many scientists, entrepreneurs, investors, and regulators have recognized these challenges and opportunities and are now actively working to transform the food system through technological and policy innovations. In particular, the development of alternative proteins to replace traditional animal proteins has been identified as one of the most important ways of improving the environmental and health impacts of the food supply. In this book, we have focused on the development of plant-based foods designed as healthy and sustainable

Fig. 10.3 Images of food products created using milk proteins produced by cellular agriculture (fermentation) rather than cows. (Images kindly supplied by Perfect Day (California) and used with permission)

alternatives to protein-rich animal-based foods like meat, seafood, eggs, and dairy products. However, other sources of alternative proteins are also likely to play an important role in the transformation away from animal-based foods, including proteins obtained from cultured meat, cellular agriculture, fungi, and insects. As an example, food products created from milk proteins produced by cellular agriculture (rather than by a cow) are shown in Fig. 10.3. Moreover, plant-derived ingredients could be used to create innovative plant-based foods that are not designed to simulate animal-based foods. Tofu, seitan, and tempeh are well-established protein-rich alternatives to animal-based products. However, completely new kinds of protein-rich foods could be created that have appearances, textures, and flavors that do not currently exist, such as spicy chewy red spheres, tangy crisp green squares, or sweet spongy purple triangles. These products could also be designed to be sustainable and healthy, *e.g.*, by including dietary fibers, nutraceuticals, vitamins, and minerals.

In the future, we believe that humans will have to greatly reduce the number of animal-based products they consume to help avoid environmental degradation and to improve their health. Science and technology will play a critical role in this transition. We hope that this book will help to stimulate further research in this important area and will lead to the development of a broader range of high-quality plant-based foods that will help to facilitate this transition.

References

Clayton ER, Specht L (2021) Introduction to plant-based meat. From https://gfi.org/science/the-science-of-plant-based-meat/

McClements, D. J. (2019). *Future foods: how modern science is transforming the way we eat*. Springer Scientific.

Parry, J., & Szejda, K. (2019). *How to drive plant-based food purchasing* (pp. 1–25). The Good Food Institute.

Tubb, C., & Seba, T. (2020). *Rethining food and agriculture: 2020–2030*. RethinkX.

Appendix: Analysis of Plant-Based Ingredients and Foods

In this Appendix, we provide information about the various methods that can be used to characterize the composition of plant-based ingredients and foods.

Introduction

Knowledge of the composition of plant-based foods and ingredients, such as their protein, fat, carbohydrate, mineral, and moisture content, is important because it determines their physicochemical, functional, and nutritional properties. Moreover, it is required for product labeling purposes. In this appendix, we provide an overview of the commonly used procedures for measuring the type and concentration of the major nutrients in plant-based foods and ingredients. We also highlight other methods that can be useful for providing information about the state of these nutrients, such as the denaturation state of proteins. It should be noted that proximate analysis methods of specific food types required for regulatory purposes are often stipulated by associations like the AOAC or ISO.

Moisture Content

The most straightforward method of measuring the moisture content of a food or ingredient is to convert it into small particles (e.g., by cutting, blending, grinding, or milling), mix and grind it with dried sand, and then dry the material in a regular convective oven or a vacuum oven until it reaches a constant mass (*e.g.,* AOAC 950.46). For an oven setup, the temperature used is usually just above 100 °C, while for a vacuum oven it can be lower due to the reduction in the boiling point of water. After letting the sample cool down in a desiccator, the dry matter is measured by weighing. It is important to ensure that the sample does not contain appreciable quantities of non-water volatile compounds and that it does not thermally degrade during drying, otherwise erroneous results will be obtained. If the temperatures in a regular convective oven are too high, then a vacuum oven should be considered.

Alternatively, the water content of samples can be analyzed using the Karl Fischer titration method, which is especially suited for low-moisture or temperature-sensitive foods (*e.g.*, plant protein powders). This method utilizes a reaction between water, sulfur dioxide, and iodine to quantify the amount of moisture in the original sample (Park, 2008):

$$2H_2O + SO_2 + I_2 \rightarrow H_2SO_4 + 2HI$$

In practice, additional reagents are included to facilitate completion and end-point detection of the reaction. Typically, the reaction requires 1 mol sulfur dioxide, 1 mol iodine, 3 mol pyridine, and 1 mol methanol per 1 mol of water. Once all the water is consumed, the presence of excess iodine is detected voltametrically using an indicator electrode in the titration unit.

Protein Analysis

Total Protein Content

Several analytical procedures are available to measure the total protein content of foods, with the most common being those that measure the total nitrogen content and those that utilize spectrophotometric assays (Mæhre et al., 2018). Spectrophotometric assays are not commonly used to determine the protein content of plant-based foods because they require calibration with an appropriate reference material, require soluble proteins, are susceptible to interfering compounds, and are not permitted as a reference method for product labeling purposes (Moore et al., 2010). In particular, the measured absorbance depends on the amino acid sequence of a protein, which varies considerably between different proteins. Although these methods are not particularly suitable for the determination of the total protein content of plant-based foods, they are useful for assessing relative changes in protein concentration, for example, in protein solubility assays (Grossmann et al., 2019).

Most laboratories use either the Dumas (*e.g.*, AOAC 992.15) or Kjeldahl (*e.g.*, AOAC 981.10 or 928.08) method to determine the protein content of plant-based foods and ingredients, which involve measuring the nitrogen content and then using a calibration factor to convert it into a protein content.

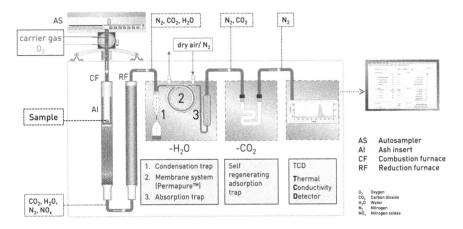

Fig. A.1 Schematic working principle of a device that measures the nitrogen content of materials by the Dumas combustion method. Protein-containing samples are burned in an oxygen-rich atmosphere. The gases are subsequently treated to obtain N_2 that is measured using a thermal-conductivity detector. The nitrogen content is then used to calculate the protein content. (Image kindly provided by C. Gerhardt GmbH & Co. KG (Koenigswinter, Germany))

Dumas Method

The *Dumas* method is based on the combustion of the food sample at temperatures between 900 and 1300 °C in an oxygen-rich atmosphere (Etheridge et al., 1998) (Fig. A.1). The released gases (O_2, CO_2, H_2O, N_2, and NO_x) are subsequently passed through a copper-filled tube, which facilitates the reduction to N_2. The remaining gases are passed through gas-/water-sensitive traps and membranes to remove the non-nitrogen containing gases. The gas mixture is finally transported through a thermal-conductivity detector that generates an electrical signal that is proportional to the nitrogen content.

Kjeldahl Method

The *Kjeldahl* method of determining the total protein content of foods is based on three steps: digestion, distillation, and titration. First, the sample is digested using boiling sulfuric acid and a catalyst, which converts the nitrogen in the samples into ammonium ions. Second, the liquid is neutralized with NaOH, which promotes the release of ammonia gas in the subsequent distillation step by the reaction of ammonium ions with hydroxide ions. The ammonia is recovered by condensation and typically reconverted into non-volatile ammonium by a weak acid, such as boric acid. Third, the amount of recovered ammonium in the borate is determined by an acid titration (Marcó et al., 2002) (Fig. A.2).

Fig. A.2 Workflow of a Kjeldahl analysis to determine the protein content in food materials. (Image kindly provided by C. Gerhardt GmbH & Co. KG (Koenigswinter, Germany))

Comparison

Both the Kjeldahl and Dumas methods are the basis for many reference methods published by associations such as the AACC, AOAC, AOACS, and ISO (Moore et al., 2010). They are commonly used to analyze the total protein content of plant-based foods and ingredients (Chiang et al., 2019; Schreuders et al., 2019). Traditionally, the Kjehldal method was used as a gold-standard reference, but the Dumas method has now become established as a reliable alternative. The Dumas method has several advantages over the Kjeldahl method, including a higher throughput (analysis time usually 3–4 min *versus* several hours), safety concerns (no handling of caustic acids, no toxic chemicals), and lower costs per sample (Moore et al., 2010). However, Dumas instruments are relatively expensive and require regular maintenance and exchange of consumables, such as copper and adsorbers, which require trained staff. Dumas instruments can detect nitrogen levels between 0.003 and 50 mg in samples with masses ranging from a few mg to a few grams, whereas the Kjeldahl method has a lower limit of quantification of around 0.02 mg nitrogen (Buchi, 2010).

Dumas and Kjehldal can be used to analyze different sample types (solids, semi-solids, and liquids). They also both need to be calibrated with a suitable standard: ethylenediaminetetraacetic acid (EDTA) or tris(hydroxymethyl) aminomethane (THAM) for Dumas; a pH calibration/indicator for Kjeldahl. Since the sample quantities used are quite low, a homogenization step (*e.g.*, blending) is usually required prior to analysis to ensure a homogenous sample is analyzed.

Amino Acid Analysis

The Dumas and Kjeldahl methods measure the total nitrogen content of a sample, which can be used to calculate the protein content using a suitable conversion factor. Since different amino acids have different nitrogen contents, the conversion factor depends on the primary sequence of the protein and therefore on protein type (Schwenzfeier et al., 2011). Several methods are available to measure the amino acid content of proteins. However, the most common method used depends on acid hydrolysis (6 M HCl) in an oxygen-free environment to release the amino acids, which are then separated using an ion-exchange HPLC system, and detected by ninhydrin post-column derivation (Mæhre et al., 2018). This method can quantify most amino acids. However, some amino acids are chemically degraded during this process and so special precautions are required.

In general, several factors should be considered to accurately determine the amino acid composition of plant proteins. First, some amino acids are prone to chemical degradation during acid hydrolysis, especially cysteine and methionine. For quantification, these amino acids should be oxidized prior to acid hydrolysis, which enhances their chemical resistance (Rayner, 1985). Tyrosine is also prone to chemical degradation during acid hydrolysis, which can be minimized by the addition of phenol (Nissen, 1992). Asparagine and glutamine are deamidated during acid hydrolysis, which results in accumulated values for asparagine/aspartic acid and glutamine/glutamic acid residues (Rutherfurd & Gilani, 2009). Second, to distinguish between proteins and free amino acids, the proteins should be separated from the amino acids by employing a selective precipitation step using a polar solvent (*e.g.*, ethanol or acetone) or trichloroacetic acid, and then the amino acid content in the supernatant is analyzed (Park et al., 2014). This value can then be subtracted from the total amino acid content to determine the protein content.

Protein Content Calculation

The choice of a correct nitrogen-to-protein conversion factor (k) is a critical step in calculating the total protein content using both the Dumas and Kjeldahl methods: Protein (%) = k × Nitrogen (%). While both methods also detect non-proteinaceous nitrogen, Kjeldahl only detects organic nitrogen compounds, NH_3, and NH_4^+, but Dumas determines all nitrogen-containing components (Sáez-Plaza et al., 2013). For most plant proteins, the conversion factors are typically between about 5 and 6 (Schreuders et al., 2019; Mariotti et al., 2008). An overview of some published factors is given in Table A.1. However, nitrogen-to-protein conversion factors have not been published for many new plant proteins, which may result in the reporting of erroneous protein contents.

For new raw materials, the conversion factor can be established based on their known amino acid composition, their anhydrous molecular weights, (Table A.2), and the measured nitrogen content of the sample. Two factors are commonly used k_b (sometimes referred to as k_p) and k_a (FAO, 2019):

- k_b is the ratio of the sum of the weights of anhydrous amino acid residues (water molecules are added during hydrolysis, *i.e.*, it is important to not use the M_W of free amino acids) to total nitrogen content (protein nitrogen + non-protein nitrogen):

$$k_b = \sum \text{anhydrous amino acids residues} / \text{total nitrogen}$$

- k_a is the ratio of the sum of anhydrous amino acid residue weights to the sum of only the nitrogen found in the detected amino acids:

$$k_a = \sum \text{anhydrous amino acids residues} / \text{protein nitrogen}$$

Table A.1 Nitrogen to protein conversion factors for important plant materials

Material	Nitrogen-to-protein conversion factors
Barley	5.45
Triticale	5.49
Oats	5.34
Rye	5.34
Millet (foxtail millet)	5.80
Millet (pearl millet)	5.47
Wheat (whole)	5.49
Wheat flour and derived products	5.52
Wheat germ	4.99
Wheat bran	4.96
Buckwheat	5.24
Rice	5.34
Corn	5.62
Sorghum	5.67
Other cereals	5.50
Soybean or soybean meal	5.50
Pea	5.36
Lupin	5.44
Dry bean	5.28
Other legumes	5.40
Mustard (yellow)	5.12
Rapeseed	5.35
Sunflower (hulled)	5.29
Vegetables, mushrooms, and leaf proteins	4.40
Rapeseed meal	5.53[a]
Sunflower meal (hulled)	5.36[a]
Flax meal	5.41[a]
Turnip rapeseed meal	5.43[a]
Rapeseed meal	5.36[a]
Flax meal	5.40[a]
Sunflower meal	5.37[a]
Safflower meal	5.33[a]
Turnip rapeseed protein isolate	5.52[a]
Rapeseed protein isolate	5.57[a]
Flax protein isolate	5.43[a]
Sunflower protein isolate	5.43[a]
Safflower protein isolate	5.49[a]
Soy proteins	5.69–5.79
Nuts	5.18–5.46[a]
Mushrooms	4.7 / 3.99[a]

Adapted from Mariotti et al. (2008) and Krul (2019)
[a]Reported as k_A

Table A.2 Amino acids (AA) weights of free (hydrolyzed) and residues (as part of a protein)

Amino acid	M_W free AA (Da)	M_W AA residue (Da)	N in AA residue (%)
Alanine	89.1	71.1	19.7
Arginine	174.2	156.2	35.9
Asparagine	132.1	114.1	24.6
Aspartic acid	133.1	115.1	12.2
Cysteine	121.2	103.1	13.6
Glutamic acid	147.1	129.1	10.8
Glutamine	146.1	128.1	21.9
Glycine	75.1	57.1	24.6
Histidine	155.2	137.1	30.6
Isoleucine	131.2	113.2	12.4
Leucine	131.2	113.2	12.4
Lysine	146.2	128.2	21.9
Methionine	149.2	131.2	10.7
Phenylalanine	165.2	147.2	9.5
Proline	115.1	97.1	14.4
Serine	105.1	87.1	16.1
Threonine	119.1	101.1	13.9
Tryptophan	204.2	186.2	15.0
Tyrosine	181.2	163.2	8.6
Valine	117.1	99.1	14.1

Modified from FAO (2019)

Each of the two factors has its limitations: k_b may not accurately report the protein content because the non-protein nitrogen content can vary from sample to sample, while k_a may overestimate the actual protein content because it does not take into account the non-protein nitrogen. For these reasons, it is recommended to average both factors to obtain a mean k factor for the nitrogen-to-protein conversion. Moreover, as prosthetic groups can also contribute to the total protein content, scientists may also take protein-associated groups into account (Mariotti et al., 2008). However, most authors consider k_a and k_b to calculate the protein content because sophisticated methods are needed to determine the associated prosthetic groups, which are not readily available in many laboratories. An overview of how the protein content can be obtained is given in Fig. A.3:

In addition, protein quality indices, such as the protein digestibility-corrected amino acid score (PDCAAS) or digestible indispensable amino acid score (DIAAS), should be considered because they vary between the different protein sources (Chap. 2). The DIAAS value is recommended by the FAO and is determined using the true ileal amino acid digestibility of indispensable amino acids in the food protein compared to a reference protein (Mathai et al., 2017). For example, the DIAAS of soy protein is 84, whereas that of wheat protein is 45 (Mathai et al., 2017). Such scores can influence the labeling of protein values in specific countries, such as the use of the packaging statement 'good source of protein' (Nosworthy et al., 2017).

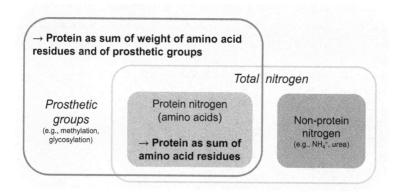

Fig. A.3 Proteins are composed of amino acids and prosthetic groups. The protein content can be determined by considering the sum of the amino acid residues and by combining this value with the weight of the associated prosthetic groups

Protein Composition

The functional performance of plant protein ingredients depends on the types and amounts of different proteins they contain. For example, soy proteins contain β-conglycinin (7S) and glycinin (11S) that have different functional attributes, *e.g.*, solubility, emulsifying, and gelling properties. The content of these fractions also varies with season (Chen, 1993). Thus, it is important to understand the protein composition of an ingredient.

Protein composition is usually determined by separating a mixture of proteins in solution using polyacrylamide gel electrophoresis. SDS-PAGE is the most common and straightforward method used for this purpose (Chen et al., 2010; Azzollini et al., 2019). These methods can also be used to monitor changes in protein properties during processing, such as the crosslinking or degradation caused by the high temperatures and shear stresses in extruders (Chen et al., 2011). During a SDS-PAGE, the proteins in a mixed sample are separated by applying them to a polyacrylamide gel (8–12%) and then applying an electric field. Proteins of known molecular weight are run under the same conditions as standards. Typically, the proteins are first dispersed in a Laemmli buffer solution, which contains sodium dodecyl sulfate (SDS), β-mercaptoethanol, and urea, and then heated at 95 °C (Chen et al., 2010, 2011; Azzollini et al., 2019; Li et al., 2018). SDS is an anionic surfactant that binds to the proteins and gives them a strong negative charge. The β-mercaptoethanol is a reducing agent that cleaves disulfide bonds. After treatment, all the proteins should have similar linear charge densities and similar extended structures. Consequently, they can be separated based on their molecular weight by passing them through the polyacrylamide gel. After electrophoresis, the gels are commonly stained with Coomassie brilliant blue to visualize the proteins.

The Laemmli buffer is useful to maximize protein solubility and to compensate for differences in the charge and conformations of the proteins. However, it does not represent the aggregation status of the original proteins because the aggregates are broken down by SDS and β-mercaptoethanol. Information about the aggregation state of the proteins can be obtained by running the gels under non-reducing conditions (without β-mercaptoethanol) or under native conditions (no additives). In addition, if the protein concentration in the bands is too low, then the proteins can be stained with other dyes, such as silver, which typically gives a 10- to 100-fold increase in sensitivity. Another issue to consider is that very small or very large protein fractions are not suitable for analysis by this type of gel electrophoresis because they either diffuse through the gel too quickly or they do not enter the gel, respectively.

The bands obtained can be quantified using image analysis. Here, a photo/scan is taken from the dyed gel and transferred into grayscale. Then either relative or absolute quantification can be carried out. These techniques involve transferring the optical data from the image into pixel intensities using image processing programs, such as ImageJ (Butler et al., 2019). This results in peak areas (Fig. A.4) for each band after brightness, contrast, background, and oversaturation adjustments (Villela et al., 2020).

Relative quantification is subsequently performed by dividing the peak area of a specific band by the total area under all the bands. The result is a percentage concentration of one individual protein in relation to the complete fraction. Absolute quantification is achieved by applying different concentrations of known proteins to the gel (*e.g.*, bovine serum albumin or egg white albumin). A standard curve is calculated by plotting the measured peak areas against the known protein concentrations and determining the linear range (Miles & Saul, 2009). The peak area should be used for this purpose, rather than the peak volume or peak intensity (Gassmann et al., 2009). Moreover, it needs to be stated that absolute quantification is only reasonable if the same protein is used as a standard because different proteins show variable staining intensities with protein dies (*e.g.*, Coomassie). More potential pitfalls of gel electrophoresis are discussed by Gassmann et al. (2009).

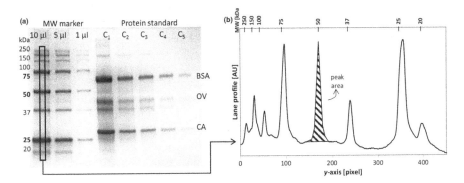

Fig. A.4 Absolute quantification using SDS-PAGE densitometry. (Adapted from (Villela et al., 2020) under CC BY 4.0 (https://creativecommons.org/licenses/by/4.0/))

Protein composition can also be characterized using proteomics. This technique is more advanced than SDS-PAGE but requires more sophisticated and expensive equipment. However, it can precisely determine the different types of proteins in a sample. In the following, we review a common approach based on 2D electrophoresis followed by a bottom-up (shotgun) tryptic approach using digestion and Nano-LC-ESI-MS/MS separation and detection. However, there are many other techniques and methods available, which are discussed by Aslam et al. (2017).

For this method, the proteins are separated using 2D gel electrophoresis, which involves two stages. In the first stage, the proteins are separated based on their isoelectric point using specialized strips that contain a defined pH-gradient. The proteins are placed on the strip and then an electrical field is applied, which causes them to move towards the oppositely charged electrode. The proteins stop moving when the external pH matches their isoelectric point, *i.e.,* the pH where the proteins have a zero net charge. In the second stage, the proteins are separated based on their molecular weight. This is achieved by treating the proteins on the strip with SDS and other reagents to denature them and give them a strong negative charge, and then placing the strip onto a polyacrylamide gel and applying an electric field, which causes the proteins to be separated based on their molecular weight. The location of the different proteins on the 2D gel can be shown by using Coomassie Blue or another suitable dye. If required, information about the identity of the different proteins can be obtained by cutting individual spots from the gel and then using mass spectrometry (often after partial hydrolysis of the protein).

Nano-Liquid Chromatography-Electrospray Ionization-Tandem Mass Spectrometry (Nano-LC-ESI-MS/MS) can be used to identify proteins recovered from such spots. Initially, a protein is partially hydrolyzed and then the peptides are separated using reversed liquid chromatography columns (typically a pre-column followed by an analytical column) and then ionized by electrospray ionization. Electrospray ionization of the peptides is performed in the ionization chamber by applying a high voltage (2–6 kV) between the spray needle and a heated capillary that leads into the mass spectrometer. The peptides are injected into the needle and the strong electrical field causes the production of an aerosol of highly charged electro-sprayed droplets. The droplets are subsequently directed to the mass spectrometer by the flow of nitrogen gas and the analytes (peptides) are released from the droplets into the orifice of the heated capillary that is located a few centimeters away from the spraying needle that leads to the mass analyzer. In the mass analyzer, the ions are separated under vacuum based on their mass-to-charge (m/z) ratio and then passed to the detector (Banerjee & Mazumdar, 2012).

Typically, a mass spectrometer system consists of a triple quadrupole setup (called tandem MS or MS/MS, quadrupoles are able to select ions based on their m/z ratio that influences their trajectories in the oscillating electric fields) in which the ions are separated based on their m/z ratio in the first quadrupole. The selected ions are then forwarded through a fragmentation chamber where they are fragmented by collision with a gas (*e.g.,* helium or argon). The resulting ionized species are then transported into the last quadrupole, which separates the resulting ions according to their m/z ratio (Domon & Aebersold, 2006; Ho et al., 2003).

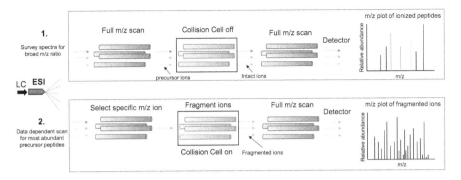

Fig. A.5 Schematic workflow of bottom-up proteomics in a triple quadrupole instrument based on tryptic peptides separated by nano liquid chromatography (LC) and being ionized by electrospray ionization (ESI) techniques. The peptides are identified by comparing the measured fragmentation patterns with theoretical (in silico) digestion patterns of a protein database. (Modified from Zhang et al., 2013)

A workflow consists of analyzing the ionized peptides first without fragmentation in a LC/MS system to obtain the m/z ratios of the peptides without fragmentation, which are also called precursor ions (Fig. A.5). Based on the m/z results, a 'product scan' is applied in which the precursor ions are separated in the first quadrupole, fragmented in the collision chamber, and the fragmented ions are detected based on their m/z ratio in the last quadrupole. This is repeated for the different precursor ions by adjusting the parameters of the quadrupole. The fragment ion spectra are then assigned peptide sequences based on database comparisons with theoretical fragmentation patterns of peptides and the peptide sequence is used to predict the protein type (Chen et al., 2020).

Protein Denaturation State

The denaturation state of proteins plays an important role in many of their functional attributes, such as their solubility, emulsifying, foaming, thickening, and gelling properties. Plant proteins are often denatured during the processes used to extract and purify them from botanical materials. Consequently, it is usually important to have information about the denaturation state of the proteins in food ingredients. The most common method for measuring the denaturation state is to use differential scanning calorimetry (DSC). In this method, the protein is dissolved or dispersed in a suitable aqueous solution of known pH and ionic strength. Typically, a protein concentration of around 1–20% is used for this purpose, depending on the sensitivity of the instrument used. Conventional DSC instruments require relatively concentrated samples (*e.g.*, 10% protein), whereas more sensitive micro-DSC instruments (baseline noise 0.015 μW in a micro DSC *vs.* 5 μW in a normal DSC) can use relatively dilute samples (*e.g.*, 1% protein).

Typically, DSC compares the heat flow between two different pans (the sample pan and the reference pan) when they are heated at the same rate. The protein solution is placed in the sample pan and the reference pan is left empty. The two pans are then placed into the DSC instrument and allowed to equilibrate to the starting temperature. The two pans are then heated from the start to the end temperature at a constant heating rate (*e.g.,* from 20 to 150 °C at 10 °C/min). During this process, the heat flow (q), i.e., the adjusted power input, required to keep the two pans at the same temperature is measured. If the initial proteins are in a native state, they unfold at their thermal denaturation temperature (T_m) and absorb energy (endothermic transition), which can be seen as a peak in the heat-flow *versus* temperature profiles (Fig. A.6). Conversely, if the initial proteins are already denatured, no peak is observed. Consequently, DSC can be used to provide valuable insights into the denaturation state and denaturation temperature of plant proteins.

Other methods can also be useful for providing information about protein denaturation and unfolding, such as fluorescence spectroscopy. The peak in a fluorescence spectrum depends on the local molecular environment of the phenolic amino acids in proteins (such as tyrosine, tryptophan, and phenylalanine). In the native state, these groups are located in the hydrophobic interior of the protein and are mainly surrounded by hydrophobic groups. However, once the protein unfolds,

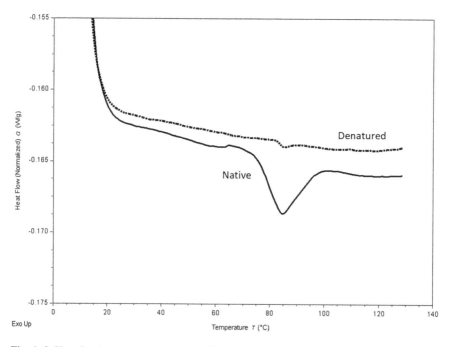

Fig. A.6 Heat flow *versus* temperature profiles for native and denatured plant proteins during heating at a constant rate in a DSC instrument. Native proteins unfold and exhibit an endothermic peak, whereas denatured proteins do not

Appendix: Analysis of Plant-Based Ingredients and Foods

these phenolic amino acids are exposed, which changes the fluorescence signal. Consequently, measurements of the fluorescence spectra *versus* temperature can be used to monitor the thermal denaturation behavior of globular plant proteins.

Fat Analysis

Total Fat Content

Fats contribute to the physicochemical, sensory, and nutritional properties of plant-based foods and so it is important to measure the type and amount present. The Soxhlet extraction method is one of the most commonly used methods for determining the total fat content of foods (Fig. A.7). Initially, the samples are usually weighed and then dried in an oven to remove the majority of the water. The samples are then placed in a thimble that is soaked several times with a nonpolar organic solvent with a low boiling point (such as hexane or petroleum ether), which solubilizes and removes the fat (*e.g.*, AOAC 960.39). After a sufficiently long time for complete

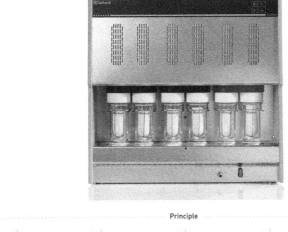

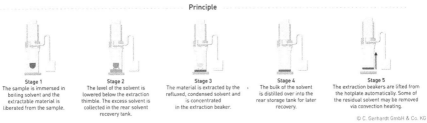

Fig. A.7 Schematic working principle of a device that automatically extracts the fat by the Soxhlet solvent extraction method. (Image shows Soxtherm and was kindly provided by C. Gerhardt GmbH & Co. KG (Koenigswinter, Germany))

extraction to occur, the fat-loaded solvent is collected and then the solvent is evaporated using a distillation apparatus. The solvent can then be reused for future analyses. The pre-weighed solvent containers are dried and the extracted fat is determined gravimetrically. To decrease the extraction time, devices are available that utilize sonication to break down the food structure or that apply high-pressure to achieve higher boiling points and consequently higher extraction temperatures and rates (Webster, 2006). In practice, homogenized samples (*i.e.,* with a pestle and silica sand) are transferred to the Soxhlet thimble after they have been dried overnight or longer above 100 °C to determine the dry matter content and remove any excess water that would interfere with the solvent extraction. If samples were acid hydrolyzed (see below) prior to the Soxhlet extraction, the filters need to be dried completely before the extraction procedure (Hilbig et al., 2019a).

If the sample has a low-fat content ($< 0.1\%$), the total fat content is difficult to measure accurately since it is a gravimetric method. Most analytical balances have a readability of around 0.1 mg, which typically results in reliable results for sample weights before solvent extraction of around 10 g. In the case of very low-fat contents, the sample size may be increased depending on the maximum thimble size available for the Soxhlet extractor. Alternatively, methods such as nuclear magnetic resonance (NMR) or near-infrared spectroscopy (NIR) can be employed (Colnago et al., 2011; Krepper et al., 2018).

If the fat is bound to a protein matrix and is not fully extracted by the solvent, then a simple Soxhlet extraction will result in underreporting of the total fat content (*e.g.,* AOAC 922.06). The bound fat can be released by hydrolysis using boiling acid (4 M HCl) prior to Soxhlet extraction, thereby improving the accuracy of the total fat content analysis, which is often referred to as the Weibull-Stoldt method (Petracci & Baéza, 2011). In this method, samples are hydrolyzed in a manual reflux heater equipped with an air cooler or an automatized hydrolyzer (Fig. A.8), the excess acid is washed out, and then the samples are dried before the solvent extraction step.

In another method, which is widely used for liquid milk (e.g., AOAC 905.02) and referred to as the Röse-Gottlieb method, a defined amount of heated sample (60–70 °C) is mixed with ammonia to hydrolyze the proteins that surround the fat globules. The mixture is precipitated with ethanol and extracted with solvents such as diethyl ether and petroleum benzine. The samples are mixed, centrifuged, and the upper solvent phase is removed and distilled to release the extracted fat, which is quantified by weighing.

Fat Composition

The fatty acid composition of lipids is usually analyzed using a gas chromatography (GC) approach coupled with flame ionization detection (GC-FID). Initially, the extracted lipids are converted into compounds with higher volatility, known as fatty acid methyl esters (FAMEs). In this operation, the fatty acids are released from the triacylglycerols and phospholipids by hydrolysis and are then esterified into FAMEs. The lipids are first solubilized in a suitable solvent (*e.g.,* methyl

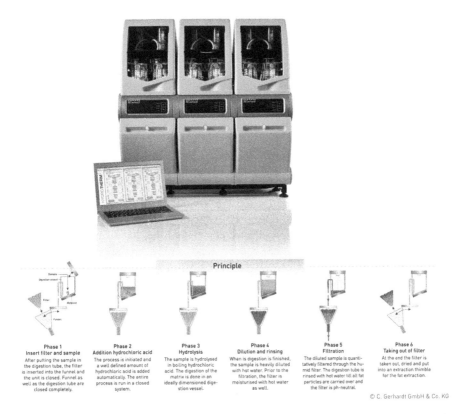

Fig. A.8 Schematic working principle of a device that automatically releases fats from food matrices by hydrochloric acid hydrolyzation. The released fats remain in the dried filter, which is used for subsequent Soxhlet extraction. (Image shows Hydrotherm device and was kindly provided by C. Gerhardt GmbH & Co. KG (Koenigswinter, Germany))

tertiary-butyl ether) and then esterified and hydrolyzed with trimethylsulfonium hydroxide or potassium hydroxide. The FAMEs are subsequently transferred into GC vials and injected into a GC column, which often uses heated hydrogen gas (H_2, 280 °C) as a carrier gas. The FAMEs are usually separated on fused silica capillary columns that have a chemically bound stationary phase consisting of polyethylene glycol or acid-modified polyethylene glycol to separate the fatty acids. The separated FAMEs are detected by FID, which combusts the eluted compounds using a hydrogen flame. The ions formed as a result of the combustion in the flame are electrically detected by measuring the current produced. This electrical signal is then converted into peaks that can be used to qualitatively and quantitatively detect and determine the fatty acid composition and their respective concentrations by analyzing fatty acid standards of known concentrations (Nielsen, 2017). Here, the position of the peaks in the chromatogram can be used to determine the identity of the fatty acids, whereas the area under the peaks can be used to determine their concentration.

Mineral Analysis

The type and concentration of minerals present in a food also influence its physicochemical, sensory, and nutritional properties. The concentration of the various different kinds of minerals in plant-based foods (such as sodium, calcium, or potassium) can be determined using atomic emission spectroscopy or ion chromatography methods.

The total mineral content of plant-based foods and ingredients is usually determined by measuring the ash content. In this method, a known mass of food (m_T) is incinerated within a muffle furnace (which combusts all the organic compounds) and the mass of the remaining ash (inorganic compounds) is measured, m_A (*e.g.*, AOAC 920.153): Ash content (%) = 100 × m_A/m_T. Typically, a wet solid sample is first homogenized with magnesium acetate (which releases oxygen and forms bubbles that increase the surface area) in dried porcelain dishes and then this mixture is dried to remove residual water. Subsequently, the samples are incinerated at a high temperature (<550 °C to not release alkali halides) to remove any organic compounds from the non-volatile minerals. The whole process usually takes several hours, and the final incineration is carried out overnight or longer to ensure complete combustion. After complete combustion, the mass of the ash is determined gravimetrically and the ash content is calculated (Hilbig et al., 2019b).

Even though the method is relatively straightforward, a number of things should be carefully considered to obtain accurate results. First, the complete removal of all organic compounds is essential for an accurate determination of the total mineral content. The final combustion should be carried out long enough, and the time needed to fully incinerate may vary from sample to sample. Usually, samples are combusted for several days but the minimum time to reach weight equilibrium (*i.e.*, the sample is fully incinerated) can be analyzed to minimize the time of the analysis. Second, the sample dishes need to be dried in an oven and cooled down in a desiccator prior to analysis to remove any residual water.

Sodium Chloride Content The sodium chloride content is another important component of food materials and plant-based foods. Many of these foods are rich in sodium chloride (*e.g.*, plant-based cheeses and meats) and the sodium (chloride) content has to be labeled on the package in many countries. There are several methodologies available to determine the content and most of them rely on measuring either the sodium or the chloride content (NaCl is 40% sodium) and then calculating back to the sodium chloride content (assuming that NaCl is the major form of Na^+ and Cl^- in the food, which might not always be the case). The most common techniques to measure the sodium chloride content are flame emission spectroscopy and chloride titration by potentiometry (Capuano et al., 2013). Flame emission spectroscopy utilizes the characteristic emission spectra of electrons in elements once they have been excited by a flame and move back to their initial energy level. The position of the peaks in the spectra is used to determine the identity of the element, while the intensity of the peaks is used to determine their concentration. Typically, the extracted sample is burned using an acetylene/air or hydrogen/air mixture at 2200 °C or 2800 °C, and sodium is detected at around 589 nm.

Moreover, the chloride content of samples can be obtained by titration with Ag^+ ions to form AgCl precipitates until chloride ions are depleted. The excess Ag^+ ions are either determined with chromate to form an orange-colored solid, silver chromate (Mohr's method) or are titrated back with a thiocyanate solution that reacts with ferric ions once surplus Ag^+ is depleted (Volhard's method). However, modern instruments typically use a potentiometric precipitation titration to detect the titration endpoint in this reaction.

Carbohydrate Analysis

Various kinds of carbohydrates are often used as functional ingredients to formulate plant-based foods, including monosaccharides, disaccharides, oligosaccharides, and polysaccharides. After ingestion, some of these carbohydrates are directly absorbed (*e.g.,* glucose), some of them are digested and then absorbed (*e.g.,* starch), and some of them are not digested or absorbed (*e.g.,* dietary fibers).

The total carbohydrate content of food samples is often calculated from the known concentrations of the other substances present, such as the protein, fat, ash, and water content:

$$C_{carbohydrate} = 100 - \left(C_{protein} + C_{fat} + C_{ash} + C_{water}\right) \quad (A1)$$

This value is therefore prone to measurement errors that may occur during the determination of the other components. However, these errors are generally quite low because the techniques used to measure protein, fat, ash, and water content are reliable (Miller et al., 2007). There are also methods available that determine the total carbohydrate content by spectrophotometric methods, such as the anthrone or phenol-sulfuric acid methods (Capuano et al., 2013; Masuko et al., 2005). These methods rely on solubilizing the carbohydrates and establishing standard curves using known compounds.

Plant-based foods may also contain significant amounts of sugars, such as monosaccharides (*e.g.,* glucose and fructose) and disaccharides (*e.g.,* sucrose and maltose). The type and concentration of these sugars can be determined by HPLC analysis. Typically, the extracted sugars are separated using anion-exchange or modified silica gel columns that have been derivatized with amino acids. The sugars are subsequently detected with pulsed electrochemical or refractive index detectors. Each sugar can be detected by reference compounds and the concentration is determined by peak integration using external standards (BeMiller, 2017).

Fiber Content Plant-based foods also contain various kinds of fibers, such as cellulose, hemicellulose, lignin, pectin, and resistant starch. The total fiber content of a food can be determined using standardized methods that are designed to simulate the human gastrointestinal tract, such as the AOAC 985.29, AOAC 2009.01, and AOAC 2017.06 methods. The food sample is passed through a simulated upper gastrointestinal tract (mouth, stomach, and small intestine) and then the amount of dietary fibers remaining is measured.

Initially, the samples are dried and converted into a powder, which is then dispersed into an aqueous solution. The digestible macronutrients (*e.g.*, digestible lipids, starches, and proteins) are then broken down by adding digestive enzymes (*e.g.*, lipases, amylases, proteases) that simulate the enzyme activity of the human gut. The total dietary fiber content is analyzed by measuring the soluble dietary fiber (*e.g.*, pectin) and insoluble dietary fiber (*e.g.*, cellulose) and summing up their contents. The soluble fiber is found as the non-digested but soluble material after protease, amylase, and glucoamylase digestion that has been separated from the insoluble parts by filtration. Similarly, the insoluble fiber is the non-digested but insoluble material after similar enzymatic treatment and filtration. In detail, the insoluble dietary fiber is obtained by filtering the enzymatically digested material and washing the retentate with water, ethanol, and acetone to remove all soluble compounds. The weight of the dried retentate equals the insoluble dietary fiber. The permeate that is solubilized during the filtration and water washing steps is subsequently treated with ethanol and the precipitate is collected by filtration with pre-washed Celite. The resulting retentate is washed with ethanol and acetone, dried and the obtained powder equals the soluble fiber (BeMiller, 2017). The low molecular weight soluble dietary fiber that remains in the permeate can be further analyzed by HPLC (McCleary et al., 2015).

The total dietary fiber can be determined in one step by precipitating all the fibers from the digested material using ethanol. The precipitates can then be collected by filtration, are washed with ethanol and acetone, and then the material is weighed. In addition, the mass of any residual fat, protein, and ash can be determined using the methods discussed earlier and is then subtracted from the mass of the precipitate to determine the dietary fiber content (Fig. A.9).

Vitamin Analysis

The type and concentration of vitamins present in plant-based foods can be analyzed using various kinds of analytical methods, including spectroscopic, chromatographic, and mass spectrometry. The interested reader is referred to textbooks for more information (de la Guardia & Garrigues, 2015; Vitamin analysis for the health and food sciences, 2016).

Rapid Composition Analysis: Near-Infrared Spectroscopy

Near-infrared (NIR) spectroscopy has been successfully used to analyze the main constituents (*e.g.*, water, fat, and protein) of conventional animal-based products for decades (Tøgersen et al., 1999; Prieto et al., 2009). NIR employs wavelengths in the infrared spectrum of about 700 to 2500 nm. The technique analyzes the absorption of near-infrared light at different wavelengths and records the molecular vibrations of all molecules containing C–H, N–H, or O–H groups (such as water, fat, and

Fig. A.9 Schematic overview of the method to determine total dietary fiber. (Modified from BeMiller, 2017)

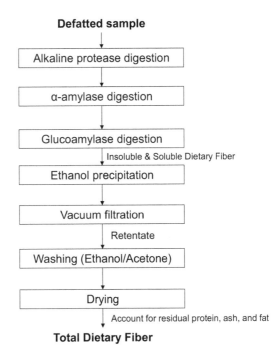

proteins). Each food is typically analyzed at specific wavelengths that are suited for the particular application. For example, the successful implementation in many studies was followed by the release of an official AOAC method using the FOSS FoodScan™ (AOAC 2007.04) in meat and meat products. The device employs near-infrared light generated by a tungsten-halogen lamp and emits monochromatic light in the spectral region between 850 to 1100 nm to analyze the sample composition in transmission mode and 400 to 700 nm in reflection mode coupled with the FOSS Artificial Neural Network Calibration Model (Anderson, 2007). The main advantages of the NIR technique are the rapid analysis time, its potential use as an in-line measurement technique, and its non-destructive nature, but it requires extensive calibration. Research is currently being carried out to use NIR measurement techniques to characterize plant-based foods.

Conclusions

In this appendix, we described some of the methods that can be used to determine the composition and properties of plant-based foods and ingredients (Table A.3). The establishment of a systematic testing regime for these products is critical to ensure that the ingredients and foods meet processing requirements and consumer expectations. There are numerous kinds of other analytical techniques that can be used to provide valuable information about plant-based foods and ingredients, but these are beyond the scope of this work.

Table A.3 Overview of common methods employed for analyzing the proximate composition

Compound	Method of choice
Protein	Kjeldahl digestion/Dumas combustion
Protein composition	Gel electrophoresis; proteomics based on soft ionization coupled with tandem MS
Fats	Soxhlet extraction, Weibull-Stoldt, Röse-Gottlieb
Fatty acid composition	Fatty acid methyl esters analyzed by GC-FID
Dry matter	Oven drying; Karl Fischer titration
Total minerals	Ashing
Specific minerals	Atomic emission spectroscopy; ion chromatography; titration methods
Carbohydrates	By subtraction, HPLC
Fibers	Enzymatic digestion and extraction

References

Anderson, S. (2007). Determination of fat, moisture, and protein in meat and meat products by using the FOSS FoodScan™ near-infrared spectrophotometer with FOSS artificial neural network calibration model and associated database: Collaborative study. *Journal of AOAC International, 90*(4), 1073–1083.

Aslam, B., Basit, M., Nisar, M. A., Khurshid, M., & Rasool, M. H. (2017). Proteomics: Technologies and their applications. *Journal of Chromatographic Science, 55*(2), 182–196. https://doi.org/10.1093/chromsci/bmw167

Azzollini, D., Wibisaphira, T., Lakemond, C. M. M., & Fogliano, V. (2019). Toward the design of insect-based meat analogue: The role of calcium and temperature in coagulation behavior of Alphitobius diaperinus proteins. *LWT - Food Science and Technology, 100*, 75–82. https://doi.org/10.1016/j.lwt.2018.10.037

Banerjee, S., & Mazumdar, S. (2012). Electrospray ionization mass spectrometry: A technique to access the information beyond the molecular weight of the Analyte. *International Journal of Analytical Chemistry, 2012*, 282574. https://doi.org/10.1155/2012/282574

BeMiller, J. N. (2017). Carbohydrate analysis. In S. S. Nielsen (Ed.), *Food analysis* (pp. 333–360). Springer. https://doi.org/10.1007/978-3-319-45776-5_19

Buchi. (2010). *How to achieve low detection and quantification limits for the nitrogen determination with Kjeldahl*. Buchi. Retrieved from: https://static1.buchi.com/sites/default/files/downloads/Low_detection_and_quantification_limits.pdf?f2c00d880fe54d1623deba992961f0849c769815. Accessed: August 2020.

Butler, T. A. J., Paul, J. W., Chan, E.-C., Smith, R., & Tolosa, J. M. (2019). Misleading westerns: Common quantification mistakes in Western blot densitometry and proposed corrective measures. *BioMed Research International, 2019*, e5214821. https://doi.org/10.1155/2019/5214821

Capuano, E., van der Veer, G., Verheijen, P. J. J., Heenan, S. P., van de Laak, L. F. J., Koopmans, H. B. M., & van Ruth, S. M. (2013). Comparison of a sodium-based and a chloride-based approach for the determination of sodium chloride content of processed foods in the Netherlands. *Journal of Food Composition and Analysis, 31*(1), 129–136. https://doi.org/10.1016/j.jfca.2013.04.004

Chen, C., Hou, J., Tanner, J. J., & Cheng, J. (2020). Bioinformatics methods for mass spectrometry-based proteomics data analysis. *International Journal of Molecular Sciences, 21*(8), 2873. https://doi.org/10.3390/ijms21082873

Chen, F. L., Wei, Y. M., & Zhang, B. (2011). Chemical cross-linking and molecular aggregation of soybean protein during extrusion cooking at low and high moisture content. *LWT - Food Science and Technology, 44*(4), 957–962. https://doi.org/10.1016/j.lwt.2010.12.008

Chen, F. L., Wei, Y. M., Zhang, B., & Ojokoh, A. O. (2010). System parameters and product properties response of soybean protein extruded at wide moisture range. *Journal of Food Engineering, 96*(2), 208–213. https://doi.org/10.1016/j.jfoodeng.2009.07.014

Chen, H.-P. (1993). *Effects of glycinin, β-conglycinin, their subunits and storage conditions on tofu sensory characteristics* (Doctor of Philosophy). Iowa State University, Digital Repository, Ames. Retrieved from: https://lib.dr.iastate.edu/rtd/10412/

Chiang, J. H., Loveday, S. M., Hardacre, A. K., & Parker, M. E. (2019). Effects of soy protein to wheat gluten ratio on the physicochemical properties of extruded meat analogues. *Food Structure, 19*, 100102. https://doi.org/10.1016/j.foostr.2018.11.002

Colnago, L. A., Azeredo, R. B. V., Netto, A. M., Andrade, F. D., & Venâncio, T. (2011). Rapid analyses of oil and fat content in agri-food products using continuous wave free precession time domain NMR. *Magnetic Resonance in Chemistry, 49*(S1), S113–S120. https://doi.org/10.1002/mrc.2841

de la Guardia, M., & Garrigues, S. (2015). *Handbook of mineral elements in food*. Wiley.

Domon, B., & Aebersold, R. (2006). Mass spectrometry and protein analysis. *Science, 312*(5771), 212–217. https://doi.org/10.1126/science.1124619

Etheridge, R. D., Pesti, G. M., & Foster, E. H. (1998). A comparison of nitrogen values obtained utilizing the Kjeldahl nitrogen and Dumas combustion methodologies (Leco CNS 2000) on samples typical of an animal nutrition analytical laboratory. *Animal Feed Science and Technology, 73*(1), 21–28. https://doi.org/10.1016/S0377-8401(98)00136-9

FAO. (2019). *Nitrogen and protein content measurement and nitrogen to protein conversion factors for dairy and soy protein-based foods: A systematic review and modelling analysis*. World Health Organization.

Gassmann, M., Grenacher, B., Rohde, B., & Vogel, J. (2009). Quantifying Western blots: Pitfalls of densitometry. *Electrophoresis, 30*(11), 1845–1855. https://doi.org/10.1002/elps.200800720

Grossmann, L., Hinrichs, J., & Weiss, J. (2019). Solubility of extracted proteins from Chlorella sorokiniana, Phaeodactylum tricornutum, and Nannochloropsis oceanica: Impact of pH-value. *LWT - Food Science and Technology, 105*, 408–416. https://doi.org/10.1016/j.lwt.2019.01.040

Hilbig, J., Gisder, J., Prechtl, R. M., Herrmann, K., Weiss, J., & Loeffler, M. (2019a). Influence of exopolysaccharide-producing lactic acid bacteria on the spreadability of fat-reduced raw fermented sausages (Teewurst). *Food Hydrocolloids, 93*, 422–431. https://doi.org/10.1016/j.foodhyd.2019.01.056

Hilbig, J., Murugesan, V., Gibis, M., Herrmann, K., & Weiss, J. (2019b). Surface treatment with condensed phosphates reduced efflorescence formation on dry fermented sausages with alginate casings. *Journal of Food Engineering, 262*, 189–199. https://doi.org/10.1016/j.jfoodeng.2019.06.014

Ho, C., Lam, C., Chan, M., Cheung, R., Law, L., Lit, L., Ng, K. F., Suen, M. W. M., & Tai, H. (2003). Electrospray ionisation mass spectrometry: Principles and clinical applications. *The Clinical Biochemist Reviews, 24*(1), 3–12.

Krepper, G., Romeo, F., de Sousa Fernandes, D. D., Diniz, P. H. G. D., de Araújo, M. C. U., Di Nezio, M. S., Pistonesi, M. F., & Centurión, M. E. (2018). Determination of fat content in chicken hamburgers using NIR spectroscopy and the successive projections algorithm for interval selection in PLS regression (iSPA-PLS). *Spectrochimica Acta Part A: Molecular and Biomolecular Spectroscopy, 189*, 300–306. https://doi.org/10.1016/j.saa.2017.08.046

Krul, E. S. (2019). Calculation of nitrogen-to-protein conversion factors: A review with a focus on soy protein. *Journal of the American Oil Chemists' Society, 96*(4), 339–364. https://doi.org/10.1002/aocs.12196

Li, T., Guo, X.-N., Zhu, K.-X., & Zhou, H.-M. (2018). Effects of alkali on protein polymerization and textural characteristics of textured wheat protein. *Food Chemistry, 239*, 579–587. https://doi.org/10.1016/j.foodchem.2017.06.155

Mæhre, H. K., Dalheim, L., Edvinsen, G. K., Elvevoll, E. O., & Jensen, I.-J. (2018). Protein determination—Method matters. *Foods, 7*(1). https://doi.org/10.3390/foods7010005

Marcó, A., Rubio, R., Compañó, R., & Casals, I. (2002). Comparison of the Kjeldahl method and a combustion method for total nitrogen determination in animal feed. *Talanta, 57*(5), 1019–1026. https://doi.org/10.1016/S0039-9140(02)00136-4

Mariotti, F., Tomé, D., & Mirand, P. P. (2008). Converting nitrogen into protein—Beyond 6.25 and Jones' factors. *Critical Reviews in Food Science and Nutrition, 48*(2), 177–184. https://doi.org/10.1080/10408390701279749

Masuko, T., Minami, A., Iwasaki, N., Majima, T., Nishimura, S.-I., & Lee, Y. C. (2005). Carbohydrate analysis by a phenol-sulfuric acid method in microplate format. *Analytical Biochemistry, 339*(1), 69–72. https://doi.org/10.1016/j.ab.2004.12.001

Mathai, J. K., Liu, Y., & Stein, H. H. (2017). Values for digestible indispensable amino acid scores (DIAAS) for some dairy and plant proteins may better describe protein quality than values calculated using the concept for protein digestibility-corrected amino acid scores (PDCAAS). *British Journal of Nutrition, 117*(4), 490–499. https://doi.org/10.1017/S0007114517000125

McCleary, B. V., Sloane, N., & Draga, A. (2015). Determination of total dietary fibre and available carbohydrates: A rapid integrated procedure that simulates in vivo digestion. *Starch - Stärke, 67*(9–10), 860–883. https://doi.org/10.1002/star.201500017

Miles, A. P., & Saul, A. (2009). Using SDS-PAGE and scanning laser densitometry to measure yield and degradation of proteins. In J. M. Walker (Ed.), *The protein protocols handbook* (pp. 487–496). Humana Press. https://doi.org/10.1007/978-1-59745-198-7_44

Miller, E. L., Bimbo, A. P., Barlow, S. M., Sheridan, B., Burks, L. B. W., & Collaborators. (2007). Repeatability and reproducibility of determination of the nitrogen content of fishmeal by the combustion (Dumas) method and comparison with the Kjeldahl method: Interlaboratory study. *Journal of AOAC International, 90*(1), 6–20. https://doi.org/10.1093/jaoac/90.1.6

Moore, J. C., DeVries, J. W., Lipp, M., Griffiths, J. C., & Abernethy, D. R. (2010). Total protein methods and their potential utility to reduce the risk of food protein adulteration. *Comprehensive Reviews in Food Science and Food Safety, 9*(4), 330–357. https://doi.org/10.1111/j.1541-4337.2010.00114.x

Nielsen, S. S. (2017). *Food analysis* (5th ed.). Springer.

Nissen, S. (1992). *Modern methods in protein nutrition and metabolism*. Academic.

Nosworthy, M. G., Neufeld, J., Frohlich, P., Young, G., Malcolmson, L., & House, J. D. (2017). Determination of the protein quality of cooked Canadian pulses. *Food Science & Nutrition, 5*(4), 896–903. https://doi.org/10.1002/fsn3.473

Park, S., Valan Arasu, M., Lee, M.-K., Chun, J.-H., Seo, J. M., Lee, S.-W., Al-Dhabi, N. A., & Kim, S.-J. (2014). Quantification of glucosinolates, anthocyanins, free amino acids, and vitamin C in inbred lines of cabbage (Brassica oleracea L.). *Food Chemistry, 145*, 77–85. https://doi.org/10.1016/j.foodchem.2013.08.010

Park, Y. W. (2008). Moisture and water activity. In *Routledge handbooks online*. https://doi.org/10.1201/9781420045338.ch3

Petracci, M., & Baéza, E. (2011). Harmonization of methodologies for the assessment of poultry meat quality features. *World's Poultry Science Journal, 67*(1), 137–153. https://doi.org/10.1017/S0043933911000122

Prieto, N., Roehe, R., Lavín, P., Batten, G., & Andrés, S. (2009). Application of near infrared reflectance spectroscopy to predict meat and meat products quality: A review. *Meat Science, 83*(2), 175–186. https://doi.org/10.1016/j.meatsci.2009.04.016

Rayner, C. J. (1985). Protein hydrolysis of animal feeds for amino acid content. *Journal of Agricultural and Food Chemistry, 33*(4), 722–725. https://doi.org/10.1021/jf00064a039

Rutherfurd, S. M., & Gilani, G. S. (2009). Amino Acid Analysis. *Current Protocols in Protein Science, 58*(1), 11.9.1–11.9.37. https://doi.org/10.1002/0471140864.ps1109s58

Sáez-Plaza, P., Navas, M. J., Wybraniec, S., Michałowski, T., & Asuero, A. G. (2013). An overview of the Kjeldahl method of nitrogen determination. Part II. Sample preparation, working scale, instrumental finish, and quality control. *Critical Reviews in Analytical Chemistry, 43*(4), 224–272. https://doi.org/10.1080/10408347.2012.751787

Schreuders, F. K. G., Dekkers, B. L., Bodnár, I., Erni, P., Boom, R. M., & van der Goot, A. J. (2019). Comparing structuring potential of pea and soy protein with gluten for meat analogue preparation. *Journal of Food Engineering, 261*, 32–39. https://doi.org/10.1016/j.jfoodeng.2019.04.022

Schwenzfeier, A., Wierenga, P. A., & Gruppen, H. (2011). Isolation and characterization of soluble protein from the green microalgae Tetraselmis sp. *Bioresource Technology, 102*(19), 9121–9127. https://doi.org/10.1016/j.biortech.2011.07.046

Tøgersen, G., Isaksson, T., Nilsen, B. N., Bakker, E. A., & Hildrum, K. I. (1999). On-line NIR analysis of fat, water and protein in industrial scale ground meat batches. *Meat Science, 51*(1), 97–102. https://doi.org/10.1016/S0309-1740(98)00106-5

Villela, S. M. A., Kraïem, H., Bouhaouala-Zahar, B., Bideaux, C., Lara, C. A. A., & Fillaudeau, L. (2020). A protocol for recombinant protein quantification by densitometry. *MicrobiologyOpen, 9*(6), e1027. https://doi.org/10.1002/mbo3.1027

Vitamin analysis for the health and food sciences. (2016). CRC Press. https://doi.org/10.1201/9781420009750.

Webster, G. R. B. (2006). *Soxhlet and ultrasonic extraction of organics in solids*. American Cancer Society. https://doi.org/10.1002/9780470027318.a0864

Zhang, Y., Fonslow, B. R., Shan, B., Baek, M.-C., & Yates, J. R. (2013). Protein analysis by shotgun/bottom-up proteomics. *Chemical Reviews, 113*(4), 2343–2394. https://doi.org/10.1021/cr3003533

Index

A
Absorption coefficient, 166
Absorption spectrum, 166
Acidification, 5
Additive manufacturing
    edible inks, 130
    3D extrusion, 127, 129
    3D printing, 127
Additives
    colorings, 59, 60
    crosslinking agents, 61
    flavorings, 59, 60
    micronutrients, 62, 63
    nutraceuticals, 62, 63
    pH controllers, 61
    preservatives, 62
    salts, 58, 59
Advanced particle technologies
    application, 97
        chemically labile, 101
        gastrointestinal tract, 102
        light scattering, 103
        macronutrient replacement, 104
        macronutrients, 103
        in plant-based foods, 101
        textural characteristics, 103
        water-dispersibility, 101
    biopolymer microgels, 100
    biopolymer particles, 100
    emulsion technology, 97
    kinds, 97
    liposomes, 99
    solid fat particles, 99
    types, 98

Affective tests, 221, 222
Agar, 72
Aggregated biopolymers, 184
Agricultural crops, 5, 6
Agricultural Revolution, 267
Agricultural sector, 10
Agriculture industry, 10
Algal sources, 52
Alpha-linolenic acid (ALA), 55, 241
Amino acid (AA), 27, 42, 328
    age groups, 234
    DIAAS, 235
    plants *vs.* animals, 230
Amino acid compositions, 429
Amino acids (AA) weights, 541
Amphiphilic charged molecules, 30
Animal-and plant-based milk products, 409
Animal-based cheeses, 471
    caseins, 444
    cheese production, 453
    constituents, 452
    cow's milk, 452
    nutrient-rich fluid, 452
    physicochemical and sensory
        properties, 444
    physicochemical properties
        color and textural, 458
        flavorful cheeses, 463
        gel-sol transition, 462
        melting behavior, 462
        melting characteristics, 463
        shreddability, 464
        textural attributes, 461
        texture, 457

Animal-based cheeses (*cont.*)
  production, 444
  properties, 452
  steric and electrostatic repulsion, 453
  sub-protein fractions, 452
Animal-based food, 1, 11, 14, 17, 19, 177, 248
Animal-based products, 16, 61
  properties, 16
Animal-derived foods, 43
Anthocyanins, 256
Anthropologists, 264
Antibiotics, 13
Antimicrobial agents, 364
Antimicrobials, 62
Antinutrients, 260
  lectins, 260
  oxalate, 260
  phylate, 260
  plants *vs*. animals, 261
  problem, 260
  tannins, 260
Antinutritional factors (ANFs), 232
Antioxidants, 57, 62
Apparent elastic modulus, 188
Appearance
  affecting factors
    scattering, 160–162
    selective absorption, 159, 160
    spatial uniformity, 156, 157
    surface gloss, 158, 159
    transmission and reflection, 157, 158
  animal-based product, 156
  color attributes, 176
  factors impacting
    chromophore type and concentration, 168, 170, 171
    refractive index contrast, 174
    size and concentration, 171, 172, 174
  measurement
    CIELAB system, 175
    colorimeters, 175
    glossmeters, 174
    image analysis program, 174
    instruments, 174
    size and concentration, 175
    spatial uniformity, 174
    UV-visible spectrophotometer, 175
  modeling and prediction
    chromophores absorption spectra calculation, 166
    colloidal dispersion, 162
    mathematical models, 162
    particles scattering characteristics calculation, 164–166
    physical processes, 162
    reflectance measurements, 164
    scattered waves, 163
    spectral reflectance calculation, 166, 167
    spherical/cylindrical particles, 162
    tristimulus color coordinates, 163, 167, 168
Aqueous extraction methods, 53
Arthrospira platensis (Spirulina), 221
Artificial sweeteners, 271
Asparagopsis, 7
Associative phase separation, 94
  applications, 95
  attraction and repulsion, 95
  biopolymer molecules, 94
  electrostatic complexes, 95
  emulsions, 95
  plant proteins, 94, 95
  protein-polysaccharide complexes, 94
  proteins and anionic polysaccharides, 94
Atomic force microscopy, 49
Automation, 10
Avian flu, 17

B
Bernoulli equation, 109
Big data, 10
Binders, 73, 74
Bioactive substances, 432
Biodiversity, 10, 11
Biological stability, 198
Biomarkers, 262
Biopolymer-based ingredients, 208
Biopolymer hydrogels, 70
Biopolymer matrix, 66, 187
Biopolymer microgels, 100
Biopolymer molecules, 67, 205
Biopolymer particles, 100
Biopolymer phases, 206
Biopolymer phase separation, 206
  segregative regime, 92
Biopolymers, 63–65, 78, 91, 155, 306
Bioreactor setup, 143
Biotechnology, 10
Blanching, 134, 135
Bologna-type sausages, 320
Bovine cheese, 490
Bovine milk, 390
Breeders, 345
Bridging effects, 363
Bridging flocculation mechanism, 421
Brownian motion effects, 411

Index   561

Buffering salts, 478
Burgers, 5
Butter, 505
    cooling and heating process, 505
    functional attributes, 506
    plant-based, 506
    production, 505
    textural characteristics, 505

C
Calcium, 255
Camembert, 455, 461–462
Caramelization, 60, 208
Carbohydrates, 262, 267, 551
    animal-derived foods, 43
    characterization, 48, 49
    ingredients, 45–47
    isolation, 45, 46
    molecular properties, 45
    organic molecules, 43
    structure, 44, 45
    synthesized, 43
Cardiovascular disease, 54, 489
Carotenoids, 59, 62, 256, 310
Carrageenan, 73
Casein micelles, 430
Caseins, 29
Cationic mineral ions, 61
Cellular agriculture, 10
Cereals, 2
Characterization
    carbohydrates, 48, 49
    lipids, 57, 58
    proteins, 41–43
Cheddar, 456
Cheddar cheese (CC), 459, 467
Cheese manufacturing
    acidification, 454
    curd processing, 454
    pasteurization, 453
    ripening, 454
    sol-gel transition, 454
    standardization, 454
Cheese manufacturing process, 457
Cheese production parameters, 453
Cheese production process, 455
Cheese ripening, 454
Chelating agents, 57
Chemical degradation, 213
Chemical derivatization, 48
Chemically reactive ingredients, 216
Chemical reactivity, 213
Chemical stability, 198

Cholesterol, 52, 58, 242, 243
Chromatography methods, 48, 58
Chromophore concentration, 168
Chromophores, 159
Chymosin, 454
Circular dichroism (CD), 42, 49
Coagulation, 454
Coalescence, 422
Cognitive dissonance, 11
Cold-set polysaccharides, 72
Collagen, 34
Colloid mills, 107
Colloidal dispersions, 161, 166, 179, 180
Colloidal particles, 419
Colloidal-polymer materials, 222
Colloids, 155
Colon, 259, 260
Colorimeters, 175
Colorings, 59, 60
Colors, 78
Commercial products, 23
Commission International de l'Eclairage
    (CIE), 175
Compression stresses, 185
Compression testing, 193, 195
Concentrated animal feeding operations
    (CAFOs), 7
Concentrated colloidal dispersions, 166
Conjugated linolenic acid (CLA), 257
Consumer awareness
    knowledge, 529
Contrast time-dependent shear thickening, 179
Controlled environmental conditions, 221
Conventional dairy cheeses, 444
Cooking loss, 206
Corn starch, 465
Couette design, 125
Covid-19, 17
Cow's cheese, 487
Cow's milk, 406, 409, 413, 414, 425, 427, 430
    casein and whey proteins, 391
    characteristics, 394
    colloidal dispersion, 393
    colloidal particles, 391
    composition and microstructure, 390
    consumption, 395
    creaming, 392
    dairy products, 395
    droplets, 392
    energy and nutrients, 395
    fat concentration, 392
    fat crystals, 394
    functional versatility, 394
    globules and casein micelles, 390

Cow's milk (*cont.*)
　interfacial layer, 392
　lightness and shear viscosity, 393
　milk fat globules, 390
　pasteurized milk, 393
　phospholipid molecules, 391
　physicochemical and sensory
　　attributes, 392
　plant-based milk analogs, 390
　plant-based milks, 390
　processing operations, 392
　profile, 393
　raw milk, 392
Crank model, 216
Critical packing parameter, 180
Crop breeding, 50
Crop production, 11
Cross constant, 178
Cross model, 178
Crossflow-filtration techniques, 113
Crosslinking, 61, 208
Crystalline milk fat globules, 411
Crystalline salt, 58
Crystallization, 75, 76
Curcumin, 59, 256
Curd formation, 484

D

Dairy-based cheeses, 484
Dairy cheeses, 456
Dairy products, 443
Delivery systems, 433
Depletion effects, 363
Descriptive sensory test, 220
Descriptive tests, 220
Diabetes, 77, 227, 245, 246, 248, 256, 261,
　　263, 267
Diacylglycerols, 50, 58
Dietary changes, 10
Dietary fibers, 77, 247
　aggregation state, 248
　benefits, 249
　binding interactions, 248
　coating/embedding, 248
　fermentation, 248
　gastrointestinal barrier properties, 248
　gut microbiome, 248
　rheological modifications, 247, 248
Dietary proteins, 18, 19
Different scanning calorimetry (DSC), 42
Diffuse scattering, 162
Digestible Indispensable Amino Acid Score
　　(DIAAS), 76, 77, 233, 235, 428

Disaccharides, 44, 246
Dispensable amino acids (DAAs), 229
Docosahexaenoic acid (DHA), 55, 241
Domesticated animals, 6
Dry milling devices, 105
Dumas and Kjeldahl methods, 538, 539
Dumas method, 537
Duo-trio test, 220
Dynamic light scattering, 210
Dynamic shear rheology measurements,
　　194, 196

E

EAT-Lancet commission, 8–10, 19
Edam cheese, 461
Edible oils, 52, 53
Education
　chemical engineering, 529
　plant-based food industry, 528
Effective medium theory (EMT), 180, 410
Egg analogs
　food manufacturers, 382
　plant-based ingredient, 383
　properties, 367
Egg products
　baked products, 381
　cooking, 380
　custards, 379
　emulsified canola oil, 379
　emulsified foods, 372
　emulsified products, 374, 378
　emulsifier-coated fat droplets, 373
　foamed foods, 380
　functional attributes, 372
　hydrophilic emulsifier, 373
　mayonnaise and salad dressings, 372, 374
　physicochemical attributes, 374
　plant-based ingredients, 380
Egg white (EW), 383
Eicosapentaenoic acid (EPA), 55, 241
Einstein equation, 180
Elastic modulus, 185, 187
Electron microscopy, 49
Electrophoresis methods, 48
Electrostatic bridging effects, 61
Electrostatic complexes, 95
Electrostatic interactions, 347
Electrostatic repulsion, 71, 362
Empirical equation, 414
Emulsification, 74, 75
　additives, 404, 405
　algal and flaxseed oils, 399
　droplets, 400

Index

emulsifier selection, 400
homogenization devices, 399
HPVHs, 405
ingredients, 399
mixtures, 403
nutrient composition, 404
oil-in-water emulsions, 398
plant-based, 403
plant-based emulsifiers, 401
plant-based oils, 399
polysaccharides, 401, 403, 404
processing operations, 404
stability, 401
temperatures, 398
water, 400
Emulsified food products, 376
Emulsified oil, 476
Emulsified plant-based foods, 74, 206
Emulsifying, 32, 38, 44, 45, 47, 59, 80
Emulsion-based delivery system, 432
Emulsion gels, 479
Emulsion technology, 97
Emulsions, 59, 98, 211
Endogenous enzymes, 35
Endomysium, 291
Environmental and sustainability, plant-based foods
  animal foods, 5–6
  biodiversity, 10, 11
  economic resources and cultural habits, 3
  efficiency of food production, 6–8
  global population, 2, 3
  human diet, 2
  inefficiency of animals, 4, 5
  planetary boundaries, 8–10
Environmental conditions, 216
Environmental sustainability and human health
  carbohydrates, 490
  CO_2-eq emissions, 487
  cow's cheese, 487
  healthy, 489
  macronutrients and micronutrients, 489
  nutritional profiles, 491
  plant-based cheese, 487, 488
  salt, 491
  sugary and starchy foods, 489
Enzymatic conversion processes, 142
Enzymatic crosslinking, 479
Enzymatic hydrolysis, 48
Enzymatic reactions, 60
Enzymatic technique, 485
Enzyme-assisted aqueous-based methods, 53
Enzyme-assisted extraction methods, 53
Enzyme crosslinking, 485

Epimysium, 291
Equilibrium effects, 216
Equilibrium partition coefficient, 213, 214
Essential oils, 62
Estimated average requirement (EAR), 233
Ethical reasons, plant-based foods, 11, 12
Evolution, 264, 266
Exogeneous hydrolytic enzymes, 35
Experimental measurements, 414
Extenders, 73, 74
Extrusion, 120, 315

F
Factory farms, 12
Factory-farmed beef production, 7
Farming practices, 7
Fat melting and crystallization, 53
Fats, 262
  in cheese products, 469
  edible, 469
  milk, 469
  physicochemical characteristics, 469
  plant oils, 470
  solid, 470
  unsaturated fatty acids, 470
  unsaturated lipids, 470
Fatty acid (FA), 50, 52, 303
Fatty acid composition, 548
Fatty acid methyl esters (FAMEs), 548
Fatty acid profiles, 57, 238
Feed conversion efficiencies, 4
Fermentation, 18, 19, 142, 494
Fermentation process, 141, 143
  enzymes, 141
  food product, 142
  microbial fermentation, 141
  plant-based milks, 141
  reactors, 141
Fermented nut pastes, 486
Fermented tofu, 482
Fertilizers, 274
Fibrous proteins, 27, 34, 42
Filament length density, 187
Filtration techniques, 112, 114
Final gel strength, 71
Fish, 8
Fish oils, 55
Fishing industry, 10
Flavanols, 38
Flavor precursors, 314
Flavorings, 59, 60
Flavors, 78
Flexible proteins, 27, 28, 33, 34

Flexitarians, 1
Flocculation, 205, 422
Fluid dairy products, 410
Fluid holding properties
  biopolymer molecules, 212
  biopolymer network, 212
  external force, 212
  water-biopolymer mixing effects, 212
  WHC, 211
Fluid-holding capacity, 65, 66
Fluorescence spectroscopy instruments, 42
Foaming, 75
Food industry, 15–18, 50
Food ingredients, 1
Food manufacturer, 17, 144
Food manufacturing sector, 10
Food preparation and consumption, 217
Food production, efficiency of, 6–8
Food protein, 4
Food quality, 15
Food quality design
  computational techniques, 525
  functional ingredients, 525
  manufacturers, 525
  plant-based ingredients, 525
Food security, 4
Food supply, 5, 531
Food systems approaches
  animal-based foods, 531
  plant-based crops, 531
  stakeholders, 531
Foods causes, 12
Fourier Transform infrared (FTIR) spectroscopy, 42
Fractionation route, 472
Fracturability, 195
Fracture strain, 188
Fracture stress, 188
Fragmentation properties, 71
Free fatty acids, 52, 58, 464
Fructo-oligosaccharides (FOS), 246
Fruits, 2
Functional ingredients, 213

G

Galacto-oligosaccharides (GOS), 246
Gas chromatography (GC), 58
Gastrointestinal fate, 76–78
  colon, 259, 260
  mouth, 258
  small intestine, 259
  stomach, 258, 259
Gastrointestinal tract (GIT), 12, 77

Gelatin, 34, 73
Gelation
  biopolymers, 96
  cold-set gels, 96
  electrostatic screening, 96
  enzymes, 96
  globular proteins, 96
  ions, 96
  pH, 96
  polysaccharides gel, 96
  solid-like textures, 95
Gelation temperature, 71
Gelling, 70–73
Gene editing, 10
Genetic engineering, 50
Genetically-engineered soybeans, 57
Global population, 2, 3
Global warming, 7, 8, 10, 11
Globular plant proteins, 362
Globular proteins, 27, 30, 32, 33, 42
Glossmeters, 174
Glucono-delta-lactone (GDL), 318
Glutenin, 34
Glycerol molecules, 50
Glycosidic bonds, 48
Good Food Institute (GFI), 14, 16–18, 341, 528, 529
Gravimetric/volumetric feeders, 118
Gravitational separation, 66, 201, 209, 417, 418
Grazing animals, 265
Greenhouse gas emissions (GHG), 5, 7, 329, 488
Greenhouse gasses, 7
Grinding
  bowl choppers, 108
  equipment, 107
  grinders, 107
  vacuum fillers, 107
Gut microbiome, 261
  biomarkers, 262
  carbohydrates, 262
  composition and function, 261
  fats, 262
  nature, 262
  proteins, 262

H

Hagen-Poiseuille equation, 129
Hammer mills, 106
Health, 1, 7, 8
  plant-based foods, 12–14
Health conscious, 57
Healthy diets, 8

Heart disease, 52, 77
Heat-set gelling agents, 32
Heat treatments, 480
Heavy metals, 274
Heme protein, 59
Hen's eggs
　appearance, 346
　aroma, 353
　characteristics, 343, 366
　chemical reactions, 348
　colloidal particles, 346, 350
　commercial scale, 345
　cooking and thermal processing, 351
　egg proteins, 366
　emulsifying and foaming, 352
　emulsion and foam stability, 353
　emulsions, 367
　energy, 342
　environmental impact, 372
　functional attributes, 348, 349
　globular proteins, 365
　instability, 347
　LDL, 352
　lipid oxidation, 348
　nutrients, 342
　nutritional and health profiles, 371
　nutritional contents, 370
　phospholipids and proteins, 344
　polypeptide chains, 349
　processed fluids, 345
　products, 345
　properties, 345
　proximate analysis, 343
　rheological properties, 346
　storage and food preparation, 371
　structure, 343
　temperature, 350
　textural attributes, 366
　thermal denaturation temperatures, 365
　thickening, gelling, or binding
　　properties, 364
　unsaturated fatty acids, 344
　volatile components, 354
　white, 344
　yolk, 343, 344
Herschel-Bulkley model, 190, 360
Herschel-Bulkley parameters, 376
High internal phase emulsions (HIPEs), 75
High performance liquid chromatography
　(HPLC), 58
Higher-quality diet, 3
High-pressure valve homogenizers
　(HPVHs), 405
High-temperature extrusion, 135

Homeostasis, 242
Hominids, 265, 266
Homogeneous transparent material, 157
Homogenization, 108, 109, 357, 417
Human diet, 2
Human gastrointestinal tract, 265
Hydrocolloids, 45, 504
Hydrocyclone, 112, 113
Hydrodynamic radius, 67, 181
Hydrogel matrix, 70
Hydrogenization, 50
Hydrolysis, 207
Hydrolysis reactions, 364
Hydrolyzing plant proteins, 314
Hydrophilic bioactives, 432
Hydrophobic additives, 404
Hydrophobic functional ingredients, 207
Hydrophobic ingredient, 217
Hydrophobic interactions, 361
Hydrophobic molecules, 429
Hydrophobic nutraceuticals, 256
Hydrophobic preservatives, 62
Hydrophobicity, 214

I
Ice cream, 500
　aqueous phase, 501
　coalescence, 501
　fat globules, 500
　freezing, 501
　ingredients, 503
　nutritional profile, 503
　plant proteins, 502
　processing step and ingredient, 500
　production, 502
　scanning electron microscope, 500
Ideal plastics, 189, 190
Ideal solids, 184–188
Image analysis program, 174
Immunoglobin E (IgE), 236
In vitro study (INFOGEST), 77
Indispensable amino acids (IAAs), 229
Infrared spectroscopy, 48
Ingredient innovation
　agricultural crops, 525, 527
　design and production, 523
　environmental benefits, 527
　environmental impact, 526, 527
　functional, 525
　governments, 530
　socioeconomic impacts, 528
　standardized analytical methods, 525
　sustainable sources, 524

Ingredients
    additives (*see* Additives)
    binders, 73, 74
    carbohydrates (*see* Carbohydrates)
    crystallization, 75, 76
    emulsification, 74, 75
    extenders, 73, 74
    fluid-holding capacity, 65, 66
    foaming, 75
    functional, 23
    gastrointestinal fate, 76–78
    gelling, 70–73
    Lipids (*see* Lipids)
    melting, 75, 76
    minimal processing approaches, 80
    minimally-processed, 79
    molecular binding interactions, 64, 65
    nutrition, 76, 77
    physicochemical, functional, and sensory attributes, 23
    plant-based foods, 23
    proteins (*see* Proteins)
    solubility, 63, 64
    thickening, 66–70
    utilization, 78, 79
Insects, 18, 19
Insoluble fraction, 36
Instability mechanisms, 198
Instrumental colorimeters, 176
Instrumental methods, 163
Instrumental texture, 222
Instrumental viscometers, 182
International Food Information Council (IFIC), 13
Iron, 253, 254, 261
Irregular shaped plant-tissue fragments, 162
Irreversible deformation and fracture, 188
Isolation
    carbohydrates, 45, 46
    lipids, 52, 53
    proteins, 34–37

J
JUST Egg Folded, 368

K
Kappa carrageenan, 73
Kjeldahl method, 537
Kubelka-Munk theory, 166, 167

L
Lactic acid fermentation, 483
Laemmli buffer, 542
Land-based animals, 289
Land use, 5
Lectins, 260
Leghemoglobin, 311
Legumes, 2
Life cycle analysis (LCA), 433, 527
    milk analogs, 434
    plant-tissue disruption approach, 434
    transportation method, 435
Light scattering, 49
Lima bean aquafaba (LBA), 382
Linear viscoelastic region (LVR), 194, 459
Lipid droplets, 74
Lipids
    animal, 49
    characterization, 57, 58
    crystallization/melting properties, 49
    ingredients, 53–57
    isolation, 52, 53
    organic solvents, 49
    sources, 49
    structure, 50–52
    unsaturated fatty acids, 49
Lipophilic bioactive agents, 261
Liposomes, 99
Livestock, 6
Livestock animals, 54
Low density lipoprotein (LDL), 239, 243
Low-moisture texturized vegetable protein, 120
Lysine, 76

M
Macronutrient balance, 227
Macronutrients, 12, 228
    amino acid profiles, 231
    carbohydrates, 244
        dietary fibers, 247–249
        oligosaccharides, 246, 247
        starches, 245
        sugars, 246, 247
    digestibility, 231
        AF, 232
        aggregation state, 231
        dietary fibres, 232
        food matrix effects, 232
        measuring, 232, 233
        protein structure, 231
    lipids, 237, 238
        cholesterol, 242, 243
        MUFA, 240
        PUFA, 241, 242
        saturated fatty acids, 238, 239

Index 567

trans fatty acids, 242
unsaturated fatty acids, 239, 240
nutritional benefits, 232
proteins, 229
  allergenicity, 236, 237
  amino acid profiles, 229
  bioactivity, 236
  DIAAS, 234, 235
  digestibility, 231
  quality, 233
Maillard reaction, 60, 208
Management innovations, 10
Mass spectrometry, 42, 48
Mastication process, 218
Mathematical/computational models, 222
Mathematical models, 155, 162, 171, 183, 188
Maxwell model, 460
Mayonnaise, 74, 376
Mayonnaise and salad dressings, 375
Meat and fish products
  consumer, 286
  key properties
    color, 321, 322
    environmental sustainability, 329
    flavor, 325
    fluids, 323, 324
    nutritional value, 326, 327
    texture, 322
  meat analog products, 287–288
  plant based meat analog, 299
  plant-based foods, 285
  plant-based meat and seafood products, 286
  plant-based meats, 331
  plant-derived ingredients, 330
  processing methods
    plant-based ground meat, 317, 318
    plant-based meat products, 318–321
    preparations, 315
    protein texturization, 315, 316
  properties
    appearance, 292, 293
    cooking loss, 294–297
    flavor profile/oral processing, 297, 298
    muscle structure/composition, 289–292
    textural attributes, 294
Meat eaters, 264
Meat-like products, 65
Meat-like structures, 93
Mechanical homogenization device, 201
Mechanical processing methods
  dry grinding, 105
  plant-based foods, 104
  size reduction, 104

Mechanical rheometers, 219
Melting, 75, 76
Membrane filtration system, 115, 116
Membrane module systems, 116
Methylcellulose, 63, 73
Microbial cells, 19
Microbial contamination, 209
Microbial fermentation, 141
Microbial plasmid, 41
Microbiological issues, 274
Micronutrients, 12, 62, 63
  definition, 249
  minerals
    calcium, 255
    iron, 253, 254
    zinc, 254, 255
  vitamins
    vitamin B12, 251
    vitamin D, 251, 252
Microorganisms, 41
Microwave-assisted extraction methods, 53
Mie theory, 161, 164, 165
Milk analogs, 405, 407, 409, 410, 416, 419, 421, 425, 426
  appearance, 408
  colloidal particles, 421
  commercial products, 427
  instability mechanisms, 416
  microbial contamination, 416
  nutrient and calorie content, 426
  polymer molecules, 421
  protein-coated particles, 421
  remnants, 431
  sedimentation, 416
  stability, 416
Milk fat globule membrane (MFGM), 391
Milk fat globules, 393
Milling operations, 107
Minerals, 76, 250
Minimal processing approaches, 80
Minimum concentration, 71
Modern food processing methods, 267
Molecular approaches
  associative separation, 91
  biopolymers, 90
  hydrophobic interactions, 90
  interactions, 90
  phase transitions, 97
  polysaccharides, 91
  proteins and polysaccharides, 90
  segregative regime, 92
Molecular binding interactions, 64, 65
Molecular interactions, 27

Molecular plant-based emulsifiers, 74
Monoacylglycerols, 50, 58
Monosaccharide, 44, 48, 246
Monosodium glutamate (MSG), 273
Monounsaturated fatty acids (MUFAs), 240
Mouth, 258
Multiple scattering, 161
Muscle tissues, 34
Mycotoxins, 274

N
Nanoemulsions, 59, 98
Nano-Liquid Chromatography-Electrospray Ionization-Tandem Mass Spectrometry (Nano-LC-ESI-MS/MS), 544
Nanotechnology, 10
National Health and Nutrition Examination Survey (NHANES), 229
Natural ecosystems, 11
Natural pigments, 171
Near-infrared (NIR) spectroscopy, 552
Network elastic deformation, 212
Net-zero emissions, 15
Neurodegenerative diseases, 55
Newtonian fluids, 125, 127
Next-generation high-quality plant-based foods, 2
Nitrogen-to-protein conversion, 541
Non-adsorbed biopolymers, 205
Non-equilibrium effects, 216
Non-heme iron, 261
Non-ideal fluid-like behavior, 190
Non-ideal plastics, 190
Non-ideal solids, 188, 189
Non-polar amino acids, 31
Non-polar substances, 213
Non-protein fractions, 37
Normalized turbidity *vs.* particle diameter, 161
Nuclear magnetic resonance (NMR), 48, 209–210
Nutraceuticals, 12, 62, 63, 256, 257
  bioavailability, 256
  composition and structure, 256
Nutrition, 14, 76, 77
Nutritional deficiencies, 62
Nutritional implications
  food industry, 526
  gut microbiome and human health, 526
  plant-based food movement, 526
Nuts, 2

O
Obesity, 77
Oil quality, 53
Oil-rich plant, 52
Oil-rich plant materials, 35
Oil-soluble bioactive substances, 432
Oil-soluble vitamins, 52
Oil-water partition coefficient, 217
Oleogelator, 73
Oligosaccharides, 44, 46, 246
Omega-3 amino acids, 239
Omega-3 fatty acid, 249
Omega-6, 55
Omega-6 amino acids, 239
Omega-9, 55
Omnivores, 1, 12
Oral processing, 218, 219
Our World in Data, 3
Ovalbumin, 32
Oxalate, 260
Oxidation, 57, 207
Oxidation reactions, 57

P
Paleolithic diets, 266
Parasites, 266
Partial coalescence, 205, 505
Particle plant-based emulsifiers, 74
Particle-particle interactions, 167
Partitioning phenomena
  equilibrium partition coefficient, 213, 214
  flavors into the headspace, 214, 215
  multiphase system, 214
Pasteurization, 453
Pea protein, 36, 300
Perimysium, 291
Pesticides, 274
pH controllers, 61
Phase separation, 93, 205
Phenolic acids, 38
Phenol-sulfuric acid method, 48
Phosphate group, 52
Phospholipids, 52, 58, 74
Physical stability, 198
Physicochemical and sensory properties
  appearance (*see* Appearance)
  fundamental factors, 155
  products, 155
  stability (*see* Stability)
  textural attributes (*see* Texture)
Physicochemical attributes
  consistency index and power index, 413
  creamy appearance, 407

Index 569

critical packing parameter, 410
dilute colloidal dispersions, 411
EMT, 410
milk analogs, 406, 407
optical properties, 407
research and development laboratories, 412
textural characteristics, 410
viscosity, 410, 411
Physicochemical phenomena, 196
Phytate, 260
Phytic acid, 260
Phytochemicals, 38
Phytostanols, 52
Phytosterols, 52
Pickering mechanism, 373
Pigments, 52
Planetary boundaries, 8–10
Plant based meat analog
 binders, 306–309
 coloring agents, 310–312
 flavoring agents, 312–314
 lipids, 302, 303, 305
 plant proteins, 299, 300
 protein-rich ingredients, 301–302
 textural attribute, 299
Plant vs. animal-based diets, 263, 264
Plant-based biopolymers, 91
Plant-based burgers, 15
Plant-based cheese, 49, 445–451, 455, 473–475, 483
Plant-based cheese applications, 466
Plant-based cheese formation, 468
Plant-based cheese ingredients
 fats, 468
 meat alternatives, 465
 polysaccharide, 465
 proteins, 467
Plant-based cheeses, 444, 477, 487–489
 fractionation route, 475
 functional ingredients, 471
 milk production, 471
 production, 470
 raw materials and processing operations, 444
 structuring principles, 471
 tissue disruption route, 471
 tofu, 444
Plant-based diet, 1, 12, 13, 76, 227
Plant-based dressings, 74
Plant-based egg analogs, 59
 aqueous phase, 357
 constituents, 354
 egg yolk, 354
 emulsifier, 357
 fat droplets, 356
 functional ingredients, 356
 globular proteins, 354
 hydrophilic emulsifier, 357
 ingredients, 355
 lipids, 354
 manufacturers, 355
 oil-soluble additives, 357
 omega-3 rich oils, 354
 packaging, 357
 physicochemical properties
  appearance, 358
  chemical reactions, 363
  colloidal particles, 361
  heating, 360
  hydrocolloids, 361
  ingredients, 364
  physical stability, 361
  properties, 359
  rheological properties, 360
  rheology, 359
  textural properties, 360
  thermal denaturation temperatures, 360
  yellowish color, 358
 thermal processing, 357
Plant-based emulsified products, 375
Plant-based food products, 58
Plant-based foods, 14, 16, 77, 90, 145, 551
 category, 23
 complex materials, 222
 environmental and sustainability (see Environmental and sustainability, plant-based foods)
 flexitarians, 1
 food industry, 15–18
 generation, 23
 health, 12–14
 healthiness, 267
  agriculture, 274
  bioactive, 269
  chemical stability, 269
  factors, 269
  food matrix compatibility, 269
  fortification, 268
  partitioning, 268
  reduced digestibility, 273
  reduced fat, 270
  reduced salt, 272, 273
  reduced sugar, 270, 272
  reformulation, 269, 270
  solubility, 268
 low viscosity fluids, 155
 next-generation high-quality plant-based foods, 2

Plant-based foods (*cont.*)
    physicochemical properties, 155
    rheological characteristics, 177
    sensory attributes, 155
    source of dietary proteins, 18, 19
    taste, 13–15
Plant-based Foods Association in the United
        States, 1
Plant-based ingredients, 182
Plant-based lipids, 49
Plant-based mayonnaise, 375
Plant-based meat, 17, 285
Plant-based meat segment, 16
Plant-based milk analogs, 144, 395, 415
    aqueous solution, 395
    colloidal dispersions, 395
Plant-based milk and cream analogs, 65
Plant-based milk production, 144, 146
Plant-based milks, 16, 17, 109, 138, 395, 396,
        443, 483
Plant-based polysaccharides, 72
Plant-based product, 23
    high-quality, 23
Plant-based product lines, 15
Plant-based sausages, 318
Plant-based solid foods, 138
Plant-based whipped creams, 75
Plant-based yogurt analogs, 78
Plant-derived ingredients, 89, 274
Plant-derived lipids, 53
Plant tissue disruption approaches
    aqueous phase, 397
    fat globules, 396
    filtration processes, 396
    fragments, 396
    material, 397
    oil bodies, 396
    physicochemical and sensory
        attributes, 396
    seeds, 396
    source, 396
    triacylglycerol-rich core, 396
Plastic consistency index, 190
Polymer chain, 415
Polymer matrix, 187
Polymer molecules, 186
Polymeric porous foods, 66
Polyphenol, 256
Polysaccharide-based cheese, 476
Polysaccharide-based ingredients, 48
Polysaccharide-based nanoparticles, 174
Polysaccharides, 34, 35, 44–48, 53, 59, 62–67,
        70, 71, 73–75, 78, 79, 404, 415, 465
    amylose and amylopectin, 466

    pasting temperature, 466
    starch, 465
    starches, 465
Polyunsaturated fatty acids (PUFAs), 54,
        241, 242
Polyunsaturated lipids, 207
Poor water-solubility, 38, 40
Potato proteins, 36, 37
Potato starch, 465
Power-law model, 69
Preservatives, 62
Primates, 265
Processing techniques, 479
Pro-oxidants, 52
Prospective Urban Rural Epidemiology
        (PURE), 239
Protein composition, 542
Protein Denaturation State, 545
Protein digestibility-corrected amino acid
        score (PDCAAS), 234, 541
Protein-coated oil droplets, 65
Protein-rich foods, 5
Protein-rich food source, 18
Protein-rich source, 19
Proteins, 262, 478
    application, 24
    characterization, 41–43
    emulsifiers, 24
    food applications, 467
    functional ingredients, 24
    gelation properties, 467
    ingredients
        agricultural commodities, 41
        application, 38
        concentrates, 37
        dietary fibers, 37
        in food industry, 37
        flours, 37
        inconsistency, 40
        lipids, 37
        microbial fermentation approaches, 41
        minerals, 38
        molecular and functional
            attributes, 37
        molecular and physicochemical
            attributes, 38
        off flavors, 38, 39
        poor water-solubility, 38, 40
        purity, 40, 41
        starches, 37
        structure, 38
        TVPs, 37
        variabilities, 37
    isolation, 34–37

Index   571

molecular, physicochemical, functional, and nutritional attributes, 24, 25
structure
   caseins, 29
   denatured states, 27
   fibrous, 27, 34
   flexible, 27
   flexible proteins, 28, 33, 34
   globular, 27
   globular proteins, 30, 32, 33
   molecular architecture, 28
   molecular characteristics, 30
   native state, 27
   primary, 24
   quaternary, 24, 28
   secondary, 24
   supramolecular, 27
   tertiary, 24, 27, 28
terrestrial sources, 24
Purity, 40, 41

Q
Quality of life, 12

R
Raman spectroscopy, 48
Randomized controlled trials (RCT), 238, 263
Rapidly digestible starch (RDS), 245
Recommended daily allowance (RDA), 229, 233
Reduce waste, 10
Reflectance spectra, 168
Reflectance vs. wavelength spectrum, 175
Refractive index contrast, 174
Relative refractive index, 157, 164
Retention and release processes, 216, 217
Retrogradation/setback, 465
Reversible heat-setting polysaccharides, 73
Rheological attributes, 415
Rheological behavior, 414
Roadkill, 12
Robotics, 10
Royal Institute of International Affairs, 11
Rural communities, 11

S
Salad dressings, 377
Salts, 58, 59
Saponins, 38
Saturated fatty acids, 53–54, 239
Scattering, 160–162

Science and technology, 532
SDS-PAGE densitometry, 543
Seaweed-derived ingredients, 73
Selective absorption, 159, 160
Self-association techniques, 479
Semi-empirical effective medium theory, 180
Semi-solid 3D networks, 53
Semi-solid plant-based foods, 61, 65, 69
Sensors, 10
Sensory attributes, 218
Sensory component, 211
Sensory evaluation, 219, 220
   affective tests, 221, 222
   descriptive tests, 220
   discrimination tests, 220
Separation and fractionation methods
   decanter processing, 111
   decanters, 110
   dynamic viscosity, 111
   manufacturing process, 110
   parameter, 111
   suspension, 111
Shear cell processing methods, 300
Shear cells, 124–126
Shear cell technology, 127
Shear modulus, 185, 189, 196
Shear stress, 177, 183, 376
Shear stress vs. shear rate, 177
Shear testing, 194, 196, 197
Shear thinning liquid, 413
Shear viscosity, 68, 177, 179, 180, 183
Short chain fatty acids (SCFAs), 245, 248, 259, 262
Sidestream, 35
Single scattering, 161
Skeletal muscle fiber, 290
Skim milk, 410
Slowly digestible starch (SDS), 245
Small intestine, 259
Sodium chloride content, 550
Soft-ripened cheese, 485
Soil quality, 7
Sol-gel and ripening processes, 487
Sol-gel transition, 471, 480, 481, 484
Solid fat content (SFC), 53, 57, 75, 76, 296, 469
Solid fat particles, 99
Solubility, 63, 64
Soluble proteins, 35
Solvent extraction methods, 52
Solvent extraction processes, 53
Soxhlet extraction, 547, 548
Soy and lupin proteins, 38
Soy protein, 35, 36, 300

Soybeans, 41
Soymilk production, 145
Spatial uniformity, 156, 157
Specific mechanical energy input (SME), 123
Stability
  biological, 198
  chemical, 198
  gravitational seperation
    creaming/sedimentation, 199, 201
    density contrast, 201
    partial coalescence, 201
    plant-tissue fragments, 200
    Stokes' law, 199, 200
    viscosity, 200
  particle aggregation
    attractive and repulsive forces, 202
    chemical degradation, 207–209
    colloidal, 205
    colloidal plant-based foods, 203
    flocculation, 205
    increased electrostatic stabilization, 204
    increased steric stabilization, 203, 204
    microbial contamination, 209
    particle size reduction, 203
    phase separation, 206, 207
    plant-based dairy analogs, 205
    plant tissue fragments, 202
    quality attributes, 202
    quantification, 209, 210
    reduce bridging effects, 205
    reduce depletion attraction, 204, 205
    reduce hydrophobic attraction, 204
  physical, 198
  resist changes ability, 198
  resistance, 198
Starch granules, 145
Starches, 45, 245, 476
Steam injection, 137
Steric repulsive forces, 419
Stokes' settling viscosity, 111
Stokes's Law, 200, 417
Stomach, 258, 259
Structural elements, 189
Structure formation methods
  barrel temperatures, 120
  cooling die, 121
  extruders, 118
  feeding zone, 120
  fibrous structures, 118
  food processing techniques, 117
  heated mass, 120
  meat analog production, 119
  meat analogs, 123
  plant-based meat production, 122
  plant proteins, 117

  polysaccharides, 123
  processing equipment, 118
  protein dispersion, 121
  protein molecules, 122
  protein processing, 123
  screw elements, 119
  SME, 123
  temperature-controlled barrel, 118
  temperature-dependent heat capacity, 124
  transition temperature, 121
  vegetable protein, 121
Sugars, 45, 246, 271
Supercritical fluid extraction methods, 53
Surface gloss, 158, 159
Sweeteners, 270, 272
Swine flu, 17
Syneresis, 206

T
Tannins, 260
Taste, 13–15
Temperature-controlled measurement cell, 182
Terrestrial sources, 24
Textural profile analysis (TPA), 193, 351, 368
Texture
  fluids
    cup-and-bob cell, 183
    instrumental viscometers, 182
    major factors impacting
      viscosity, 179–182
    power indices, 183
    rotational viscometers, 182
    shear viscosity, 177–179, 183
    sophisticated instruments, 183
  practical considerations, 197, 198
  solids
    analytical techniques, 193
    compression testing, 193, 195
    disadvantages, 193
    elastic and viscous properties, 183
    environmental conditions, 184
    fracture properties, 184
    ideal plastics, 189, 190
    ideal solids, 184–188
    muscle protein networks, 184
    non-ideal plastics, 190
    non-ideal solids, 188, 189
    rheological parameters, 184
    shear testing, 194, 196, 197
    structural elements, 184
    viscoelastic materials, 191–193
Texture Technologies Corporation, 195
Texturized vegetable protein (TVP), 37, 285
Theoretical models, 187

Index 573

Thermal processing methods, 134
    antinutrients, 135
    antinutritional factors, 136
    blanching, 134
    plate heat-exchangers, 139
    thermal treatment, 137
    trypsin inhibitors, 135, 137
Thickening, 66–70
Thickening agent, 68
Thickening power (TP), 66–67
Thin layer chromatography (TLC), 58
Time-dependent shear thinning behavior, 179
Tissue disruption route, 480, 486
    colloidal dispersions, 480
    lupin seeds, 481
    proteins, 482
    sol-gel transition, 480
    soymilk, 482
Tocopherols, 62
Traditional animal-based food, 170
Trans fatty acids, 50, 54, 242
Translational diffusion coefficient, 216
Transportation sector, 10
Triacylglycerols, 50, 53–55, 58
Triangle test, 220
Tribology instruments, 219
Tribometers, 183
Trimethylamine N-oxide (TMAO), 262
Tubular heat exchangers, 140
Tyrosine, 539

U
UHT processes, 145
UHT treatment, 138, 139
Ultrafiltration, 35
Ultra-high temperature (UHT) processing, 137
Ultrasound-assisted extraction methods, 53
Umami, 222
Unhealthful plant-based diets, 13
Unsaturated fatty acids, 53–55, 57, 239, 240
UV-visible spectrophotometer, 175

V
Van der Waals attraction, 361
Vegan cheese, 458
Vegetables, 2
Vegetarian baked products, 381
Viscoelastic emulsion gels, 477
Viscoelastic filled-emulsion gel, 486
Viscoelastic materials, 191–193
Viscoelastic rheological properties
    Maxwell model, 460
    Voigt-Kelvin model, 461

Viscosity enhancement, 417
Vitamin B12, 251
Vitamin D, 52, 251, 252, 432–433
Vitamin D_2, 252
Vitamins, 76, 250
Voigt-Kelvin model, 461

W
Water holding capacity (WHC), 65, 66, 211
Water use, 5
Water-in-oil-in-water (W/O/W), 97
Water-insoluble polysaccharides, 64
Water-insoluble proteins, 63
Water-solubility, 32, 63
Wet milling devices, 105
Wet-milling approach, 106
Wheat proteins, 36, 300
Whipped cream, 503
    fat droplets and network formation, 503
    plant-based, 504
    proteins, 504
Whipped cream and ice cream, 394
Whipping cream, 504
Whipping process, 380
Winterization, 52
World Economic Forum, 15
World Wildlife Fund, 10

X
X-ray diffraction analysis, 42
X-ray tomography, 210

Y
Yields and efficiency, 10
Yogurt, 491
    acidification, 493
    bacteria, 494
    casein micelles, 492
    fermentation approaches, 493
    organic acids, 494
    plant-based, 493, 499
    sensory profile, 499
    shear stress, 492
    sol-gel transition, 492
    standardization and formulation, 493
    textural attributes, 499
    3D protein network, 492
Young's modulus, 185

Z
Zinc, 254, 255
Zoonotic disease transmission, 11, 13

CPSIA information can be obtained
at www.ICGtesting.com
Printed in the USA
LVHW080401190622
721595LV00001B/1